P9-BJL-318

Principles and Techniques of Electron Microscopy

Principles and Techniques of Electron Microscopy

BIOLOGICAL APPLICATIONS

Volume 7

M. A. HAYAT
Professor of Biology
Kean College of New Jersey
Union, New Jersey

VNR VAN NOSTRAND REINHOLD COMPANY
NEW YORK CINCINNATI ATLANTA DALLAS SAN FRANCISCO
LONDON TORONTO MELBOURNE

Van Nostrand Reinhold Company Regional Offices:
New York Cincinnati Atlanta Dallas San Francisco

Van Nostrand Reinhold Company International Offices:
London Toronto Melbourne

Library of Congress Catalog Card Number: 70-129544
ISBN 0-442-25691-4

Manufactured in the United States of America

Published by Van Nostrand Reinhold Company
450 West 33rd Street, New York, N.Y. 10001

Published simultaneously in Canada by Van Nostrand Reinhold Ltd.

15 14 13 12 11 10 9 8 7 6 5 4 3 2 1

Library of Congress Cataloging in Publication Data

Hayat, M A
Principles and techniques of electron microscopy.

Includes bibliographies.
Vols. 6 edited by M. A. Hayat.
1. Electron microscope–Collected works. I. Title.
[DNLM: 1. Microscopy, Electron. QH205 H413p H2VO]
QH212.E4H38 578'.4'5 70-129544
ISBN 0-442-25691-4 (v. 7)

Contributors to This Volume

Gary D. Burkholder

A. J. Gibbs

L. T. Ellison

M. S. Isaacson

M. Geuskens

D. G. Jones

M. V. Nermut

A. J. Rowe

Klaus-Rüdiger Peters

F. Ruzicka

PREFACE

This is the seventh volume of a multi-volume series on the principles and techniques employed for studying biological specimens with the aid of an electron microscope. Since its inception in 1970, the series has successfully reflected the growth of electron microscopy in instrumentation as well as in methodology. There was a pressing need to keep the readers abreast of the remarkable expansion of the field in recent years and the ever growing importance of its contributions to the understanding of many problems in biological and medical sciences. This treatise serves as an international authoritative source in the field, and is designed to cover important new developments systematically. The treatise departs from the tradition that books on methodology present only the contemporary consensus of knowledge. It is written by scholars, and when they have anticipated the potential usefulness of a new method, they have so stated. The authors have not hesitated to include ideas in progress. The treatise should serve as a guide and survey, which can save a newcomer the tedious search for information scattered in biological journals.

This volume has developed over the years through the joint effort of ten distinguished author-scientists. As a result, a most comprehensive compilation of methods developed and used by a large number of competent scientists has been achieved. The book contains new viewpoints with particular regard to current problems. Areas of disagreement and potential research problems have been pointed out. It is hoped that the readers would become aware that correct interpretation of the information retrieved from electron micrographs is dependent upon an understanding of the principles underlying the methodology and instrumentation.

The basic approach in this volume is similar to that in the previous six volumes in that the methods presented have been tested for their reliability, and are the best of those currently available. The instructions for the preparation and use of various solutions, media, stains, and apparatus are straightforward and complete, and should enable the worker to prepare his or her specimens without outside help. It is suggested that prior to undertaking the processing, the entire procedure should be read and necessary solutions and other media prepared. Each chapter is provided with an exhaustive list of references with complete titles. Full author and subject indexes are included at the end of the book.

It is encouraging to know that the previous six volumes have been received favorably. It is my impression that this volume will also fulfill its purpose: to provide an understanding of the usefulness, limitations, and potential applications of special methods employed for studying the structure, composition, size, number, and location of cellular components, and to provide details of current improvements in the instrumentation.

M. A. HAYAT

CONTENTS

2 FREEZE-DRYING FOR ELECTRON MICROSCOPY

M. V. Nermut

3 IMAGE RECONSTRUCTION OF ELECTRON MICROGRAPHS BY USING EQUIDENSITE INTEGRATION ANALYSIS

Klaus-Rüdiger Peters

4 G-BANDING OF CHROMOSOMES

F. Ruzicka

6 OPTICAL ANALYSIS AND RECONSTRUCTION OF IMAGES

A. J. Gibbs and A. J. Rowe

7 MIRROR ELECTRON MICROSCOPY

R. S. Gvosdover and B. Ya. Zel'dovich

8 ELECTRON MICROSCOPY OF BANDED MAMMALIAN CHROMOSOMES
Gary D. Burkholder

9 EQUIDENSITOMETRY: SOME NEUROBIOLOGICAL APPLICATIONS
L. T. Ellison and D. G. Jones

Contents to

Volume 1, M. A. Hayat

Volume 2

Volume 3

Volume 4

Volume 5

Volume 6

Volume 8

Principles and Techniques of Electron Microscopy

1. SPECIMEN DAMAGE IN THE ELECTRON MICROSCOPE

M. S. Isaacson

Department of Physics and Enrico Fermi Institute, The University of Chicago, Chicago, Illinois

INTRODUCTION

For over two hundred years the biologist has used the light microscope to unravel the structures of macro-organisms. With the advent of the electron microscope in the early 1930s, it became apparent that here was a promise of being able to deduce structures of biological objects much smaller than those which had been observed in light microscopy (Marton, 1934). Therefore, the possibility existed (in principle, at least) of being able to deduce function starting at some basic molecular level. Of course, solving structures at the molecular level could be obtained using X-ray diffraction. However, the necessity of being able to produce large crystals, the size of the unit cell, and other practical difficulties limited the types of biological objects which could be effectively studied by the X-ray techniques (e.g., DeRosier, 1971; McPherson, 1976). Electron microscopy thus offered the potential for being able to unravel biological structures too complicated for study by other means.

It is worthwhile noting that progress in biological electron microscopy has fallen far behind recent developments in instrumental techniques and application of these techniques to inorganic materials insofar as obtaining information near the resolution limit of the instruments is concerned (Echlin and Fendley, 1973). This is evidenced, to some extent, by the fact that while spatial resolution of 2 to 3 Å is comfortably achievable and visualization of single heavy atoms in the electron microscope is a reality (Crewe, 1971a; Henkelman and

Ottensmeyer, 1971; Hashimoto *et al.*, 1971; Thon and Willasch, 1972; Wall *et al.*, 1974b), spatial information obtained concerning biological specimens is generally no better than ~15 Å (ten times lower than the resolution limit of existing instruments).

Technically, one reason for this limitation has been the preparative procedures which have been necessary to produce suitable contrast for visualization of structure in the conventional transmission electron microscope (CTEM) using the bright field mode of operation (Horne, 1973; Amos, 1974; DeRosier, 1971; Nermut, 1972). It was mainly the advent of these preparative techniques which led to the advances in biological electron microscopy in the fifties and sixties. However, with new microscope methods of obtaining increased contrast using dark field in the CTEM and with the development of the high resolution scanning transmission electron microscope (STEM) (Crewe and Wall, 1970b; Crewe, 1971b), the need for the old contrast-enhancing techniques (such as negative staining) can be eliminated. Thus, the hope of removing the 15-Å barrier in biological electron microscopy remains a distinct possibility.

Furthermore, there is hope that the full utilization of the STEM to produce superior contrast and obtain many types of information simultaneously (Crewe, 1971b; Crewe and Wall, 1970a; Crewe *et al.*, 1975; Isaacson *et al.*, 1974b) will have a great effect upon increasing the resolution attainable with biological objects. However, there are few such microscopes in existence today, and there are still new specimen preparation techniques to be developed. Moreover, while the necessary electron irradiation dose in the STEM is less than that needed in an equivalent CTEM for a given amount of information (Langmore *et al.*, 1973b; Crewe, 1973; Wall *et al.*, 1974a), one is led to the inevitable conclusion that insofar as biological electron microscopy is concerned, the fundamental limit to biological structure determination in the electron microscope is the damage incurred by the specimen because of the action of the incident electron beam (Echlin and Fendley, 1973; Glaeser, 1971; Isaacson *et al.*, 1973; Breedlove and Trammel, 1970; Stenn and Bahr, 1970a). (This statement assumes, of course, that all technical problems such as specimen preparation are mastered.)

As will be seen in the following sections, the situation is not quite as discouraging as it may seem. It is clear that the radiation-damage limit is becoming of increasing concern to electron microscopists, as evidenced by the fact that there have probably been more symposia and workshops on the subject in the last few years than there had been in the first three decades of electron microscopy. One should take heart that even though radiation damage poses the fundamental limit in structure determination in biological electron microscopy, there are several avenues which still exist for overcoming this limit, and many of these trails have only just begun to be explored.

This chapter was written with the explicit goal in mind of summarizing what is known at present about radiation damage as it pertains to biological electron microscopy. Perhaps by a clearer understanding of the nature of the damage process, we might be better able to devise new techniques and schemes which would allow us to reduce its deleterious effect on the problem of biological structure determination. The chapter will be divided into essentially three basic sections comprising a brief discussion of electron scattering and how the scattering process relates to potential damage; a survey of experimental measurements of radiation damage relating to electron microscopy and a discussion of these measurements insofar as what we can learn about the damage mechanisms; and finally a survey of techniques which have been proposed (or might be considered) to reduce radiation damage and a discussion of the particular merits (if any) of these techniques.

The radiation damage literature as it pertains to electron microscopy is by no means extensive. Moreover, whatever experimental measurements exist have, in many instances, to be looked upon with a very discerning eye, since the quantitations and experimental methods of many published measurements are not described in sufficient pertinent detail for one to be able to draw sound conclusions with respect to the field as a whole. I should hope that if nothing else, this chapter will instill in electron microscopists the need to quantitate carefully their data and methods with regard to radiation damage, so that they can more efficiently utilize the literature and therefore draw conclusions in the minimum possible time which could lead them on the correct path in pushing back the radiation-damage limit in biological electron microscopy.

THEORETICAL CONSIDERATIONS CONCERNING ELECTRON SCATTERING

Before we begin discussing experimental results concerning radiation damage, possible mechanisms and implications for electron microscopy, it is necessary to give a brief review of the nature of electron scattering. It is, of course, the electrons being scattered from the specimen which give us our information as well as cause the ultimate destruction of the information which we seek. Let us look into the parameters involved in specimen–electron scattering. The discussion in this section is by no means aimed at completeness, but rather is a survey into the pertinent quantities involved in electron microscopy. Readers interested in more thorough discussions should consult the reviews by Scott (1965) and Zeitler (1965) aimed at electron microscopists or the more detailed descriptions by Mott and Massey (1965) or Massey and Burhop (1969).

There are basically three types of interaction which a fast incident electron can experience in traversing a thin specimen of the type used in electron microscopy (Fig. 1.1). (The word "thin" as employed here means that the incident electron is transmitted through the film with an energy loss which is smaller than its initial kinetic energy.) The electron can traverse the specimen losing a negligible amount of energy and be elastically scattered from the atomic nuclei. The elec-

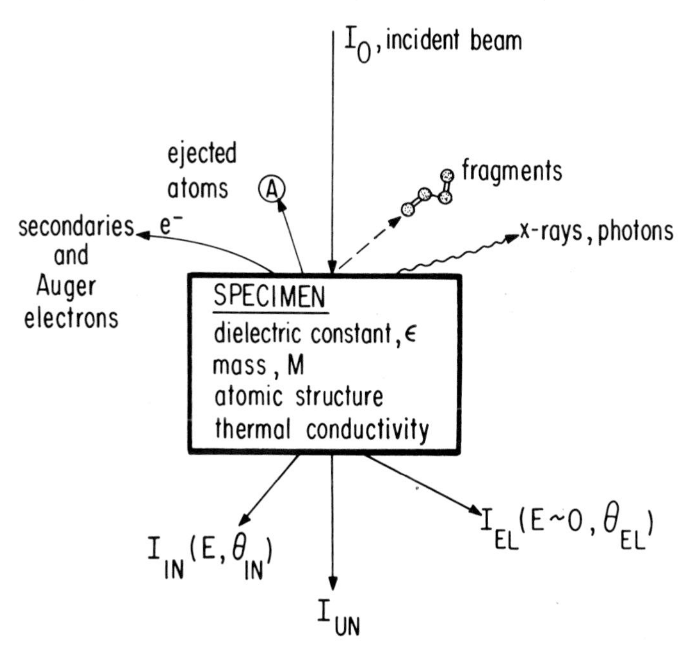

Fig. 1.1 Schematic diagram illustrating the various quantities involved which determine the ultimate structural information obtainable from a specimen in the electron microscope.

tron can be inelastically scattered from the atomic electrons and impart energy to the specimen by either exciting the atoms in the specimen to higher energy states or ionizing them. The electron can also traverse the specimen without interacting at all. We obviously do not consider the latter in terms of specimen damage.

Elastic interactions are generally useful for high resolution imaging since the nuclei present almost point objects to the electrons (at resolutions greater than 1.5 Å, of course) (Crewe *et al.*, 1975). These interactions are generally not thought of as producing damage to the specimen. However, as we shall show, if the angle of scattering is sufficiently large, the conservation of energy and momentum requires a certain amount of energy to be imparted to the atom itself, and this energy can exceed that which binds the atom to the specimen, resulting in the atom's "recoiling" and moving. The probability of such an occurrence is generally small and usually ignored. However, under certain conditions, i.e., high voltage microscopy, this is not necessarily true.

Inelastic interactions generally can result in some molecular or chemical change in the specimen. Moreover, in general one cannot use all the inelastic electrons for high resolution imaging (Isaacson *et al.*, 1974a; Rose, 1973). This fact is very important when we consider electron microscopy of organic or biological materials, since both theory and experiments show that one expects more inelastic than elastic collisions in materials of low atomic number (Brünger and Menz, 1965; Wall, 1972a; Isaacson, 1975b). Moreover, those inelastic collisions which ionize the atoms in the specimen can produce electrons with sufficient kinetic energy to produce damage via further inelastic scattering.

Elastic Scattering

We would like to summarize here the formulae pertinent to calculations of the probability of elastic scattering events and with some assumptions derive the probability for an "elastic" event to result in a displaced atom. The differential cross section for the elastic scattering of electrons (cross section/unit solid angle) is defined as $d\sigma/d\Omega = |f(s)|^2$, where $f(s)$ is called the electron scattering amplitude and $s = (1/\lambda) \cdot \sin \theta/2$ is the scattering parameter for electrons of wavelength λ to be scattered through an angle θ by the atom potential transferring a momentum $2 \cdot h \cdot s$ to the atom (h is Planck's constant).

This $f(s)$ is a complex quantity possessing both amplitude and phase, and exact calculations of it are difficult. In the limit that the interaction between the electron and atom is weak, one can use the first Born Approximation to calculate $f(s)$. This $f(s)$ is real and can be scaled with voltage since $f(s)/\gamma$ is voltage-independent ($\gamma = [1 - \beta^2]^{-1/2}$ where $\beta = v/c$). Values of elastic scattering differential and total scattering cross sections using the first Born Approximation and different atomic models have been given in the literature (e.g., Zeitler, 1965; Langmore *et al.*, 1974). However, the first Born Approximation is not valid for many operational conditions used in electron microscopy (Zeitler and Olsen, 1967). In fact, it is rigorously valid only for $Z/(137\,\beta) \ll 1$. As an example, for Hg atoms irradiated by 100 keV electrons, $Z/(137\,\beta) \cong 1.07$. So total electron scattering cross sections obtained using the first Born Approximation have to be corrected for its failure. As it turns out, this correction takes the form of a multiplicative factor which is less than or equal to unity (Zeitler and Olsen, 1967; Langmore *et al.*, 1974). Experimental measurements of electron scattering cross sections agree fairly well with theoretical predictions obtained using the first Born Approximation and this correction (Crewe *et al.*, 1974).

While the above results are accurate, they do not allow simple expressions for electron scattering. A simple, though not as accurate, expression of the electron scattering amplitude starts from a model of the atom due to Wentzel (1927), who assumed the atomic potential to be an exponentially screened coulomb potential. The result is that

$$f(s) = \left[\frac{\gamma}{8\pi^2 a_0}\right] \frac{Z}{s^2 + s_0^2} \tag{1.1}$$

where $s_0 = (4\pi a)^{-1}$, a is the effective atomic radius, a_0 is the Bohr radius and Z is the nuclear charge. Integration of $|f(s)|^2$ over all scattering angles then yields the total elastic cross section in the Wentzel model:

$$\sigma_W = \frac{\lambda_c^2}{\pi\beta^2}\left[\frac{a}{a_0}\right]^2 Z^2 \tag{1.2a}$$

The value of the differential cross section, $|f(s)|^2$, and the total elastic cross section of Eq. (1.2a) then depends only upon the choice of a. If we choose the effective atomic radius as $a = a_0 Z^{-1/3}$, we obtain the expression for the differential cross section and the total elastic cross section due to Lenz (1954). Equation (1.2a) then becomes in the Lenz-Wentzel model

$$\sigma_{LW} = \frac{\lambda_c^2}{\pi\beta^2} Z^{4/3} \tag{1.2b}$$

where $\lambda_c = 2.426 \times 10^{-2}$ Å is the Compton wavelength of the electron. Because of their simplicity, these Lenz-Wentzel expressions are often used in electron

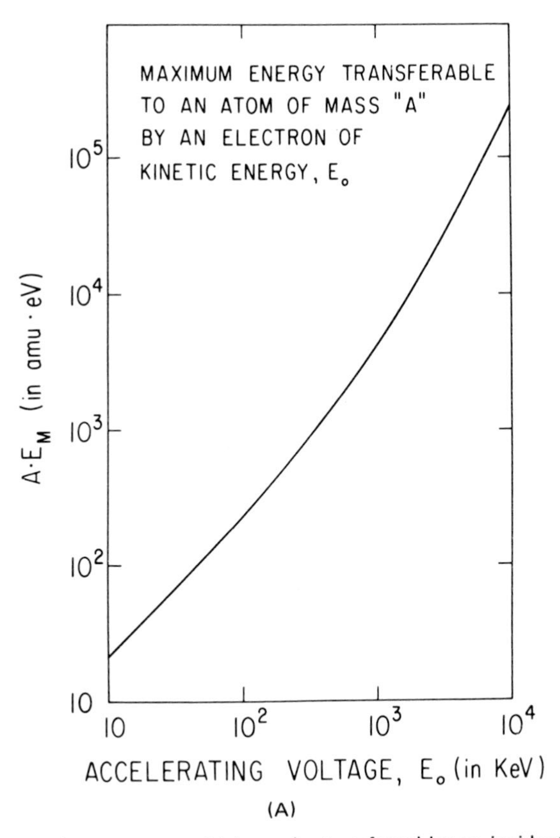

Fig. 1.2 A. The maximum energy which can be transferred by an incident electron of kinetic energy E_0, in an elastic nuclear collision with an atom of mass A. $E_{max} = 2(m/m_pA) \cdot E_0(E_0 + 2mc^2)/mc^2$, where m is the electron rest mass, m_p is the proton rest mass and c is the velocity of light ($M_A = m_p \cdot A$). B. The threshold energy of the incident electron, E_t, necessary just to produce a displacement of an atom of mass A in an elastic nuclear collision. The term E_t is that energy such that the maximum transferable energy shown in Fig. 1.2A is equal to the displacement energy, E_d. The arrows indicate the threshold energies for various atoms assuming $E_d = 1$ eV.

microscopy even though they give somewhat misleading results (the total cross sections tend to be low for high Z and the differential cross sections have the wrong angular dependance for high Z).

The process whereby elastic nuclear collision could result in specimen damage occurs if the energy transferred to the atom (due to a very large momentum transfer) is greater than the binding energy of the atom to the substrate or within the structure under investigation. Using simple kinematic arguments (conservation of momentum and energy) one can show (e.g., Corbett, 1966) that the recoil kinetic energy imparted to an atom by an electron scattered through an angle θ is:

$$E_{\text{recoil}} = E_M[\sin(\theta/2)]^2 \tag{1.3}$$

where $E_M = 2(m/M_A)E_0(E_0 + 2mc^2)/mc^2$ is the maximum energy which can be transferred by an electron of kinetic energy E and mass m to an atom of mass M_A (see Fig. 1.2A).

Note that usually the recoil energy is small, since the bulk of elastic scattering occurs at moderately small angles. However, for large incident electron energies and light atoms, E_M can be nonnegligible (e.g., for 1 MeV electrons and carbon atoms, $E_M \cong 350$ eV!). Therefore, we would like to calculate the cross section for displacing an atom in a specimen.

This can be accomplished by finding the cross section for elastic scattering through angles greater than the minimum scattering angle for which the recoil

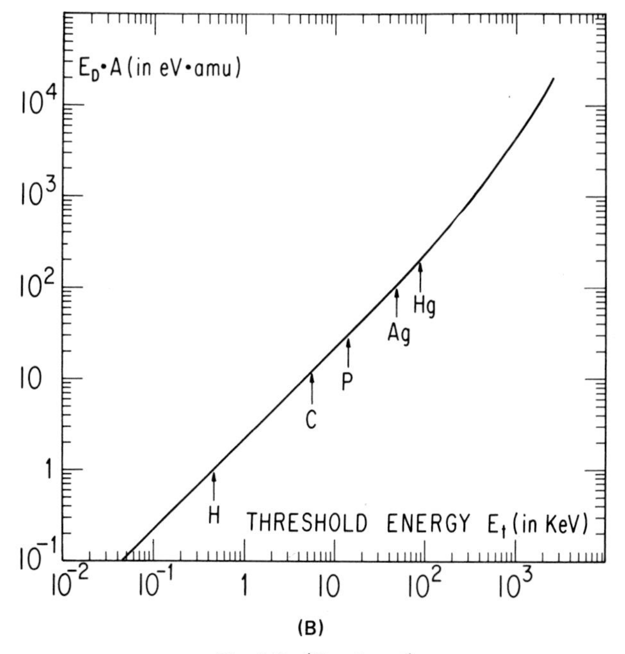

(B)

Fig. 1.2 *(Continued)*

energy is sufficient to displace the atom from its structure. We call this angle the critical angle, θ_c, and it is obtained from Eq. (1.3) when $E_{\text{recoil}} = E_d$, the displacement energy necessary to move the atom. Therefore,

$$\sin(\theta_c/2) = \sqrt{\frac{E_d}{E_M}} = \lambda s_c \tag{1.4}$$

where s_c is the critical scattering parameter (i.e., the minimum value of s for which displacement occurs). Hence, the cross section for atom displacement becomes:

$$\sigma_{\text{disp}} = \int_{\theta_c}^{\pi} \frac{d\sigma_{\text{el}}}{d\Omega} 2\pi \sin\theta d\theta$$

Although the Wentzel model for the atom is not really correct, it allows us easily to calculate cross sections for displacement which will, at least, allow us to see which parameters are important. Therefore, from Eq. (1.1) we obtain:

$$\frac{d\sigma_{\text{el}}}{d\Omega} = \left[\frac{\gamma}{8\pi^2 a_0}\right]^2 \frac{Z^2}{[s_0^2 + s^2]^2} \tag{1.6}$$

Inserting this into Eq. (1.5) and integrating yields:

$$\sigma_{\text{disp}} = \frac{1}{\pi}\left[\frac{\gamma\lambda Z}{4\pi a_0}\right]^2 \left[\frac{1}{s_0^2 + s_c^2} - \frac{1}{s_0^2 + (1/\lambda)^2}\right] \tag{1.7}$$

Using the total Wentzel elastic cross section given in Eq. (1.2a) and realizing that $\lambda\gamma = \lambda_c/\beta$, we get from Eq. (1.7):

$$\frac{\sigma_{\text{disp}}}{\sigma_{\text{el}}} = \left[1 + \left(\frac{s_c}{s_0}\right)^2\right]^{-1} - [1 + (1/s_0\lambda)^2]^{-1} \tag{1.8}$$

Obviously, this expression is meaningless if the maximum energy which can be transferred to the atom is less than the displacement energy (since then $s_c > 1/\lambda$). In that case, $\sigma_{\text{disp}} = 0$. We call the kinetic energy of the electron for which $E_m = E_d$ the threshold energy,

$$E_t = mc^2 [(1 + 918 \cdot A \cdot E_d/mc^2)^{1/2} - 1] \tag{1.9}$$

This is the minimum incident electron kinetic energy needed to displace the atom. We have plotted this in Fig. 1.2B as a function of $A \cdot E_d$. The points indicated by arrows are the value of the threshold energy for an atom bound by 1 eV (i.e., $E_d = 1$ eV). We shall discuss the implications of the threshold energy later, but it is obvious from Fig. 1.2 that even in medium-energy electron microscopy ($E_0 \sim 25$–100 keV) the threshold energy is sufficient to displace low-Z atoms bound by 1 eV (i.e., hydrogen) and in some cases even higher-Z atoms. At 1 MeV the situation becomes even more serious.

We can rewrite Eq. (1.8) to allow us to see more easily the relative energy dependences involved in the ratio $\sigma_{disp}/\sigma_{el}$. This ratio is the quantity of interest rather than σ_{disp} alone since it tells us the probability of an elastic event causing atomic displacement.

$$\frac{\sigma_{disp}}{\sigma_{el}} = \left[1 + \frac{E_d}{E_M}(1/s_0\lambda)^2\right]^{-1} - [1 + (1/s_0\lambda)^2]^{-1} \tag{1.10}$$

This becomes very simple when $\lambda \ll 1/s_0 = 4\pi a$, i.e., when the wavelength of the incident electron becomes much smaller than atomic dimensions. Then Eq. (1.10) simplifies to:

$$\frac{\sigma_{disp}}{\sigma_{el}} \cong \left(\frac{\lambda}{4\pi a}\right)^2 \left[\frac{E_M}{E_d} - 1\right] \tag{1.11}$$

(valid if $(4\pi a/\lambda)^2 \gg E_M/E_d$).

Admittedly, the above description of the displacement cross section is a simple one since we start from a Wentzel picture of the atom and ignore the fact that for large angle scattering such a model gives Rutherford scattering which we know is not quite correct (Corbett, 1966). However, various approximate corrections to non-Rutherford scattering have been given in the literature, and based on the work by McKinley and Feshbach (1948) they have been shown to agree within 20% of the exact solution for lighter elements. If we use their expression for the correction to Rutherford scattering (R_{MF}) and apply it to our Wentzel elastic cross section, the corrected displacement cross section becomes:

$$\sigma_{MF} = \int_{\theta_c}^{\pi} \frac{d\sigma_{el}}{d\Omega} \cdot R_{MF} \cdot 2\pi \sin\theta d\theta \tag{1.12}$$

We find small differences between cross sections from Eq. (1.12) and those obtained using Eq. (1.10) (see Figs. 1.3A–C). So for practical purposes at estimating knock-on damage, Eq. (1.10) is perfectly adequate.

It should be noted that the Z-dependence of the Lenz-Wentzel cross section given in Eq. (1.2B) which was obtained using the atomic radius $a = a_0 Z^{-1/3}$ is not quite correct. To a better approximation, it was found (Langmore *et al.*, 1973b) that if one used an $a = 0.9\, a_0 Z^{-1/4}$ one obtained a Wentzel cross section of

$$\sigma_{el} \cong \frac{3}{4}\frac{\lambda_c^2}{\pi\beta^2} Z^{3/2} \tag{1.13}$$

which agreed within 30% of the total elastic scattering cross sections obtained using Hartree-Fock-Slater scattering amplitudes in the first Born Approximation. The expressions for the displacement cross sections evaluated in Fig. 1.3 A, B, C were therefore obtained using $a = 0.9\, a_0 Z^{-1/4}$.

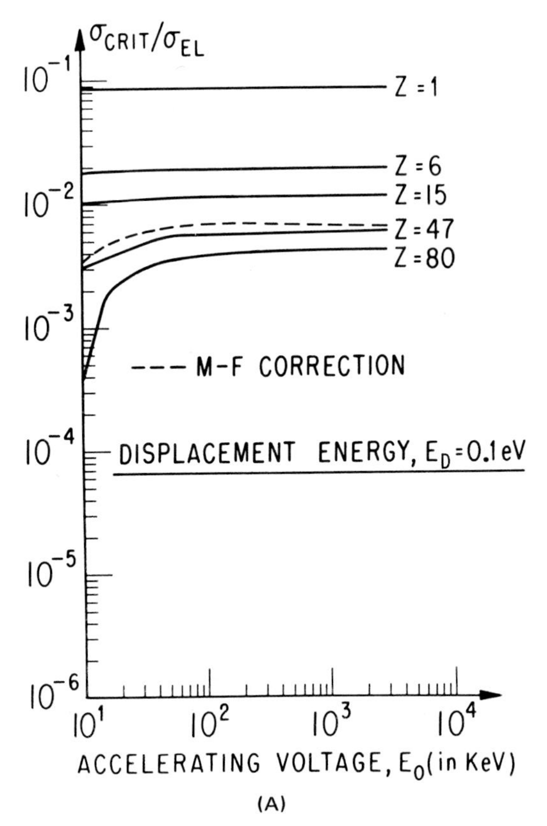

(A)

Fig. 1.3 The ratio of the knock-on cross section ($\sigma_{crit} = \sigma_{disp}$) to the elastic scattering cross section (σ_{el}) calculated using Eqs. (1.10) and (1.13), plotted as a function of the kinetic energy of the incident electron (i.e., the microscope accelerating voltage). The dashed curves correspond to the knock-on cross sections being calculated using the McKinley-Feshbach correction to Rutherford scattering as indicated in Eq. (1.12). (A) Displacement energy, $E_d = 0.1$ eV. (B) Displacement energy, $E_d = 1.0$ eV. (C) Displacement energy, $E_d = 10.0$ eV. It should be noted that for solids, $E_d \sim 20$ eV for most materials.

Inelastic Scattering

In this section we would like to summarize the formulae pertinent to inelastic scattering, both for the probability for the occurrence of all energy loss events and for the probability of particular energy loss events. The latter is of importance since we would like to relate particular energy loss occurrences with events which "damage" the specimen.

It is known from the dielectric theory (Daniels *et al.*, 1970) that the probability for a fast electron of velocity v to be inelastically scattered once into an infinitesimal element of solid angle $d\Omega$ around a scattering angle $\vec{\theta}$ (from its

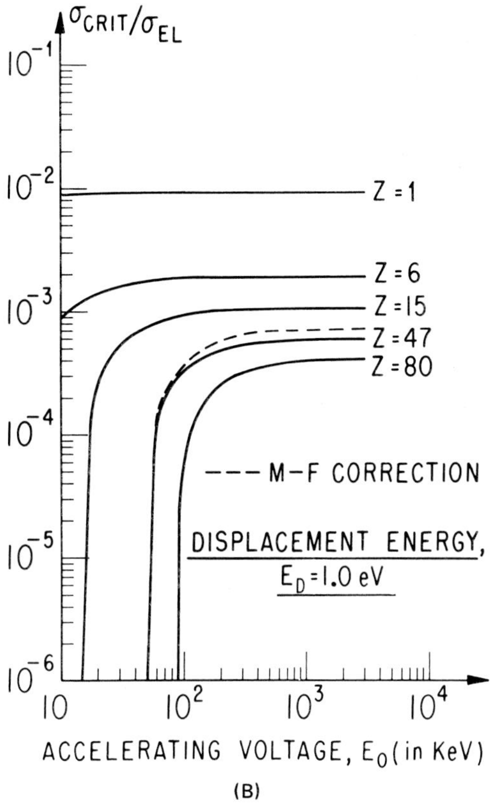

(B)

Fig. 1.3 *(Continued)*

incident direction), and to lose an amount of energy E while traversing a specimen of thickness T is given as:

$$P_1(E, \vec{\theta}, T)\, dEd\Omega = \frac{T}{\pi^2 a_0 mc^2 \beta^2} \left[-\mathrm{Im} \left(\frac{1}{\epsilon(E, \vec{\theta})} \right) \right] \frac{dEd\Omega}{\vec{\theta}^2 + \theta_E^2} \qquad (1.14)$$

In this equation, $\beta = v/c$, c is the velocity of light, m is the electron rest mass, a_0 is the Bohr radius, $\theta_E = E/pv$ where p is the incident relativistic electron momentum and $\epsilon(E, \vec{\theta})$ is the complex dielectric constant of the specimen. Classically, θ_E is related to the minimum momentum which must be transferred in electron scattering from a free electron for an energy loss E. $-\mathrm{Im}\ 1/\epsilon$ is usually called the energy-loss function (Pines, 1969) because of its proportionality with the energy loss probability P_1, and it is this quantity which one measures. The differential scattering cross section for the above process is:

$$\frac{d^2 \sigma_{\mathrm{in}}}{dEd\Omega}(E, \vec{\theta}) = \frac{1}{NT} P_1(E, \vec{\theta}, T) \qquad (1.15)$$

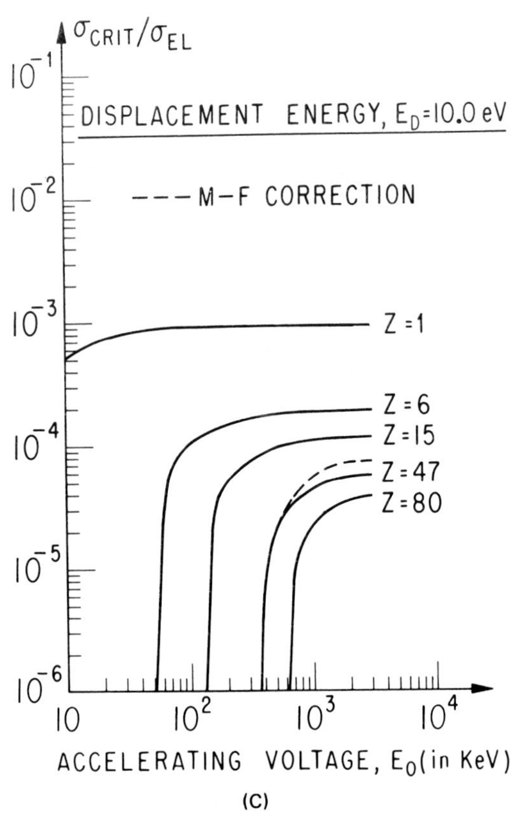

(C)

Fig. 1.3 *(Continued)*

where N is the molecular density of the specimen. Note that the above relations hold for both isolated atoms and molecules as well as for solid specimens, the nature of the specimen appearing only in $1/N\ [-\text{Im}\ 1/\epsilon]$.

It should be noted that $-\text{Im}\ 1/\epsilon$ is independent of the incident electron energy and that one can obtain, upon integration of Eq. (1.15) over all possible scattering angles [remembering that the maximum scattering angle for an energy loss E in the integration is usually taken to be $\theta_{max} = \sqrt{2\theta_E}$ (Mott and Massey, 1965)], the differential inelastic scattering cross section per unit energy loss:

$$\frac{d\sigma_{in}}{dE} = \frac{1}{\pi a_0 mc^2 \beta^2 N}\left[-\text{Im}\frac{1}{\epsilon}\right] h(\beta) \tag{1.16}$$

Here $h(\beta) \cong \ln(2/\theta_E)$ is a slowly increasing function of the incident electron energy, and we have assumed that the dielectric constant is relatively constant over the aperture angles generally used in electron microscopy. Apart from the logarithmic term, $d\sigma/dE$ has the same energy dependence as the cross section for elastic scattering.

In order to understand how inelastic scattering relates to specimen damage in the electron microscope, the probability for various types of energy-loss events in typical biological specimens must be known. While this probability (the characteristic electron energy loss spectrum) differs from one material to the next, the gross features of all biological (or organic) objects are sufficiently similar so that we can see the trends from one spectrum. As an example, consider the nucleic acid base, adenine ($C_5N_5H_5$). The characteristic energy loss spectrum for a 500-Å-thick film of adenine supported on a 30-Å-thick carbon substrate is shown in Fig. 1.4, extending from 0 to 500 eV energy loss of the incident electrons. The structure in this spectrum has been described in the literature in some detail (Isaacson, 1972a; 1972b), and we merely summarize the features here. The spectrum for adenine is similar in general to that of all biological molecules thus far studied (Crewe *et al.*, 1971a; Johnson, 1972; Isaacson, 1972a, b; Lin, 1974), in that it exhibits various peaks between 0 and 10 eV, some structure between 10 and 20 eV, a rather broad peak at ~20 eV, and sharp peaks in the region greater than 280 eV energy loss.

The spectral region less than 10 eV (>1240 Å) corresponds to $\pi \rightarrow \pi^*$ excitations in the molecule, the minimum at ~10 eV corresponding to the onset of π electron ionization. The fraction of total inelastic scattering occurring in this region is only .08. The broad peak centered around 20 eV corresponds mainly to outer shell ionization of the molecule and $\sigma \rightarrow \sigma^*$ molecular transitions, transitions from σ molecular orbitals to superexcited σ^* orbitals with the excitation energies greater than the first ionization potential (Johnson and Isaacson, 1973; Platzman, 1962).

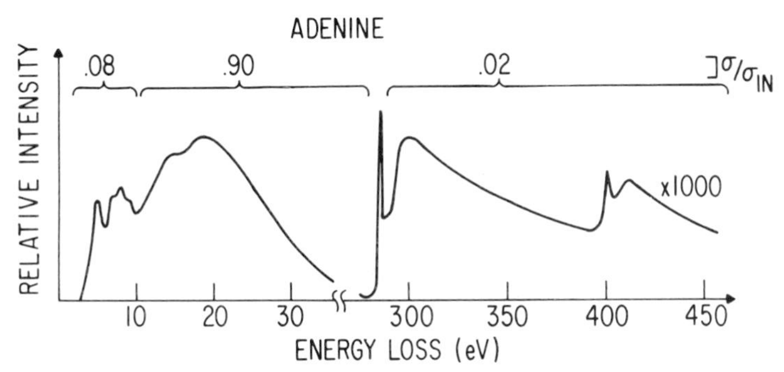

Fig. 1.4 The characteristic electron energy loss spectrum for an approximately 500-Å-thick film of the nucleic acid base, adenine ($C_5N_5H_5$), supported on a 30-Å-thick carbon substrate. The horizontal scale is the amount of energy lost by incident 25 keV electrons in traversing the film. The electron intensity was obtained by detecting only those electrons scattered in the forward direction. The numbers over the brackets indicate the fraction of the total inelastic scattering cross section which occurs in the respective energy loss regions (from Isaacson, 1975b). The peaks near 285 and 395 eV correspond to the *K*-shell excitation of the carbon and nitrogen atoms, respectively, while the region less than 50 eV corresponds mainly to valence shell excitations and ionizations.

The sharp peaks in the energy loss region greater than 280 eV correspond to the K-shell excitation and ionization of the constituent atoms within the molecule. In adenine, the carbon peak at 285 eV and the nitrogen peak at 395 eV can be seen. Although it is not evident on the scale shown, the sharp peaks contain fine structure that is due to excitation from K-levels to bound excited states, while the broader peak and tail correspond to the inner K electrons being ionized and ejected with some kinetic energy. The fraction of total inelastic scattering that is due to the K-shell electrons is only .02 for adenine. This K-shell ionization results in an Auger effect for low-Z materials with the result that there is a high probability of a carbon, nitrogen, or oxygen atom with two vacancies. This violent change in the molecule can lead in some cases to atomic displacement (Durup and Platzman, 1961), a result which could have implications for beam damage. This change will be considered further when we concern ourselves with possible specimen damage mechanisms.

Before this section is concluded, we need to evaluate the total probability (or total inelastic cross section) for inelastic collisions in biological specimens, since we want to determine how efficiently we can extract information from the specimen as well as to determine the possibilities of beam damage by inelastic scattering.

In order to obtain rigorously the total inelastic scattering cross section, we must integrate Eq. (1.16) over all energy losses. For biological materials, this can be performed, at present, only for the nucleic acid bases, since we have experimental values of $-\text{Im}\ 1/\epsilon$ for these (Isaacson, 1972a; 1972b; Johnson, 1972). However, we can obtain a general expression applicable to most materials by performing a few approximations using the experimental inelastic scattering results for the nucleic acid bases and some other materials as a guide. One can show that upon integration of Eq. (1.16) over all energy losses, the total inelastic scattering cross section is given by:

$$\sigma_{\text{in}} \cong H \frac{1}{\beta^2} \{\ln (2/\theta_{\overline{E}}) - \beta^2 + \ln \gamma\}, \tag{1.17}$$

where all the properties of the specimen (i.e., atomic weight, density, etc.) are contained in the function H, and the terms in the brackets contain the entire energy dependence of the incident electron. $\gamma = (1 - \beta^2)^{-1/2}$, $\theta_{\overline{E}} = \overline{E}/pv$, and $\overline{E}$ is the average energy loss per collision of the incident electrons. Note that σ_{in} does not decrease as fast as $1/\beta^2$. This means that the ratio of inelastic to elastic scattering (Eq. (1.13)) is not constant with the energy of the incident electrons but rather increases slightly as the electron energy is increased.

We would now like to find a general expression for H so that σ_{in} for any material can be evaluated with reasonable accuracy. To do this we rely on a general argument concerning the nature of elastic and inelastic scattering. We know that the cross section for an incident electron to be scattered from a charged particle of charge $n \cdot e$ is proportional to $(ne \cdot e)^2$, where e is the electronic charge (i.e., Rutherford scattering).

If we neglect atomic screening, then the elastic cross section for an electron to be scattered from a nucleus of atomic number Z is proportional to $(Ze \cdot e)^2$. Similarly, the cross section for an electron to be scattered from another electron is proportional to $(e \cdot e)^2$. In inelastic scattering, the incident electrons are scattered by all the atomic electrons so that the cross section for inelastic scattering (neglecting atomic screening again) is then proportional to $Z \cdot (e \cdot e)^2$; i.e., it is the sum of the electron scattering from each atomic electron. Therefore, in this approximation the ratio of elastic to inelastic scattering is $\sigma_{el}/\sigma_{in} \propto Z$. One can show, in fact, that even if we include atomic screening, the ratio is still roughly proportional to Z, exclusive of a logarithmic term in β^2 (Scott, 1965; Lenz, 1954). Therefore, utilizing this fact and combining Eqs. (1.17) and (1.13), we obtain:

$$\sigma_{in}(Z) = \text{constant} \cdot \frac{1}{\beta^2}\sqrt{Z}\ \{\ln(2/\theta_{\overline{E}}) - \beta^2 + \ln\gamma\} \tag{1.18}$$

For energies up to 1 MeV, the terms in the bracket reduce to $\ln(2/\theta_{\overline{E}})$ to ~10% accuracy. The constant can be evaluated empirically using experimental data for the nucleic acid bases (Isaacson, 1972a, b; Johnson, 1972). The result is that for biological materials, the total inelastic scattering cross section is given as:

$$\sigma_{in} \cong \frac{3}{4}\frac{\lambda_c^2}{\pi\beta^2}\sum_i Z_i^{1/2}\ [\ln(2/\theta_{\overline{E}})] \tag{1.19}$$

where Z_i is the atomic number of the ith atom in the molecule. It is assumed, of course, that we can simply sum up the cross sections of the constituent atoms.

The above argument can be carried one step further to see that Eq. (1.19) is valid for higher-atomic-number materials. There are few inelastic scattering measurements for electrons in the tens of keV range. We can, however, extrapolate the inelastic scattering cross sections from measurements of $-\text{Im}\ 1/\epsilon$ for Cu, Ag, and Au (Daniels, 1969; Daniels *et al.*, 1970) and Si (Philipp and Ehrenreich, 1967). We have indicated these data along with measured values of the inelastic cross section for 50 keV electrons in C and Ge (Brünger and Menz, 1965) in Fig. 1.5. If we take the averaged value of $\ln(2/\theta_E)$ rather than $\ln(2/\theta_{\overline{E}})$ in Eq. (1.19) we get:

$$\sigma_{in} \cong \frac{3}{4}\frac{\lambda_c^2}{\pi\beta^2} Z^{1/2}\ \langle\ln(2/\theta_E)\rangle \tag{1.20}$$

which is plotted in Fig. 1.5 and appears to yield reasonable agreement with the data.

It should be pointed out, however, that one does expect the shell structure of the atom to manifest itself in the inelastic scattering cross section in the sense that the cross section will decrease as one proceeds across each row of the Periodic Table (in the direction of increasing Z). Therefore, although Eq. (1.20)

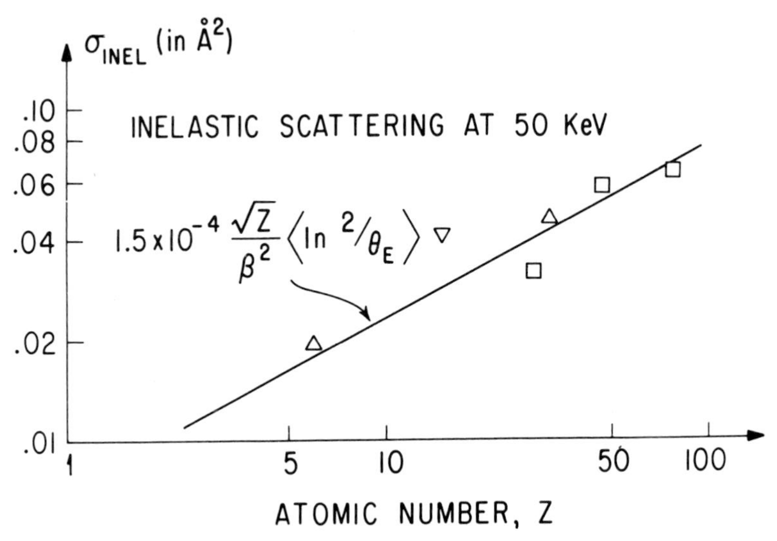

Fig. 1.5 The inelastic scattering cross section (in units of Å^2) as a function of the atomic number (Z) for 50 keV electrons. The solid line represents Eq. (1.20) of the text, while the open squares and triangles are from experimental data in the literature. The error in these experimental data is ~20%. Symbols: Δ from Brünger and Menz (1965); ∇ extrapolated from $-\text{Im } 1/\epsilon$ measurements of Philipp and Ehrenreich (1967); □ extrapolated from $-\text{Im } 1/\epsilon$ measurements of Daniels (1969) and Daniels *et al.* (1970). β is the ratio of the incident electron velocity to that of light.

forms a reasonable (and simple) basis for describing the general trend in the inelastic cross section, it is not very accurate in predicting the relative values of elements within a row and should only be used as a guideline. Recent values of the inelastic scattering cross section for elements in the first few rows ($Z \leqslant 18$) of the Periodic Table have been calculated by Inokuti *et al.* (1975) for 5 keV electrons and more accurately reflect the atomic shell structure than our Eq. (1.20). However, for the purposes of this chapter Eq. (1.20) will suffice since it is simple and represents a fit to the scant experimental data which exist.

Eq. (1.20) has been evaluated for a variety of biological molecules in Table 1.1 for 100 keV electrons assuming the average energy loss, $\overline{E} \cong 37$ eV for all the molecules [37 eV is the experimentally determined value for the nucleic acid bases (Isaacson *et al.*, 1973)]. Also included in this table are the partial cross sections for K-shell excitations (σ_k) obtained by assuming that (Isaacson and Crewe, 1975):

$$\sigma_k \cong 640 \frac{\lambda_c^2}{\pi\beta^2} \frac{Z_k}{E_k^3} (Z - .3)^4 \ln (1.2/\theta_{E_k}) \tag{1.21}$$

where E_k is the K-shell ionization energy in eV, Z_k is the number of K-shell electrons, and Z is the atomic number. The value $(Z - .3)$ approximates the effective nuclear charge.

Table 1.1 Inelastic Scattering Cross Sections at 100 keV*

Material	*Composition*	σ_{in} (in Å^2)	σ_k(in 10^{-3} Å^2)
Polyethylene	C_2H_4	.041	0.504
Polyester	$C_5H_4O_2$	.101	1.49
Teflon	C_2F_4	.077	0.84
Polyvinyl chloride	C_2H_2Cl	.051	0.52
Polyamide	$C_{12}H_{22}N_2O_2$	.28	3.6
Adenine	$C_5N_5H_5$	.141	2.23
Guanine	$C_5N_5OH_5$	.153	2.34
Cytosine	$C_4N_3OH_5$	.118	1.64
Thymine	$C_5N_2O_2H_6$	.134	1.84
Uracil	$C_4N_2O_2H_4$	.113	1.58
Histidine	$C_6N_3O_2H_9$	.168	2.26
Phenylalanine	$C_9NO_2H_{11}$	.187	2.66
Tyrosine	$C_9NO_3H_{11}$	.202	2.79
Tryptophan	$C_{11}N_2O_2H_{12}$	.229	3.35
Glycine	$C_2NO_2H_5$	.084	0.907
Acridine orange	$C_{17}N_3H_{20}$	.331	4.60
Cu-phthalocyanine	$C_{32}N_8H_{16}Cu$	.55	9.45
Paraffin	$C_{32}H_{66}$	.665	8.06
Tetracene	$C_{18}H_{12}$	.258	4.54
Coronene	$C_{24}H_{12}$	.326	6.05
Adenosine	$C_{10}N_5O_2H_{11}$	.250	3.62
Indigo	$C_{16}O_2N_2H_{10}$	.276	3.61
Pentacene	$C_{22}H_{14}$	.313	5.90
Valine	$C_5NO_2H_{11}$	.145	1.66

*Obtained from the equations in the text.

EXPERIMENTAL DATA ON MOLECULAR RADIATION DAMAGE

Types of Damage Measurements and Their Validity

Measurements of radiation damage in thin organic samples of the type used for electron microscopy have been the subject of numerous experimental investigations, particularly during the last ten years. A variety of different indicators have been used to assay damage to the specimen. These have included the total loss of mass, the loss of specific elements, the loss (or change) of crystalline structure and the change in the optical spectra (infrared, visible, ultraviolet, and characteristic electron energy loss). One can classify these experiments into two basic categories: those which are performed by simulating electron microscope conditions during irradiation, while the assay itself is performed outside the microscope, and those in which both the irradiation and the assay are performed entirely within the electron microscope. The types of experiments which belong to the first category are those in which the mass loss incurred by the specimen because of irradiation is measured by directly weighing the material before and after irradiation, and optical absorption measurements where large-area thick

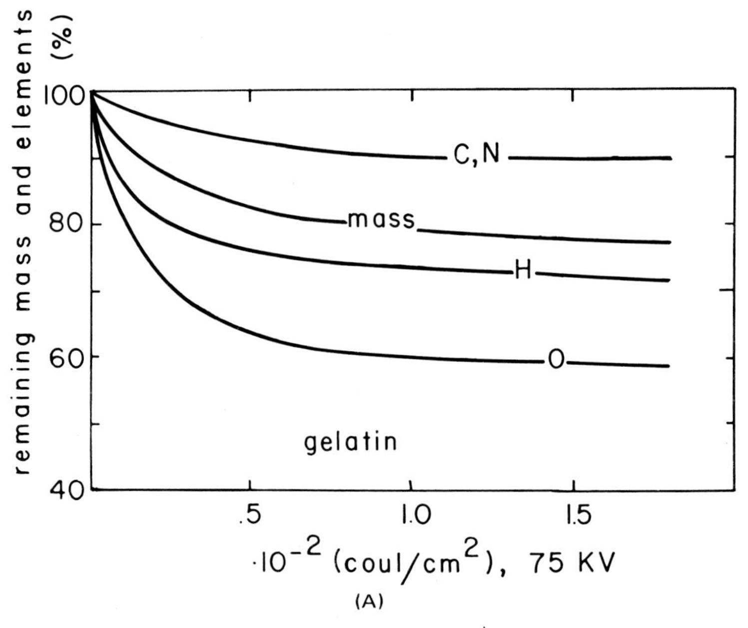
remaining mass and elements (%)
100
80
60
40
C,N
mass
H
O
gelatin
.5
1.0
1.5
·10⁻² (coul/cm²), 75 KV

(A)

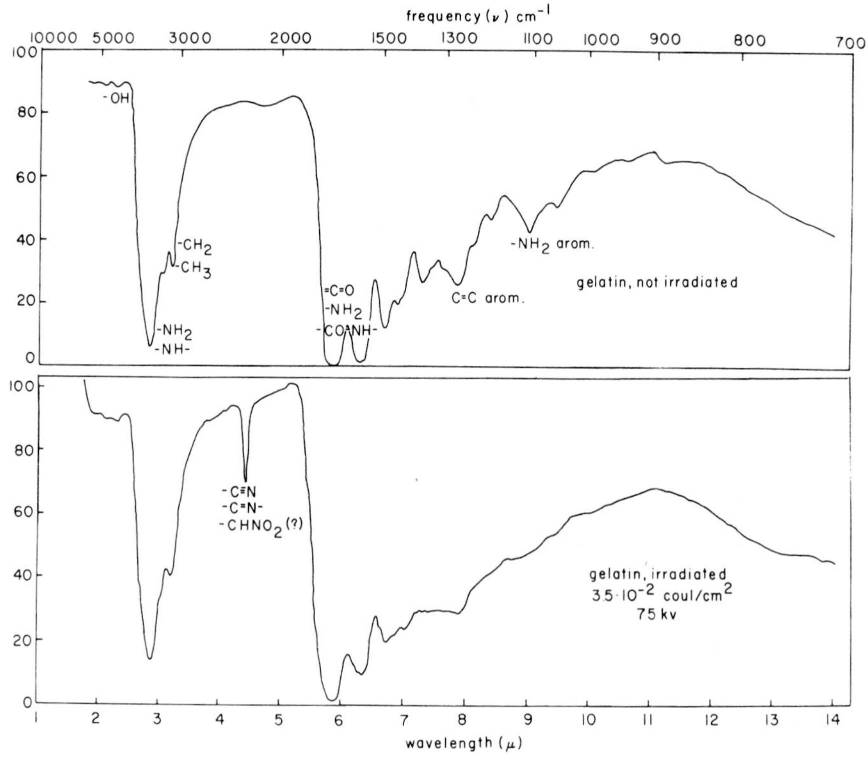
frequency (ν) cm⁻¹
10000 5000 3000 2000 1500 1300 1100 1000 900 800 700
-OH
-CH2
-CH3
-NH2
-NH-
=C=O
-NH2
-CO-NH-
C=C arom.
-NH2 arom.
gelatin, not irradiated
-C≡N
-C=N-
-CHNO2 (?)
gelatin, irradiated
3.5·10⁻² coul/cm²
75 kv
wavelength (μ)

(B)

films are analyzed for changes in optical transmission or absorption at various wavelengths.

The main disadvantage of the above-mentioned methods is that the specimens used must of necessity extend over fairly large areas and therefore be relatively thick ($\sim$1-20 μ) compared to the usual electron microscope specimen ($\lesssim$1,000 Å). This results in uncertainty in that the fraction of the total incident energy deposited in the specimens might be quite different from that under normal microscope conditions, and in that there may be damage due to heating which might make extrapolation back to thin specimens ambiguous.

An example of the two types of measurements mentioned above is shown in Fig. 1.6 (Bahr *et al.*, 1965). Figure 1.6A shows the measurements of the loss of mass incurred by irradiating a 10-μ-thick specimen of gelatin with 75 keV electrons. It has been determined by directly weighing the material and by chemical methods. Figure 1.6B shows the effects of irradiation on the infrared absorption spectrum of gelatin. In both cases, a considerable change due to the action of the electron beam can be seen.

Another type of measurement belonging to the first category is that of autoradiography. In this method, radioactive atoms (such as H^3, P^{32}, C^{14}, etc.) are incorporated into the specimen and the radioactivity (by autoradiography) of regions of the specimen, which have been irradiated by electrons in the electron microscope, is measured. A comparison of the activity with unirradiated regions then yields a measure of the radioactive atoms lost and hence the relative fraction of labeled atoms which have been lost because of the electron beam (Thach and Thach, 1971; Dubochet, 1975). Such measurements can be difficult to interpret if there is any diffusion of atoms occurring on the surface of the specimen grid.

Let us now consider the types of measurements which can be performed entirely within the electron microscope. These measurements more closely approximate conditions under which the electron microscope is used, although the results must be carefully interpreted since operating conditions vary widely from one microscope to another. There are basically three different types of such measurements which have been performed.

One can obtain the characteristic electron energy loss spectrum of the incident electrons which have been inelastically scattered upon traversing the specimen, the specimen being of the order of 20-500 Å thick. This requires, of course, that the microscope be equipped with an energy analyzer. Since the energy loss spectra can be correlated with the optical absorption (Daniels *et al.*, 1970),

---

Fig. 1.6 Electron radiation damage of 1 mg/cm^2 films of gelatin using 75 keV incident electrons (from Bahr *et al.*, 1965). (A) The remaining mass and elementary composition measured by direct weighing and chemical determination. (B) The infrared absorption spectra of nonirradiated (top) and irradiated (bottom) gelatin. Note the appearance of an absorption band in the irradiated material at $\sim$4.5 μm, which is attributed to double and triple bonds between carbon and nitrogen that are due to the loss of oxygen from the molecule.

these measurements can be related to optical measurements on much thicker films of the order of 1-10 μ (Isaacson *et al.*, 1973).

Since the low-lying (less than 10 eV loss) energy loss spectrum in biological materials is determined by the π molecular orbital levels of the molecule, one can obtain information concerning molecular structure damage by observing the changes in this spectrum as a function of irradiation dose. On the other hand, if one monitors the change in the energy loss spectrum in the region of K-shell excitation (>280 eV loss) as a function of irradiation dose, then one obtains information concerning the relative chemical compositional change as well as changes in the charge distribution in the molecule (Isaacson *et al.*, 1973).

One can also obtain electron diffraction patterns of small crystals or polycrystalline films. The decrease in certain diffraction peaks or the change in relative spacings as a function of electron irradiation provide us information on crystalline damage (or change in the weak bonding forces).

A third damage indicator which has been used is the change in bright field and dark field image intensities as a function of incident irradiation dose (loss of contrast method). The bright field method has been predominantly utilized in the literature. The use of this technique as an indication of mass loss can be seen more easily if we realize that to a good approximation, the number of unscattered electrons, n_{un}, is related to the number of incident electrons, n_0, by Beer's law (Zeitler and Bahr, 1965) as:

$$n_{un}/n_0 = \exp(-N\sigma T), \tag{1.22}$$

where N is the molecular density, T is the specimen thickness, and σ is the total molecular cross section for scattering outside the objective aperture in a CTEM or outside the detector in the STEM. Since $N = \rho/m_m$ where ρ is the specimen density and m_m is the mass per molecule, NT is proportional to ρT, the mass per unit area of the specimen. Therefore, changes in n_{un} are reflected in changes in the mass of the specimen to a first approximation (σ/m_m is approximately a constant for a given experimental set-up). Selective loss of one type of atom before the others will not affect the results since the majority of scattering comes from atoms of approximately the same atomic number (C, N, and O). A more detailed discussion of the loss of bright field contrast method to determine loss of mass is given by Lin (1974).

The method of assaying mass loss by monitoring the change in the dark field image intensity is eventually the same as for bright field except that as mass is lost, the image intensity decreases since

$$n_{sc}/n_0 = 1 - \exp(-N\sigma_{sc} T), \tag{1.23}$$

where n_{sc} is the number of scattered electrons (elastic or inelastic) and σ_{sc} is the cross section for scattering into the dark field detector or aperture. This method has an advantage over simply using the bright field change in contrast, since by monitoring both the elastic and inelastic scattering intensities, one can also determine the change in the effective atomic number of the specimen (Isaacson *et al.*, 1974b; Langmore, 1975).

Examples of the types of damage indicators which one can obtain directly within the electron microscope are shown in Figs. 1.7 through 1.10. The results shown in these Figures were obtained at the University of Chicago using an electron gun scanning microscope with a field emission source and equipped with a high resolution magnetic spectrometer for energy analysis of the transmitted electrons. The entire apparatus has been described in detail before (Crewe *et al.*, 1969, 1971b). The important points to be noted are the following: (1) the microscope chamber operates at $\sim 3 \times 10^{-10}$ torr, thus eliminating specimen contamination due to residual gases in the chamber; (2) the system can produce an electron probe as small as 100 Å in diameter with beam currents up to 2×10^{-10} amp; (3) the system can be used to obtain characteristic energy loss spectra, energy filtered scanning diffraction patterns, total unscattered beam current, and bright field (or dark field) micrographs, all from the same specimen in the same environment within the same microscope.

Figure 1.7 shows low-lying energy loss spectra of fast electrons which have traversed two different thin films composed of biological molecules supported on 30-Å-thick carbon substrates. Figure 1.7A shows the spectrum of 25 keV electrons which have traversed an ~500-Å-thick adenine film, while Fig. 1.7B shows the spectrum of 25 keV electrons which have traversed a film of the aromatic amino acid, L-phenylalanine, of similar thickness. Note that after an incident electron dose of 0.7 coul/cm^2 in Fig. 1.7A and 0.2 coul/cm^2 in Fig. 1.7B, the spectra have been altered to a great extent. Irradiation with doses greater than these causes no further change in the spectra. The dose at which the structure disappears (or more correctly, the stabilization occurs) is a relatively subjective quantity; a more quantitative method of assaying the damage is to characterize the damage by the dose necessary to decrease the intensity of a given peak to $1/e$ of its initial value at zero dose. This dose is called $D_{1/e}$, the characteristic damage dose, and it is useful to use a similar definition for both mass loss and diffraction damage.

Figure 1.8 shows the electron-beam-induced changes in the characteristic electron energy loss spectrum observed at the onset of excitation of the K-shells of the carbon and nitrogen atoms in adenine. Note that even after a dose greater than .75 coul/cm^2, there is only a very slight change in the nitrogen K-shell intensity.

Figure 1.9 shows an energy filtered electron diffraction line scan of 17.5 keV electrons which have traversed a 400-Å-thick film of the nucleic acid base thymine supported by a 30-Å-thick carbon substrate. The damaged pattern was obtained after a total incident dose of .1 coul/cm^2. This damaged structure remained the same even after exposure to much higher doses indicating "stabilization."

Figure 1.10 shows bright field micrographs obtained in an STEM using 17.5 keV

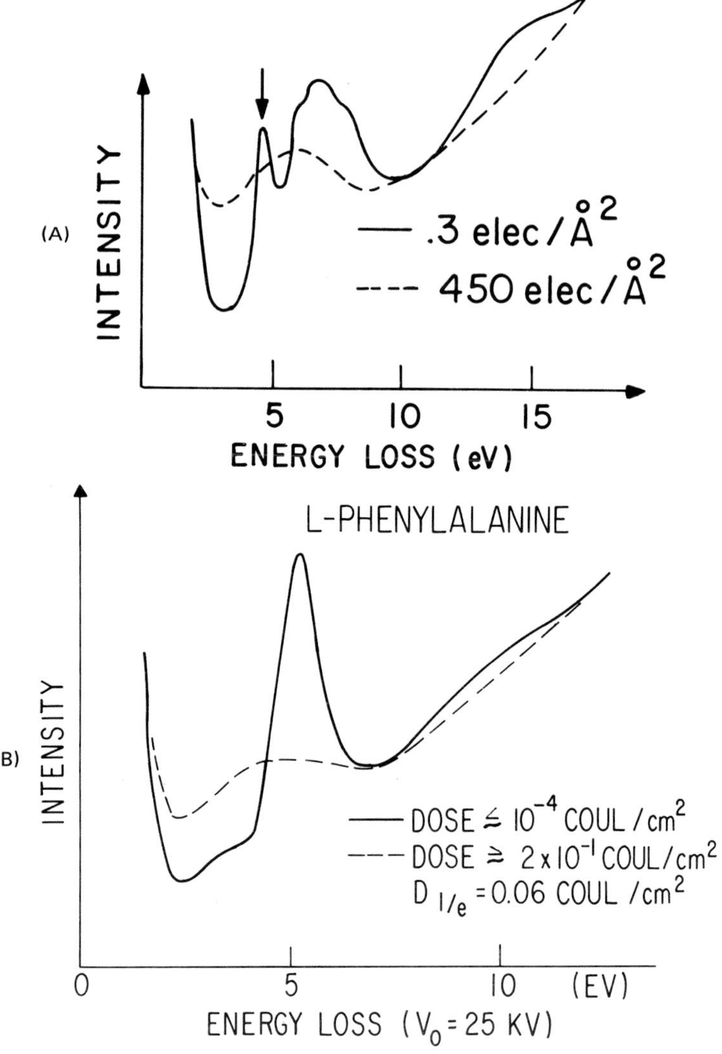

Fig. 1.7 Electron-irradiation-induced changes in 0 to 15 eV energy loss structure of the characteristic energy loss spectra of ~500-Å-thick biological films supported on 30-Å-thick carbon substrates. The incident electron energy was 25 keV.

(A) Adenine, from Isaacson (1972b).

(B) L-Phenylalanine, , from Lin (1974).

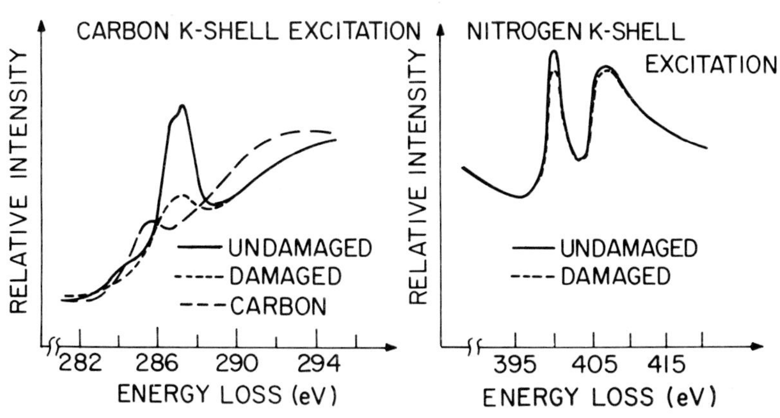

Fig. 1.8 Electron-irradiation-induced changes in the structure observed at the onset of electron energy loss due to *K*-shell excitation of the nitrogen and carbon atoms in adenine. The undamaged curve is for a total incident dose less than 2×10^{-4} coul/cm^2 ($\frac{1}{8}$ electron/Å^2), while the damaged curve is for a dose greater than .75 coul/cm^2 (470 electrons/Å^2). The *K*-shell excitation spectrum for a thin carbon film is shown for comparison. The incident electron energy was 25 keV (from Isaacson *et al.*, 1973).

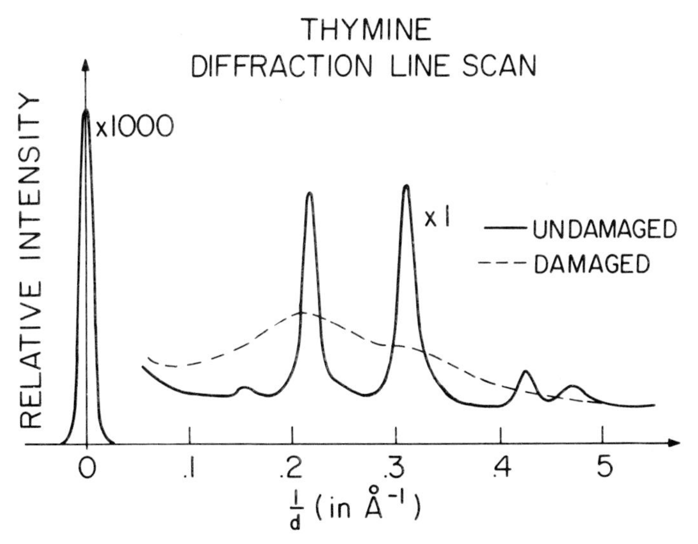

Fig. 1.9 Electron diffraction line scan for an ~400-Å-thick thymine ($C_5N_2O_2H_6$) film supported on a 30-Å-thick carbon substrate. The line scan was obtained by filtering out electrons which had lost more than 1 eV in energy in traversing the film. The incident electron energy was 17.5 keV. The undamaged curve was obtained with a total dose of less than 5×10^{-4} coul/cm^2 while the damaged curve was obtained after a dose of 10^{-1} coul/cm^2. (from Isaacson *et al.*, 1973).

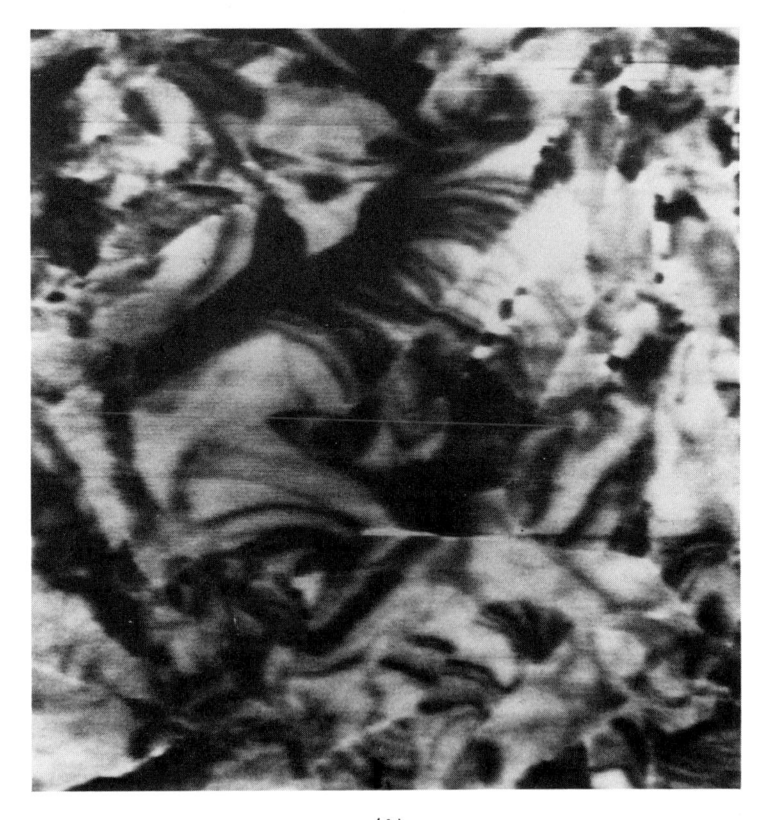

(A)

Fig. 1.10 Bright field micrographs of an ~300-Å-thick cytosine film supported on a 30-Å-thick carbon substrate. The micrographs were obtained using all electrons scattered within a 12.0-mrad cone about the forward direction. The incident electron energy was 17.5 keV, and the full horizontal scale is 6 μm. (A) Micrograph taken with a total dose less than 3×10^{-4} coul/cm^2. Note the extinction contours on the various crystallites (which are ~1 μm in size). (B) Micrograph of the same area as A, taken after a dose of $\sim 5 \times 10^{-2}$ coul/cm^2. This dose is greater than the $D_{1/e}$ for diffraction damage of 9×10^{-3} coul/cm^2 and is therefore sufficient to destroy the crystal structure. The crystal grain boundaries have not changed, but most of the extinction contours have disappeared.

electrons. The specimen is a 500-Å-thick film of the nucleic acid base cytosine supported on a 30-Å-thick carbon substrate. The micrographs have been taken using a detector aperture subtending a half angle of 12 mrad at the specimen. Figure 1.10A was taken with a dose $< 10^{-4}$ coul/cm^2, and the crystalline grain boundaries and crystal extinction contours are quite evident. Figure 1.10B is the same area taken after a dose of 5×10^{-2} coul/cm^2. The effect of the beam damage has been to destroy the crystal structure and therefore the crystalline extinction contours.

(B)

Fig. 1.10 *(Continued)*

Figure 1.11 shows typical dose-response measurements on L-histidine indicating the three different fundamental indicators which can be used within the microscope. In this figure can be seen as a function of irradiation dose the low-lying energy loss spectra (curve 3), the loss of mass (curve 2), and the decrease in intensity of the 3.6 Å line in the diffraction pattern (curve 1). All these indicators were obtained from the same specimen in the same environment. Note that the loss of crystallinity occurs at much less dose than do the other damage indicators (Isaacson *et al.*, 1974c). This will be discussed further in the next few sections.

A summary of a wide variety of experimental results concerning radiation damage in electron microscopy is given in Tables 1.2, 1.3, and 1.4. In these Tables we have listed experiments which assay damage outside of the electron microscope as well as those experiments which assay damage within the microscope. Moreover, the experiments tabulated have been performed utilizing inci-

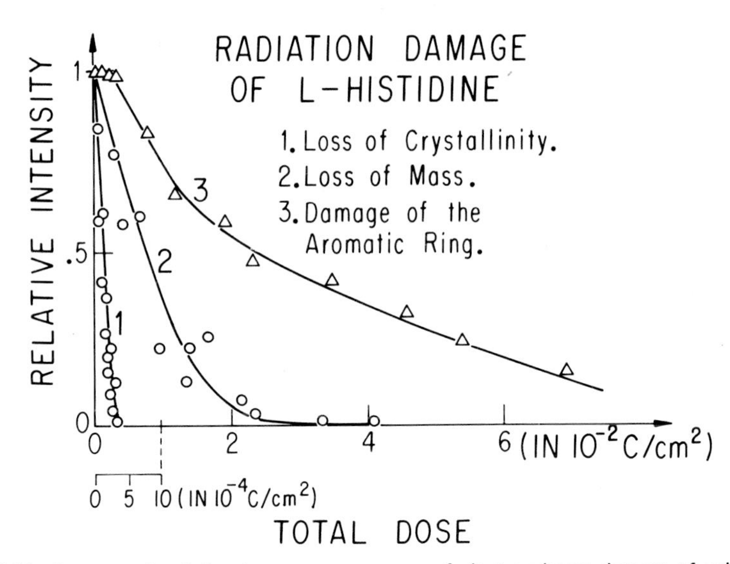

Fig. 1.11 An example of the dose-response curves of electron beam damage of a thin film of the aromatic amino acid L-histidine supported on a 20-Å-thick carbon substrate. The data were obtained by P. Lin using the system described by Crewe *et al.* (1971b). The 10^{-4} coul/cm^2 scale refers only to curve 1. Curve 3 refers to the damage as measured by the change in intensity of the 0–10 eV structure in the energy loss spectra. The dose rate at which the measurements were performed was ~10^{-5} amps/cm^2 and the incident electron energy was 25 keV.

dent electron energies from 17.5 to 100 keV and under a wide range of conditions. The results have also been obtained under various conditions of current density of the illuminating beam. However, since the measurements of Glaeser (1971), Isaacson *et al.* (1973), Lin (1973, 1974), and Langmore (1975) indicate that there is no effect of current density on radiation damage in the range from 10^{-6} to 10^4 amp/cm^2 for the types of specimens considered in these Tables, the current density as a parameter when comparing results from different researchers (at least for now) is disregarded.

We would like to be able to extract from these results some information regarding the relative radiation stability of different materials. In order to do this, we must be aware of the dependence of electron beam damage on the kinetic energy of the incident electrons and of the different criteria of "damage" chosen by different researchers.

There are basically two different criteria which are used in assaying radiation damage (characterizing a "dose" for damage). The one most easily obtained is called the "end point dose" or D_{ep}. This is the incident electron dose at which "stabilization" of the structure under investigation occurs. (By structure we mean an absorption spectrum, a diffraction pattern, or simply the mass.) Increasing the dose beyond the D_{ep} produces no further apparent change. While this

Table 1.2 Experimental Measurements of Electron Beam Damage at Room Temperature: Characteristic Energy Loss and Optical Absorption

Material	*Damage dose [in electrons/Å²]*	E_0 *[in keV]*	*Method*	*Reference*
Adenine	119 (a)	25	0–10 ev characteristic electron energy loss spectra	Isaacson *et al.* (1973)
Guanine	100 (a)	25		
Cytosine	88 (a)	25		
Thymine	32.5 (a)	25		
Tryptophan	77 (a)	25	0–10 ev characteristic electron energy loss spectra	Lin (1974)
Phenylalanine	37 (a)	25		
Histidine	23.2 (a)	25		
Di-glycine	6.9 (a)	25		
DNA	11.1 (a)	25	0–10 ev characteristic electron energy loss spectra	Isaacson *et al.* (1973)
Poly-A	24 (a)	25		
E. coli ribosomes	12.5 (a)	25		
Acridine orange	71.5 (a)	25		
Copper-phthalocyanine	111 (a)	25		
Copper-phthalocyanine	~650 (a)	60	Optical transmission at 5,000 Å	Reimer (1961)
Phthalocyanine	500 (a)	60		
Indigo	250 (a)	60	Optical transmission at 4,000 Å	Reimer (1961)
Tetracene	31.3 (b)	60	Optical absorption maxima between 4,000 Å and 5,000 Å	Reimer (1961)
Pentacene	31 (b)	60	Optical absorption maxima between 4,000 Å and 6,500 Å	Reimer (1961)
Polyethylene	25 (b)	80	Characteristic electron energy loss spectra	Ditchfield *et al.* (1973)

Abbreviations: a—dose at which the intensity of a peak decreases to $1/e$ of the difference in intensity between 0 and ∞ dose: $D_{1/e}$; b—dose necessary to completely destroy the characteristic peaks being monitored: D_{ep}.

quantity often appears in the literature because of the simplicity in obtaining it, it is fairly subjective in the sense that no two researchers will obtain the exact same end point dose. In fact, researchers performing the same experiment under identical conditions can often obtain end point doses differing by almost a factor of two Glaeser *et al.*, 1971a; Grubb and Groves, 1971).

A more quantitative measure of radiation damage is obtained by defining a particular point on the dose-response curve of the material. The dose-response curve is a plot of some measurable quantity of the specimen versus the incident

Table 1.3 Experimental Measurements of Electron Beam Damage at Room Temperature: Loss of Mass

Material	*Damage dose [electrons/Å²]*	*TFML*	E_0 *[in keV]*	*Method*	*Reference*
Nucleic acid bases	>10³ (a)	<.10	25	Bright field contrast in a STEM	Isaacson *et al.* (1973)
DNA	>10⁴ (a)	<.10	29	Dark field contrast in a STEM	Wall (1972a)
DNA	~15 (a)	~.40	28.6	Dark field contrast in a STEM	Langmore (1975)
Glycine	2.5 (a)	.87	25	Bright field contrast in a STEM (using increase in number of unscattered electrons)	Lin 1974
Di-glycine	1.25 (a)	.30	25		
L-Histidine	5.6 (a)	.48	25		
L-Tyrosine	5.3 (a)	.45	25		
L-Tryptophan	5.4 (a)	.45	25		
L-Phenylalanine	2.2 (a)	.80	25		
L-Tryptophan	0.10 (a)	.20	70	Direct weighing of 0.5 mg/cm² thick films	Stenn and Bahr (1970b)
L-Phenylalanine	0.13 (a)	.80	70		
Poly-L-Phenylalanine	0.18 (a)	.30	70		
Collagen	0.125 (a)	.10	70		
Insulin	0.18 (a)	.15	70		
Polyethylene	0.62 (a)	.05	75	Direct weighing of 1 mg/cm² thick films	Bahr *et al.* (1965)
Polyester	5.5 (a)	.30	75		
Teflon	1.9 (a)	.85	75		
Polyvinylchloride	0.13 (a)	.40	75		
Polyamide	3.7 (a)	.16	75		
Gelatin	2.5 (a)	.22	75		
Polymethacrylate	6.25 (b)	.40	60	Bright field contrast in a CTEM	Reimer (1965)
fD phage	62.5 (c)	.50	29	Dark field contrast in a STEM	Wall (1972a)
Chromatin	~5 (a)	~.40	30.5	Dark field contrast in a STEM	Langmore (1975)
TMV	~1.6 (a)	>.40	25.5		
Blowfly larva Salivary gland Section	28.2 (a,d)	.30	30	Decrease in emitted continuum X-rays	Hall and Gupta (1974)
PCMPS	5.0 (e,f)	>.80	28.6	Decrease in ratio of elastic to inelastic scattering	Langmore (1975)

(a) Dose at which $1/e$ of the total fractional mass loss (TFML) occurs
(b) End point dose (D_{ep}); no more mass loss occurs after this dose.
(c) Dose at which half of the mass is lost.
(d) Section cut frozen and dried; 4 microns thick.
(e) Parachloromercuriphenyl sulfonic acid; dose at which the ratio signal falls to $1/e$ of the total change.
(f) Assays only the loss of mercury.

Table 1.4 Experimental Measurements of Electron Beam Damage at Room Temperature: Electron Diffraction

Material	*Damage dose [in electrons/Å²]*	*E_0 [in keV]*	*Method*	*Reference*
Cytosine	5.6 (a)	17.5	Decay of 4.75 and 6.52 Å peaks	Isaacson *et al.* (1973)
Copper-phthalocyanine	6.2 (a)	17.5	Decay of 5.7 Å peak	Isaacson *et al.* (1973)
Copper-phthalocyanine	300 (b)	17.5	Disappearance of 12.5 Å peak	Isaacson *et al.* (1973)
Copper-phthalocyanine	625 (b)	60	Disappearance of 12.5 Å peak	Reimer (1965)
Phthalocyanine	62 (b)	60	Disappearance of 5.7 Å peak	Reimer (1965)
Tetracene	62.5 (b)	60	Decay of 4.6 Å peak	Reimer (1961)
Indigo	81 (b)	60	Disappearance of crystalline pattern	Reimer (1961)
Tetracene	100 (a)	100	Decay of 3.23 and 3.96 Å peaks	Siegel (1972)
Paraffin	3.8 (a)	100	Decay of 4.15 Å peak	Siegel (1972)
Polyethylene	3.1 (b)	100	Disappearance of crystalline pattern	Kobayashi and Ohara (1966)
Polyethylene	3.0 (c)	100	Decay of [110] peaks	Thomas and Ast (1974)
Polyethylene	4.7 (b)	100	Disappearance of crystalline pattern	Thomas *et al.* (1970)
Polyoxymethylene	2.5 (b)	100	Disappearance of crystalline pattern	Thomas *et al.* (1970)
Histidine	.07 (a)	25	Decay of 3.7 Å peak	Lin (1974)
Valine	.82 (b)	80	Disappearance of crystalline pattern	Glaeser (1971)
Adenosine	6.25 (b)	80	Disappearance of crystalline pattern	Glaeser (1971)
Thymine	73 (b)	1000	Disappearance of crystalline pattern	Dupouy (1974)
Catalase (stained with uranyl acetate)	310	80	Diffraction pattern degraded to 25 Å order	Glaeser (1971)
Coronene	150 (b)	100	Disappearance of crystalline pattern	Salih (1974)
Coronene	30 (d)	100	Decay of peaks within 1.15 Å annulus	Claffey and Parsons (1972)

(a) Dose at which the intensity of a peak decreases to $1/e$ the difference in intensity between 0 and ∞ dose: $D_{1/e}$.
(b) Dose necessary to completely destroy the crystalline diffraction pattern: D_{ep}.
(c) Dose necessary to reduce intensity of [110] peaks to 60% of the initial intensity.
(d) Dose necessary to reduce the intensity of the peaks to ¾ of the initial intensity.

electron dose which was imparted to the specimen before the measurement was made. A particularly useful characteristic dose is the $D_{1/e}$ dose. This is the dose necessary for the intensity of particular peaks in the diffraction pattern, optical absorption or characteristic energy loss spectra, or for mass lost, to decrease to $1/e$ of the difference between that quantity at zero dose and at infinite dose. One can show that for a "single hit" model of radiation damage the $D_{1/e}$ dose (in electrons/Å^2) is the inverse of the cross section, σ_D, for damage to the specimen (e.g., Isaacson *et al.*, 1973). This model gives an exponential decay to the dose-response curve. Although the dose-response curves for electron microscopy radiation damage studies are not purely exponential, this model is quite adequate for preliminary comparative purposes.

In comparing a particular experiment in which a $D_{1/e}$ dose was measured to one in which only the end point dose, D_{ep}, was obtained under the same conditions, one can only accurately state that $D_{1/e} < D_{\text{ep}}$. However, to a good approximation, the D_{ep} is approximately equal to the dose at which the quantity measured falls to one tenth to one hundredth of the difference between that at zero and infinite dose. Then for an exponential decay of the dose-response curve beyond a dose of $D_{1/e}$, we would get $D_{\text{ep}} \cong (2.3 \text{ to } 4.6) \cdot D_{1/e}$. This simple expression allows us to obtain, reasonably well, some correspondence between the two different measurements.

The other quantity which is frequently different for radiation damage experiments is the energy of the incident electrons, and we would like to be able to compare results obtained at different energies. There have been several studies in the literature which attempted to elucidate the energy dependence of radiation damage in the electron microscope (Kobayashi and Sakaoku, 1965; Kobayashi and Ohara, 1966; Bahr *et al.*, 1966; Grubb and Groves, 1971; Thomas *et al.*, 1970; Dupouy, 1974). A summary of some of these studies is presented in Fig. 1.12.

In Fig. 1.12 we have plotted the end point dose (D_{ep}) for loss of the crystalline electron diffraction pattern at a given incident electron energy normalized to that quantity at 100 keV incident energy. This ratio is plotted as a function of the inverse of the energy-dependent term of the inelastic scattering cross section in Eq. (1.18). It can be seen that the experimental points lie along a straight line of slope 1, indicating that to a first approximation radiation damage is proportional to the inelastic scattering cross section. This would be expected if the bulk of the damage was due to inelastic collisions of the incident electrons with the specimen. Of course, at the higher electron energies, knock-on collisions become increasingly important as an avenue for radiation damage. The main point of Fig. 1.12 is that to a good approximation the dose for radiation damage should be inversely proportional to $\beta^2/\{\ln [(2\beta^2 \gamma mc^2)/E] - \beta^2 + \ln \gamma\}$.

It should be pointed out that very often in the literature the energy dependence of the dose for radiation damage is compared to the energy dependence of the stopping power cross section, dE/dx, obtained from variations of the Bethe-Bloch formula (e.g., Glaeser, 1971). The energy dependence of this formula (for example, Rohrlich and Carlson, 1954) is within a few percent of that given in Eq. (1.18). So it makes little difference with regard to the energy dependence of damage whether one compares experiments with stopping power calculations or the inelastic scattering cross sections.

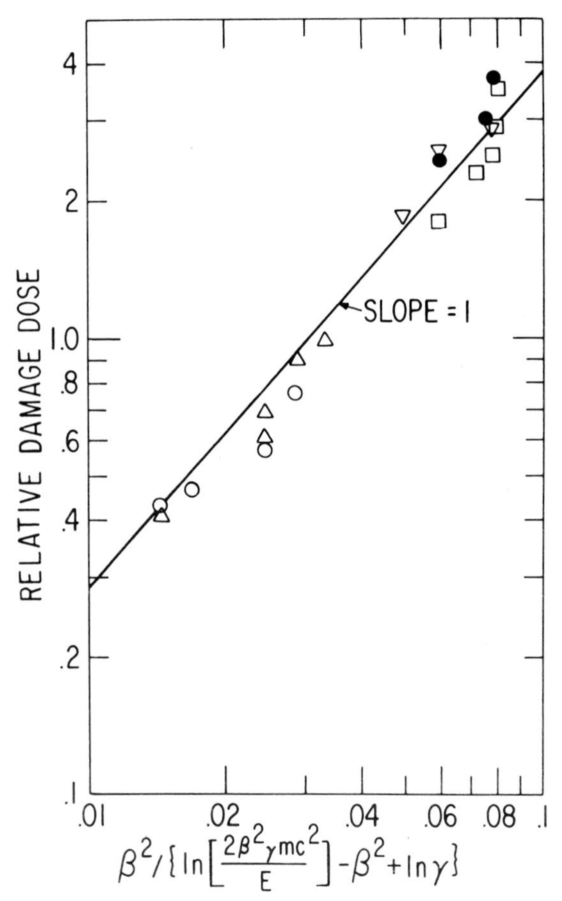

Fig. 1.12 The energy dependence of electron beam damage. The relative damage dose is given as the ratio of the end point dose for diffraction damage at a given incident electron energy to the end point dose at 100 keV. (All the 100 keV data are not indicated for the sake of clarity.) The horizontal scale is the inverse of the energy-dependent part of the inelastic scattering cross section given in Eq. (1.18) (where we have assumed an average energy loss, $\overline{E}$ equal to 37 eV [the experimental value for the nucleic acid bases determined by Isaacson *et al.* (1973)]. If the damage dose were inversely proportional to the inelastic scattering cross section (i.e., the damage cross section proportional to σ_{in}), then the points of the plot should lie on a line with unit slope. For orientation purposes, $\beta^2/\{\ln [2\beta^2\gamma mc^2/\overline{E}] - \beta^2 + \ln\gamma\} = .033$ for 100 keV electrons. Symbols: ○ polyethylene and polyoxymethylene (Grubb and Groves, 1971); △ cytosine (Isaacson); ▽ polyethylene (Kobayashi and Ohara, 1966); □ polyethylene and polyoxymethylene (Thomas *et al.*, 1970); ● thymine (Dupouy, 1974).

Comparison of Different Materials

It is obvious in Tables 1.2 through 1.4 that after voltage dependence effects and the different damage criteria are taken into account there is still a considerable variability in the ability of different materials to withstand electron beam destruction. As a general observation, however, the aliphatic (or saturated bond) compounds are more sensitive to radiation damage than the aromatic compounds by about an order of magnitude. This finding appears to be consistent with radiolytic measurements (Hart and Platzman, 1961; Charlesby, 1960), even though the dose rate in the electron beam measurements is two to three orders of magnitude greater than that used in conventional radiolytic studies. As a general note, if one assumes a specimen density of 1.3 grams/cm^3 and 310 eV/μ as the energy loss per unit path length at 100 keV, then for an incident beam current density of 10^{-3} amp/cm^2, the specimen would receive a dose rate of $\sim 10^8$ rads/second. (The quantity 10^8 rads absorbed by a one-gram specimen is enough energy to convert a gram of ice to steam!)

As Lin (1974) has shown, the G values obtained in electron microscope studies* are consistent with those obtained from conventional radiation chemistry, and it makes little difference whether the impinging radiation consists of fast electrons or high energy γ rays (Dertinger and Jung, 1970). This fact is of use in our study of specimen damage since it allows us to utilize the large amount of information on radiation damage which is known from radiation chemistry methods.

Another general point worth noting concerning the results in Tables 1.2 through 1.4 is that for the most part, radiation damage occurs in three stages: first, loss of crystallinity; then, loss of mass; and, finally, destruction of the optical absorption or characteristic energy loss spectra (Isaacson, 1975a). However, in some cases (for instance, tetracene) this order does not hold, and so one must be extremely cautious in trying to determine the extent of specimen damage on the basis of only one type of damage measurement.

One final observation is that there is a difference of radiation sensitivity between groups of similar molecules. For example, among the nucleic acid bases, the order of radiation stability is that $A > G > C > T$ with thymine being the most radiation-sensitive. Insofar as destruction of the characteristic electron energy loss spectra of aromatic amino acids is concerned, TRY > TYR > PHE > HIS. This fact will be discussed further in a later section.

*The G value is defined as the number of molecules damaged per 100 eV deposited in the specimen. We can calculate G values from results in Tables 1.2–1.4 by using the equation

$$G = \frac{100}{f\bar{E}} \frac{\sigma_D}{\sigma_{in}}$$

where f is the fraction of energy actually deposited in the specimen.

Effects of Low Temperatures

Periodically, it has been suggested that if the specimen were kept at low temperatures, the molecular disorder which invariably accompanies radiation damage might be prevented (Reimer, 1965; Glaeser, 1971; Box, 1973). The general rationale for this argument is that if disorder is caused by the motion of ionized molecular fragments (which are the result of the primary radiation process), then perhaps this motion could be restricted at low temperatures. Therefore, even though the fundamental radiation damage processes might not be temperature-dependent, the broken fragments would remain in approximately the same place, thus allowing some "structural" determination to be made.

Table 1.5 Low Temperature Observations of Specimen Damage by Electron Irradiation

Specimen	*Temperature*	*Low temperature to room temperature improvement*	*Method*
Polyethylene			
(Kobayashi and Sakaoku, 1965	90°K	~1	Diffraction
(Grubb and Groves, 1971)	18°K	~3	Diffraction
(Venables and Basset, 1967)	10°-20°K	~2	Diffraction
Paraffin			
(Siegel, 1972)	~4°K	~2	Diffraction
Tetracene			
(Siegel, 1972)	~4°K	~4	Diffraction
Hexadecachloro-Cu-phthalocyanine			
(Harada *et al.*, 1972)	100°K	~1.7	Diffraction
Nucleic acid bases			
(Isaacson *et al.*, 1973)	90°K	~1.4	0-10 eV characteristic energy loss
Adenosine			
(Glaeser *et al.*, 1971a)	10°-20°K	~1.5	Diffraction
φX-174 DNA			
(Hotz and Muller, 1968)	4.2°K	~2	Inactivation
L-Valine			
(Glaeser *et al.*, 1971a)	10°-20°K	~1	Diffraction
Lysozyme			
(Fluke, 1966)	100°K	~5	Inactivation
Ribonuclease			
(Fluke, 1966)	100°K	~4	Inactivation
Hyaluronidase			
(Vollmer and Fluke, 1967)	66°K	~8	Inactivation
Catalase			
(Setlow, 1952)	100°K	~2	Inactivation
Invertase			
(Pollard *et al.*, 1952)	100°K	~3	Inactivation

Few controlled experiments have been made, however, at low temperatures. We have listed in Table 1.5 a selection of some of the measurements appearing in the literature in which the radiation sensitivities at room temperature and at low temperature are compared. The types of radiation damage assayed here are the loss of electron diffraction pattern, the change in the characteristic electron energy loss spectra, and biological inactivation. The last assay must of necessity be performed upon warming the sample up to room temperature.

The main feature to be noticed in this tabulation is that except for one inactivation measurement, the improvement in radiation stability (i.e., the increase in $D_{1/e}$) at low temperature compared to that at room temperature is never more than a factor of five. For the purely electron microscope observations, the improvement is less even when the measurements were performed near liquid helium temperatures.

It should be mentioned that presently there is little published quantitative data on the effect of mass loss at low temperatures, although one might expect a significant reduction in mass loss at low temperatures because of the fact that one is asking not for preservation of order, but only that the material does not vaporize. In fact, recent cryogenic measurements on moss loss of thin films of amino acids upon irradiation with 17 keV electrons (Ramamurti, *et al.*, 1975) and one measurement of *E. Coli* irradiated by 100 keV electrons (Dubochet, 1975) seem to support this expectation.

DISCUSSION OF MOLECULAR DAMAGE MECHANISMS

Upon looking at the data which we have presented in the preceding section, one gets a first impression that it is an impossible task to sift through the results to obtain trends which might allow one to understand the basic mechanism of radiation damage in the electron microscope. As we mentioned before, one must first scale the damage dose with incident electron voltage and then reconcile the different definitions of damage before one can actually compare damage between different materials. We have attempted to do this for some selected materials and have presented the results in Tables 1.6a–c, where we have scaled all doses to an equivalent $D_{1/e}$ dose at 25 keV. We chose 25 keV as the comparison voltage since we have most experimental results at this incident electron energy. The data has been presented in the form of a cross section for damage, σ_D, where $\sigma_D = 1/D_{1/e}$ if $D_{1/e}$ is expressed in incident electrons per Å^2.

It was mentioned before that there is a fair variability of radiation resistance among the nucleic acid bases and the aromatic amino acids, and that is shown quite clearly in Table 1.6a. It is thought that the molecular parameter that determines a molecule's ability to resist damage due to ionizing radiation is the resonance energy per π electron (Pullman and Pullman, 1963), the resonance energy being defined as the energy that the molecule has over that which it would have if all its bonds were localized. The larger the resonance energy per π

Table 1.6a Summary of Beam Damage Determined From Energy Loss Spectra*

Molecule	$\sigma_{D,E}$ [*in* 10^{-3} Å^2]	$\sigma_k/\sigma_{D,E}$	$\sigma_{in}/\sigma_{D,E}$
Adenine (a)	8.4	.88	55.5
Guanine (a)	10.0	.77	50.5
Cytosine (a)	11.4	.48	34.2
Thymine (a)	30.8	.20	14.4
Poly A (a)	41.5	.33	26.5
DNA (a)	90.0	.15	11.7
L-Histidine (b)	43.0	.17	12.9
L-Phenylalanine (b)	27.0	.32	22.8
L-Tyrosine (b)	18.0	.51	37.0
L-Tryptophan (b)	13.0	.85	58.0
Acridine orange (a)	14.0	1.09	78.0
Copper phthalocyanine (a)	9.0	3.45	199.0
Indigo (c)	8.8	.41	31.4
Polyethylene (d)	340.0	.05	0.4

*All values are referred to 25 keV incident electron energy and have been taken from Table 1.2. Original data not obtained at 25 keV have been scaled with voltage as described in the text.

(a) Isaccson *et al.* (1973).
(b) Lin (1974).
(c) From optical transmission data of Reimer (1961).
(d) Extrapolated from Ditchfield *et al.* (1973).

electron, the more delocalized are the π electrons in the molecule. Thus it becomes easier to have the excitation energy deposited by incident ionizing radiation dissipated over the entire π electron pool rather than over one particular bond resulting in less bond damage.

From theoretical calculations (Pullman and Pullman, 1963), one finds that the relative resonance energy per π electron for the nucleic acid bases follows the order $A > G > C > T$ (i.e., adenine has a greater resonance energy per π electron than guanine, etc.). In order to compare this with experimental determinations of radiation damage, we should compare the theory with damage assayed by observing the change in the 0–10 eV energy loss structure, since this measures changes in the π electron structure of the molecule. Therefore, in Fig. 1.13A, Isaacson *et al.* (1973) have plotted the $D_{1/e}$ dose for 0–10 eV energy loss damage (in electrons/Å^2) against the relative resonance energy per π electron calculated by Pullman and Pullman (1963). It is seen that relative resonance energy per π electron does indeed appear to determine the molecular radiation stability in this case.

Let us now obtain the same type of plot for the aromatic amino acids. Unfortunately, the relative resonance energy per π electron for all the aromatic amino acids are not available from a self-consistent source, so one must make certain

Table 1.6b Summary of Beam Damage Determined by Loss of Mass*

Molecule	$\sigma_{D,M}$ [*in* Å^2]	$\sigma_k/\sigma_{D,M}$	$\sigma_{in}/\sigma_{D,M}$
Adenine (a)	~0	∞	∞
Guanine (a)	~0	∞	∞
Cytosine (a)	~0	∞	∞
Thymine (a)	~0	∞	∞
DNA (b)	$\lesssim 10^{-4}$	∞	∞
DNA (c)	~.07	~.35 (f)	~26 (f)
L-Histidine (d)	0.18	0.041	3.08
L-Phenylalanine (d)	0.45	0.020	1.37
L-Tyrosine (d)	0.19	0.049	3.51
L-Tryptophan (d)	0.18	0.061	4.20
Glycine (d)	0.40	0.007	0.69
TMV (c)	~.62	~.012 (g)	~.6 (g)
Polyethylene (e)	0.84	0.002	0.16

*All values are scaled to 25 keV incident electron energy and have been taken from Table 1.3.

(a) Isaacson *et al.* (1973).
(b) Wall (1972a, b).
(c) Langmore (1975).
(d) Lin (1974).
(e) Bahr *et al.* (1965).
(f) Values are per base pair along strand.
(g) Values are per amino acid (measured on single particles).

Table 1.6c Summary of Beam Damage Determined from Electron Diffraction*

Molecule	$\sigma_{D,D}$ [*in* Å^2]	$\sigma_k/\sigma_{D,D}$	$\sigma_{in}/\sigma_{D,D}$
Cytosine (a)	0.13	0.0370	2.94
Adenosine (b)	1.37	0.0086	0.60
L-Histidine (c)	14.3	0.0005	0.04
L-Valine (c)	10.3	0.0005	0.05
Phthalocyanine			
[5.7 Å line] (a)	0.12	0.258	15.0
[12.5 Å line] (a, d)	~0.01	2.83	181
Indigo (e)	0.08	0.045	3.42
Polyethylene (f)	2.2	0.0002	0.02

*All values are scaled to 25 keV incident electron energy and have been taken from Table 1.4.

(a) Isaacson *et al.* (1973).
(b) Glaeser (1971).
(c) Lin (1974).
(d) Reimer (1965).
(e) Reimer (1961).
(f) Extrapolated from Kobayashi and Ohara (1966).

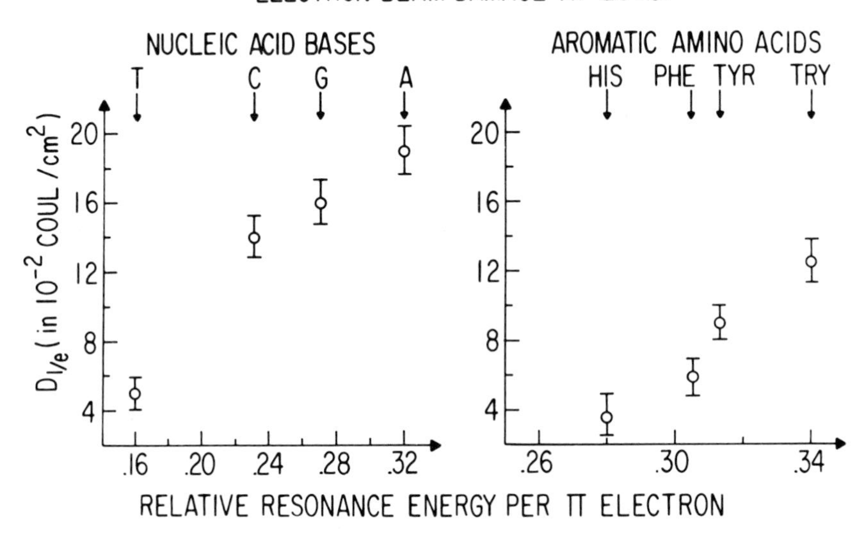

Fig. 1.13 A plot of the characteristic dose for damage of the 0 to 10 eV structure in the electron energy loss spectra as a function of the relative resonance energy per π electron for: (A) (left) four common nucleic acid bases (Isaacson *et al.*, 1973); (B) (right) four aromatic amino acids (Lin, 1974). The data are shown for irradiations of thin films of the molecules at room temperature using 25 keV electrons.

assumptions in obtaining this quantity for those materials (Lin, 1974). It can be assumed that the relative resonance energy per π electron follows the order TRY > TYR > PHE > HIS. This sequence agrees with the experimental measurements by Lin for the damage of the 0–10 eV energy loss spectra of these amino acids, and the results are plotted in Fig. 1.13B.

Another point to note concerning the electron beam damage assayed by observing the 0–10 eV energy loss structure is the relative sensitivity of a base alone compared to a base when it is incorporated in its nucleotide. We see from Table 1.6a that adenine is more radiation-resistant than poly A. This order of stability is observed for all the bases in radiolysis studies (Muller, 1966). It would appear that the attachment of the sugar phosphate grouping to the bases increases the π bonding structural damage. However, since no electron beam damage studies have been performed on the sugars and phosphates alone, it is not clear whether or not the aromatic bases actually reduce the damage to the sugar-phosphate groups.

In addition, we see from Table 1.6a that poly A is more radiation-resistant than is DNA. This might be accounted for if one realizes that since the aromatic part of DNA consists of a combination of four bases, one would expect it to be more radiation-sensitive than a nucleotide whose aromatic part consisted only of the most radiation-resistant base, adenine.

It is interesting to note that the cross section for the 0–10 eV energy loss structure damage, $\sigma_{D,E}$, is approximately constant (within a factor of three or four) for all aromatic materials in Table 1.6a. Since this type of damage assays damage to the π electron bonds, this confirms radiolysis measurements which show no order of magnitude difference in the radiation sensitivity of aromatic molecules (Hart and Platzman, 1961).

Now let us consider what information can be obtained on radiation damage using the second type of assay—that of the measurement of mass lost by the specimen because of irradiation. In some sense, mass loss measurements are most important with regard to electron microscopy. Mass loss is the most damaging of all structural degradations, since when material leaves the specimen one cannot even obtain as crude a structural determination as the molecular weight.

Experimentally, there are several observations to be made upon inspection of the selected results in Table 1.6b. First, from contrast measurements on thin films of the nucleic acid bases, it is found that within experimental error there is negligible loss of mass at 25 keV (i.e., $\sigma_{D,M}$ was not measurable) (Isaacson *et al.*, 1973).

Measurements by Wall (1972a) on the elastic and inelastic scattering intensities from unstained strands of DNA in a high resolution STEM indicated minimal mass loss upon irradiation with 29 keV electron doses exceeding 10^3 electrons/Å^2. However, more recent measurements by Langmore (1975) using a similar technique on unstained aggregates of DNA strands showed finite mass loss with a $\sigma_{D,M} \cong 7 \times 10^{-2}$ Å^2. The difference between these two measurements lies possibly in the ambiguity of the amount of salt bound to the DNA. One point to be speculated upon here is that if Langmore's mass loss data are correct and DNA loses ~40% of its total mass upon irradiation, this would be consistent with the nucleic acid bases remaining and all the other components leaving since the fractional mass of DNA composed of the bases is ~40%. Since no mass loss data exist on ribose and phosphate components, this is only conjecture at this time.

Using the contrast method, Lin (1974) found that half of the mass or more could be lost from thin films of aromatic amino acids upon irradiation with 25 keV electrons, and the $\sigma_{D,M}$ varied from .18 to .45 Å^2. This observation appears to be consistent with preliminary results of $\sigma_{D,M} \cong .62$ Å^2 from single particles of tobacco mosaic virus which showed greater than 62% mass loss upon irradiation with 25.5 keV electrons by Langmore (1975). Moreover, both these measurements are consistent with earlier measurements of mass loss on thick samples of different aliphatic-type polymers (Table 1.3) (Bahr *et al.*, 1965; Stenn and Bahr, 1970b).

In addition, the characteristic damage dose for mass loss (defined as the $D_{1/e}$ dose) results in damage cross sections for mass loss, $\sigma_{D,M}$, which are an order of magnitude larger than for destruction of the 0–10 eV energy loss spectra, $\sigma_{D,E}$

(compare the second columns of Tables 1.6a and 1.6b). Furthermore, the $\sigma_{D,M}$ for the materials containing amino acids are larger than for those containing nucleic acid bases.

Some insight into the trends observed in these radiation damage experiments can be gained by comparing these damage cross sections with the scattering cross sections for K-shell excitation, σ_k, and those for total inelastic scattering, σ_{in}. These comparisons are given in Tables 1.6a and 1.6b. If we realize that the cross section for all valence shell excitations between 10 eV and 280 eV, σ_{10-280}, is approximately nine-tenths of the total inelastic cross section (i.e., $\sigma_{10-280} \approx 0.9\ \sigma_{\text{in}}$), then we find that:

(1) The ratio of the cross section for K-shell excitation to the 0–10 eV energy loss damage cross section is of the order of unity for the nucleic acid bases, the aromatic amino acids, and the aromatic compounds (i.e., $\sigma_k/\sigma_{D,E} \sim 1$). However, the ratio $\sigma_{10-280}/\sigma_{D,E}$ is approximately 20 for the same materials.

(2) The ratio of the cross section for valence shell excitations greater than 10 eV to the cross section for mass loss is of the order of unity (i.e., $\sigma_{10-280}/\sigma_{D,M} \sim 1$) for all the compounds listed in Table 1.6b except the nucleic acid bases and DNA. For the nucleic acid bases, this ratio is essentially infinite since there is negligible mass loss. For DNA, this ratio is greater than 200 for Wall's measurement and of the order of unity for Langmore's measurement.

We can draw some tentative conclusions from the above observations with the aid of some known experimental data from the literature. First, it has been shown that strand breakage of DNA does not occur upon photon irradiation until the energy of the incident photons is greater than 10 eV (Wirths and Jung, 1972). Therefore, it would appear that for DNA, at least, radiation damage does not occur for 0–10 eV photon irradiation or for 0–10 eV energy loss collisions. This seems more plausible if we note that the threshold for electron emission upon photon irradiation of adenine, poly A, and DNA occurs also at ~10 eV, which is also the first π ionization potential of the molecule (Wirths and Jung, 1972). Therefore, 0–10 eV collisions are non-ionizing ones, and we would not expect such soft collisions to produce damage. If this is true, then damage must be initiated by collisions in which the incident electron loses more than 10 eV.

It should be further noted that the structure in the 0–10 eV energy loss spectra is primarily due to $\pi \to \pi^*$ transitions generally associated with nonlocalized bonding in the molecules. Therefore, changes in this spectra upon irradiation most certainly reflect changes in the nonlocal bonding structure (or π charge densities) and hence changes in the aromatic components of the molecules. Since the probability for a K-shell excitation for most of the molecules listed in Table 1.6a is the same order of magnitude as the probability for a change in the 0–10 eV energy loss structure, one might expect that this K-shell excitation could have some relation to an event which is damaging to the aromatic component of the molecule (Isaacson, 1975b). In fact, since a K-shell ionization leads with a high probability to a carbon, nitrogen, or oxygen with two L-shell vacancies, one might expect the atomic system and its environment (and therefore the molecular system) to undergo a violent rearrangement.

The fact that the cross section for all valence shell excitations greater than 10 eV is ~20 times larger than the 0-10 eV energy loss structure damage cross section implies that these excitations are not very dominant in causing damage to the π electron bonding structure. Moreover, if one assumes that the chemical bond energies for the molecules studied are ~4 eV (Parikh, 1973), then one can calculate the cross sections for atomic displacement by elastic nuclear collisions as presented earlier for C—H, C—N, etc., bonds. At 25 keV, there is not sufficient energy available in an elastic collision to break more than one bond. Thus, if we assume that only atoms bound to one another can be displaced, one finds that $\sigma_{\text{disp}}/\sigma_{D,E} \sim 10^{-3}$ for all the materials listed in Table 1.6 (except polyethylene).

The situation is, in fact, even less favorable for elastic nuclear collision damage if we realize that a 4 eV displacement for these light atoms in a solid is quite optimistic. Experimental measurements seem to indicate displacement energies in graphite of ~20 eV (Corbett, 1966) (one needs energy not only to break the chemical bonds, but also to move the atom to a different position). With such values for displacement energies, $\sigma_{\text{disp}}/\sigma_{D,E}$ becomes completely negligible at 25 keV and less than 10^{-3} even at 100 keV. The implication here is that knock-on damage by direct elastic nuclear collisions is not a dominant effect even at incident electron energies up to 100 keV. However, at higher voltages, this effect can be expected to become increasingly important, since, for example at 1 MeV, the maximum energy which can be transferred to a carbon atom is ~350 eV compared to only 4.6 eV at 25 keV, so that one can simultaneously break more than one bond and get a multiplicative effect.

As we briefly mentioned earlier there is another mechanism available for displacing atoms (Durup and Platzman, 1961). Instead of a single-step process as in an elastic collision, there is a two-step sequence: energy is first imparted to the electronic system of the molecule (i.e., a K-shell ionization) and then during the reorganization of the nuclei in the drastically altered electronic environment (two electrons missing via Auger effect), the atom is displaced. This is exactly the way in which molecules are dissociated by light absorption. It should be noted that the minimum energy necessary for atomic displacement by this method is generally less than for displacement by a direct elastic collision, so that this process would be more energetically favorable. Therefore, it would appear that K-shell excitation could result in atomic displacement by the above two-step process and in doing so disrupt the relatively radiation-resistant aromatic molecular components.

Insofar as mass loss is concerned, we have seen that the probability for mass loss in the amino acids listed in Table 1.6 as well as the materials with aliphatic components listed in Table 1.3 is approximately the same order of magnitude as that for an inelastic collision resulting in a 10 to 280 eV loss of energy by the incident electron. Since no ionization occurs at energies less than 10 eV, and the 10–280 eV range consists of ionizations and excitations to superexcited bound states (Platzman, 1962; Johnson and Isaacson, 1973), one can image mass loss damage to be due, in general, to the valence shell ionizations, since they are two orders of magnitude more probable than the K-shell excitations. These outer-shell ionizations could be capable of disrupting the localized bonds in aliphatic-type materials.

For instance, the aromatic amino acids are molecules which consist of an aromatic component and an aliphatic component. The damage cross section for mass loss, $\sigma_{D,M}$, is an order of magnitude greater than that for the damage of the 0–10 eV energy loss structure, $\sigma_{D,E}$. Therefore, we can assume that damage to these molecules proceeds first through destruction of the aliphatic component due to the valence shell ionizations. This results, in many cases, in the loss of only aliphatic component of the total mass, while at the $D_{1/e}$ dose for mass loss the aromatic component remains relatively unharmed as determined by the 0–10 eV energy loss spectra (Lin, 1974). Such a picture is consistent with negligible mass loss of the nucleic acid bases since they consist entirely of aromatic structures. It is also consistent with the damage data of polyethylene, which, since it has no aromatic component, has $\sigma_{D,M} \sim \sigma_{D,E}$.

The fact that there is negligible mass loss upon electron irradiation of the nucleic acid bases implies that the resulting damaged material is not composed entirely of carbon. This is substantitated by chemical assays of electron beam-damaged materials, where it is found that nitrogen and oxygen never completely disappear (Bahr *et al.*, 1965). Further substantiation for the nucleic acid base adenine can be seen in Fig. 1.8. The characteristic energy loss spectra in the region of the K-shell excitation of the carbon and nitrogen atoms in adenine are shown in this figure. One can see that the damaged K-shell fine structure of the carbon atoms is different than that of amorphous carbon, and the K-shell excitation structure of the nitrogen atoms indicates negligible loss of nitrogen (Isaacson, 1975b).

Finally, let us concentrate our attention on Table 1.6c which gives selected damage cross sections for destruction of specimen crystallinity as judged by the change in the electron diffraction pattern. For the aromatic molecules we find the ratio of the inelastic scattering cross section, σ_{in}, to that for damage to the diffraction pattern, $\sigma_{D,D}$, is always greater or equal to unity, while for the materials which are predominantly aliphatic, this ratio is $\sim 10^{-2}$. In addition, the materials for which electron energy loss measurements have been performed indicate that $\sigma_{D,D} > \sigma_{D,E}$.

The fact that $\sigma_{D,D} > \sigma_{D,E}$ appears to be true for systems which contain more than carbon and hydrogen. For pure hydrocarbons the situation can sometimes be reversed (in tetracene, for instance, see Tables 1.2 and 1.4). This fact has been noted by Reimer (1965), who pointed out that one can be misled in interpreting radiation damage data if one only relies on one type of damage assay.

Since the materials of biological interest are not pure hydrocarbons, we shall assume for our discussion that $\sigma_{D,D} > \sigma_{D,E}$ is valid.

According to Table 1.6, when one irradiates nonhydrocarbon aromatic components, one needs at least one valence shell excitation to disrupt the bonds holding the molecule in its position in the crystal. By and large, however, the molecule itself is relatively unscathed since there is negligible mass loss and its π electron structure remains intact for more than one order of magnitude more dose.

On the other hand, the cross section for damage of the electron diffraction patterns of compounds containing aliphatic components appears always to be one to two orders of magnitude greater than the inelastic scattering cross section. This implies that either each primary inelastic event causes many crystal bond breakages or that some damage is caused by secondary reactions (such as by energetic secondary electrons and reactive molecular fragments). Such effects would tend to show up in departure of the dose-response curves from pure exponentials. As we shall see later, this has been observed in some cases and so it would appear that one must also deal with these secondary events in treating the entire radiation damage problem.

One should note that for the aliphatic compounds listed here, $\sigma_{D,D}$ is greater than for the aromatic compounds. This might presumably be due to the fact that the aromatic compounds studied are generally planar while the aliphatic compounds are not. One therefore requires less energy to rotate bonds in the aliphatic molecules than the aromatic ones since the planar compounds are more sterically hindered. We might thus expect nonplanar crystal systems to damage more easily than planar systems.

In all the cases that we have considered in Table 1.6, we have $\sigma_{D,D} > \sigma_{D,M}$ and $\sigma_{D,D} > \sigma_{D,E}$, so that at doses which damage the crystal structure, the molecule as a whole is relatively unchanged. Therefore, electron beam damage as assayed by diffraction generally gives a lower limit on the dose necessary for structural preservation of biological molecules.

In closing this section, I should emphasize that one still needs much more quantitative data on electron beam damage in order to determine conclusively the mechanism of damage. The views I offer here are therefore speculative, being consistent only with the scarce existing experimental results. The relation between the damage cross sections evaluated here and the ability to obtain structurally significant high resolution images of biological objects will be discussed later.

DAMAGE OF BIOLOGICAL SPECIMENS

Cells

Living processes exhibit a wide range of resistance to the degrading action of ionizing radiation (e.g., Ducoff *et al.*, 1971; Engel and Adler, 1961; Cleveland

and Deering, 1971). For several decades, the interesting question has been asked as to whether or not it is possible to examine biological function in living cells with the electron microscope at a spatial resolution level which exceeds that of the light microscope. The problem encountered in the electron microscope is that living function can be halted by the very electrons needed to photograph it.

In order to appreciate the range of radiation sensitivities of living organisms, we have compiled in Table 1.7 some selected values from the literature of doses necessary to damage some aspect of biological function. The data have been scaled with energy to an equivalent 100 keV incident electron energy. To appreciate how these doses to damage function compare to the doses necessary to damage certain aspects of structure, one should compare Table 1.7 with Tables 1.2 through 1.4. It is seen that, in most cases, the doses necessary to destroy function are much less than those necessary to destroy the structure assayed in the methods presented earlier.

One can see from Table 1.7 that the damaging dose for killing *Bacillus subtillus* spores is only 4×10^{-4} electrons/Å^2, and so it would appear that for observing living organisms in the electron microscope, the future would not appear extremely bright. However, one must realize that a spore is ~1 μ thick, so that an incident 100 keV electron would undergo ~10 inelastic collisions and half as

Table 1.7 Radiation Damage of Living Organisms*

Structure	$D_{1/e}$ [*in elec*/Å^2]	*Method*	*Reference*
E. coli	6.2×10^{-6}	Irradiated in wet state; survival	Bhattacharjee (1961)
TMV	1.7×10^{-4}	Irradiated in wet state; inactivation	Ginoza (1967)
Ribonuclease	1.9×10^{-4}	Irradiated in wet state; inactivation	Jung and Schüssler (1966)
Bacillus subtillus spores	3.7×10^{-4}	Irradiated in an electron microscope; survival	Siegel (1970)
Bacillus subtillus spores	8.7×10^{-4}	Irradiated in dry state; transforming DNA	Tanooka and Hutchinson (1965)
Ribonuclease	2.2×10^{-3}	Irradiated in dry state; inactivation	Günther and Jung (1967)
Bacillus megaterium	1.9×10^{-3}	Irradiated in an electron microscope; survival	Nagata and Fukai (1974)
Valine T-RNA	4.7×10^{-3}	Irradiated in dry state; binding ability	Fawaz-Estrup and Setlow (1964)
Hyaluronidase	1.4×10^{-2}	Irradiated in dry state; inactivation	Vollmer and Fluke (1967)
Lysozyme	1.7×10^{-2}	Irradiated in dry state; inactivation	Fluke (1966)

*All results have been scaled to 100 keV incident electron energy.

many elastic collisions in traversing the spore. If one were only interested in information at the 1,000 Å resolution level, then almost 100 electrons could traverse the 1,000 Å diameter area with only a 20% probability of killing the spore (i.e., 400 electrons would result in a 63% probability of death). Therefore, we could obtain over 10^3 collisions which could be useful for imaging purposes. In other words, if one does not necessarily require high resolution imaging, there might exist regions of spatial resolution in which electron microscopy of living organisms might prove to be a viable technique.

Over the last twenty years, several researchers have tried to examine the technique of observing living organisms in the electron microscope. Most of the early work, however, was concerned solely with visualizing objects such as spores and cells (Dupouy *et al.*, 1960; Locquin, 1954; Stoyanova *et al.*, 1960). This work was of necessity very qualitative in nature, and so one could not be certain whether the organisms were, in fact, viable after the microscopy was performed.

Spores have generally been the test object which researchers use in tests of electron microscopy of living organisms, since as we can see from Table 1.7 they are moderately radiation-resistant. Moreover, they have the additional advantage in that they are not very much affected by being dehydrated for a short period of time.

Nagata and co-workers have recently examined various spores in the dry state and assayed cellular viability by determining whether the spores advanced into a vegetative state after exposure and observation under the electron beam. Their criterion of viability was whether the exact spores irradiated doubled in size. It was found that spores of *Bacillus megaterium* irradiated with 100 keV electrons were prevented from advancing into a vegetative state with a dose of 6.2×10^{-3} elec/Å^2 (Nagata and Ishikawa, 1971). However, in a later paper which was more quantitative, it was found that for the same type spores irradiated with 100 keV electrons, there was 100% survival if the incident dose was less than 9.4×10^{-4} elec/Å^2, but only 37% survival if the dose was 1.25 elec/Å^2 (Nagata and Fukai, 1974). In another paper, Nagata and Ishikawa (1972) observed green pond organisms (*Scendesmus obliques*) in a wet chamber using 750 keV electrons and found that a dose of 6.25 elec/Å^2 was needed to destroy the specimen.

More extensive observations on cells using stages in which the specimen is kept in a hydrated environment have been made by Parsons and co-workers (Parsons, 1970; Parsons *et al.*, 1972; Parsons *et al.*, 1974). In several experiments they examined whole cell mounts of wet cells using 1 MeV electrons and claimed that internal cytoplasmic structures could be visualized. In their most recent paper they showed micrographs which claimed to visualize mitochondria in the cytoplasm and claimed resolutions of 500–1,000 Å for cellular components. No attempt was made to assess cellular viability, and it was mentioned that no movement of the cytoplasmic organelles was observed. For the interested reader, a review of attempts at electron microscopic observation of aqueous biological specimens is given by Joy (1973). A more recent review on this subject is presented by Allinson (1975). It is clear at present that it has not yet been demonstrated whether visibility of living organisms or wet unstained cells in the electron microscope can exceed that of the light microscope.

It is pertinent to point out that there is a difference between the radiation damage of specimens by ionizing radiation depending upon whether the specimens are in a hydrated or dehydrated state. This is due to the secondary (or indirect) effects of damage which usually lead to the production of small diffusible radicals. Thus, where water is present, invariably very reactive OH^- radicals are formed. The G value for OH^- production in water is ~2.3 (Dertinger and Jung, 1970) (i.e., 2.3 OH^- produced per 100 eV deposited). Since the average energy lost in an inelastic fast electron collision in biological molecules is ~30-40 eV, this implies that almost every inelastic collision produces a reactive OH^- radical in solution.

In some cases, these indirect OH^- reactions can have an appreciable effect upon radiation damage. For instance, it was found that ribonuclease irradiated with ^{60}Co γ rays was almost two orders of magnitude more radiation-resistant (to inactivation) when it was in a dry state than when it was in a 5 mg/ml solution (Günther and Jung, 1967; Jung and Schüssler, 1966). Insofar as irradiation of cells is concerned, Bhattacharjee (1961) found that irradiation of *E. coli* bacteria with X rays resulted in a $D_{1/e}$ dose for survival of about four times greater for those cells in the dry state compared to the wet cells.

However, one must be cautious in assuming that OH^- radical production in solution will always result in radiation damage in solution being greater than damage caused to dehydrated objects. Damage in solution is very concentration-dependent, and it can be shown that while the fraction of damaged molecules (per undamaged molecules) per unit dose is less in the dry state than the hydrated state for dilute concentrations, the situation is reversed for substances in high-concentration solutions (Bacq and Alexander, 1961).

Thin Sections

It became obvious to researchers in the 1950s that if one wanted to obtain information on biological organisms in the electron microscope, then one would have to resort to some of the techniques familiar in light microscopy such as embedding the biological object in some plastic and then cutting reasonably thin sections. At the normally used electron microscope voltages (~100 keV), sections ~200-1,000 Å thick were required to permit reasonable penetration by the electron beam.

The crucial question is that of whether one can preserve spacings between various elements of the biological objects upon embedding so that what occurs in the embedded thin section is a faithful representation of the spacings in the living object. That is, can the molecular architecture be preserved upon embedding and sectioning? There has evolved a wealth of literature on specimen damage caused by the specimen preparation techniques of thin section fixation and embedding, and this part of damage will not be covered in this article. The interested reader is referred to the recent reviews by Hayat (1970) and Luft (1973) on the subject. What we want to consider here is the damage caused by the effect of the electron microscope beam on thin sections.

There are essentially three potential types of specimen damage which can occur when one is dealing with thin plastic sections. They include radiation damage in the sense of electron beam degradation of chemical bonds as discussed in the preceding sections; and in addition to this type, there is also the possibility of damage due to specimen heating and charging. These last two effects can, of course, be considered for other specimens besides thin sections. However, since thin sections are generally composed of plastic embedding materials which are usually poor conductors both electrically and thermally, the potential damage due to heating and charging is increased.

Let us first consider the radiation damage of thin sections. In order to accomplish this we must investigate the materials which are representative of those used as embedding compounds. Methacrylate was used very often in early electron microscope work with thin sections (Luft, 1973). However, because of its extreme sensitivity under the action of the electron beam and the tissue damage caused by the polymerization of this compound, it has generally been replaced by polyester and epoxy resins.

We have not listed any of these embedding plastics in the damage Tables 1.2–1.4, because the electron beam damage measurements of these materials were made under quite unspecified conditions, and, in some instances, damage due to specimen heating could not be distinguished from pure radiation damage (Reimer, 1965). All of these measurements on embedding materials dealt with mass loss as being the most obvious manifestation of damage, since it was (and still is) quite common to notice an increased electron transparency of a thin embedded section during the first few seconds of exposure to the electron beam. In fact, this mass loss was used as an advantage, since it meant that one did not necessarily have to cut ultrathin sections. The thick sections gradually became thin upon irradiation by the electron beam (Van Dorsten, 1954).

The most common feature of all mass loss measurements of embedding material was that 30 to 70% of the total mass could be lost upon prolonged exposure to an electron beam using a 60 keV electron beam with current densities of 10^{-4} to 10^{-5} amp/cm^2 at the specimen. Reimer (1959), using the bright field contrast method of measurement, found mass losses of 40 to 50% for methacrylate, 20% for Araldite, and 13% for Vestopal. The total dose necessary for all the mass lost was $\sim 10^{-2}$ coul/cm^2 for all three materials. He also noted that the mass loss values that he measured were lower than those measured using somewhat higher current densities at the specimen (Reimer, 1965). For instance, Lippert (1962) found mass loss of 60% for methacrylate, 30% for Vestopal and 30% for Araldite upon irradiation using current densities of 10^{-4} amp/cm^2 at the specimen. Zelander and Ekholm (1960) observed a 56% loss of mass after electron beam exposure to methacrylate sections using a "normal beam current density" at the specimen. They measured the changes in optical path length to determine loss of mass. Using a similar instrumental technique on ~2,000-Å-thick sections, Cosslett (1960) measured 50% mass loss in methacrylate, 20 to 50% in Araldite and 30 to 50% in Vestopal. In that paper, the current densities were listed only as "high" or "low."

The apparent increase in mass loss due to higher current densities is in contrast to the results quoted in the earlier section where we found no dependence of radiation damage on current density over a very wide range. This current-density dependence of mass loss in plastic sections is probably due to the fact that plastic sections are poor thermal conductors and so there are thermal heating effects as well as radiation damage effects at work. In addition, the thin-section measurements were generally made by illuminating a fairly large area of the specimen, and this would also tend to increase the specimen temperature (see below). It is worthwhile observing that Reimer (1959) in fact noted a small decrease in mass loss of plastic sections if they were coated with carbon (a good thermal conductor).

As a historical note, it is interesting that in Araldite and Vestopal the remaining mass left after electron beam damage is approximately the same as the carbon content in the original material (~70%). This observation has probably led the term "carbonization" to appear frequently in the literature and has erroneously led readers to assume that only carbon remains in a biological specimen after irradiation by an electron beam.

The important point to be stressed here is that all existing measurements at room temperature confirm a substantial loss of mass of embedding material in thin sections in the electron microscope. This loss of embedding material is probably one of the chief obstacles in being able to perform dry mass determination in sectioned materials by the method of Zeitler and Bahr (1959, 1965), since, in general, the object structure and the embedding material damage differently under the action of the electron beam. Figure 1.14 (Zeitler and Bahr, 1965) illustrates dramatically the loss of the embedding medium Araldite from a thin section after irradiation with 75 keV electrons. The section was irradiated and then subsequently shadowed with platinum at an angle of 7°. The selective mass loss of the Araldite leaves the tissue structure standing out as ridges or elevations. Perhaps when more quantitative measurements are available on the mass loss of embedding materials, one will be able to quantitatively determine the dry mass of objects in thin sections.

In some respects, the mass loss of embedding material in thin sections might possibly prove beneficial. That is, if one could selectively damage the embedding material to cause loss of mass of it without a significant mechanical displacement of the components of the embedded object under scrutiny, then one could increase the contrast of the object against the embedding background and

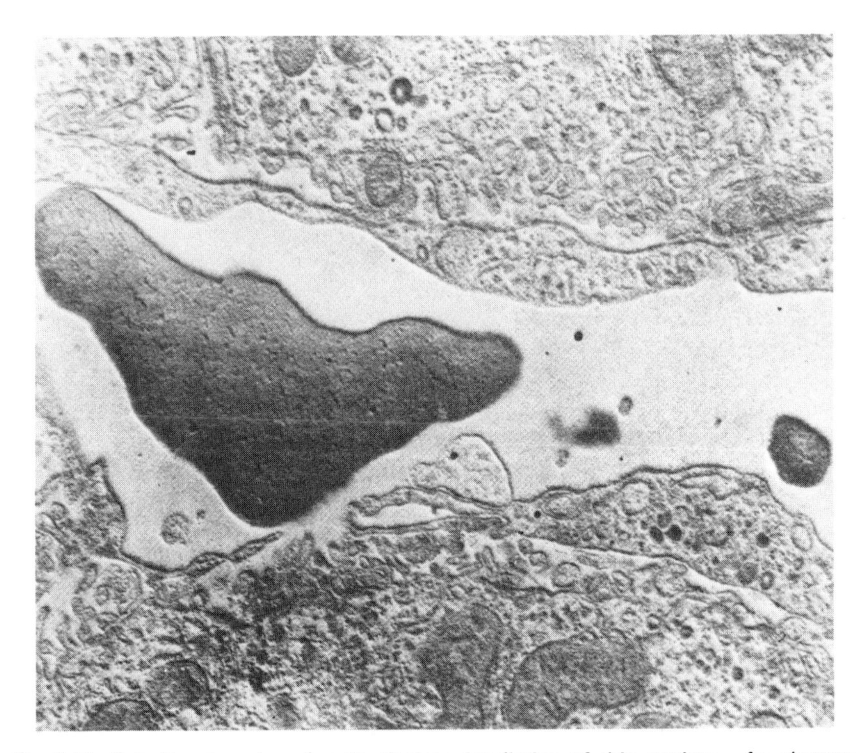

Fig. 1.14 Selective mass loss due to electron irradiation of thin sections. An electron micrograph of a thin section of osmium tetroxide-fixed, Epon-embedded rat liver tissue. The section has been irradiated with 75 keV electrons using a dose-rate of 3×10^{-3} amp/cm^2 and subsequently shadowed with platinum at an angle of 7°. Note the selective removal or shrinkage of the Epon, leaving tissue structures standing out as ridges or elevations. Full horizontal scale is 7.15 μm. (*From Zeitler and Bahr, 1965.*)

possibly allow oneself to visualize sectioned objects without resorting to staining techniques.

Let us now consider the possibilities of specimen heating and charging as they might pertain to specimen damage. We have seen in the case of thin plastic sections how the existing experimental results indicate that specimen heating might have affected the total damage. To calculate precisely the temperature that a specimen reaches because of the action of an incident electron beam is an extremely complex problem depending upon local geometry and other not so general conditions. What we would rather like to calculate here is a fairly simple general expression for specimen heating which under certain assumptions would be useful in obtaining order-of-magnitude estimates of the temperature.

Heat gets deposited into a specimen as a result of inelastic collisions of the incident beam electrons. For specimens in which most of the incident electrons are transmitted, the rate of heat deposition is given by

$$\frac{dQ}{dt} = \frac{dE}{dx} \cdot f_q \cdot T \cdot I_b \tag{1.24}$$

where dE/dx is the energy loss per unit path length of the incident electron, f_q is the fraction of energy lost by the incident particle which actually gets deposited in the specimen (some energy escapes in the form of secondary electrons and photons), T is the specimen thickness, and I_b is the incident beam current. For most thin specimens used in electron microscopy, the predominant mechanism for heat removal from an irradiated area is by heat conduction through the specimen support film, or if the specimen is a thin film (such as a section), through the film itself.

We shall consider the case of a thin specimen where there are few multiple collisions. We take a specimen of thickness T, thermal conductivity k, mounted over a metallic grid with hole openings of radius r_G and illuminate it (in the center of a grid hole) with a cylindrical electron beam of radius r_B. If we assume that the energy from the incident beam is uniformly deposited throughout the cylindrical irradiated volume and that the front and back of the specimen are at the same temperature, then the heat flow is purely radial. Then, if we regard the grid as a heat sink, one can show that the temperature rise, $\Delta\Theta$, in the center of the irradiated volume is given by:

$$\Delta\Theta = \frac{dQ}{dt} \cdot \frac{1}{2\pi kT} \left[\ln(r_G/r_B) + \frac{1}{2} \right] \tag{1.25}$$

Since $dE/dx = \bar{E} N \sigma_{\text{in}}$, where N is the molecular density and is equal to ρ/AM_p, where ρ is the specimen density, A is the molecular weight and M_p is the proton mass, we get using Eqs. (1.19), (1.24), and (1.25), that the temperature rise in the center is:

$$\Delta\Theta \cong \frac{3}{4} \frac{\lambda_c^2}{\pi\beta^2} Z^{1/2} \left\langle \ln\left(\frac{4E_0}{E}\right) \right\rangle \frac{\bar{E}\rho}{AM_p} \cdot \frac{N_b}{2\pi kJ} \left[\ln(r_G/r_B) + \frac{1}{2} \right] \tag{1.26}$$

where N_b is the beam current in electrons/sec, J is the mechanical equivalent of heat, and we have used $f_q \sim 1$.

If we note that $\rho Z^{1/2}/A \sim \frac{1}{2}$ gm/cm^3 throughout the Periodic Table, then Eq. (1.26) simplifies to:

$$\Delta\Theta \cong 3 \times 10^{-17} \frac{\bar{E} N_b}{\beta^2 k} \left\langle \ln\left(\frac{4E_0}{E}\right) \right\rangle \left[\ln(r_G/r_B) + \frac{1}{2} \right] \tag{1.27}$$

where $\Delta\Theta$ is in °C, E and E_0 are in eV, and k is in cal/cm/sec/°C. In addition, for most materials $\bar{E} \cdot \langle \ln(4E_0/E) \rangle \sim 400$ from $E_0 = 10$ keV to 1 MeV, so that within a factor of 2 or 3 over the Periodic Table we have reduced Eq. (1.26) to:

$$\Delta\Theta \approx 1.2 \times 10^{-14} \frac{N_b}{\beta^2 k} \left[\ln(r_G/r_B) + \frac{1}{2}\right] \tag{1.28}$$

Note that in the CTEM, the number of electrons/sec striking the specimen is given by

$$N_b(\text{CTEM}) = M^2 j_{\text{pl}} \pi r_B^2 / f_c \tag{1.29}$$

where M is the magnification, j_{pl} is the current density on the photographic plate, and f_c is the fraction of incident electrons which is used to form the image.

Equation (1.28) also is valid for the STEM, since even at low magnifications the scanning speed of the spot is only $\sim 10^2$ cm/sec, which is much less than the speed of sound. Therefore, the temperature rise can be calculated as if no scanning were taking place. It is useful to compare the temperature rise in the CTEM and STEM. Substituting Eq. (1.29) in (1.28) gives us:

$$\frac{\Delta\Theta(\text{STEM})}{\Delta\Theta(\text{CTEM})} \approx \frac{N_b(\text{STEM})}{M^2 j_{\text{pl}} \pi r_{BC}^2 / f_c} \frac{\left[\ln(r_G/r_{BS}) + \frac{1}{2}\right]}{\left[\ln(r_G/r_{BC}) + \frac{1}{2}\right]} \tag{1.30}$$

where r_{BS} is the beam radius in the STEM and r_{BC} is the radius of the illuminated area in the CTEM.

To make a reasonable comparison, we must compare similar conditions. As an example, consider the case where one wants a high resolution image. According to Valentine (1966), this requires a minimum magnification of $M = 10^5$ in the CTEM. If we assume a photographic plate area of 100 cm^2 and that the illuminated area of the specimen fills up the plate, then the radius of illumination is $r_{BC} = .56\ \mu$. For the STEM, we assume a probe diameter of 2 Å so that $r_{BS} = 10^{-4}\ \mu$. If we assume a 40 μ grid hole diameter, we then get from Eq. (1.30):

$$\frac{\Delta\Theta(\text{STEM})}{\Delta\Theta(\text{CTEM})} \approx 3 \frac{N_b(\text{STEM})}{M^2 j_{\text{pl}} \pi r_{BC}^2 / f_c} \tag{1.31}$$

Since the STEM probe size and the CTEM illumination diameter both appear in the logarithm in Eq. (1.30), the constant of Eq. (1.31) is relatively insensitive to their exact sizes.

In the CTEM, we need $\sim 10^{-12}$ amp/cm^2 at the photographic plate for an acceptable image. Zeitler (1968) and Valentine (1966) estimate that 2×10^{-11} amp/cm^2 is needed at the photographic plate for an optical density of unity with a 1 sec exposure. Therefore for the CTEM, $M^2 j_{\text{pl}} \pi r_{BC}^2 = 6 \times 10^8$ elec/sec to record a high-resolution image. For a bright field image, $f_c \sim 1$, so this is the number of electrons incident on the specimen. However, for a dark field image in a CTEM using 100 keV electrons, $f_c \sim 0.5$-.20 for most specimens (Langmore *et al.*, 1974), and so the number of incident electrons would have to be 3 to 10×10^9 elec/sec. For the STEM, an adequate high-resolution image (bright or dark field) can be obtained with a probe current of 10^{-11} amp (Crewe, 1973) or $N_b = 6 \times 10^8$ elec/sec. Therefore, under these conditions, we get:

$$\frac{\Delta\Theta(\text{STEM})}{\Delta\Theta(\text{CTEM})} \approx 3f_c \tag{1.32}$$

That is, under bright field conditions, the specimen in the STEM will get slightly hotter than that in the CTEM, whereas under dark field conditions the situation is reversed.

Referring back to Eq. (1.28) we can estimate the temperature rise in a uniform specimen in the electron microscope for high resolution conditions. One finds that the temperature rise can go from fractions of a degree for metals ($k \sim 0.1$-1 cal/cm/sec/°C) to tens of degrees for poor thermal conductors such as some plastics ($k \sim 10^{-4}$-10^{-3}). Therefore, even in the case where the embedding medium of a thin section had a thermal conductivity of 10^{-4} cal/cm/sec/°C, we would not expect a drastic temperature rise under optimum viewing conditions. Of course, by greatly increasing the incident beam current or increasing the illumination area in the CTEM, one could get larger temperature rises (Leisengang, 1954).

In the above considerations we have assumed that the incident beam does not irradiate the metal grid support of the specimen. If this occurs, almost all the beam energy is converted into heating the support, and quite high specimen temperatures can result. For instance, Reimer and Christenhusz (1965) have shown that one could melt indium films in the electron microscope with moderate currents on the specimen if part of the beam irradiated the grid holder.

The apparent heating effects that were seen in the early mass loss measurements on plastic films were probably not due to the direct heating as discussed here. They were more likely due to the fact that the specimen grid was partly illuminated by the incident beam, resulting in a large temperature rise, or that because of the thick specimens used ($>$ 2,000 Å), multiple scattering increased the heat production.

Let us finally consider the effect that charging has upon the destruction of the specimen. As we mentioned before, for the thin specimens used in transmission electron microscopy, most of the incident electrons are transmitted through the specimen. The only method in which the specimen can accumulate charge is by means of the secondary electrons which escape from the specimen. The ratio of the number of secondary electrons escaping to the number of incident electrons is called the total secondary emission coefficient, δ. (We loosely apply the term secondary electron to mean any electron other than the incident one which leaves the specimen.) In general, for thin specimens δ is a function of the specimen thickness T.

In any increment of time, dt, the amount of charge per unit area, dq, imparted to a specimen that is due to an incident beam of area a and current I_b is given as:

$$dq = I_b \delta dt/a \tag{1.33}$$

If we assume that the specimen is supported over a metallic grid which is at ground potential, then we can imagine that this charge decays to ground as a function of time t as

$$dq(t) = [I_b \cdot \delta \cdot dt/a]\, e^{-t/\rho'\epsilon'} \tag{1.34}$$

where ρ' is the resistivity of the specimen and ϵ' its permitivity. Now if the illuminated area has accumulated an amount of charge per unit area $q(t)$ after a time t, then an electrostatic field has been set up in that area, which by Gauss' law is given by:

$$\mathcal{E} = \frac{1}{4\epsilon'} \frac{q\sqrt{a}}{T} \tag{1.35}$$

Therefore, that area of the specimen experiences an electrostatic stress equal to:

$$\frac{dF}{da} = \frac{1}{2}\epsilon' \mathcal{E}^2 = \frac{1}{32\epsilon'} \frac{q^2 a}{T^2} \tag{1.36}$$

where all the units are given in the MKS system.

We would like to determine what this stress is in the electron microscope. First, consider the CTEM. In this case, all elements of the field of view are being illuminated at the same time, so that to calculate the total charge per unit area after an exposure τ_c, we merely integrate Eq. (1.34) from $t = 0$ to τ_c. The total accumulated charge per unit area, Q_c, is therefore:

$$Q_c = \frac{I_b \delta \rho' \epsilon'}{a} [1 - e^{-\tau_c/\rho'\epsilon'}] \tag{1.37}$$

In the case of the STEM, we must calculate the deposited charge in a different fashion since each element of the field of view is illuminated sequentially in time. That is, after the beam has left the first picture element, the charge per unit area imparted is:

$$q_1 = I_b \delta t_{\text{pic}}/a_b \tag{1.38}$$

where t_{pic} is the time of the picture element and a_b is the beam cross section (we assume that the beam cross section equals the area of a picture element). This charge decays, so that when the beam returns to this picture element after one scan of time τ, only the amount $q_1 \exp(-\tau/\rho'\epsilon')$ is left. The beam then imparts more charge given by $I_b \delta t_{\text{pic}}/a_b$, so that after the beam leaves the picture element for the second time, that area has accumulated a charge of:

$$q_2 = \left[\frac{I_b \delta t_{\text{pic}}}{a_b}\right] [1 + e^{-\tau/\rho'\epsilon'}] \tag{1.39}$$

One can extend the argument and show that after $M + 1$ scans, the total charge per unit area accumulated by a picture element is:

$$q_{M+1}(t) = \left[\frac{I_b \delta t_{\text{pic}}}{a_b}\right] \sum_{m=0}^{M} e^{-m\tau/\rho'\epsilon'} e^{-(t-M\tau)/\rho'\epsilon'} \tag{1.40}$$

where $M\tau \leqslant t < (M + 1)\tau$ (Shaffner and Van Veld, 1971). Note that the above argument assumes that the beam is scanned in a random fashion so that one need not concern oneself with accumulated charge draining away from adjacent areas. The solution to the consecutive line scan raster is much more complicated. Suffice it to say that if the beam is scanned in a line raster fashion, the total accumulated charge per scan would be larger than that obtained here.

In the case of a good conductor, where $\rho'\epsilon' \ll 1$, we see that Eq. (1.40) reduces to:

$$q_{M+1}(t) \approx \frac{I_b \delta t_{\text{pic}}}{a_b} e^{-(t-M\tau)/\rho'\epsilon'}, M\tau \leqslant t < (M + 1)\tau \tag{1.41}$$

On the other hand, for a poor conductor, where $\rho'\epsilon' \gtrsim 1$, Eq. (1.40) reduces to:

$$q_{M+1}(t) \approx \frac{I_b \delta t_{\text{pic}}}{a_b} \left[\frac{\rho'\epsilon'}{\tau}\right] [1 - e^{-M\tau/\rho'\epsilon'}] e^{-(t-M\tau)/\rho'\epsilon'} \tag{1.42}$$

for $M\tau \leqslant t < (M + 1)\tau$.

Notice that where $t \approx M\tau$, then Eq. (1.42) is equivalent to Eq. (1.37) for the CTEM with $\tau_c = M\tau$ since $t_{\text{pic}}/\tau = a_b/a$ and a is the total area scanned. Moreover, when $\rho'\epsilon' \ll 1$, then Eq. (1.41) and Eq. (1.37) both approach zero. That is, in any electron microscope, no charge is built up on a good conductor.

Let us see what the above equations tell us concerning carbon substrates and thin plastic sections. For carbon, $\rho' \approx 30\Omega me$ and $\epsilon' \approx 27 \times 10^{-12}$ $\text{coul}^2/\text{nt} \cdot me^2$ which gives $\rho'\epsilon' \approx 8.1 \times 10^{-10}$ sec, and so for any reasonable parameters there is little electrostatic stress on carbon substrates, the substrate quickly reaching equilibrium with the ground metal grid.

On the other hand, consider a thin plastic section. Generally, for plastics $\rho' \sim 10^{11}\ \Omega me$. If we take ϵ' to be roughly that for carbon, then $\rho'\epsilon' \sim 2.7$ sec. They are poor conductors. Therefore for the CTEM the electrostatic stress exerted on the entire irradiated area is given as:

$$\left[\frac{dF}{da}\right]_c = \frac{1}{32\epsilon' a T^2} [I_b \delta \rho'\epsilon']^2 [1 - e^{-\tau_c/\rho'\epsilon'}]^2 \tag{1.43}$$

while for the STEM after M scans it is given as:

$$\left[\frac{dF}{da}\right]_s \approx \frac{1}{32\epsilon' a T^2}[I_b \delta \rho' \epsilon']^2 [1 - e^{-M\tau/\rho'\epsilon'}]^2 \tag{1.44}$$

If we consider a 1,000-Å-thick plastic section which has a secondary emission coefficient $\delta \approx 10^{-1}$ (Isaacson and Lin, 1975), then if we let $\tau_E = \tau_c$ for the CTEM and $\tau_E = M\tau$ for the STEM, Eqs. (1.43) and (1.44) become:

$$\frac{dF}{da} \approx 8.6 \times 10^{21} \frac{I_b^2}{a}[1 - e^{-\tau_E/2.7}]^2 \tag{1.45}$$

where all units are in the MKS system. Let us consider a 10-μ-square illuminated area. If we use 10^{-11} amp beam current in the STEM, then after six ten-second scans the electrostatic stress on that area is $\sim 6 \times 10^9$ newton/me^2 which exceeds the tensile strength of most plastics ($\sim 10^8$ nt/me^2). Therefore, under these conditions we would expect the irradiated area in the STEM to tear apart. In fact, we have noticed exactly this effect on regions of self-supporting thin sections which are not stained. Stained regions stand up better under the beam, presumably because they are slightly better conductors and the stress is proportional to the square of resistivity. Moreover, if the sections are coated with a thin layer of evaporated carbon they do not rip apart under the same conditions because carbon is a sufficiently good conductor to carry off the excess charge.

Note that from Eq. (1.43) we would also predict self-supporting thin sections to tear apart in the CTEM under the same conditions. However, since the photographic plate has more resolution than a STEM recording system (5,000 lines to 1,000-2,000 lines), the illuminated area can be an order of magnitude larger in the CTEM. Moreover, under usual operating conditions in the CTEM one illuminates a larger region of the specimen than is actually recorded on the film. For instance, at current densities of 10^{-3} amp/cm^2 on the specimen, the electrostatic stress would be less than 10^8 nt/me^2 so that under such conditions one might not expect the section to tear up under the illumination.

It is fairly clear by now that if one attempts to minimize the dose which the specimen receives in order to extract an image, the main damage to sections is probably not due to charging or heating, but rather is due to chemical bond degradation because of the ionizing character of the incident beam. In other words, radiation damage is still the fundamental problem for observation of thin sections and cells.

THEORETICAL LIMITS ON STRUCTURAL IMAGING IN ELECTRON MICROSCOPE

In order to be able to minimize radiation damage of sensitive specimens in the electron microscope, one obviously wants to record an image using the least possible number of incident electrons. In the next section, we shall discuss some of the possible methods with which this can be achieved. What we shall discuss now is the theoretical limit of the minimum number of incident electrons which is necessary to form a statistically meaningful image.

The ability to detect an object of dimension Δ depends upon the signal-to-noise ratio of the measurement. In the electron microscope, there are two sources of noise which can occur in the imaging process, excluding the noise inherent in the recording process itself. There are the statistical fluctuations of the number of electrons in adjacent areas of the image which are due to the statistical fluctuations in the number of incident electrons (beam noise). In addition, there are fluctuations due to differences in scattering from adjacent areas of the substrate because the number of substrate atoms varies from area to area (substrate noise) (Isaacson *et al.*, 1974b). For the purposes of this section, we shall not consider substrate noise. This is a reasonable omission since there are different substrates which can be used that exhibit varying amounts of noise (Langmore *et al.*, 1973a; Muller and Koller, 1972; Vollenweider *et al.*, 1973). And there also exist the possibilities of obtaining single crystal supports which are relatively noise-free (White *et al.*, 1971; Hashimoto *et al.*, 1971; Baumeister and Hahn, 1974; Riddle and Siegel, 1971).

We would like to determine the minimum number of incident electrons per unit area necessary to record an image. In the discussion to follow, we assume that the only limiting process in detecting the presence of an object of size Δ is the statistical fluctuation of the number of electrons used to form the image. One can define the image contrast, C, to be the ratio of the spatial variations in image intensity of an object, δI, to the local average of intensity, $\bar{I}$. That is, $C = \delta I/\bar{I}$. If it were possible to illuminate the specimen with an unlimited number of electrons, this would be the intensity fluctuations due to the object which we would record, and we call this its "inherent contrast." Of course, because of the radiation damage of the specimen, the number of incident electrons must be limited. The result is a statistical fluctuation in the number of electrons from one image point to the next, and this effect produces "spatial noise." One can think of this noise as spurious or "noise contrast" which is given by $C_N = \delta N/\bar{N}$, with $\bar{N}$ being the average number of electrons per unit area in the image.

It becomes obvious that if one is to visualize (or detect) the object, its inherent contrast must exceed the noise contrast due to limiting the number of incident electrons. That is, we must have $C \geqslant C_N$. Following the general criterion for visual perception (Rose, 1948), this noise contrast must be less than the inherent contrast by a minimum signal-to-noise factor, κ, so that we get:

$$C \geqslant \kappa/\sqrt{\bar{N}} \tag{1.46}$$

where we have assumed that $\delta N \cong \sqrt{\bar{N}}$. A minimum signal-to-noise of five was determined by Rose (1948) with regard to television problems and has been used in the literature in calculations of electron microscope detectability limits

(Isaacson *et al.*, 1973; Glaeser, 1971). However, with modern techniques it is not unreasonable to expect values of $\kappa < 5$.

Let us consider what Eq. (1.46) implies concerning the minimum dose necessary for visualizing an object of linear dimension Δ. If N is the number of electrons per resolution element in the image and if f' is the fraction of incident electrons which actually contribute to that resolution element, then $N = f' \cdot D \cdot \Delta^2$, where D is the total number of incident electrons per unit area (the incident dose). (For optimal recording of detail of size Δ, we want the area of a resolution element to be Δ^2.) It can be deduced from Eq. (1.46) that for an object of size Δ with an inherent contrast C to be detected (or resolved), the incident dose can be no less than the minimum dose, D_{min}, given by:

$$D_{min} = \frac{\kappa^2}{f' C^2 \Delta^2} \tag{1.47}$$

In drawing some conclusions from this formula, one must recognize that there are several factors involved and therefore general statements must be cautiously interpreted. We must appreciate the fact that the minimum dose depends upon the type of specimen, and each specimen (or specimen detail) has its own inherent contrast. The minimum dose depends upon the accelerating voltage since both f' and C are voltage-dependent. It depends upon the particular mode of operating the electron microscope (for instance, bright field and dark field operation yield different values of f' and C). The minimum dose can depend upon the type of instrument used.

Given these interdependences, it is still instructive to plot D_{min} as a function of detectable object size Δ for certain parameters and to compare this with the various indicators of damage which have been experimentally determined. At least, then, one can get a feeling for the type of problem which lies ahead of us in obtaining high resolution information concerning biological structures with the electron microscope. Various comparisons of this type have appeared in the literature (Isaacson *et al.*, 1973; Glaeser, 1974), and the values of $f'C^2$ used have ranged from 2.5×10^{-3} to 2×10^{-2}. It is is beyond the scope of this chapter to go into exact details of determining $f'C^2$, but if for our purposes we take $f'C^2 = 10^{-2}$ to be a happy compromise (for instance, $C = .5$ and $f' = .04$ or $C = .14$, $f' = .5$), we obtain the plot in Fig. 1.15 for different values of the minimum signal-to-noise, κ. The damage doses for different materials listed on the right are all scaled to 100 keV incident electron energy.

It is interesting to note that diffraction damage of L-histidine would imply a resolution limit or detectable object size of ~100 Å for $\kappa = 5$. But mass loss damage would only limit one to ~15 Å for $\kappa = 5$ and 2.5 Å for $\kappa = 1$. For aromatic compounds with negligible mass loss, the resolution limit is even smaller. The point is that while radiation damage is a real problem, there are regions of moderate resolution which the minimum dose is less than or equal to some damaging dose. In particular, if one were only interested in determining

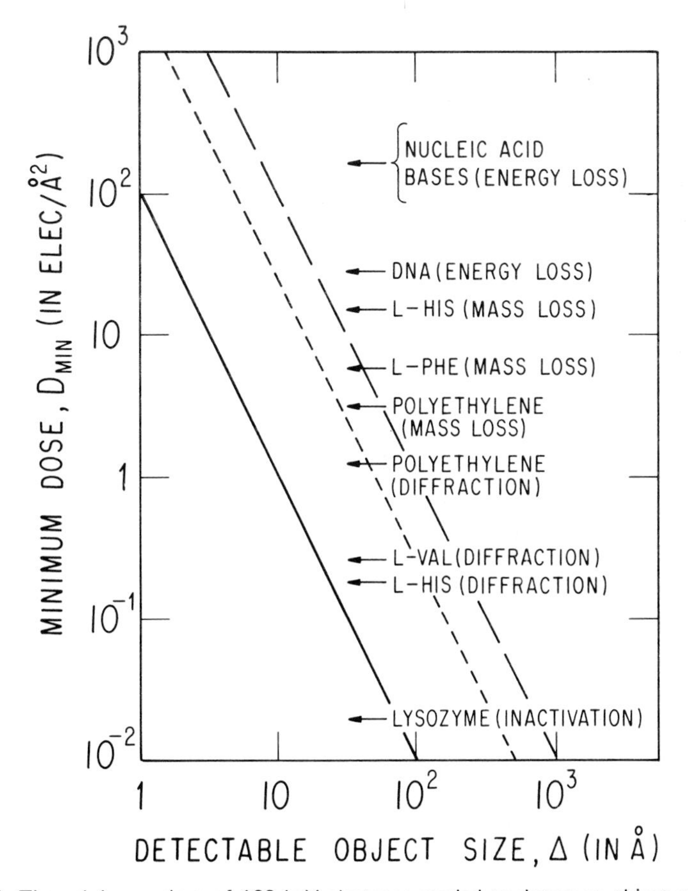

Fig. 1.15 The minimum dose of 100 keV electrons needed to detect an object of size Δ. The details of the calculation are given in the text. The arrows indicate the $D_{1/e}$ doses necessary to damage the molecules shown using the various damage indicators. This minimum dose assumes that the substrate is noiseless and is limited only by the statistical fluctuations of the incident beam. The curves have been obtained from Eq. (1.47).

—— —— —— $f'C^2 = 2.5 \times 10^{-3}$, $\kappa = 5$.

-------------- $f'C^2 = 1.0 \times 10^{-2}$, $\kappa = 5$.

———————— $f'C^2 = 1.0 \times 10^{-2}$, $\kappa = 1$.

dry molecular weights of biological particles, one might expect to make such measurements with $\kappa = 5$ for individual particles as small as a few tens of Å in size. And with some of the possible techniques discussed in the next section, it is certainly not unreasonable to expect the structure of biological objects to be studied at resolutions of 5-10 Å in the near future with minimal radiation damage.

IMPROVING THE RADIATION DAMAGE LIMIT: POSSIBILITIES

Minimal Exposure Techniques

Once it is realized that radiation damage in the electron microscope is the fundamental limitation on structural determination of biological objects, it becomes obvious that techniques to reduce this damage are a necessity. Apart from devising ways to increase the number of electrons necessary to damage a sensitive specimen, it is essential to minimize the number of electrons incident upon the specimen which are needed to record an image. One method of achieving this goal is to focus and correct astigmatism on a different area of the specimen other than the area to be imaged. The specimen area to be photographed is then exposed to the incident beam only for the time actually required for recording the image.

This technique is called the method of "minimum exposure" and it has been described in the literature for the CTEM by Williams and Fisher (1970a,b). In this method, the illuminating electron beam is focused to a sufficiently small spot on the specimen so that there is sufficient brightness on the fluorescent screen to permit one to focus and correct astigmatism. The next step is to move the illumination so that the illuminated area either falls off the fluorescent screen or is positioned at the corner of the screen. This is performed by either a small motion of the projector lens or with a small DC deflection. When the screen is lifted to expose the photographic plate, the illumination is spread to cover a new area of the specimen that has now been projected over the photographic plate. In this manner, the area photographed is only irradiated during plate exposure.

One uses a relatively similar technique in the STEM. In this case, all focusing and astigmatism correction is performed over a field of view adjacent to some field of interest, and at the moment before recording, a DC deflection is applied to the electron beam to bring an unirradiated area into the scanned field of view. The main disadvantage of these minimum exposure techniques is that one must "photograph blindly" and is never sure until after exposure whether the area photographed was interesting.

However, even with this difficulty, the technique has achieved very good results which could not be obtained under normal beam exposure techniques. The results with the CTEM have mainly been obtained with negatively stained specimens (Williams and Fisher, 1970a,b; Richards *et al.*, 1973; Unwin, 1974; Horne, 1973). However, even in these cases the differences between minimum and normal exposures are remarkable. In addition, the results obtained by Unwin (1974), Glaeser (1971), and Williams (1970a,b; Richards *et al.*, 1973) seem to provide evidence that even negative stain is affected by the electron beam and moves appreciably through the organic material which it has encapsulated. Although the nature of the process is not clear, the stained object does contain structural information which is usually destroyed under normal conditions. So in many instances, it is the radiation damage of the stain rather than the stain itself which limits the useful resolution (Glaeser, 1971; Unwin, 1974; Dubochet, 1973a).

The results with the STEM are preliminary and have been performed mainly with unstained biological specimens (Wall, 1972b; Langmore, 1975). However, it does appear that considerable loss of mass can be avoided if one does not insist on focusing directly upon the object to be measured (photographed). It seems that because of the simplicity, there is no excuse for not operating electron microscopes under minimum exposure techniques if one wants to obtain the most useful, signigicant structural detail of biological objects.

Another method which has been suggested to minimize the number of electrons necessary to record an image has been the use of image intensifiers. However, this method is somewhat erroneous since the minimum detectable object size depends only upon the minimum dose given in Eq. (1.47). If one uses a lower dose than that, the information becomes statistically insignificant and no amount of "intensification" used subsequently will improve the situation.

The only real way that an image intensifier system could allow one to utilize fewer electrons for recording an image would be if it were more efficient at electron detection than a fluorescent screen-photographic plate combination. For electron microscopy in the range of conventional voltages (~100 keV) the photographic plate is essentially a perfect detector in the sense that every incident electron is detected (Zeitler and Hayes, 1965; Zeitler, 1968; Valentine, 1966). The detective quantum efficiency (*DQE*, the square of the ratio of the signal to noise in the developed image to that in the incident electron image) is ~0.25-1 over a wide density range (Hamilton and Marchant, 1967), at these voltages. Moreover, when one is sufficiently dark-adapted, the same is true of the fluorescent screens used in the CTEM's. Therefore, it appears that at conventional electron microscope voltages, image intensifiers are of little use in minimizing the number of electrons per exposure. However, because of the sensitivity of fluorescent screens and photographic plates for higher-voltage electrons (Bahr *et al.*, 1966; Jones and Cosslett, 1970; Cosslett *et al.*, 1974), improved image intensifiers might be useful for electron exposure minimization at these higher voltages.

High Voltage Electron Microscopy

Many papers have appeared in the literature advocating the use of high voltage electron microscopes (incident electron energies greater than several hundred keV) to minimize some of the effects of radiation damage (e.g., Cosslett, 1969; Dupouy, 1968; Ris, 1970). Such papers were based upon the fact that the damaging dose for destruction of crystalline diffraction patterns increased over a factor of two from 100 keV to 1 MeV. What was usually neglected was the fact that while the damage cross section was getting smaller with increasing voltage (the voltage dependence being just that for inelastic scattering–see Fig. 1.12), the elastic cross section was decreasing more rapidly since it depended only upon $1/\beta^2$. The net result is that there is no real gain in the number of scattering events per damage (inelastic) event. At present, this fact is generally accepted in the literature. However, because there still is some confusion (e.g., Thomas, 1973; Hama, 1973; Humphreys, 1975), it is most instructive to look more closely at the trend of radiation damage as the energy of the incident electron is increased.

A convenient quantity to evaluate in order to determine this more clearly is the ratio of the total elastic cross section to the inelastic cross section as a function of the incident electron energy. This quantity gives us the number of high resolution scattering events per inelastic event. We have plotted this in Fig. 1.16 as $\sigma_{el}/\sigma_{in}/Z$ where σ_{in} is taken from Eqs. (1.18) and (1.20) and σ_{el} is given by Eq. (1.13) corrected for the failure of the first Born Approximation at lower voltages (Langmore *et al.*, 1973b). It is evident that this ratio decreases from 100 keV upwards in energy, so that one can expect no improvement in the number of scattered events to damaging events when going to higher incident electron energies.

The situation is, however, not quite as simple as this, since there are other quantities which must also be considered. For instance, the contrast changes as a function of incident electron energy; the efficiency of collection of electrons for imaging [f' in Eq. (1.47)] also is a function of voltage, and it is a different function for different microscopes. For the details of the changes in these quantities with voltage, the reader is referred to Langmore *et al.*, (1973b), Zeitler and Thomson (1970), Thomson (1975), Wall *et al.* (1974a), and Siegel (1971). We merely want to indicate the trends here with regard to damage.

A useful figure of merit for assessing the improvement in useful events per damage events is the ratio of the contrast to the inelastic scattering cross section. For dark field elastic scattering, the contrast is proportional to the elastic scattering cross section and so the ratio plotted in Fig. 1.16 indicates what we need to know: there is a slight increase in effective damage in dark field when going from 100 keV to 10 MeV. Actually, since the ultimate resolution improves with increasing voltage (barring "engineering" difficulties), the dark field contrast increases slightly with voltage. Moreover, in the CTEM, the collection efficiency for elastically scattered electrons increases with voltage approaching that of the STEM at 1-2 MeV (Langmore *et al.*, 1973b; Thomson, 1975). However, the ratio $\eta_{el} \cdot \sigma_{el}/\sigma_{in}$ (where η_{el} is the collection efficiency for elastic scattering) can never be greater than σ_{el}/σ_{in}, so the net result is that for dark field microscopy there is little reduction in radiation damage to be expected at high voltages.

Insofar as bright field scattering contrast is concerned, since the characteristic elastic scattering angle shrinks faster with voltage than the optimum aperture size, we get more scattering within the bright field aperture as we increase the

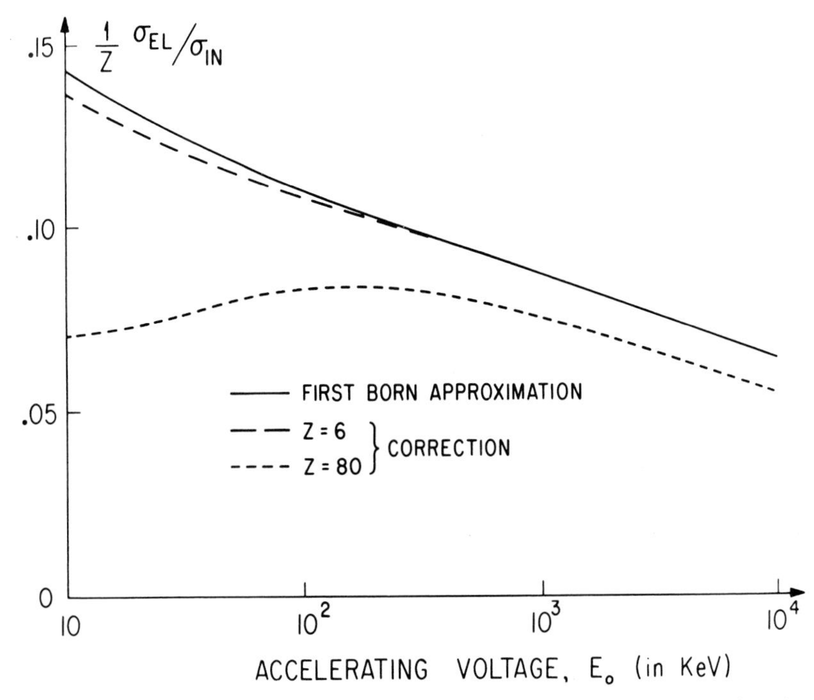

Fig. 1.16 The ratio of the elastic to the inelastic scattering cross sections as a function of the energy of the incident electron. The inelastic cross section was calculated using Eqs. (1.18) and (1.20). The solid curve was obtained using the first Born Approximation expression for the elastic scattering cross section given in Eq. (1.13). The dashed curves were obtained using the correction for the failure of the first Born Approximation [due to Zeitler and Olsen (1967)] given by Langmore *et al.* (1973b):

$$\sigma_{el} = \frac{3}{4}\frac{\lambda_c^2}{\pi\beta^2} Z^{3/2}\left[1 - \frac{.23\,Z}{137\,\beta}\right]$$

where Z is the atomic number and β is the ratio of the electron velocity to that of light.

voltage. As a result, the bright field scattering contrast decreases with voltage (a factor of approximately four when going from 100 keV to 10 MeV for a mercury atom). Therefore, for this contrast mode, we again find little improvement in radiation damage at high voltages.

For the phase contrast mode of operation, Zeitler and Thomson (1970) have found that single atom contrast is relatively constant (increasing by ~50% for Au) over the energy range from 100 keV to MeV, so that over this energy range little improvement in radiation damage would be expected. On the other hand, the single atom phase contrast increases by almost a factor of two from 1 MeV to 10 MeV. Since the inelastic scattering cross section only increases by 30% in this range, one might hope for some small reduction in radiation damage at 10

MeV (if one neglects the increased probability for knock-on collision damage). However, the phase contrast calculations are performed by keeping the product of spherical aberration of the objective lens times the electron wavelength constant with voltage. The result is that the resolution is getting better with increasing voltage, and therefore the bulk of the increased contrast is derived from this increased resolution. If one were to obtain this small improvement, the microscope would have to operate at its theoretical electron optical resolution limit (~1 Å). At present, there is no existing high voltage electron microscope which operates at a resolution equal to its theoretical limit, the performance being limited by electrical and mechanical instability.

One point which sometimes gets lost in the literature among advocates of high voltage electron microscopy for minimizing radiation damage is the fact that as we increase the voltage, the maximum transferable energy, E_M, increases (Fig. 1.2A). Therefore, even though the ratio $\sigma_{\text{disp}}/\sigma_{\text{el}}$ for low atomic number materials does not substantially increase from 100 keV to 10 MeV (Fig. 1.3), it becomes energetically possible to break more than one bond. For instance, for carbon at 10 MeV, $E_M \approx 20$ keV! In addition, because this maximum energy is much greater than any displacement energy, we can get a multiplicative effect in which the displaced atom displaces others. For all known models of multiple displacements, Parikh (1974) has stated that for $E_M \gg E_D$, the displacement energy, the number of displacements per primary knock-on collision is:

$$\bar{\nu} \approx \frac{1}{2}[1 + \ln(E_M/2E_D)] \tag{1.48}$$

This would give a value of $\bar{\nu} \approx 4$ for a displacement energy of 10 eV using 10 MeV electrons. Therefore, knock-on collisions may pose a damage problem at extremely high energies.

I have neglected to mention the question of the improvement in contrast that is due to smaller chromatic aberration smearing at higher voltages in the CTEM. While this is in fact true, the use of energy filters at conventional voltages to filter out the inelastically scattered electrons (Castaing and Henry, 1962; Henkleman, 1973; Henkleman and Ottensmeyer, 1973; Rose and Plies, 1974; Watanabe and Uyeda, 1962) accomplishes the same contrast improvement with less difficulty. Moreover, the problem is completely eliminated by the use of STEM techniques. Therefore, the improvement in contrast because of less chromatic aberration smearing as a means of reducing radiation damage is no longer a clear advantage of utilizing high voltage electron microscopes.

While there are many reasons for advocating the use of high voltage electron microscopes [such as increased penetration power which would enable one to view thicker samples, the ultimate theoretical resolution of 1 Å or better, and many applications in materials science (Cosslett, 1974)], it appears to this author that there is no substantial evidence to indicate that a reduction in radiation damage is one of those reasons.

Scanning Transmission Electron Microscopy

There has been some confusion in the literature as to whether or not the STEM can be operated utilizing a lower incident electron dose per micrograph than the CTEM. In fact, recent remarks by Scherzer (1970, 1972) have been widely misinterpreted as indicating that the STEM causes more specimen damage than the CTEM. The point that is generally missed is that he confined his attention to instruments with resolutions better than 0.4 Å, where his conclusions are most probably true. For resolutions lower than 0.4 Å, it has been shown that for any mode of contrast, the STEM is more efficient than the CTEM in utilizing electrons for imaging purposes.

Therefore, the minimum dose necessary for detection is smaller for the STEM (Crewe, 1973; Langmore *et al.*, 1973b; Wall *et al.*, 1974a; Isaacson *et al.*, 1973; Rose, 1974). In addition, the STEM is capable of recording several signals simultaneously so that one can, in principle, extract a maximum amount of information per incident electron (Crewe *et al.*, 1975). Therefore, for all practical purposes (resolution greater than ~1/2 Å), the STEM causes less specimen damage than the CTEM. The reader who is interested in the detailed comparisons between the two types of instruments is referred to the above-mentioned references or to the review by Zeitler and Isaacson (1976) in this series.

Signal Averaging

Even with the use of minimum exposure techniques, radiation damage occurs in the CTEM (Glaeser, 1971; Dubochet, 1973a,b; Unwin, 1972, 1974) and the STEM (Langmore, 1975; Isaacson *et al.*, 1974b; Wall, 1972b). However, if one could obtain micrographs with an incident dose less than the minimum dose given in Eq. (1.47), one could reduce specimen damage even further. The problem is that the micrographs would consist of statistically noisy images. And so the question becomes one of whether there exist possibilities (in certain cases) in which information on the object can be recovered from noisy images.

The obvious technique is to record many noisy micrographs of identical objects in identical orientations. One could then simply superimpose the noisy images (with proper orientation) and improve the signal-to-noise of the micrograph as the square root of the number of superpositions. In fact, this is just what one does in recording an electron diffraction pattern of a periodic object. In this case, one obtains information from identical unit cells utilizing a much lower dose than the "minimum dose" which would be necessary to record a direct image of just one of the unit cells. As an example, consider the nucleic acid base, cytosine. The $D_{1/e}$ dose necessary for damage of the diffraction pattern of cytosine is ~23 elec/Å^2 at 100 keV. Yet to record an image at 1 Å resolution we would require more than 1,000 elec/Å^2 so that the resulting image would be of a structurally damaged object. However, one can easily record an electron diffraction pattern of a few thin single crystals of cytosine (approximately a few microns square in area) at 100 keV with an incident electron dose of less than .1 elec/Å^2. Such a diffraction pattern is shown in Fig. 1.17, and one can see reflections out to 1 Å.

Therefore, for periodic structures at least, one can suppose that there would exist possibilities of superimposing repeating units of statistically noisy data to obtain a statistically well-defined image. The simplest techniques of spatial averaging of repeating structures are at least a decade old. Techniques have included either rotational or translational superposition using photographic techniques (Markham *et al.*, 1963), computer techniques (Nathan, 1971), or techniques of filtering certain spatial frequencies in the Fourier transform of the images (Klug and DeRosier, 1966; Hoppe *et al.*, 1968; DeRosier and Klug, 1968; DeRosier, 1971; Hoppe *et al.*, 1973). However, in most of the examples which appear in the literature, the averaging was accomplished using statistically well-

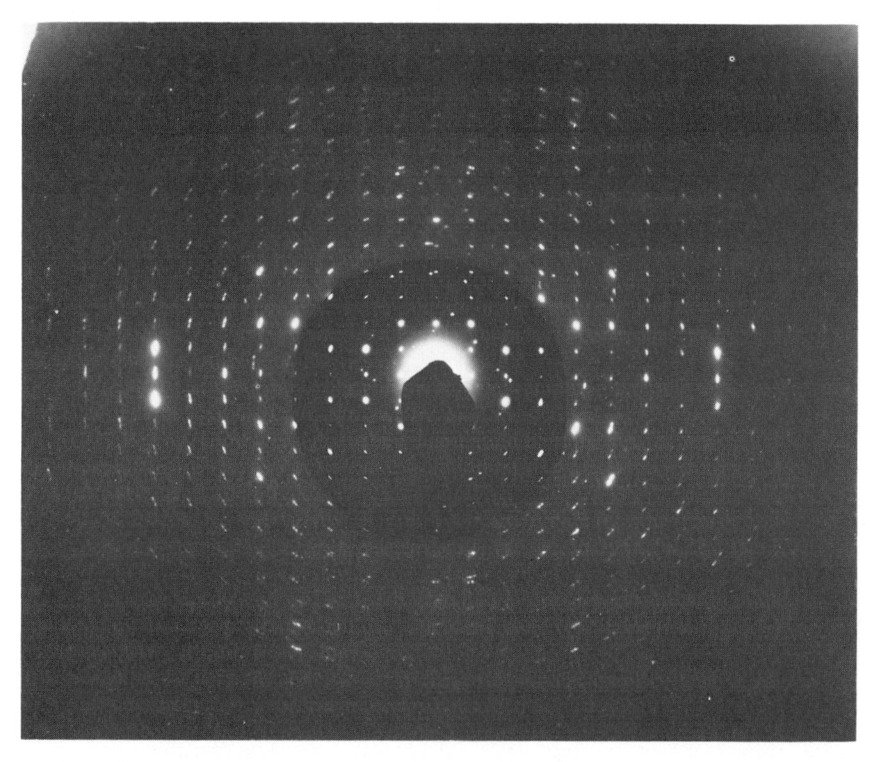

Fig. 1.17 Selected area electron diffraction pattern of a polycrystalline thin (400-Å-thick) film of the nucleic acid base, cytosine, supported on a 30-Å-thick carbon substrate. The selected area corresponded to about a 3-μ diameter area at the specimen, and the incident electron energy was 100 keV. The dose necessary to record the micrograph was less than 0.1 elec/Å^2. Cytosine has an orthorhombic crystal structure and the pattern shown is a projection of the *ab* plane (a = 13.041 Å, b = 9.494 Å, c = 3.815 Å are the lattice constants of the unit cell). One can easily see reflections out to 1 Å or better.

defined images. The averaging was performed because the repeating structures were all slightly different because of damage in the specimen preparation procedures.

The possibility of using these spatial averaging techniques in electron microscopy to obtain information about repeating structures from statistical noise has been discussed by Glaeser *et al.* (1971b) and Frank (1973, 1975), each advocating a different method. According to the method of Glaeser *et al.* (1971b), one obtains the power spectrum of the noisy image in order to obtain the repeating distances (i.e., lattice dimensions). This information is then used to superimpose the image by lattice distance translation (or rotation). The feasibility of this technique has not been tested on actual electron micrographs, but they have demonstrated it by computing a noisy two-dimensional periodic function obtained by sampling it in a random fashion. The resulting "noisy image" shows no periodicity. However, after obtaining the power spectrum and by properly translating and superimposing the noisy image, they were able to reconstruct the original structure.

Frank (1975) claims that correlation techniques using the autocorrelation function are superior to power spectrum techniques in being able to retrieve signals from statistical noise. Hoppe (1974) also suggests that correlation functions will be an important tool in interpreting noisy micrographs, since they are significant in images which look uninterpretable because of noise. However, as of the present, no experimental data on low level detection using correlation functions have been published.

One could ask the question as to why one does not simply utilize electron diffraction for obtaining low dose information concerning periodic structures rather than resort to these spatial averaging techniques of direct image micrographs. The reason is that one needs crystals of a reasonable size to do this (single crystals several microns in size), and this is not always possible; one is sometimes hampered by multiple scattering in interpreting electron diffraction patterns. Also, one gets only intensity information from electron diffraction patterns, whereas with direct images one can obtain both the amplitude and the phase.

In the case of aperiodic objects, the ability to spatially average noisy structures is much more difficult than for repeating objects. For these objects, the techniques discussed above cannot be readily applied unless one can develop methods for orienting the image of the object in the presence of noise, so that all properly oriented objects can be superimposed.

The ability (Horne, 1975) to produce two-dimensional "rafts" of single asymmetric units (such as viruses) offers the possibility of being able more easily to orient noisy images of aperiodic objects. This method has not yet been tried as a means of minimizing the necessary incident dose and may well hold promise in reducing radiation damage. In addition, it is also possible to perform some Fourier filtering of aperiodic objects in the form of high pass or low pass filters which could be used to eliminate some noise in the image (Frank, 1973).

It appears that the method of reducing radiation damage by extracting signals from the noise has hardly been touched upon. Much might be gained in electron microscopy at reduced damage levels by further exploiting such techniques.

Low Temperatures

In an earlier section a suggestion was made that some of the molecular disorder accompanying electron irradiation damage might be reduced if the specimens were kept at low temperatures. However, since there were so few experimental measurements of this temperature effect, and there was contradictory evidence among them, it was not conclusive whether low temperatures actually reduced damage by any significant amount.

There is evidence, however, that there is a "latent dose" which is tolerable to a material without a detectable change in the indicator used to assay damage. In fact, Siegel (1972) has found that for paraffin and tetracene, this "latent dose" increases as the temperature is decreased. Moreover, the increase of the characteristic dose, $D_{1/e}$, for diffraction damage upon cooling the specimen is mainly due to the increase of this latent dose, D_{ld}. We have also found a latent dose effect at room temperature for the nucleic acid bases for damage as assayed by the characteristic energy loss spectra. This is shown in Fig. 1.18 for adenine, where we see that $D_{\mathrm{ld}} \approx \frac{1}{6} \cdot D_{1/e}$. Obviously, there is a necessity for improving upon the simple, single-hit, single-target ideas which we have used to attempt to understand radiation damage in electron microscopy.

The existence of a "latent dose" seems to be indicative of some secondary damage process. If this process were temperature-dependent, one might hope that cooling could delay the process by freezing in the broken molecular fragments. The main point to be made is that one does have some evidence to believe that radiation damage might be reduced at low temperatures. The few temperature-dependent measurements which do exist (Table 1.5) certainly do not allow one to state conclusively that low temperatures will not reduce damage. Since there are no quantitative temperature-dependent mass loss measurements in the literature and few diffraction measurements, it would appear that quantitative measurements of the temperature dependence of various damage indicators on different materials are essential.

Other Possibilities

There are other techniques which might possibly be used to minimize the structural degradation of biological objects that is due to the incident electron beam in the electron microscope. Some of these techniques do not actually reduce damage to the unstained specimen itself, but rather they convert the naked unstained organic particle into inorganic material which reflects some inherent morphology of the original structure. This inorganic "replica" is more resistant to beam damage than the original object, and so one, in essence, re-

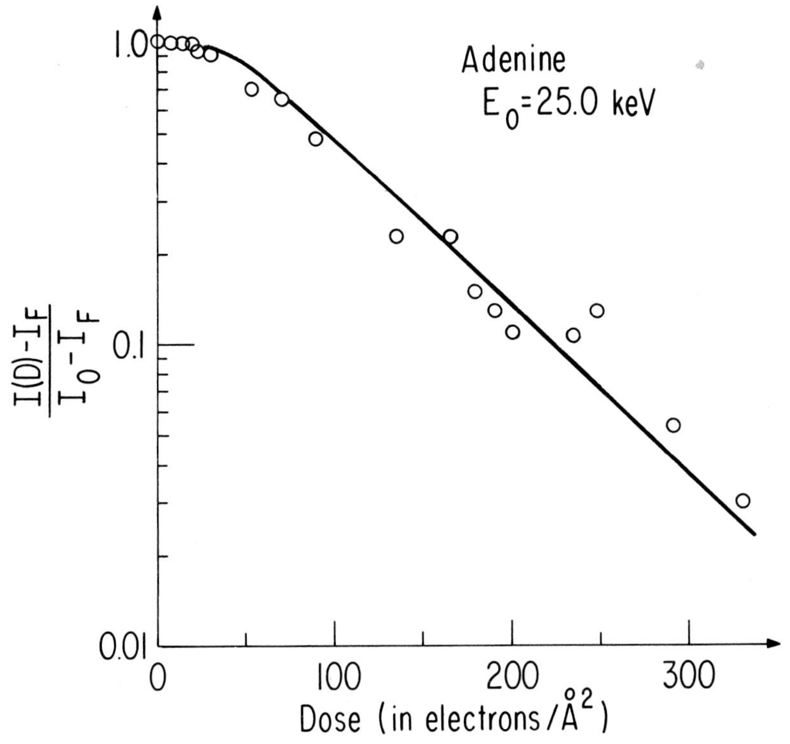

Fig. 1.18 The dose-response curve for damage to the 5 eV peak in the energy loss spectrum of adenine. The term $I(D)$ is the intensity of the 5 eV peak at a dose D, while I_0 is the intensity of that peak at zero dose, and I_f is the intensity at infinite dose. Note that the latent dose, $D_{ld} \approx \frac{1}{6} D_{1/e}$. The incident electron energy was 25 keV and the specimen was at room temperature.

duces the effects of beam damage. There are, however, certain specimen preparation problems associated with these techniques which tend to limit the useful resolution attainable.

As it was mentioned in the Introduction, the technique of negative staining has been used for almost twenty years as a method of increasing the contrast in the CTEM. In general use, the useful structural resolution attainable is only 20–30 Å (DeRosier, 1971; Haschemeyer and Meyers, 1972). The advantage of a negative stain is that it literally embeds the organic particle in a nearly amorphous inorganic coat. At an incident electron dose which would exceed the damage dose for the particle itself, the inorganic coat can be relatively unharmed by the incident beam. In many cases, the resolution attainable using negative staining "replication" techniques is limited by the crystallinity of the stain.

However, it has recently been shown that in some instances the resolution limit was imposed by the alteration of the stain that was due to the incident electron flux (Glaeser, 1971; Unwin, 1972; Williams and Fisher, 1970a,b;

Horne, 1973). It is known that negative stain can be mobile under the action of an electron beam and can get recrystallized in some cases (Unwin, 1974; Dubochet, 1973a). The exact nature of the damage is not really known, but it has been demonstrated that the redistribution of stain can reduce resolution and that a stabilization of the stained structure occurs which eventually limits the obtainable structural detail to ~25-30 Å (Glaeser, 1971; Unwin, 1974).

One can obtain useful structural information concerning negatively stained biological particles at 15 Å resolution if one uses minimum exposure techniques. Moreover, it appears that certain stains consisting of equal parts of metal ions and uranyl salts are more radiation-resistant than the standard uranyl stains (Unwin, 1972). In addition, the crystallinity of such stains seems to be of smaller size than the standard stains, so that Unwin (1972) has been able to show optical diffraction patterns of micrographs of negatively stained catalase crystals which showed distinct reflections out to 10-12 Å. The problem with negative staining techniques at present is that the mechanism of staining is not well understood. It would appear that further research on negative staining materials is necessary.

There is also the possibility of utilizing "negative stains" of low density material solely as embedding agents to prevent the original unstained structure from collapsing. Such stains would be useful in dark field microscopy where there is adequate contrast available with staining and the low density material might have less "granularity" than heavy metal stains. Dubochet (1973a) has shown some preliminary examples of some low density stains of 0.6% NaPT diluted in 1.2% sucrose, and 1% solutions of sodium silicate. There are also several laboratories working on BeF_2 as an embedding medium. So, it would appear that there is hope for obtaining useful structural information at levels below 10 Å resolution using negative staining-embedding techniques.

Another widely used "replication" technique for measuring the contrast in the CTEM is that of metal shadowing. It generally yields only resolutions of 20-30 Å, but there are certain cases where 15 Å has been claimed (Abermann *et al.*, 1972). The advantage of shadowing again is that the metal coat is more resistant to the incident electron beam than the organic object underneath.

The main problem in high resolution shadowing is in minimizing the granularity (or crystallinity) of the shadowing material. The mechanism of nuclei formation (the formation of atom clusters) is not well understood in spite of the extensive research on the subject in the literature of thin film technology. If we think of negative staining and shadowing as merely techniques to encase the object in a radiation-resistant coat, then it appears that further research into these techniques might yield methods which would possibly enable one to extract useful structural information concerning biological objects below the 10 Å level.

A third technique for obtaining structural information of biological particles at a high resolution is that of selective heavy atom staining (Beer and Zobel,

1961; Hayat, 1975). Unlike negative staining and shadowing this is not a replication technique in the sense that the biological particle remains uncoated. The technique consists of chemically attaching single heavy atoms of heavy atom complexes to specific sites on a biological molecule and then visualizing the relative locations of these markers in the electron microscope. As long as the heavy atoms do not change their relative positions (Langmore *et al.*, 1974), this technique is not dependent on beam alterations of the original structure. Unfortunately, if the markers are attached at regions which do undergo damage, then the technique is almost as prone to radiation damage as the original specimen. The main advantage of the technique is that the heavy atom markers provide more inherent contrast than unstained specimens, and so at the high resolution levels at which these markers would be useful one can reduce the minimum dose necessary for detection.

One particular application of this technique being actively pursued by several laboratories is that of selectively attaching heavy atoms to bases along strands of nucleic acid (Langmore *et al.*, 1972, Langmore and Crewe, 1974; Whiting and Ottensmeyer, 1974; Beer, 1974). The main problem encountered to date is that of producing a good selective stain which is not volatile in the beam or in the vacuum of the microscope. A detailed discussion on positive staining has been recently presented by Hayat (1975).

FUTURE OF BIOLOGICAL ELECTRON MICROSCOPY

It is pertinent to present a brief discussion on the future of biological electron microscopy insofar as its specimen damage limitations are concerned. Unfortunately, this is not possible because all the evidence for or against it is not available. It is obvious that we have just begun to understand the specimen damage mechanisms in the electron microscope. It is also clear, however, that researchers are now aware of the fact that radiation damage is the fundamental limit in structural determination in biological electron microscopy, and they are attempting to surmount the barrier.

We have tried to present an overview of the experimental evidence concerning radiation damage in the electron microscope to gain some preliminary understanding of the damage process, and to explore briefly some of the possibilities for reducing the effect of beam damage in the structural determination of sensitive organic specimens. It seems that the radiation damage limit is not insurmountable. Rather than having to try to increase the dose for radiation damage or decrease the minimum detectable dose by several order of magnitude, we find that in many cases there exists only one order of magnitude difference between the damage dose and the minimum detectable dose. By gaining factors of two or three from each of a few of the potential techniques discussed in the previous section it is not inconceivable that one might eventually be able to obtain useful structural information at ~2 Å level, with minimal beam damage.

It would appear that more research is needed on radiation damage of biological materials (particularly at low temperatures), techniques of retrieving information from statistically noisy micrographs, and new specimen preparation techniques applicable to minimum damage imaging. Only after these problems have been explored will the biologist be able either to dismiss electron microscopy as a viable tool for structure research below 20 Å, or be able to reap fully the fruits of structure research at the 2-20 Å level which electron beam technology has the potential to provide.

This work was supported by the U.S. Atomic Energy Commission, Division of Biology and Medicine. I would like to thank Drs. A. V. Crewe, D. Johnson, J. P. Langmore, S. D. Lin, and E. Zeitler for numerous helpful arguments and discussions. I would also like to thank E. Zeitler and J. Langmore for critically reading the manuscript.

References

Abermann, R., Salpeter, M. M., and Bachmann, L. (1972). High resolution shadowing. In: *Principles and Techniques of Electron Microscopy; Biological Applications*, Vol. 2 (Hayat, M. A., ed.), p. 197. Van Nostrand Reinhold Company, New York.

Allinson, D. L. (1975). Environmental devices in electron microscopy. In: *Principles and Techniques of Electron Microscopy: Biological Applications*, Vol. 5 (Hayat, M. A., ed.). Van Nostrand Reinhold Company, New York and London.

Amos, L. A. (1974). Image analysis of macromolecular structures. *J. Microscopy* (London) **100**, 143.

Bacq, Z. M., and Alexander, P. (1961). *Fundamentals of Radiobiology*. Pergamon Press, Oxford.

Bahr, G. F., Johnson, F. B., and Zeitler, E. (1965). The elementary composition of organic objects after electron irradiation. In: *Quantitative Electron Microscopy* (Bahr, G. F., and Zeitler, E., eds.), p. 377. Williams and Wilkins Co., Baltimore.

Bahr, G. F., Zeitler, E. H., and Kobayashi, K. (1966). High voltage electron microscopy. *J. Appl. Phys.* **37**, 2900.

Baumeister, W., and Hahn, M. (1974). Suppression of lattice periods in vermiculite single crystal specimen supports for high resolution electron microscopy. *J. Microscopy* (London) **101**, 111.

Beer, M. (1974). Electron microscopic studies of base sequencing in DNA. *Proc. 32nd Ann. EMSA Meeting* (St. Louis), p. 300.

Beer, M., and Zobel, C. R. (1961). Electron stains II: Electron microscope studies on the visibility of stained DNA molecules. *J. Mol. Biol.* **3**, 717.

Bhattacharjee, S. B. (1961). Action of X-irradiation on *E. coli*. *Rad. Res.* **14**, 50.

Box, H. C. (1973). Cryoprotection of irradiated specimens. *Proc. 31st Ann. EMSA Meeting* (New Orleans), p. 480.

Breedlove, J. R., Jr., and Trammel, G. T. (1970). Molecular microscopy: Fundamental limitations. *Science* **170**, 1310.

Brünger, W., and Menz, W. (1965). Wirkungsquérschnitte für Elastische und Inelastische Elektronenstreuung an Amorphen Kohlenstoff und Germaniumschnitten. *Z. Phys.* **124**, 271.

Castaing, R., and Henry, L. (1962). Filtrage magnetique des vitesses en microscopie electronique. *C. R. Acad. Sci.* (Paris) **225**, 76.

Charlesby, A. (1960). *Atomic Radiation and Polymers*. Pergamon Press, Oxford.
Claffey, W. J., and Parsons, D. F. (1972). Electron diffraction study of radiation damage in coronene. *Phil. Mag.* **25**, 635.
Cleveland, R. F., and Deering, R. A. (1971). Division delay of resistant and sensitive strains of a cellular slime mold. *Rad. Res.* **47**, 292.
Corbett, J. W. (1966). Electron radiation damage in semiconductors and metals. In: *Solid State Physics, Supplement 7* (Seitz, F., and Turnbull, D., eds.), p. 6. Academic Press, New York and London.
Cosslett, A. (1960). Some applications of the ultraviolet and interference microscopes in electron microscopy. *J. Roy. Micros. Soc.* (London) **79**, 263.
Cosslett, V. E. (1969). High voltage electron microscopy. *Quart. Rev. Biophys.* **2**, 95.
Cosslett, V. E. (1974). Current developments in high voltage electron microscopy. *J. Microscopy* (London) **100**, 233.
Cosslett, V. E., Jones, G. L., and Camps, R. A. (1974). Image viewing and recording in high voltage electron microscopy. In: *High voltage Electron Microscopy* (Swann, P. R., Humphreys, C. J., and Goringe, M. J., eds.) Academic Press, London, p. 147.
Crewe, A. V. (1971a). A scanning microscope that can see atoms. *Scientific American* **224**, 26.
Crewe, A. V. (1971b). High resolution scanning microscopy of biological specimens. *Phil. Trans. Roy. Soc. London* **B261**, 61.
Crewe, A. V. (1973). Considerations of specimen damage for the transmission electron microscope, conventional versus scanning. *J. Mol. Biol.* **80**, 315.
Crewe, A. V., Isaacson, M., and Johnson, D. (1969). A simple scanning electron microscope. *Rev. Sci. Instr.* **40**, 241.
Crewe, A. V., Isaacson, M., and Johnson, D. (1971a). The characteristic energy loss of fast electrons in the nucleic acid bases. *Nature* **231**, 262.
Crewe, A. V., Isaacson, M., and Johnson, D. (1971b). A high resolution electron spectrometer for use in transmission scanning electron microscopy. *Rev. Sci. Instr.* **42**, 411.
Crewe, A. V., Langmore, J., and Isaacson, M. (1975). Resolution and contrast in the scanning transmission electron microscope. In: *Techniques in Electron Microscopy and Microprobe Analysis* (Siegel, B., and Beaman, D., eds.). John Wiley and Sons, New York, p. 47.
Crewe, A. V., Langmore, J. P., Isaacson, M. S., and Retsky, M. (1974). Understanding single atoms in the STEM. *Proc. 8th Int. Cong. Electron Micros.* (Canberra), **1**, 260. Australian Academy of Sciences, Canberra.
Crewe, A. V., and Wall, J. (1970a). Contrast in a high resolution scanning transmission electron microscope. *Optik* **30**, 461.
Crewe, A. V., and Wall, J. (1970b). A scanning microscope with 5 Å resolution. *J. Mol. Biol.* **48**, 375.
Daniels, J. (1969). Determination of optical constants of palladium and silver from 2 to 90 eV from energy loss measurements. *Z. Phys.* **277**, 234.
Daniels, J., von Vestenberg, C., Raether, H., and Zeppenfeld, K. (1970). Optical constants of solids by electron spectroscopy. In: *Springer Tracts in Modern Physics*, Vol. 54 (Hohler, G., ed.), p. 77. Springer-Verlag, Berlin.
DeRosier, D. J. (1971). The reconstruction of three-dimensional images from electron micrographs. *Contemp. Phys.* **12**, 437.
DeRosier, D. J., and Klug, A. (1968). Reconstruction of three-dimensional structures from electron micrographs. *Nature* **217**, 130.
Dertinger, H., and Jung, H. (1970). *Molecular Radiation Biology*. Springer-Verlag, Berlin.
Ditchfield, R. W., Grubb, D. J., and Whelan, M. J. (1973). Electron energy loss studies of polymers during radiation damage. *Phil. Mag.* **28**, 1267.
Dubochet, J. (1973a). High resolution dark-field electron microscopy. *J. Microscopy* (London) **98**, 334.

Dubochet, J. (1973b). High resolution dark field electron microscopy. In: *Principles and Techniques of Electron Microscopy, Biological Applications*, Vol. 3 (Hayat, M. A., ed.), p. 115. Van Nostrand Reinhold Company, New York and London.

Dubochet, J. (1975). Carbon-loss during irradiation of T4 bacteriophages and *E. coli* bacteria in electron microscope. *J. Ultrastruct. Res.*, **52**, 276.

Ducoff, H. S., Crossland, J. L., and Vaughan, A. P. (1971). The lethal syndromes of insects. *Rad. Res.* **47**, 299.

Dupouy, G. (1968). Electron microscopy at very high voltages. In: *Advances in Optical and Electron Microscopy*, Vol. II (Barer, R., and Cosslett, V. E., eds.), p. 167. Academic Press, London.

Dupouy, G. (1974). Megavolt electron microscopy. In: *High Voltage Electron Microscopy* (Swann, P. R., Humphreys, C. T., and Goringe, M. J., eds.), p. 441. Academic Press, London.

Dupouy, G., Perrier, F., and Durrieu, L. (1960). L'observation de la matiere vivante au moyen d'un microscopie electronique fonctionmant sous tres haute tension. *C. R. Acad. Sci.* (Paris) **251**, 2836.

Durup, J., and Platzman, R. L. (1961). Role of the auger effect in the displacement of atoms in solids by ionizing radiation. *Disc. Faraday Soc. London*, **31**, 156.

Echlin, P., and Fendley, J. A. (1973). The future of electron microscopy in biology. *Nature* **244**, 409.

Engel, M. S., and Adler, H. I. (1961). Catalase activity, sensitivity to hydrogen peroxide and radiation response in the genus *Escherichia*. *Rad. Res.* **15**, 469.

Fawaz-Estrup, F., and Setlow, R. B. (1964). Inactivation of soluble ribonucleic acid by electron irradiation. *Rad. Res.* **22**, 579.

Fluke, D. J. (1966). Temperature dependence of ionizing radiation effect on dry lysozyme and ribonuclease. *Rad. Res.* **28**, 677.

Frank, J. (1973). Computer processing of electron micrographs. In: *Advanced Techniques in Biological Electron Microscopy* (Koehler, J. K., ed.), p. 215. Springer-Verlag, Berlin.

Frank, J. (1975). Digital correlation methods in electron microscopy (to be published).

Ginoza, W. (1967). The effects of ionizing radiation on nucleic acids of bacteriophages and bacterial cells. *Ann. Rev. Microbiol.* **21**, 325.

Glaeser, R. M. (1971). Limitations to significant information in biological electron microscopy as a result of radiation damage. *J. Ultrastruct. Res.* **36**, 466.

Glaeser, R. M. (1974). Radiation damage and resolution limitations in biological specimens. In: *High Voltage Electron Microscopy* (Swann, P. R., Humphreys, C. J., and Goringe, M. J., eds.), p. 370. Academic Press, London.

Glaeser, R. M., Cosslett, V. E., and Valdre, U. (1971a). Low temperature electron microscopy: Radiation damage in crystalline biological materials. *J. Microscopie* (Paris) **12**, 133.

Glaeser, R. M., Kuo, I., and Budinger, T. F. (1971b). Method for processing of periodic images at reduced levels of electron irradiation. *Proc. 29th Ann. EMSA Meeting* (Boston), p. 466. Claitor's Pub. Division, Baton Rouge, La.

Grubb, D. J., and Groves, G. W. (1971). Rate of damage of polymer crystals in the electron microscope: Dependence on temperature and beam voltage. *Phil. Mag.* **24**, 815.

Günther, W., and Jung, H. (1967). Der Einfluss der Temperatur auf die Strahlenempfindlichkeit von Ribonuclease. *Z. Naturf.* **22B**, 313.

Hall, T. A., and Gupta, B. L. (1974). Beam-induced loss of organic mass under electron microprobe conditions. *J. Microscopy* **100**, 177.

Hama, K. (1973). High voltage electron microscopy. In: *Advanced Techniques in Biological Electron Microscopy* (Koehler, J. K., ed.), p. 275. Springer-Verlag, Berlin.

Hamilton, J. F., and Marchant, J. C. (1967). Image recording in electron microscopy. *J. Opt. Soc. Amer.* **57**, 232.

Harada, Y., Taoka, T., Watanabe, M., and Ohara, M. (1972). Effect of accelerating voltage and specimen temperature on radiation damage of hexadecachlorocopper-phthalocyanine. *Proc. 30th Ann. EMSA Meeting* (Los Angeles), p. 686.

Hart, E. J., and Platzman, R. L. (1961). Radiation chemistry. In: *Mechanisms in Radiobiology*, Vol. I. (Errera, M., and Fossberg, A., eds.), p. 91. Academic Press, New York and London.

Haschemeyer, R. H., and Meyers, R. J. (1972). Negative staining. In: *Principles and Techniques of Electron Microscopy: Biological Applications*, Vol. 2 (Hayat, M. A., ed.), p. 101. Van Nostrand Reinhold Company, New York and London.

Hashimoto, H., Kumao, A., Hino, K., Yatsumoto, Y., and Ono, A. (1971). Images of thorium atoms in transmission electron microscopy. *Jap. J. Appl. Phys.* **10**, 1115.

Hayat, M. A. (1970). *Principles and Techniques of Electron Microscopy: Biological Applications*, Vol. 1. Van Nostrand Reinhold Company, New York and London.

Hayat, M. A. (1975). *Positive Staining for Electron Microscopy*. Van Nostrand Reinhold Company, New York and London.

Henkelman, R. M. (1973). An energy selecting electron microscope for biological applications. Ph.D. thesis, The University of Toronto.

Henkelman, R. M., and Ottensmeyer, F. P. (1971). Visualization of single heavy atoms by dark field electron microscopy. *Proc. Nat. Acad. Sci.* (U.S.A.), **68**, 3000.

Henkelman, R. M., and Ottensmeyer, F. P. (1973). Energy filtration of electron microscope images. *Proc. 31st Ann. EMSA Meeting* (New Orleans), p. 288.

Hoppe, W. (1974). Towards three-dimensional "Electron Microscopy" at atomic resolution. *Naturwissenschaften* **61**, 239.

Hoppe, W., Langer, R., Knesch, G., and Poppe, C. (1968). Protein-Kristall Strukturanalyse mit Elektronenstrahlen. *Naturwissenschaften* **55**, 333.

Hoppe, W., Bussler, P., Feltynowski, A., Nunsmann, N., and Hirt, A. (1973). Some experience with computerized reconstruction methods. In: *Image Processing and Computer-Aided Design in Electron Optics* (Hawkes, P. W., ed.), p. 92. Academic Press, London.

Horne, R. W. (1973). Contrast and resolution from biological objects examined in the electron microscope with particular reference to negatively stained specimens. *J. Microscopy* (London) **98**, 286.

Horne, R. W. (1975). A negative staining carbon film technique for studying viruses in the electron microscope. *J. Ultrastruct. Res.* **51**, 233.

Hotz, G., and Muller, A. (1968). The action of heat and ionizing radiation on the infectivity of isolated ϕX-174 DNA. *Proc. Nat. Acad. Sci.* (U.S.A.) **60**, 251.

Humphreys, C. (1975). High voltage electron microscopy. In: *Principles and Techniques of Electron Microscopy: Biological Applications*, Vol. 6 (Hayat, M. A., ed.). Van Nostrand Reinhold Company, New York and London.

Inokuti, M., Saxon, R., and Dehmer, J. (1975). Total cross-sections for inelastic scattering of charged particles by atoms and molecules. VIII. Systematics for atoms in the first and second row. *Int. J. Rad. Phys. Chem.*, **7**, 109.

Isaacson, M. (1972a). The interaction of 25 keV electrons with the nucleic acid bases adenine, thymine, and uracil. I. Outer shell excitation. *J. Chem. Phys.* **56**, 1803.

Isaacson, M. (1972b). The interaction of 25 keV electrons with the nucleic acid bases adenine, thymine, and uracil. II. Inner shell excitation and inelastic scattering cross-sections. *J. Chem. Phys.* **56**, 1813.

Isaacson, M. (1975a). The characteristic energy loss of electrons in biological molecules. *Proc. 4th Int. Conf. on Vacuum Ultra Violet Radiation Physics*, p. 826. Pergamon-Vieweg, Braunschweig.

Isaacson, M. (1975b). Inelastic scattering and beam damage of biological molecules. In: *Techniques in Electron Microscopy and Microprobe Analysis* (Siegel, B., and Beaman, D., eds.), Chapter 14. John Wiley and Sons, New York, p. 247.

Issacson, M., Johnson, D., and Crewe, A. V. (1973). Electron beam excitation and damage of biological molecules: Its implication for specimen damage in electron microscopy. *Rad. Res.* **55**, 205.

Isaacson, M., Langmore, J., and Rose, H. (1974a). Measurement of the non-localization of inelastic scattering by electron microscopy. *Optik* **41**, 92.

Isaacson, M., Langmore, J., and Wall, J. (1974b). The preparation and observation of biological specimens for the scanning transmission electron microscope. In: *Scanning Electron Microscopy/1974* (Johari, O., and Corvin, I., eds.), p. 19. IITRI, Chicago.

Isaacson, M., Lin, S. D., and Crewe, A. V. (1974c). Radiation damage and energy analysis in the STEM. *Proc. 8th Int. Cong. Electron Micros.* (Canberra), **2**, 680.

Isaacson, M., and Crewe, A. V. (1975). Electron microspectroscopy. *Ann. Rev. Biophys. and Bioeng.* **4**, 165.

Isaacson, M., and Johnson, D. (1975). Low Z elemental analysis using energy loss electrons. *J. Ultramicroscopy* **1**, 33.

Isaacson, M., and Lin, S. D. (1975). Secondary emission of thin films. (to be published).

Johnson, D. (1972). The interaction of 25 keV electrons with guanine and cytosine. *Rad. Res.* **49**, 63.

Johnson, D., and Isaacson, M. (1973). Cytosine reflectance measurements using electron energy loss spectra and synchrotron radiation. *Optics Comm.* **8**, 406.

Jones, G. L., and Cosslett, V. E. (1970). Sensitivity and resolution of photographic emulsions to electrons (60–700 keV). *Proc. 7th Int. Cong. Electron Micros.* (Grenoble), **2**, 349. Société Francaise de Microscopie Électronique, Paris.

Joy, R. T. (1973). The electron microscopical observations of aqueous biological specimens. In: *Advances in Optical and Electron Microscopy*, Vol. 5 (Barer, R., and Cosslett, V. E., eds.), p. 297. Academic Press, New York and London.

Jung, H., and Schüssler, H. (1966). Zur Strahleninaktivierung von Ribonuclease. I. Auftrennung der Bestrahlungsprodukte. *Z. Naturf.* **21B**, 224.

Klug, A., and DeRosier, D. J. (1966). Optical filtering of electron micrographs. Reconstruction of one-sided images. *Nature* **212**, 29.

Kobayashi, K., and Ohara, M. (1966). Voltage dependence of radiation damage of polymer specimens. *Proc. 6th Int. Cong. Electron Micros.* (Kyoto), **1**, 579.

Kobayashi, K., and Sakaoku, K. (1965). Irradiation changes in various accelerating voltages. In: *Quantitative Electron Microscopy* (Bahr, G. F., and Zeitler, E., eds.), p. 1097. Williams and Wilkins Co., Baltimore.

Langmore, J. P. (1975). Studies of unstained and selectively stained biological molecules using scanning transmission electron microscopy. Ph.D. dissertation, The University of Chicago.

Langmore, J. P., Cozzarelli, N. R., and Crewe, A. V. (1972). A base-specific single heavy atom stain for electron microscopy. *Proc. 30th Ann. EMSA Conf.* (Los Angeles), p. 184.

Langmore, J. P., and Crewe, A. V. (1974). Progress towards the sequencing of DNA by electron micrsocopy. *Proc. 32nd Ann. EMSA Meeting* (St. Louis), p. 376.

Langmore, J. P., Isaacson, M., and Crewe, A. V. (1974). The study of single heavy atom motion in the STEM. *Proc. 32nd Ann. EMSA Meeting* (St. Louis), 378.

Langmore, J., Wall, J., Isaacson, M., and Crewe, A. V. (1973a). Carbon support films for high resolution electron microscopy. *Proc. 31st Ann. EMSA Conf.* (New Orleans), p. 76.

Langmore, J., Wall, J., and Isaacson, M. (1973b). The collection of scattered electrons in dark field electron microscopy. I. Elastic scattering. *Optik* **38**, 385.

Leisengang, S. (1954). Zur Erwarmung Elektronenmikroskopischer Objekte bei Kleinem Strahlquerschnitte. *Proc. 3rd Int. Cong. Electron Micros.* (London), p. 176.

Lenz, F. (1954). Zur Streuung Mittelschneller Elektronen im Kleinste Winkel. *Z. Naturf.* **23A**, 185.

Lin, S. D. (1973). Electron radiation damage on thin films of phenylalanine. *Proc. 31st Ann. EMSA Conf.* (New Orleans), p. 484.

Lin, S. D. (1974). Electron radiation damage of thin films of glycine, di-glycine and aromatic amino acids. *Rad. Res.* **59,** 521.

Lippert, W. (1962). Mass-thickness changes of plastic films in the electron microscope. *Optik* **19,** 145.

Locquin, M. (1954). L'observation d'objets vivant au microscope electronique. *Proc. 3rd Int. Cong. Electron Micros.* (London), p. 448.

Luft, J. H. (1973). Embedding media–old and new. In: *Advanced Techniques in Biological Electron Microscopy* (Koehler, J. K., ed.), p. 1. Springer-Verlag, Berlin.

Markham, R., Frey, S., and Hills, G. J. (1963). Methods for enhancement of image detail and accentuation of structure in electron microscopy. *Virology* **20,** 88.

Marton, L. (1934). Electron microscopy of biological objects. *Bull. Acad. Roy. Soc. Belg.* **20,** 439.

Massey, H. S. W., and Burhop, E. H. S. (1969). *Electronic and Ionic Impact Phenomena*, Vol. 1 (2nd edition). Oxford at the Clarendon Press, Oxford.

McKinley, W. A., Jr., and Feshbach, H. (1948). The coulomb scattering of relativistic electrons by nuclei. *Phys. Rev.* **74,** 1759.

McPherson, A. (1976). The analysis of biological structure with X-ray diffraction techniques. In: *Principles and Techniques of Electron Microscopy: Biological Applications*, Vol. 6 (Hayat, M. A., ed.). Van Nostrand Reinhold Company, New York and London.

Mott, N. F., and Massey, H. S. W. (1965). *The Theory of Atomic Collisions* (3rd Edition). Oxford at the Clarendon Press, Oxford.

Muller, A. (1966). The formation of radicals in nucleic acids, nucleoproteins and their constituents by ionizing radiation. *Progr. Biophys. Mol. Biol.* **17,** 99.

Muller, M., and Koller, T. (1972). Preparation of aluminum oxide films for high resolution electron microscopy. *Optik* **35,** 287.

Nagata, F., and Ishikawa, I. (1971). High voltage electron microscopy of living materials. *Proc. 29th Ann. EMSA Meeting* (Boston), p. 464.

Nagata, F., and Ishikawa, I. (1972). Observation of wet biological materials in a high voltage electron microscope. *Jap. J. Appl. Phys.* **11,** 1239.

Nagata, F., and Fukai, K. (1974). Irradiation effects and penetration of non-metallic materials at higher voltages. In: *High Voltage Electron Microscopy* (Swann, P. R., Humphreys, C. J., and Goringe, J. J., eds.), p. 379. Academic Press, London.

Nathan, R. (1971). Image processing: Enhancement procedures. In: *Advances in Optical and Electron Microscopy*, Vol. 4 (Barer, R., and Cosslett, V., eds.), p. 85. Academic Press, London.

Nermut, M. V. (1972). Negative staining of viruses. *J. Microscopy* (London) **96,** 351.

Parikh, M. (1973). An analysis of dissociation of molecules in a high resolution electron microscope. *Proc. 31st Ann. EMSA Conf.* (New Orleans), p. 486.

Parikh, M. (1974). Molecular dissociation in the electron microscope. University of California, Electronics Research Laboratory Memorandum No. ERL-M431.

Parsons, D. F. (1970). Problems in high resolution electron microscopy of biological materials in their natural state. In: *Some Biological Techniques in Electron Microscopy* (Parsons, D. F., ed.), p. 1. Academic Press, New York.

Parsons, D. F., Matricardi, V. R., Subjeck, J., Uydess, I., and Wray, G. (1972). High voltage electron microscopy of wet whole cancer and normal cells: Visualization of cytoplasmic structures and surface projections. *Biochim. Biophys. Acta.* **290,** 110.

Parsons, D. F., Uydess, I. and Matricardi, V. R. (1974). High voltage electron microscopy of wet whole cells. *J. Microscopy* (London) **100,** 153.

Philipp, H. R., and Ehrenreich, H. (1967). Ultraviolet optical properties. In: *Optical Prop-*

erties of III-IV Compounds: Semiconductors and Semi-metals, Vol. 3 (Willardson, R. K., and Beer, A. C., eds.), p. 93. Academic Press, New York.

Pines, D. (1969). *Elementary Excitations in Solids*. W. A. Benjamin Press, New York.

Platzman, R. L. (1962). Superexcited states of molecules. *Rad. Res.* **17**, 419.

Pollard, E., Powell, W. F., and Reaume, S. H. (1952). The physical inactivation of invertase. *Proc. Nat. Acad. Sci.* (U.S.A.) **38**, 173.

Pullman, B., and Pullman, A. (1963). *Quantum Biochemistry*, p. 209. Interscience, New York.

Ramamurti, K., Crewe, A. V., and Isaacson, M. (1975). Low temperature mass loss of thin films of L-phenylalanine and L-tryptophan upon electron irradiation. A preliminary report. *J. Ultramicroscopy* 1, p. 156.

Reimer, L. (1959). Quantitative Untersuchungen zur Massenabnahme von Einbettungsmittelm (Methacrylat, Vestopal und Araldit) unter Elektronenbeschuss. *Z. Naturf.* **14B**, 566.

Reimer, L. (1961). Veranderungen organischer Farbstoffe im Elektronenmikroskop. *Z. Naturf.* **16B**, 166.

Reimer, L. (1965). Irradiation changes in organic and inorganic objects. In: *Quantitative Electron Microscopy* (Bahr, G. F., and Zeitler, E., eds.), p. 1082. Williams and Wilkins Co., Baltimore.

Reimer, L., and Christenhusz, R. (1965). Determination of specimen temperature. In: *Quantitative Electron Microsocpy* (Bahr, G. F., and Zeitler, E., eds.), p. 420. Williams and Wilkins Co., Baltimore.

Richards, K. E., Williams, R. C., and Calendar, R. (1973). Mode of DNA packing within bacteriophage heads. *J. Mol. Biol.* **78**, 255.

Riddle, G. H. N., and Siegel, B. M. (1971). Thin pyrolytic graphite films for electron microscope substrates. *Proc. 29th Ann. EMSA Meeting* (Boston), p. 266.

Ris, H. (1970). High voltage electron microscopy in biology. *Proc. 28th Ann. EMSA Meeting* (Houston), p. 12.

Rohrlich, F., and Carlson, B. C. (1954). Positron-electron differences in energy loss and multiple scattering. *Phys. Rev.* **93**, 38.

Rose, A. (1948). Television pickup tubes and the problem of vision. *Adv. Electronics* **I**, 131.

Rose, H. (1973). To what extent are inelastically scattered electrons useful in the STEM? *Proc. 31st Ann. EMSA Meeting* (New Orleans), p. 286.

Rose, H. (1974). Phase contrast in scanning transmission electron microscopy. *Optik* **39**, 416.

Rose, H., and Plies, E. (1974). Entwurf eines fehlesarmen magnetischen energie-Analysators. *Optik* **40**, 336.

Salih, S. M., and Cosslett, V. E. (1974). Reduction in electron irradiation damage to organic compounds by conducting coatings. Phil. Mag. **20**, 225.

Salih, S. M., and Cosslett, V. E. (1974). Some factors influencing radiation damage in organic substances. *Proc. 8th Int. Cong. Electron Micros.*, (Canberra), **2**, 671.

Scherzer, O. (1970). Die Strahlenschadigung der Objekte als Grenze fue die hochauflosende Elektronmikroskopie. *Wiss. Buensen-Gesell. Phys. Chemie* **74**, 1154.

Scherzer, O. (1972). Requirements for imaging carbon atoms in biomolecules. *Optik* **33**, 501.

Scott, W. T. (1965). The theory of small angle multiple scattering of fast charged particles. *Rev. Mod. Phys.* **35**, 231.

Setlow, R. B. (1952). The radiation sensitivity of catalase as a function of temperature. *Proc. Nat. Acad. Sci.* (U.S.A.) **38**, 166.

Shaffner, T. J., and Van Veld, R. D. (1971). Charging effects in the scanning electron microscope. *J. Phys.* **E4**, 633.

Siegel, G. (1970). The influence of low temperature on the radiation damage of organic compounds and biological objects by electron irradiation. *Proc. 7th Int. Cong. Electron Micros.* (Grenoble), **2**, 221. Société Francaise de Microscopie Électronique, Paris.

Siegel, G. (1972). The influence of very low temperature on the radiation damage of organic crystals irradiated by 100 keV electrons. *Z. Naturf.* **27A**, 325.

Siegel, B. M. (1971). Current and future prospects in electron microscopy for observations of biomolecular structure. *Phil. Trans. Roy. Soc. London* **B261**, 5.

Stenn, K., and Bahr, G. F. (1970a). Specimen damage caused by the beam of the transmission electron microscope, a correlative reconsideration. *J. Ultrastruct. Res.* **31**, 526.

Stenn, K., and Bahr, G. F. (1970b). A study of mass loss and product formation after irradiation of some amino acids, peptides, polypeptides and proteins with an electron beam of low current density. *J. Histochem. Cytochem.* **18**, 574.

Stoyanova, I. G., and Nekrasova, T. A. (1960). Study of live microorganisms in electron microscope by gas microchamber method. *Sov. Phy. Doklady* **5**, 1117.

Stoyanova, I. G., Nekrasova, T. A., and Zaides, A. L. (1960). Study of collagen in wet state in gas microchamber of electron microscope action of ionizing radiation on collagen. *Sov. Phy. Doklady* **5**, 209.

Tanooka, H., and Hutchinson, F. (1965). Modifications of the inactivation by converging radiations of the transforming activity of DNA in spores and dry cells. *Rad. Res.* **24**, 43.

Thach, R. E., and Thach, S. S. (1971). Damage to biological samples caused by the electron beam during electron microscopy. *Biophys. J.* **11**, 204.

Thomas, G. (1973). Some current and future trends of high voltage transmission electron microscopy. *Proc. 31st Ann. EMSA Meeting* (New Orleans), p. 2.

Thomas, L. E., Humphreys, C. J., Duff, W. R., and Grubb, D. J. (1970). Radiation damage of polymers in the million volt electron microscope. *Rad. Effects* **3**, 89.

Thomas, L. E., and Ast, D. G. (1974). Image intensification and the electron microscopy of radiation-sensitive polymers. *Polymer.* **15**, 37.

Thomson, M. G. R. T. (1975). Resolution and image signal-to-nosie ratio: A comparison between the conventional transmission electron microscope and the scanning transmission electron microscope. In: *Techniques in Electron Micrsocopy and Microprobe Analysis* (Siegel, B., and Beaman, D., eds.) p. 29. John Wiley & Sons, New York.

Thon, F., and Willasch, D. (1972). Imaging of heavy atoms in dark field electron microscopy using hollow cone illumination. *Optik* **36**, 55.

Unwin, P. N. T. (1972). Negative staining of biological materials using mixture salts. *Proc. 5th Eur. Cong. Electron Micros.* (Manchester), p. 232.

Unwin, P. N. T. (1974). Electron microscopy of stack disk aggregate of tobacco mosaic virus protein. II. The influence of electron irradiation on the stain distribution. *J. Mol. Biol.* 87, 657.

Valentine, R. C. (1966). The response of photographic emulsions to electrons. In: *Advances in Optical and Electron Microscopy*, Vol. 1 (Barer, R., and Cosslett, V. E., eds.), p. 180. Academic Press, New York and London.

Van Dorsten, A. C. (1954). Electron irradiation of specimens. *Proc. 3rd Int. Conf. Electron Micros.* (London), p. 172.

Venables, J. A., and Bassett, D. C. (1967). Electron microscopy of polyethylene below 20°K. *Nature* **214**, 1107.

Vollenweider, H. J., Koller, T., and Kubler, O. (1973). Aluminum-beryllium alloy films: specimen supports for high resolution electron microscopy. *J. Microscopie* (Paris) **16**, 247.

Vollmer, R. T., and Fluke, D. J. (1967). Temperature dependence of ionizing radiation effect on dry hyaluridase. *Rad. Res.* **31**, 867.

Wall, J. (1972a). Mass and mass loss measurements on DNA and fd phage. *Proc. 30th Ann. EMSA Conf.* (Los Angeles), p. 186.

Wall, J. (1972b). A high resolution scanning electron microscope for the study of single biological molecules. Ph.D. dissertation, The University of Chicago.

Wall, J., Isaacson, M., and Langmore, J. (1974a). The collection of scattered electrons in dark field electron microscopy. II. Inelastic scattering. *Optik* **39,** 359.

Wall, J., Langmore, J., Isaacson, M., and Crewe, A. V. (1974b). Scanning transmission electron microscopy at high resolution. *Proc. Nat. Acad. Sci.* (U.S.A.) **71,** 1.

Watanabe, H., and Uyeda, R. (1962). Energy selecting electron microscope. *J. Phys. Soc.* (Japan). **17,** 569.

Wentzel, G. (1927). Dispersion of corpuscular rays as diffraction appearances. *Z. Physik.* **40,** 590.

White, J. R., Beer, M., and Wiggins, J. W. (1971). Preparation of smooth graphite support films for high resolution electron microscopy. *Micron* **2,** 412.

Whiting, R. F., and Ottensmeyer, F. P. (1974). Studies of a model DNA sequence by dark field electron microscopy. *Proc. 32nd Ann. EMSA Meeting* (St. Louis), p. 384.

Williams, R. C., and Fisher, H. W. (1970a). Electron microscopy of tobacco mosaic virus under conditions of minimal beam exposure. *J. Mol. Biol.* **52,** 121.

Williams, R. C., and Fisher, H. W. (1970b). Electron microscopy with minimal beam damage. *Proc. 28th Ann. EMSA Meeting* (Houston), p. 304.

Wirths, A., and Jung, H. (1972). Single-stranded breaks induced in DNA by vacuum ultraviolet radiation. *Photochem. Photobiol.* **15,** 325.

Zeitler, E. (1965). Theory of elastic scattering of electrons. In: *Quantitative Electron Microscopy* (Bahr, G. F., and Zeitler, E., eds.), p. 36. Williams and Wilkins Co., Baltimore.

Zeitler, E. (1968). Resolution in electron microscopy. In: *Advances in Electronics and Electron Physics*, Vol. 25 (Marton, L., ed.), p. 277. Academic Press, New York and London.

Zeitler, E., and Bahr, G. F. (1959). Contributions to quantitative electron microscopy. *J. Appl. Phys.* **30,** 940.

Zeitler, E., and Bahr, G. F. (1965). Determination of dry mass in populations of isolated particles. In: *Quantitative Electron Microscopy* (Bahr, G. F., and Zeitler, E., eds.), p. 217. Williams and Wilkins Co., Baltimore.

Zeitler, E., and Hayes, J. R. (1965). Electrography. In: *Quantitative Electron Microscopy* (Bahr, G. F., and Zeitler, E., eds.), p. 586. Williams and Wilkins Co., Baltimore.

Zeitler, E., and Olsen, H. (1967). Complex scattering amplitudes in elastic electron scattering. *Phys. Rev.* **162,** 1439.

Zeitler, E. H., and Thomson, M. G. R. (1970). Scanning transmission electron microscopy. I. *Optik* **31,** 258.

Zeitler, E., and Isaacson, M. (1976). Scanning transmission electron microscopy. In: *Principles and Techniques of Electron Microscopy: Biological Applications*, Vol. 8 (Hayat, M. A., ed.). Van Nostrand Reinhold Company, New York (in press).

Zelander, T., and Ekholm, R. (1960). Determination of the thickness of electron microscopy sections. *J. Ultrastruct. Res.* **4,** 413.

2. FREEZE-DRYING FOR ELECTRON MICROSCOPY

M. V. Nermut

National Institute for Medical Research, Mill Hill, London

INTRODUCTION

Small biological structures (e.g., subcellular fractions, membranes, viruses and macromolecules) are often studied as a whole using the techniques of shadowing and positive or negative staining. Air-drying is included in all these techniques in routine work. This type of preparation is not free of artifacts, and it appears useful to deal with the artifacts at the beginning of this chapter. This chapter describes one of the most successful techniques for preserving the native state of biological structures.

Preparatory Artifacts

The biological structure can be harmfully influenced before mounting it on an electron microscope grid, i.e., during washing, fixation, dehydration, embedding, cutting, etc. The damage can also occur on the grid by positive or negative staining (specific staining artifacts) and particularly by air-drying. Finally, the damage can occur after drying on the grid, i.e., by shadowing or replication and by the electron beam during observation in the electron microscope.

Washing is usually carried out with distilled water, which may be harmful to osmotically sensitive structures (mitochondria, bacterial protoplasts, and some viruses) and has to be replaced by volatile buffers such as ammonium acetate, carbonate, or succinate. Fixation is supposed to stabilize the structure, but this is not always the case, and the results of fixation depend upon the chemical composition of the structure and the type of fixative used (Hayat, 1970).

Positive stains such as uranyl acetate bind to certain groups in the biological structure causing some degree of selective staining. It is necessary to know this when interpreting high resolution electron micrographs of macromolecules. Similarly, negative stains can interact with the biological structure either because of their chemical nature [e.g., potassium phosphotungstate (PTA) is harmful to membranes, whereas ammonium molybdate (AM) or uranyl acetate (UA) are not (Muscatello and Horne, 1968, Munn, 1968)] or because of the pH of the staining solution (Glauert and Lucy, 1969; Nermut *et al.*, 1970; Nermut, 1972a).

Most harmful, however, is the effect of air-drying which causes reorientation of particles because of the flow of liquid, aggregation, collapse, and disruption by the surface tension forces. For example, in the first studies of the bacteriophage-bacterial cell interaction, the phage particles were often found oriented with their heads towards the cell. Using the technique of critical point-drying, Anderson (1953a) was able to show that the phages attach themselves to the cells by their tails. This has been confirmed by other authors and with other techniques (e.g., agar filtration or ultrathin sectioning).

The false orientation of small particles on the grid is not the only effect of air-drying. Because of the surface tension forces the biological structure is subjected to enormous stresses which cause its collapse or even disruption (Fig. 2.1). Anderson (1956a) calculated that a stress exerted on a column

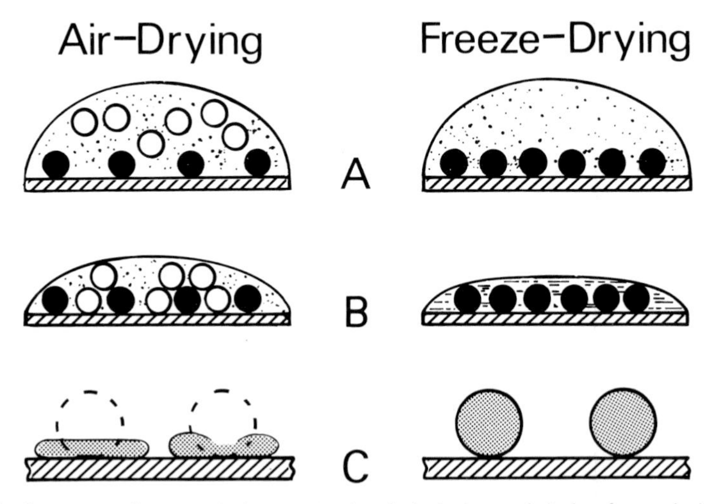

Fig. 2.1 Sequence of events during conventional air-drying and during freeze-drying on a grid. (A) Both adsorbed and unadsorbed particles are present in the capillary layer of liquid before air-drying. A monolayer of adsorbed particles is present on the grid as prepared for freeze-drying. (B) During air-drying the unadsorbed particles aggregate or overlap; this is not the case in freeze-drying. (C) Collapse or even disruption of particles is caused by surface tension forces during air-drying. The three-dimensional structure of particles is well preserved after freeze-drying.

200 Å in diameter reaches 2,000 pounds/square inch. The collapse of bacteria and particularly of viruses has been documented by many authors, and some examples will be presented later in this chapter.

Several authors in Vol. 2 of this series have dealt with shadowing artifacts and the reader is referred to them (Aberman *et al.*, 1972; Henderson and Griffith, 1972; Koehler, 1972).

In most cases one can avoid or substitute for the harmful agent (e.g., fixatives or negative stains), but in routine electron microscopy, drying cannot be avoided. The question is therefore how to dry without exposing the structures to the deleterious effects of surface tension forces. This can be achieved in three different ways:

(1) The surface tension is lowered by substituting organic solvents (ethanol, amyl acetate, ether, etc.) for water. This is a simple method and is widely used, for instance, in connection with the monomolecular film technique (Kleinschmidt *et al.*, 1962). This film technique involves spreading of filamentous molecules such as nucleic acids. It can also be employed with other macromolecules (Dubochet, 1973), but the effect of the organic solvent used on the structure must first be checked. This method cannot be used with lipid-containing structures such as enveloped viruses which are damaged by such a treatment. On the other hand, drying adenoviruses from absolute ethanol yields results comparable to those obtained by freeze-drying (Nermut, 1973a). Drying with nitrous oxide (N_2O), which has a very low surface tension, has also been used with success (Koller and Bernhard, 1964).

(2) Using critical point-drying (Anderson, 1951, 1953b, 1965a,b), water is replaced by organic solvents (acetone, ethanol, or amyl acetate), and these in turn are replaced by carbon dioxide or another compound having a suitable critical point. This technique has been described in detail in Vol. 3 of this series by Hayat and Zirkin (1973).

(3) Freeze-drying technique was introduced to biological electron microscopy by Wyckoff (1946), and further developed and improved by Williams (1952, 1953a). Originally, the specimen was sprayed onto a precooled specimen stage with grids, and the ice was removed by sublimation in vacuum with a cool trap. This procedure has been sporadically used by various authors, but until recently no device with an exact control of temperature was available. After the freeze-etch units were introduced on the market an opportunity arose to use them for freeze-drying of small structures. This procedure has the advantage that neither fixatives nor organic solvents are needed, and the required duration is reasonably short.

THEORETICAL ASPECTS OF FREEZE-DRYING

There are three main steps in the freeze-drying procedure—rapid freezing, sublimation of ice, and visualization of the structure with the electron microscope (shadowing, replication, positive staining, and negative staining). However, only the first two steps are important from the theoretical point of view. The problems of rapid freezing have been discussed thoroughly by Rebhun (1972) and by several authors dealing with the freeze-etching technique (Koehler, 1968; Bullivant, 1973; Moor, 1973a). The main point here is the suppression of ice crystal formation. It has been clearly shown (Moor, 1973a) that high freezing rates can be obtained and practically applied only in the case of thin layers (small objects), so that with larger cells or pieces of tissue there is a need for antifreeze agents. Using freeze-drying technique on a grid this is not necessary, and high freezing rates are easily achieved. Thin films (~10–20 μ high) containing small particles (viruses, cell fractions, macromolecules) can be vitrified because the freezing rate of such thin specimens in LN_2 is fairly high (~10^5 K/sec) (Moor, 1973a). This is the case in the freeze-dry technique as it is used in our laboratory (adsorption procedure). Therefore, rapid freezing does not impose any problem in the technique of freeze-drying on the grid.

Drying is a more complex physical phenomenon, and its theoretical background has been discussed by several authors (Meryman, 1960; Rowe, 1960; Stephenson, 1960). In principle, drying is based on diffusion of molecules of water along a concentration gradient. This is created by removing the escaped water molecules from the surrounding space by continuous pumping and by introducing a place of lower temperature than that of the specimen for vapor condensation.

According to the kinetic theory of gases the temperature of a substance is a function of the average speed of the molecules composing it. In the reversed form this means that the thermal motion of molecules is slower at a lower temperature. Therefore, freeze-drying (carried out at -80°C) is much slower than air-drying. However, to preserve the native structure of the biological material we have to carry out the drying at a temperature below the recrystallization point, i.e., ~-60°C or even less. As the partial pressure of water at -100°C is ~10^{-5} torr (Dunlop *et al.*, 1972), the need for a high vacuum (10^{-6} torr) is obvious.

In practice, sublimation of water molecules is achieved by introducing a cool trap in the vacuum chamber so that the movement of water molecules is directed from the specimen (-80°C) to the cool trap (~150°C to ~-190°C). According to Moor (1970), the "etching depth" is ~90 nm/min when the specimen stage is at -100°C, and every 10°C difference increases (or decreases) it by a factor of 10. As the sublimation from deeper layers becomes slower, it may be safer to count with 500 Å only. This would give a depth of ~5 μ/min at a specimen stage temperature of -80°C. Therefore, a drying time of 20 to 30 min is sufficient when the capillary layer of water left on the grid is ~10–20 μ high. This is usually the case with small biological structures such as viruses, cell fractions, and macromolecules.

However, the water molecules adsorbed directly to the structure do not sublime during "etching" (Moor, 1973b). They can form a layer 10–100 Å high on the biological surfaces and can increase substantially the size of small particles. High vacuum during drying helps to keep down the amount of adsorbed water molecules. Nevertheless, our practical experience with freeze-drying of adenoviruses and with ferritin showed that this fact does not cause serious difficulties.

PRACTICAL ASPECTS OF FREEZE-DRYING

Two more aspects are important for the practice of freeze-drying: (1) how the sample is applied to the grid; (2) how the structure is prepared for observation in the electron microscope. The procedures as used by several authors up to now can be divided into two groups: those using spraying to apply the sample onto the precooled grid (spray freeze-drying) and those applying the specimen onto the grid by adsorption (adhesion) at room temperature and freezing it afterwards.

Spray Freeze-drying

This is the classical procedure introduced by Williams (1952, 1953a), who sprayed a highly concentrated suspension of bacteriophages onto a precooled copper block covered with grids. This technique has been described in different reviews (Horne, 1965; Mühlethaler, 1973) and need not be repeated here. A modernized version of this procedure was described (e.g., by Steere, 1973), and its main steps are given below.

The specimen stage of a freeze-etch unit (e.g., Denton or Balzers) is precooled to -180°C in a vacuum. The suspension of particles such as viruses is then sprayed either onto grids placed on the specimen stage after ventilating the vacuum chamber or on a disc of mica (fixed on the specimen stage from the beginning) using a stream of dry nitrogen to prevent condensation of water. This is a little dangerous when working with human pathogenic viruses. For such cases the adsorption technique described below is recommended.

The vacuum chamber is then pumped down and drying is achieved by increasing the specimen stage temperature to -80°C and by cooling down to -150°C or less a suitable cool trap in the vacuum chamber. The vacuum obtained under these conditions is usually $\sim 10^{-6}$ torr and drying lasts for ~15 min. After this period the specimen is shadowed with Pt-C, and if mounted on mica it can be replicated with carbon. Small structures (simple viruses, ribosomes) can be observed directly as pseudoreplicas (Steere, 1973), but replicas of larger viruses such as pox virus are better cleaned with a bleach or sulfuric acid.

Certain attention should be paid to spraying, which has to be moderate but sufficiently strong to bring the particles into contact with the support before they freeze. The droplets should be as fine as possible. This reduces the duration necessary for drying and prevents overlap of structures. A microdroplet of tobacco mosaic virus is shown in Fig. 2.2.

Fig. 2.2 A microdrop of tobacco mosaic virus sprayed onto a precooled specimen stage (−150°C) of a Balzers freeze-etch unit and freeze-dried. The concentration was fairly high, ~40 mg protein/ml. Shadowed with Pt-C. ×12,500.

Adsorption Freeze-drying

This procedure has been used independently by Nanninga (1968) and Nermut and Frank (1971) and further elaborated by Nermut *et al.* (1972) and Nermut (1972a). The specimen is applied to the grids or mica by adsorption, washed thoroughly, and then frozen and dried in a high vacuum. This modification allows not only shadowing and replication but also positive or negative staining, and it proved very useful particularly in studies of viruses and macromolecules (Gelderblom *et al.*, 1972; Nermut, 1973b; Lange *et al.*, 1973; Witter *et al.*, 1973). It will therefore be described in more detail.

Specimen Preparation. A purified sample of particles should be suspended in distilled water or in a volatile buffer such as 0.1-0.3 M ammonium acetate,* carbonate, or succinate (Backus and Williams, 1950; Williams, 1953a). The concentration of particles such as viruses and ribosomes should be $\sim 10^8$ per ml (or 10 mg protein per ml). This increases the amount of particles in the viewing field (convenient for statistical evaluation or counting) and improves the retention of negative stain (when used) on the supporting film. Although

*Ammonium acetate can only be used if the specimen stage temperature during drying is −80°C or less as the melting point of its eutectic concentration is below −70°C (Williams, 1953a).

carbon-coated Formvar or parlodion films are routinely used, thin carbon films are useful for high resolution work. These are transferred from drop to drop by means of a wire loop (Fernández-Morán *et al.*, 1966), or washing is done underneath the grid using the "one-drop-technique" (Nermut, 1973a). Carbon-coated grids are preferably treated by glow discharge (for ~5 min) to render them more hydrophilic.

Two conditions are important to obtain satisfactory freeze-dried specimens: formation of a monolayer of particles on the grid and removal of organic substances and salts.

A monolayer of particles is achieved either by means of repeated washing (diffusion technique) or by spreading. In the first case the grid is placed on a drop of the suspension under study for the necessary adsorption time, which varies from 1 to 10 min and depends upon the concentration of particles in the suspension. The possibility of prolonging the adsorption time means that even low-concentration suspensions can be used. The capillary layer of liquid which remains on the grid after removal from the suspension contains already adsorbed particles as well as those still floating in the liquid. These settle down during drying and cause overlap which prevents observation in the electron microscope (Fig. 2.1). Such particles are removed by diffusion into drops of water during successive washings, each lasting for 1-2 min (Fig. 2.3).

Spreading of particles on the surface of distilled water or a volatile buffer (Kleinschmidt technique) is a better technique, but it requires more concentrated suspensions (e.g., 10–20 mg protein per ml in the case of influenza virus). Although cytochrome c is usually necessary for satisfactory spreading, it can be omitted when another protein (trypsin or phospholipase C) is present in suspension or when membrane vesicles are spread. We always achieved better preparations of sarcoplasmic reticulum vesicles by spreading than by conventional adsorption, and we use this technique even for routine (air-dry) negative staining. Although the concentration of particles has to be high, the total amount of suspension used is ~0.05 ml. For example, 0.05 ml virus suspension is mixed with 0.01 ml or even less cytochrome c (0.1% solution in 0.1 M ammonium acetate, pH 7.0) and spread on distilled water in a Petri dish of 6 cm in diameter

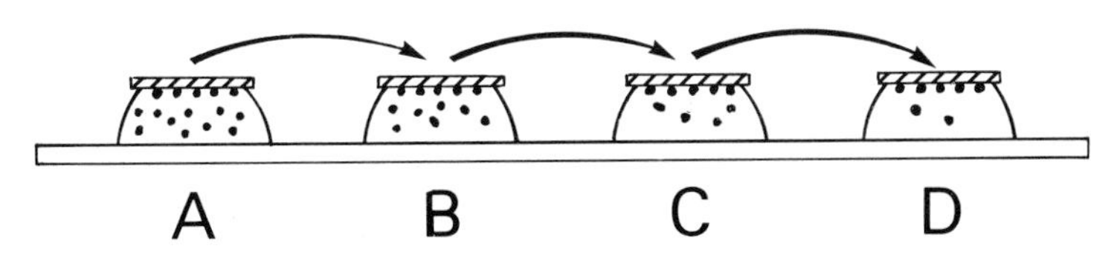

Fig. 2.3 Formation of a monolayer of particles by repeated washing. An electron microscope grid with adsorbed particles (A) is transferred three times on a drop of water or volatile buffer (B, C, D). The unadsorbed particles (transferred with the capillary layer of water) diffuse into the drop of water and are substantially diluted. Negative staining can be carried out after the last wash (=D).

(Nermut, 1972a). After a few minutes the film is touched with a grid, which is washed twice with distilled water and used for freeze-drying. When pure carbon films are used, they all are first deposited onto the monomolecular film and then collected by means of a wire loop.

Extensive washing is a prerequisite for revealing fine structural details which could be obscured if the frozen capillary layer contains salts or organic substances. These substances do not evaporate and are deposited on the surface of the biological structures, increasing their size and obscuring fine details of their surface. Moreover, the organic substances form a film on the specimen or at least the so-called strands (Fig. 2.4) which prevent shadowing and observation in the electron microscope. Usually successive washing on three drops of distilled water or volatile buffer (for 1-2 min on each drop) is necessary in case of the diffusion technique, whereas one or two washings are sufficient when

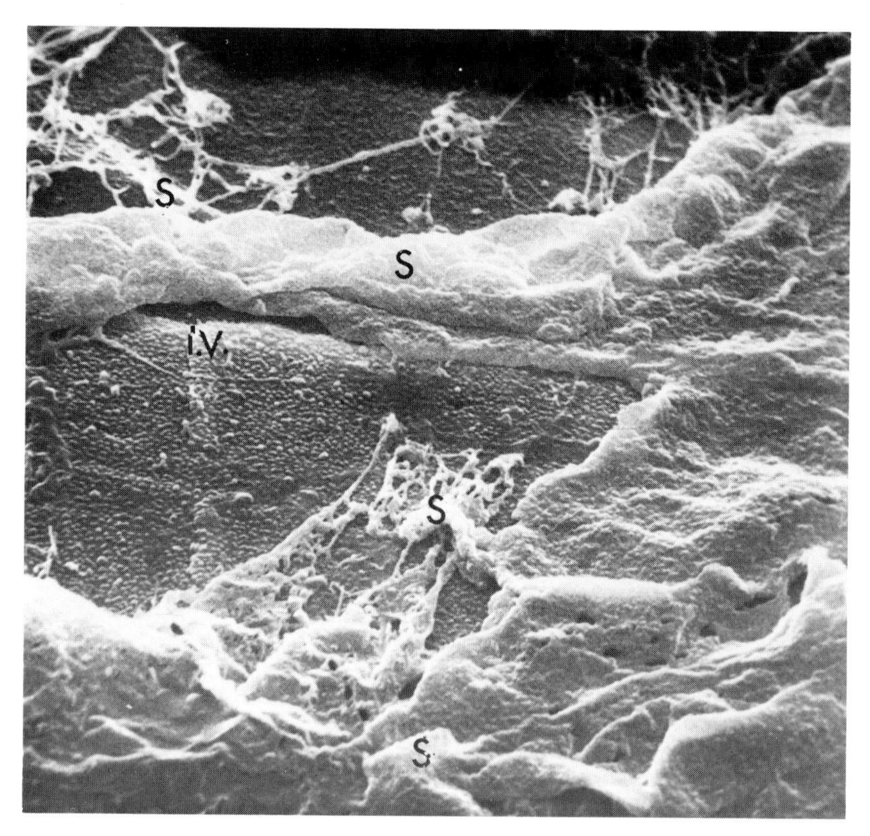

Fig. 2.4 A scanning electron micrograph of a grid with freeze-dried influenza virus (IV). Repeated washing was used before freeze-drying. Note the strands (S) covering a considerable area of the viewing field. ×5,000. (*Courtesy of Mr. K. Fecher, Cambridge Instrument Comp., Dortmund, Germany.*)

spreading is used. In the later case the "strands" are virtually absent (Nermut, 1973c).

Procedures

The advantage of the adsorption technique is that it can be combined not only with unidirectional shadowing, but also with replication, rotary shadowing, and positive or negative staining.

Freeze-drying with Shadowing. The basic procedure (Nermut *et al.*, 1972) is carried out as follows: a grid with adsorbed and washed particles (prepared either by the diffusion or the spreading technique) is held in a forceps over a Dewar bottle filled with liquid nitrogen (LN_2). The excess fluid is then drained off with a piece of filter paper (Fig. 2.5), so that a thin film is left on the grid. With hydrophilic grids there is no danger of drying for ~2 sec. The grid is dipped quickly into the LN_2 as deep as possible and kept there for at least 10 sec. Thereafter, it is quickly transferred onto a precooled to -150°C and with Freon 22 defrosted specimen stage (we use a 4-specimen stage of the Balzers freeze-etch unit). Only metal cover plate is used to fix the grids on the stage (the screwcap is not used) during operation.

The chamber is pumped down, the specimen stage temperature is adjusted to -80°C and the knife arm serving as a cool trap is cooled below -150°C. After that it is moved into position over the specimen stage and left for ~30 min. With larger viruses or lipid-containing structures it is useful to bring the specimen stage temperature up to zero for a short duration and then back to -80°C or even lower for shadowing. This is usually done with Pt-C or Ta/W (in an elec-

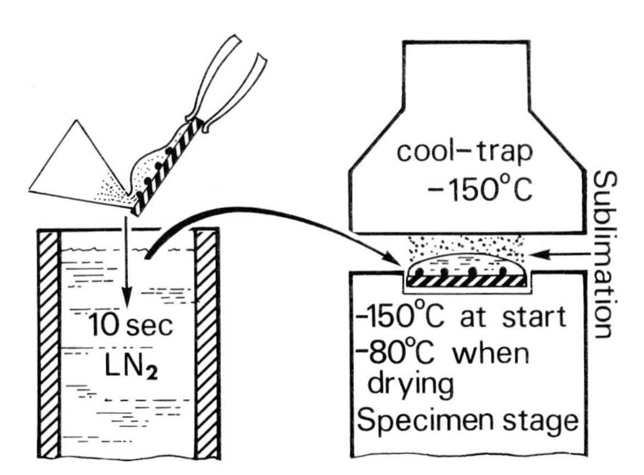

Fig. 2.5 The freeze-drying procedure. After the last wash (or negative staining) the excess liquid is drained off with a filter paper and the grid is dipped quickly into LN_2. Then it is rapidly transferred onto the precooled specimen stage of a freeze-etch unit and dried in a vacuum.

tron gun) at an angle of 45°, 35°, or 30°, depending upon the size of the structure studied and purpose. Before the vacuum chamber is ventilated, the specimen stage is warmed up to +30°C to prevent water condensation on the grids.

Freeze-drying with Rotary Shadowing. This procedure is carried out in substantially the same manner as described above, except that the Pt-C shadowing is replaced by rotary shadowing. This must be done after the whole procedure has been accomplished and the grids have been transferred onto a rotary table either in the same unit or in another evaporator. This move is usually not harmful if it is made quickly.

Freeze-drying Followed by Replication. This procedure should be employed when fine structure on the surface of larger viruses and bacteria, plant, or animal cells is studied.

(1) Replication of Freeze-dried Viruses or Small Cell Fragments. The procedure is fundamentally the same as above, but the sample is applied either to a piece of freshly cleaved mica (Nanninga, 1968; Nermut, 1973b,c) or another supporting material, e.g., a coverslip or discs of hard paper coated with Formvar (Brown *et al.*, 1972). Shadowing with Pt-C is followed by evaporation of a carbon layer of different thickness. Small particles usually do not need a very thick carbon replica, and evaporation for ~5 sec or less increases resolution (compare with 8-10 sec recommended for evaporation of carbon after freeze-etching). It is not necessary to warm the specimen stage to 30°C before ventilating the vacuum chamber, and floating (on bleach or 30% sulfuric acid) is usually easier from cold mica. One or two hours cleaning is usually sufficiently long to remove viruses or cell fragments from the replicas. They are subsequently washed with distilled water and mounted on plain 400-mesh grids.

(2) Replicas of Freeze-dried Cells. Surface of bacteria, animal or plant cells, and tissues can also be studied with freeze-drying when followed by replication. However, it is more difficult to obtain satisfactory replicas than in the previous instance (1). One of the reasons for this difficulty is the much higher profile of freeze-dried cells which are not collapsed as after conventional air-drying. To lower the profile one has to form a closely packed sheet of cells. This can be done, for instance, by smearing a piece of freshly cleaved mica with a pellet of cells (bacteria, red blood cells, etc), or by using a monolayer of tissue culture cells on a coverslip or by cutting a small piece of a tissue such as from leaves (Steere and Haga, 1970). In the case of suspensions (bacteria or cells) it is necessary to replace beforehand the culture medium by a volatile buffer of a suitable molarity and pH or by distilled water when possible to prevent the loss of fine structural details by deposited organic substances or salts. Cell monolayers or pieces of tissues can be rinsed briefly with 0.3 M ammonium acetate (pH 7.0). In all cases, care is taken to drain off excess liquid but not

to dry the cells. The piece of mica or the coverslip with cells is then quickly frozen in LN_2 and transferred onto a precooled specimen stage of the freeze-etch unit as described above.

Drying is carried out longer than with viruses (usually 1-2 hr), but it is neither necessary nor advisable to dry the cells entirely; it is sufficient to sublime the ice layer covering the cells (Fig. 2.6). After the shadowing and carbon evaporation, the replicas are immediately floated on bleach or sulfuric acid as described in the previous section (1).

Freeze-dry Negative Staining. Negative staining can be done either before or after freeze-drying. The latter procedure is useless from the point of view of preservation of native structure and prevention of drying artifacts. It is justified in those cases where staining of internal structures is desirable. The dry particles take up negative stain very quickly, so that the internal organization is revealed. The result may be a complex two-side image, but in a few cases reasonable information has been obtained.

In the adsorption technique (Nermut and Frank, 1971; Nermut *et al.*, 1972) negative staining is carried out after washing, i.e., immediately before freezing.

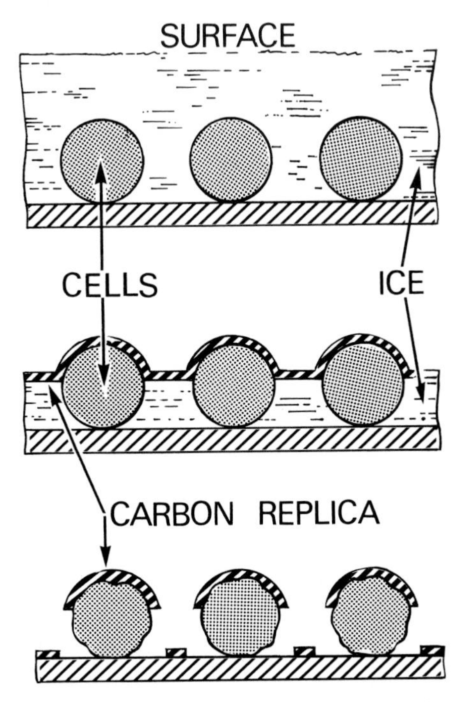

Fig. 2.6 Freeze-drying of cells. Drying should be carried out as long as necessary (for 1-2 hr) to uncover the upper half of cells. The replica is more continuous than when all ice is sublimed off, and a possibility of shrinkage of cells is fairly reduced.

The excess of negative stain is removed with a filter paper and the grid is quickly dipped into LN_2 (Fig. 2.5). Further procedure is the same as for shadowing, but this is omitted, and the specimen stage is warmed to +30°C before opening of the vacuum chamber. The cool trap (knife arm in the Balzers unit) is also warmed to 0°C. The grids appear covered with a white powder—the dried negative stain. This has to be removed immediately by gentle blowing with dry air, nitrogen, or a fine brush. This is a relatively difficult operation and grids may be entirely obscured (dark) if too much powder is left on them or may be emptied if the stream of air is too strong, which will break the film. However, with a certain skill a yield of ~50% satisfactory grids can be achieved.

The grids have to be observed with an electron microscope as soon as possible because the white powder is hygroscopic. They should otherwise be kept in a desiccator *in vacuo*. In fact, a certain collapse of freeze-dried influenza viruses was noticed by scanning electron microscopy (unpublished results), when the grids were left in the air for several hours.

Freeze-drying of Positively Stained Preparations. This procedure can also be followed, but its importance is limited to filamentous molecules like nucleic acids or to special purposes in virus research.

Evaluation of Freeze-drying Techniques

The two techniques of freeze-drying described here—the spray technique and the adsorption technique—differ in one principal point, i.e., the method in which the suspension of particles is applied to the grid. In the spray technique, particles (in a volatile buffer or distilled water) are sprayed onto a precooled specimen stage (at -150°C). It is difficult to estimate the cooling rate, but it is probably satisfactory (judging from the results obtained with viruses) if the droplets are very small. A practical disadvantage is that a highly concentrated virus suspension (10^8-10^{10} particles/ml) in a volatile buffer is needed. This is not easy to achieve if a small amount of purified virus suspension in a phosphate or Tris-buffer is supplied. One has to use a relatively long procedure, i.e., a microdialysis or high speed centrifugation before starting freeze-drying. The adsorption technique (in both variations) is simple from this point of view, as washing can be done on the grid.

It has already been mentioned that superposition of particles is not suitable for electron microscopic observation, and this occurs quite often with the spray technique (Fig. 2.7). A monolayer of particles (as produced by the adsorption technique) increases the yield in well-oriented particles for any statistical work (Fig. 2.8). On the other hand, a random orientation of particles in the spray preparations helps one to see virus particles from different angles and to understand more easily their shape without tilting the specimen in the electron microscope.

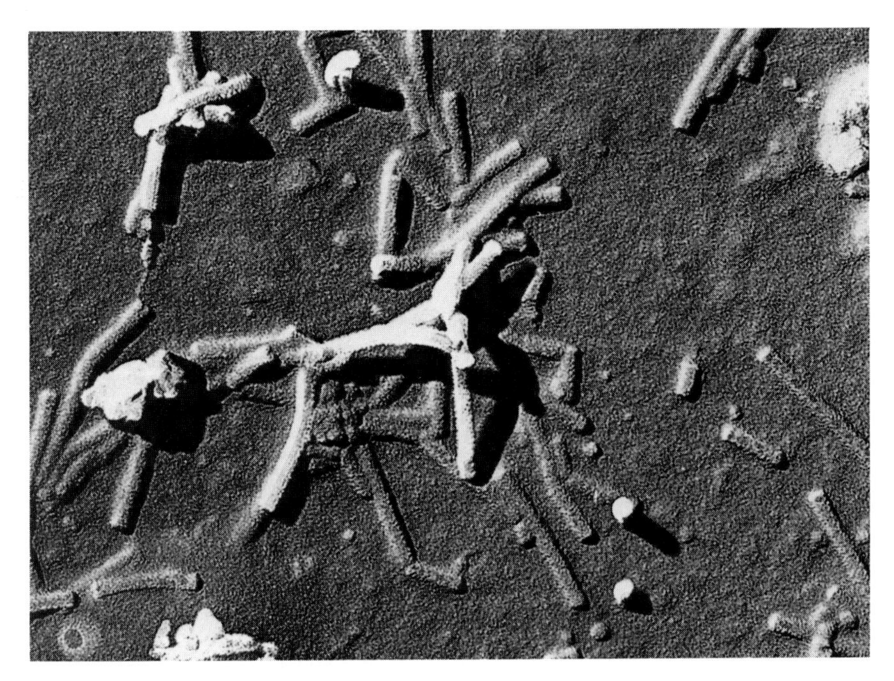

Fig. 2.7 Tobacco mosaic virus prepared by spray-freeze-drying and Pt-C shadowing. Some particles lie flat on the film, others are attached with one end only. ×100,000.

The spray technique is generally considered to be quantitative in comparison with the adsorption, which can be fairly selective (Dubochet and Kellenberger, 1972). However, this advantage seems to be lost when the droplets are rapidly frozen so that the particles touch the support film only gently or not at all. Heavier particles such as bacteria will fall down on the grid by gravitation, but lighter ones such as macromolecules or even tails of bacteriophages can be blown away by the stream of air during pumping or by a "high rate of sublimation" (Williams and Fraser, 1953). Broader contact of particles with the supporting film (as occurs with the adsorption technique) prevents such a loss.

Spraying of a mixture of particles with a negative stain was widely used at the beginning of the era of negative staining, and it is still used for quantitative work. However, it has not been described in combination with freeze-drying. In our hands it did not yield satisfactory results, as the droplets were too electron-dense for the proper observation of particles in the electron microscope. Spraying is of course quite dangerous in the case of pathogenic viruses or bacteria, particularly when a stream of nitrogen is used to prevent water condensation on the specimen stage (Steere, 1973). Spraying is also difficult to use in studies of cells or tissues.

The adsorption technique offers the following advantages: One can use differ-

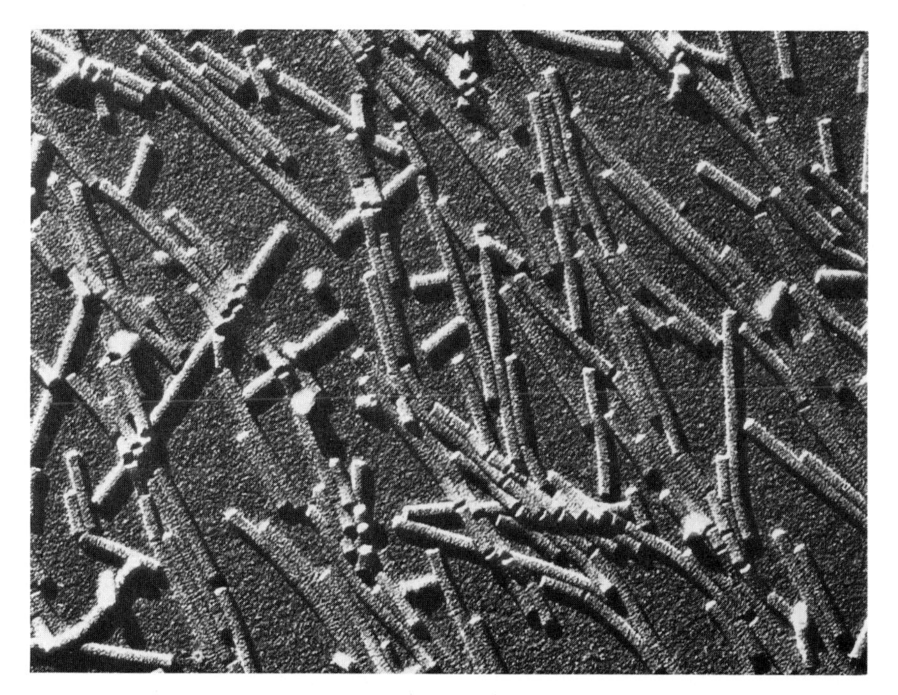

Fig. 2.8 Tobacco mosaic virus prepared by adsorption-freeze-drying and shadowed with Pt-C. The majority of virus particles are attached to the film by their full length. ×100,000.

ent concentrations of particles and in different media (buffers, culture media, gradients), as washing is done on the grid. With low concentrations, adsorption from a drop is employed, and the duration of adsorption can be fairly long. With high concentrations, spreading on the surface of water yields the best and very probably also quantitative results. It can be easily combined with replication, rotary shadowing, and negative or positive staining. It can be used to study macromolecules, viruses, bacteria, and cell surfaces. The critical moments in the procedure are quick freezing after draining off the drop of water from the grid and a quick transfer from LN_2 onto the specimen stage. However, skill can be achieved after some practice.

BIOLOGICAL APPLICATIONS OF FREEZE-DRYING

The technique of freeze-drying on the grid has proved useful in various fields, and several examples of its use will be demonstrated. This should not be regarded as an exhaustive survey of literature, but more as a guideline for the proper application of the technique and its different procedures. All the examples quoted below are from preparations made with adsorption freeze-drying.

Freeze-drying of Macromolecules

The surface tension forces acting on the biological structure during drying amount to tons per square inch and are relatively strong in case of small specimens. For example, a flagellum 200 Å in diameter stretched across a gap of 2 μ would be subjected to a stress of ~325 tons/square inch (Anderson, 1956a). Their distorting effect of course depends upon the rigidity of the biological structure. Soft lipid-containing structures collapse readily, whereas more rigid complexes (icosahedral viruses) withstand these effects better (see below).

Macromolecules usually occur as single units or as complexes of two, four, or six subunits or even as crystals. Some of them are isodiametric, others are rodlike or filamentous. They may therefore be influenced to a different degree by air-drying. Only a few examples of single macromolecules, crystals, and filamentous molecules prepared by air-drying and freeze-drying will be described here.

Adenovirus capsid (surface shell) is built up of capsomers which can be easily isolated and purified. The majority of them are the hexons (surrounded by six neighbors in an icosahedral lattice) which consist of three subunits of molecular

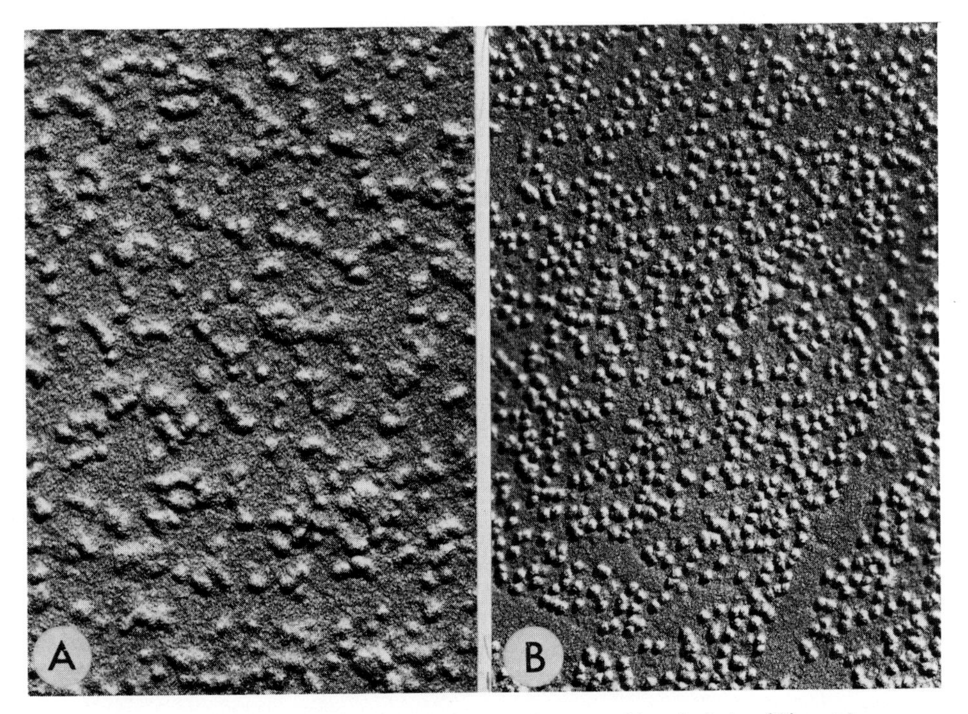

Fig. 2.9 Capsomers (hexons) of adenovirus type 5 prepared by air-drying (A) and freeze-drying (B). Shadowed with Pt-C. ×100,000.

weight of ~120,000 (Horwitz *et al.*, 1970). They were considered to be spherical in shape by some authors (Valentine and Pereira, 1965). Figure 2.9A,B shows hexons prepared by air-drying or freeze-drying and shadowing with Pt-C. It is clear that freeze-dried hexons are higher than the air-dried ones; their shadows are angular (pointed or blunt). Careful analysis of the shape of the shadows showed that hexons are presumably triangular prisms, certainly not spherical bodies (Nermut, 1975).

Larger macromolecular complexes such as ribosomes or pyruvate dehydrogenase are particularly suitable for freeze-drying (Nanninga, 1968) or freeze-etching (Bachmann and Schmitt-Fumian, 1973).

Crystalline complexes are usually more stable than single macromolecules. However, this depends very much upon the nature and strength of intermolecular

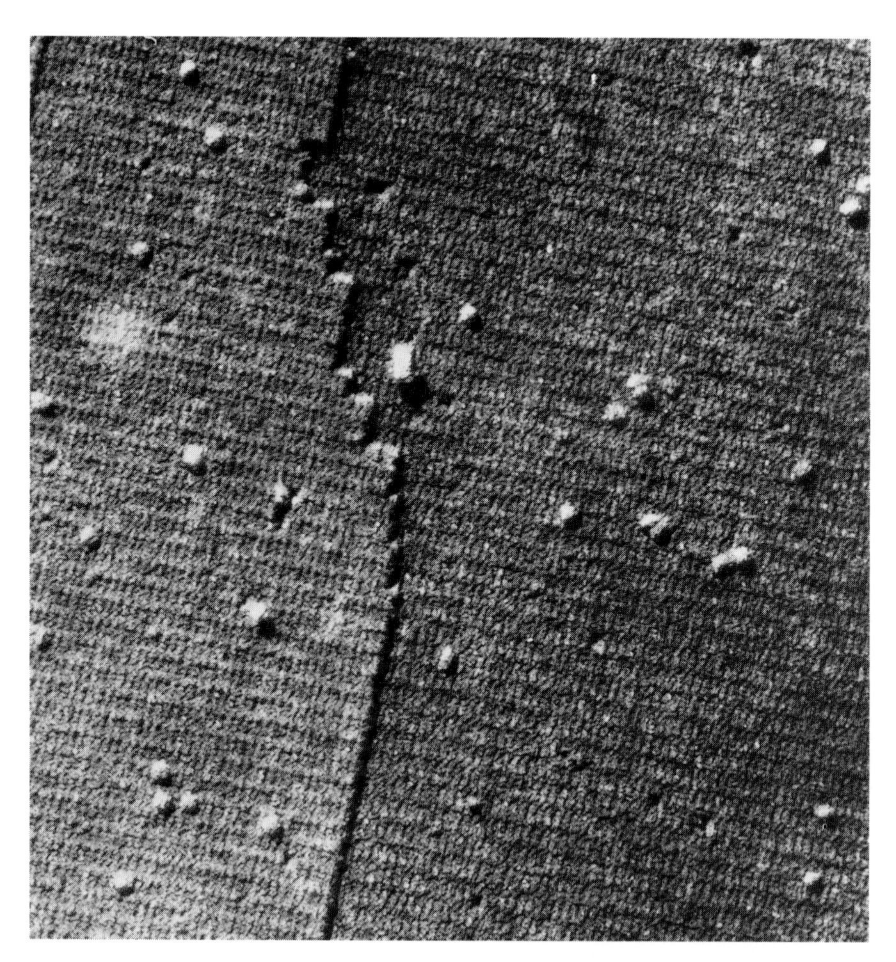

Fig. 2.10 Crystalline beef liver catalase freeze-dried and Pt-C shadowed. ×200,000.

bonds. An example of a strong, rigid complex is crystalline catalase. It is characterized by two different spacings *a* and *c* (Wrigley, 1968), which can be easily measured by optical diffraction. We have done this with both glutaraldehyde-fixed and unfixed catalase crystals prepared by conventional negative staining (air-drying) as well as by freeze-drying. The differences were statistically not significant. However, when the unfixed crystals were air-dried first and then stained negatively with silicotungstate or ammonium molybdate, they were entirely disintegrated. This was not the case with uranyl acetate (Nermut and Ward, unpublished results). High resolution shadowing with Ta/W or Pt-C (using the new Balzers electron gun) was also used, and could confirm that Ta/W has finer graininess (~10–15 Å) but less contrast than Pt-C (Fig. 2.10).

To compare the effects of drying on filamentous molecules, polyoma virus DNA was prepared by spreading (Kleinschmidt technique) and air-drying from distilled water, by air-drying from absolute alcohol, and by freeze-drying. Air-drying from water resulted in aggregation of DNA filaments into bundles or thick strands (although cytochrome c was used), but there was no substantial difference between ethanol-drying and freeze-drying. This finding is understandable as the surface tension of ethanol is three times lower than that of water, i.e., 22.75 dyne/cm^2 at 20°C as compared with surface tension of water of 72.75 dyne/cm^2.

Freeze-drying in Virus Research

This is probably the main field of application of this technique, because viruses by size and chemical composition belong to structures most sensitive to the surface tension. There are two main groups of viruses: those covered by a lipid membrane (enveloped viruses) and those built up of proteins and nucleic acid only (icosahedral and helical viruses).

Enveloped viruses have frequently been described as "pleomorphic." However, it has been shown recently in several laboratories that their pleomorphism is an artifact of purification and particularly of preparation techniques for electron microscopy. Figures 2.11A and B show influenza virus prepared by conventional negative staining and by negative staining followed by freeze-drying. It is obvious that the pleomorphism of this virus is caused by air-drying. The diameter of air-dried particles (where measurable) is ~30% larger than that obtained after freeze-drying (Williams, 1953b; Nermut and Frank, 1971).

Another typical example of drying artifacts is provided by oncogenic RNA-viruses such as murine leukemia virus and mammary tumor virus. All of them display the so-called tails, i.e., long protrusions or other pleomorphic appearances, when prepared by air-dried negative staining with PTA, STA, or AM. This is not the case with uranyl acetate (Nermut, 1972a,b), which is known to stabilize lipids in the virus membrane. Neither is this the case with freeze-drying (Nermut *et al.*, 1972, Nermut, 1973b). The artifactual nature of the "tails" was recognized

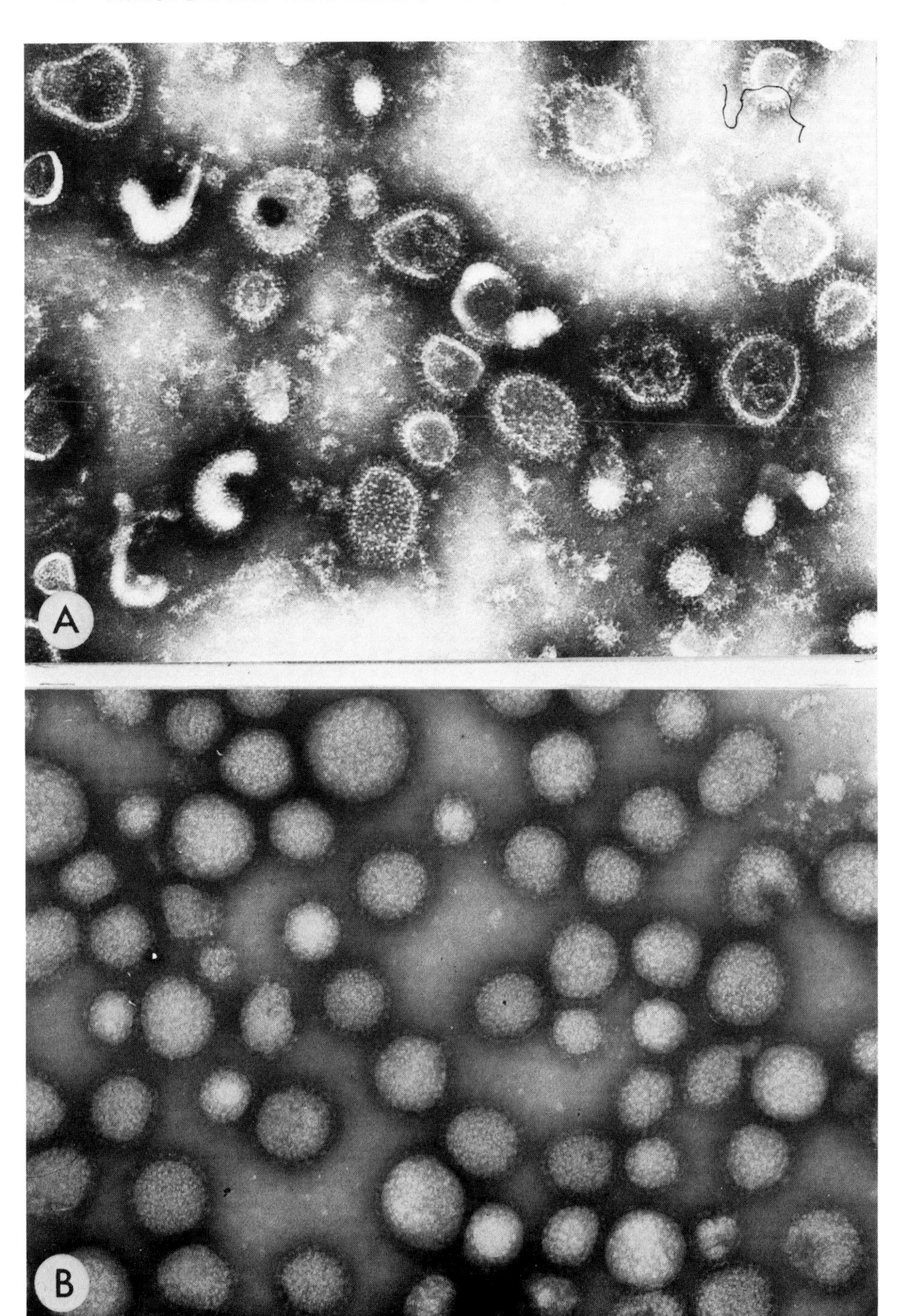
A
B

at least by some authors using fixatives before negative staining (de Harven and Friend, 1964; Luftig and Kilham, 1971), but it has until recently been believed that the group of leukemia viruses (the so-called C-type particles) lack surface projections which are very clear and distinct on mammary tumor viruses (B-type particles) (Nowinski *et al.*, 1970; de Harven *et al.*, 1973).

Using negative staining with uranyl acetate and the techniques of freeze-drying and freeze-etching, several authors (Luftig and Kilham, 1971; Nermut *et al.*, 1972; Witter *et al.*, 1973) were able to show that "knobs" are present on C-type particles (Fig. 2.12 and 2.13). Moreover, by using freeze-drying, it has been shown that the "cores" of murine leukemia viruses as well as avian myeloblastosis virus are cubical and covered with small subunits in a regular hexagonal pattern (Nermut *et al.*, 1972; Gelderblom *et al.*, 1972; Lange *et al.*, 1973) (Fig. 2.14). This has been previously seen in only a few cases (Zeigel and Rauscher, 1964; Dourmashkin and Simons, 1961).

Vaccinia virus belongs to the large complex enveloped viruses, and its surface is covered with small beads arranged in the so-called "paired ridges." These beads were seen after negative staining and more clearly after freeze-etching (Medzon and Bauer, 1970). However, for freeze-etching one needs a highly concentrated virus suspension, and the yield in particles displaying their surface (even after deep etching) is not very high. The technique of freeze-drying with replication provides preparations with an abundance of virus particles with a well-resolved surface pattern (Fig. 2.15).

Icosahedral viruses are less sensitive to air-drying, especially in the presence of a negative stain (Johnson and Horne, 1970). However, drying from water only causes a considerable disruption of particles (Fig. 2.16A). This picture is in great contrast to that seen with freeze-dried viruses in Fig. 2.16B. In the latter case and in Fig. 2.17 (an avian adenovirus) the virions show convincingly their icosahedral shape and display the capsomers on their surface. The presence of fibers (sticking out of the corners) was dependent upon the preparation of virus suspension (purification, storage, etc.). We were able to find them frequently after negative staining and freeze-drying.

Rotary shadowing is usually used to reveal filaments such as nucleic acids or regular patterns for optical diffraction. "Cores" of adenoviruses prepared by treatment with deoxycholate (Russell *et al.*, 1971) display loops of DNA running out of them. These loops were not seen after freeze-drying and unidirectional shadowing with Pt-C (Fig. 2.18A), but were distinct after freeze-drying and rotary shadowing with uranyl oxide (Fig. 2.18B). Therefore, it is concluded that the loops are not drying artifacts, but were probably loosened from the cores by deoxycholate.

Fig. 2.11 Influenza virus A_2 prepared by air-dried negative staining with 4% silicotungstate pH 6.5 (A) and by freeze-dried negative staining with ammonium molybdate pH 6.5 (B). Note the uniformity in shape after freeze-drying. ×100,000. (*Figure 11B from Nermut, 1972a, by permission of the* Journal of Microscopy.)

Fig. 2.12 Murine leukemia virus (Friend) freeze-dried and Pt-C shadowed. Note that some particles are partly devoid of "knobs" (surface glycoprotein macromolecules). X132,000. *(Electron micrograph by courtesy of Dr. H. Frank, Tübingen.)*

One of the first applications of freeze-drying was to study the T_1-T_7 bacteriophages (Williams and Fraser, 1953). It became obvious that the form of the phage head could be better studied on uncollapsed particles. However, the technique has not been used for that purpose as would be desirable. Very satisfactory results were achieved with deep freeze-etching (Bayer and Remsen, 1970; Bayer and Bocharov, 1973), but freeze-drying is superior to freeze-etching

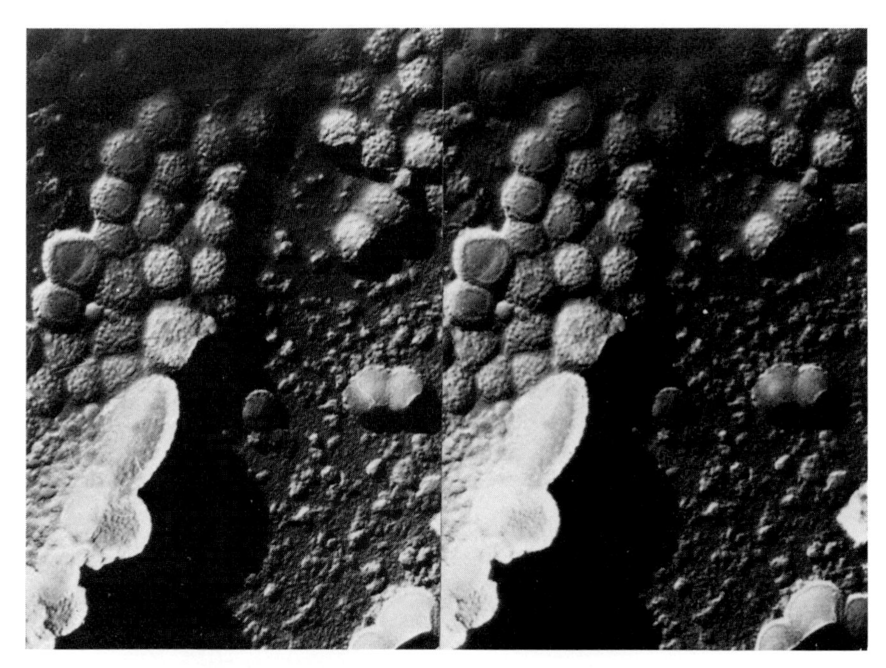

Fig. 2.13 Stereo-pair of a freeze-dried Friend virus. Pt-C shadowing. ×80,000. (*Courtesy of Dr. H. Frank, Tübingen.*)

Fig. 2.14 A core of Friend virus (*arrow*) showing an angular shadow and small hexagonally arranged subunits. Freeze-dried and Pt-C shadowed. ×220,000. (*Courtesy of Dr. H. Frank.*)

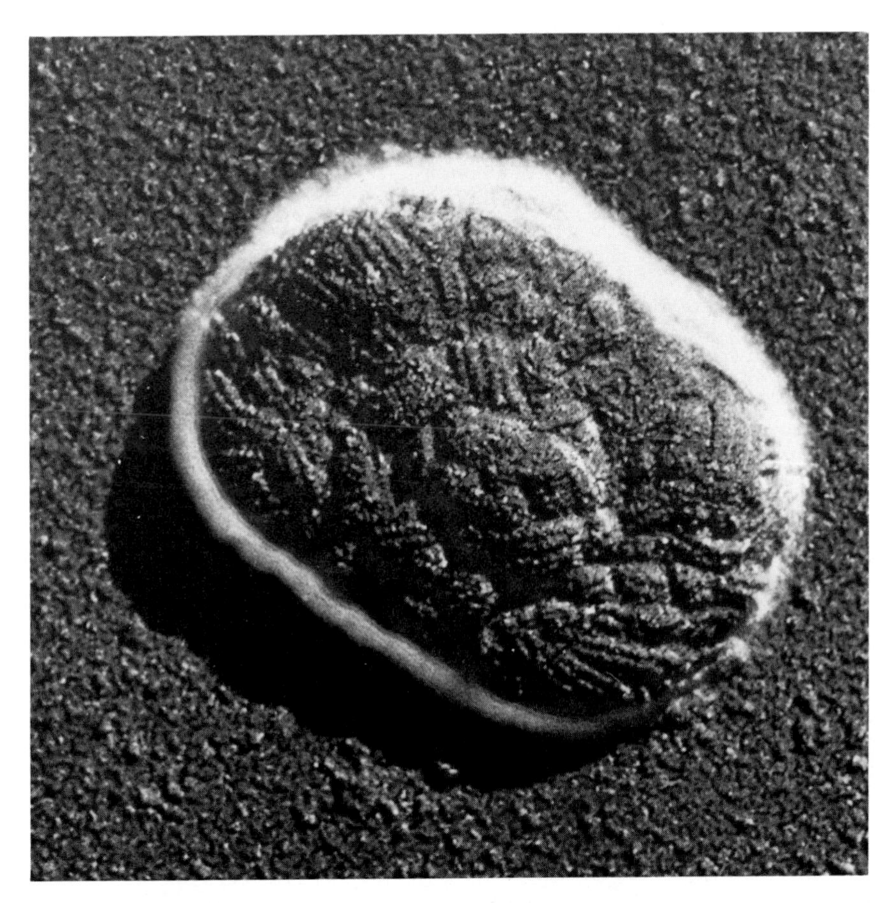

Fig. 2.15 Carbon replica of a freeze-dried vaccinia virus displaying the so-called paired ridges with little spherical subunits (~30–40 Å in diameter). Shadowed with Pt-C. X240,000.

Fig. 2.16 Adenovirus type 5 air-dried (A) and freeze-dried (B). Shadowed with Pt-C. The degree of disintegration by air-drying depends on the length of storage of purified virus. Here the virus was stored for a few weeks so that many free capsomers were seen even after freeze-drying. X100,000.

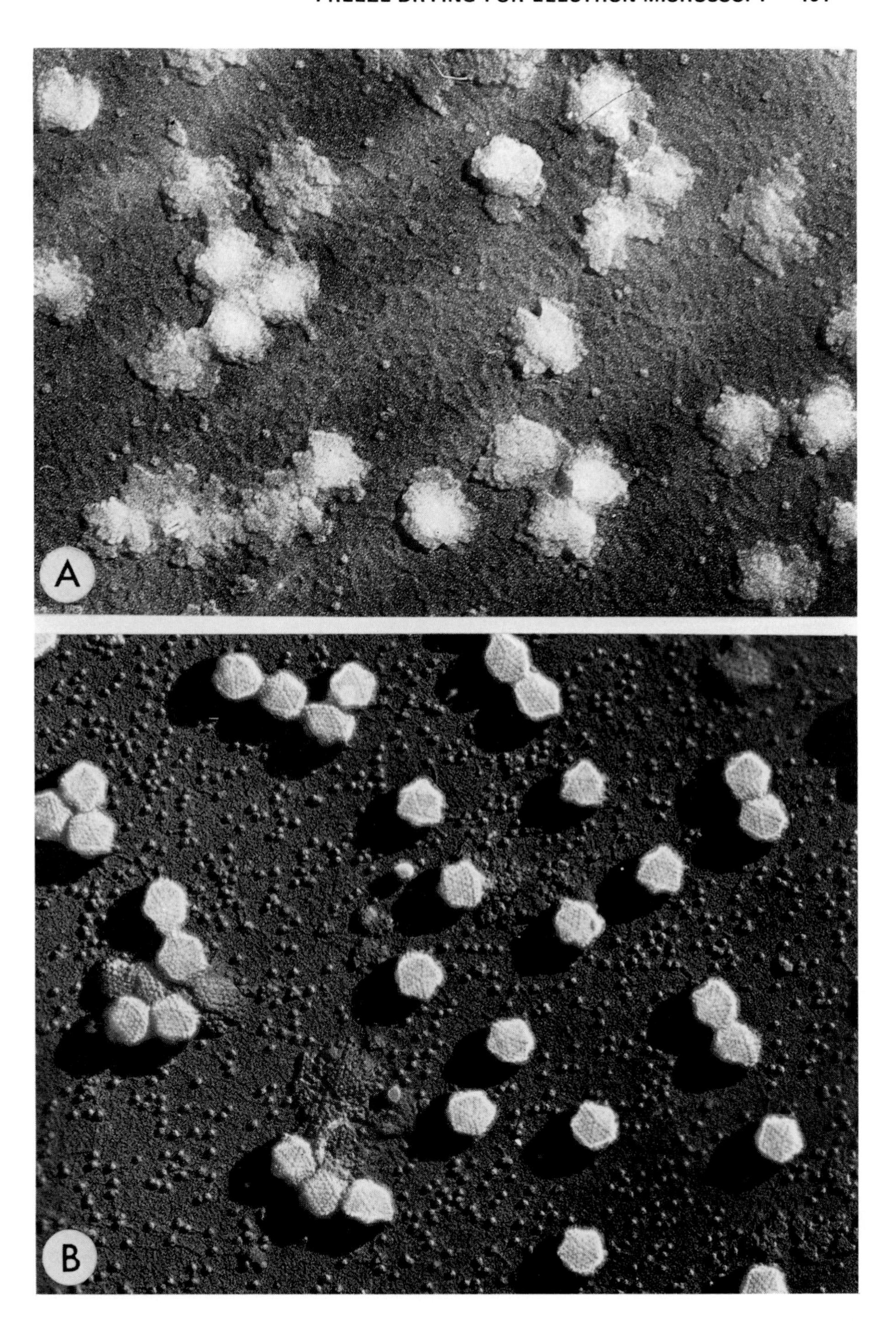
A
B

Fig. 2.17 An avian adenovirus (CELO-virus) freeze-dried and shadowed with Pt-C. The icosahedral form of the virus is clearly demonstrated, and individual capsomers can be easily distinguished. ×200,000.

because of absence of fracturing as well as a higher yield in particles available for observation than after deep-etching. Recently, H. Frank (personal communication) was able to show a conspicuous difference in shape of an air-dried and freeze-dried T_5-bacteriophage (Fig. 2.19A,B) suggesting that its head might be an octahedron.

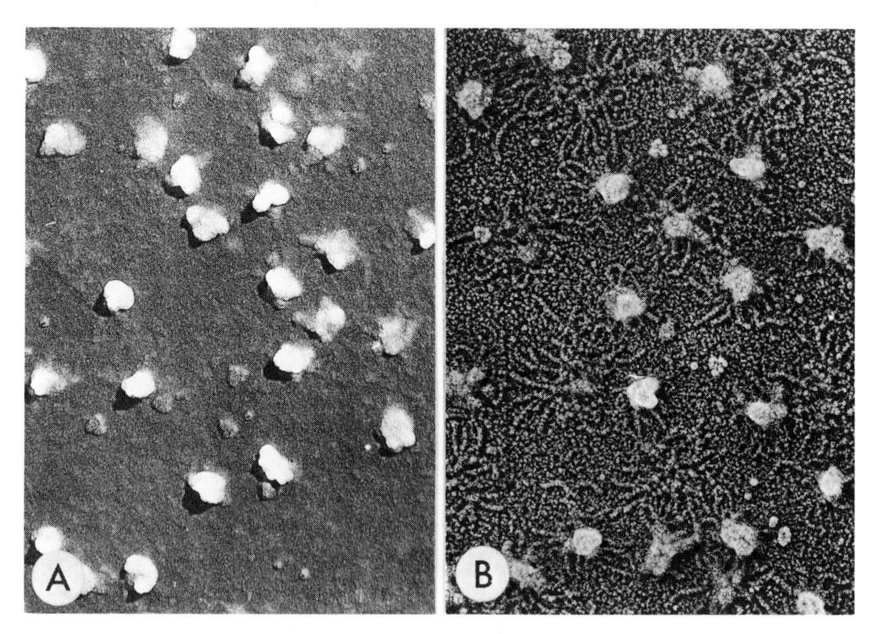

Fig. 2.18 "Cores" of adenovirus type 5 obtained by treating the virus with sodium deoxycholate. Freeze-dried and shadowed with Pt-C (A). Rotary shadowing of the same material revealed filaments of DNA around almost every core (B). ×75,000.

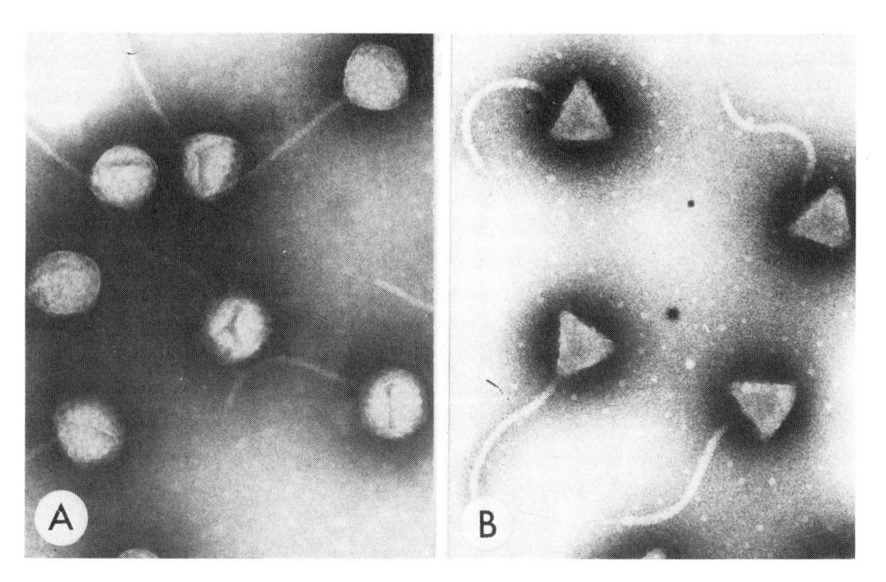

Fig. 2.19 *E. coli* bacteriophage (T_5) stained negatively with PTA and air-dried (A) or freeze-dried (B). Note the fundamental differences in shape of the head. The freeze-dried image suggests that the head is an octahedron. ×132,000. (*Courtesy of Dr. H. Frank.*)

Cell Membranes and Lipid Vesicles

Another interesting field for applying freeze-drying is cell membranes, lipid vesicles, etc. Figure 2.20 shows a red blood cell ghost prepared by freeze-drying. It is evident that the specimen did not collapse, and it is sufficiently rigid to maintain the three-dimensional structure, even under the hydrostatic pressure. A certain folding or wrinkling could often be seen, and this problem will be discussed below. On the other hand, vesicles of rabbit sarcoplasmic reticulum are usually collapsed even in the liquid medium, because most of them are considerably disrupted. Only a few cast a long shadow (Fig. 2.21). However, reconstituted vesicles of sarcoplasmic reticulum are fairly high after freeze-drying because they are closed and often "multi-lamellar" (Fig. 2.22A). However, the most interesting features are the surface projections (presumably the external part of the ATPase), which are clearly seen even on single sheets of ATPase. They are not seen after air-drying (Fig. 2.22B,C).

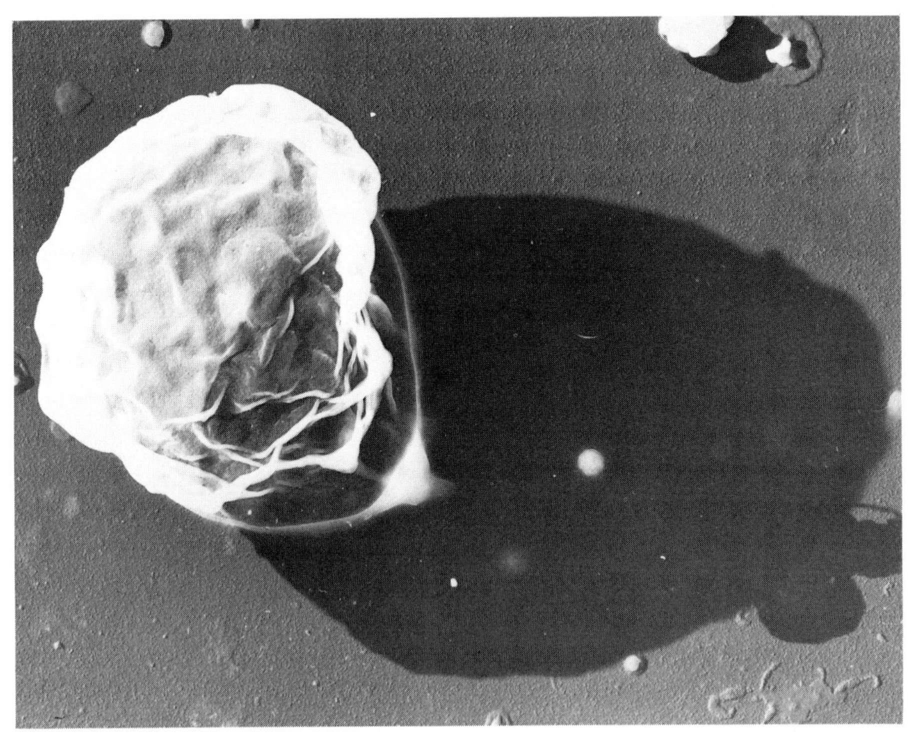

Fig. 2.20 A red blood cell ghost obtained by osmotic lysis and spread with cytochrome c on 0.001 M $MgSO_4$. Freeze-dried and shadowed with Pt-C. Note the light folding of the surface. ×25,000.

Fig. 2.21 Preparation of rabbit sarcoplasmic reticulum, freeze-dried and shadowed. Most vesicles are flat and covered with "projections." ×100,000.

Surface Structures of Freeze-dried Cells

Bacteria and eucaryotic cells are fairly electron-dense after freeze-drying and can be observed in the electron microscope only as replicas. For demonstration, cells of *B. polymyxa* have been chosen because of their conspicuous rectangular pattern covering the cell wall (Nermut and Murray, 1967). After 1 hr of freeze-drying the replicas revealed a regular pattern on the surface of many bacterial cells (Fig. 2.23A,B).

Cultivation of animal cells in monolayers provides a suitable opportunity to study cell surfaces with the freeze-drying technique. A freeze-dried monolayer of cells can be observed either in a scanning electron microscope or after replication in a transmission electron microscope. As an example of the latter technique, a replica of a cell from chick embryo fibroblast monolayer is shown in Fig. 2.24.

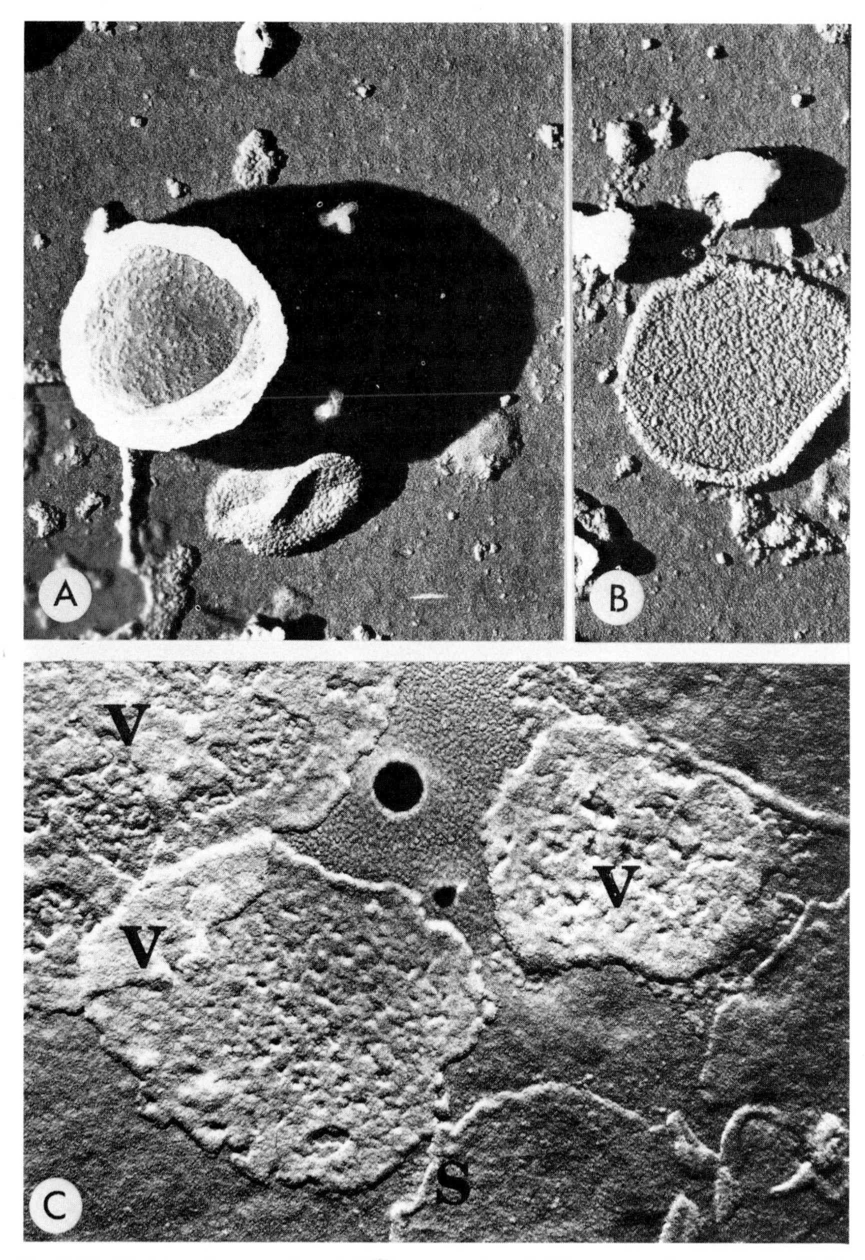

Fig. 2.22 Vesicles of reconstituted ATPase spread on 0.15 M ammonium acetate, pH 8.0, briefly washed on water, then treated for 10 min with 1% uranyl acetate, washed, and freeze-dried. Shadowed with Pt-C (electron-gun). Vesicles (A) are not collapsed and bear surface projections. Single sheets (B) lie flat and are densely covered with ATPase projections. Air-dried vesicles (C) are collapsed and no distinct projections are seen. V = vesicles; S = single sheet. ×70,000 (A and B); ×50,000 (C).

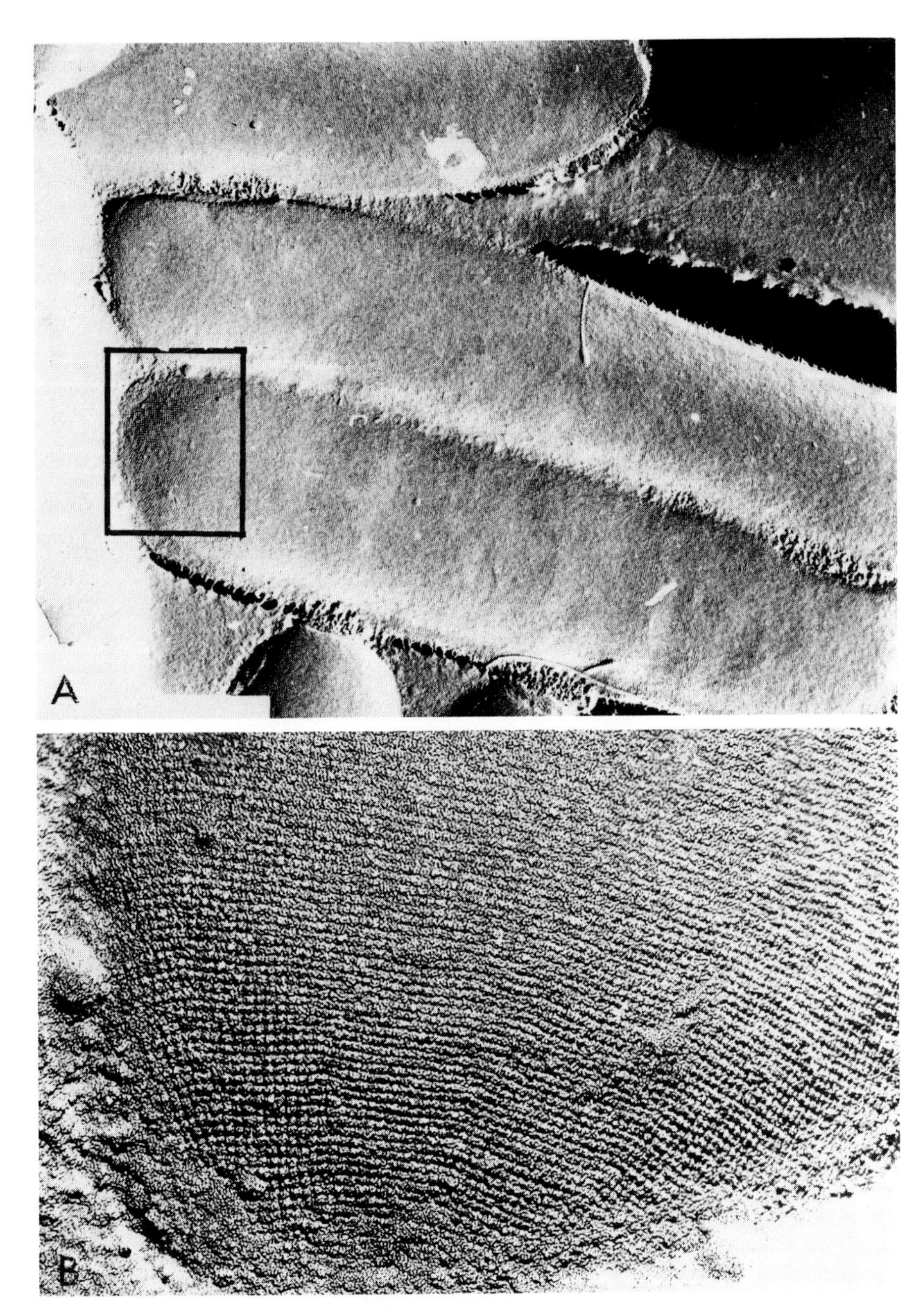

Fig. 2.23 Cells of *B. polymyxa* freeze-dried and replicated. The replicas are well preserved when the bacteria are closely packed (A). A higher magnification of the labeled area shows that the cell surface was really revealed after $1\frac{1}{2}$ hr of drying (B). ×30,000 (A); ×150,000 (B).

Fig. 2.24 A cell from a monolayer of chick embryo fibroblasts freeze-dried and carbon-replicated. Note the long processes originating at the upper part of the cell and also fibrillar structures right of the cell. The cell surface is mainly smooth with a fine granularity on the central area. ×16,500.

Artifacts of Freeze-drying

No technique is free of artifacts, and it would be illusory to believe that we shall ever be able to observe cells in a living state without influencing them by our preparation or observation technique. However, we hope to be able to reduce both the preparation and the observation artifacts to a reasonable minimum in the future. The contribution of freeze-drying and critical-point-drying is that they remove the deleterious effects of the surface tension forces on the biological structure–the so-called drying artifacts. When appraising these techniques, we have to ask whether or not they introduce their own specific changes in the living matter. There is no doubt that the dry state obtained as a result of freeze-drying is not a native state of the biological structure. The dry state is already something unnatural; however, it cannot be avoided at the present state of electron microscopy.

When considering the possibilities of artifacts, we have to start with freezing. There is no doubt that large ice crystals can cause morphological as well as functional changes in the structure. However, if the freezing rate is sufficiently high (10^5 K/sec), a monolayer of cells or particles on a grid or cells sprayed directly into a suitable coolant (Bachmann and Schmitt-Fumian, 1973) can be vitrified without any observable damage to them.

More problems seem to accompany the process of drying and the existence in a dry state. During sublimation of water molecules (from ice) conditions are created which could bring about an escape of small macromolecules from the specimen by a strong stream of water molecules. Anderson (1954, 1956a) calculated that small particles up to ~100 Å in diameter could escape the influence of gravitational forces and leave the specimen, provided they were embedded in ice only without any previous contact with the support film or another structure. Anderson (1956a) considered drying at -50°C (at the specimen), which would mean a difference of ~100°C or more between the cool trap and the specimen. The conditions used in the Balzers freeze-etch unit are -80°C at the specimen and -190°C or less at the cool trap; i.e., they are very similar.

Williams and Fraser (1953) reported that the tails of the bacteriophage T_5 were lost when drying was carried out at -40°C, but they were present when drying was accomplished at -65°C. It seems therefore plausible to believe that some of the knobs absent from the Friend virus preparations (Fig. 2.12) were lost by the above-mentioned mechanism, supposing that they were bound to the virus membrane by weak bonds only. It is generally known that "knobs" of murine leukemia viruses are very easily lost, for example, during a purification procedure. Similar phenomenon was recently reported by Bayer and Bocharov (1973), who observed loss of capsomers in bacteriophage lambda during deep-etching. This was prevented by fixation with formaldehyde. It should be added that forces responsible for absorption (e.g., van der Waals forces) or other type of bonds are usually sufficiently strong to prevent escape of small macromolecules from the specimen.

Structures larger than viruses are distinctly influenced by gravity which is manifested not only by their presence on the grid in a dry state but probably also by the observed folding or even shrinkage of their surface. In one of the early studies, Nei (1962) observed shrinkage and wrinkling of *E. coli* cells after freeze-drying at -60°C (which may be below the recrystallization point). As mentioned above, some kind of folding or distortion of the ideally spherical shape has also been observed with the red cell ghosts or large ATPase vesicles (Figs. 2.20, 2.22). Similarly Pinto da Silva (1973) described depressions on the surface of erythrocytes after 10 min deep-etching. This shows that the cell membrane might collapse to a certain extent when it is no longer supported by a liquid medium from both sides.

Some of the folding could be attributed to the effects of shadowing and/or of the electron beam. However, a detailed study of the top part of such structures and of cell replicas revealed that the folds exist before shadowing and only minor changes occur during observation in an electron microscope. Thus the freeze-dry studies can provide important information concerning the rigidity of various biological structures (cell walls, plasma membranes, etc.) in a dry state.

FREEZE-DRYING VERSUS CRITICAL-POINT DRYING

It appears to be useful to compare the two main procedures for preventing drying artifacts [i.e., the freeze-drying and the critical-point-drying (CPD) techniques]. Both procedures try to prevent the movement of a phase boundary across the specimen during the transition of water from a liquid (or solid) state into a gaseous state, and both first fix the specimen to avoid structural changes. In the case of freeze-drying this occurs by rapid freezing leading to vitrification of water, and undoubtedly this is superior to chemical fixation used in the CPD techniques. Of course, the rate of freezing must be sufficiently high to prevent ice crystal formation even without added cryoprotectives (10^5 K/sec). This is definitely the case with the absorption technique described above and very probably also with the spray technique, although this situation might be more difficult to assess.

Drying (sublimation of ice) occurs in a high vacuum under conditions previously described where the partial pressure of water molecules in the chamber is lower than the saturation pressure on the specimen, and the water molecules condense on a place of a lower temperature than that of the specimen, the so-called cool-trap. The situation is relatively simple with macromolecules, cell fragments, or open membrane vesicles, where the ambient ice can sublime easily. However, it is more complicated with large enveloped viruses, bacteria, and cells of higher organisms. The complete drying would mean transport of water molecules through the surface membranes or even more complex envelopes, and this process would be fairly long. With the exception of viruses this process is not necessary, and replicas of the revealed cell surface usually provide sufficient information (Figs. 2.23, 2.24).

The water molecules are thought to leave the cells through the hydrophilic protein channels associated with the intramembranous particles (Pinto da Silva, 1973). Nothing is known at present regarding the transport of water through a virus membrane which has no internal proteins (e.g., influenza virus). Perhaps the glycoprotein spikes may mediate the water transport. However, transport of nonelectrolytes through cell membranes as well as liposomes has been described by several authors (Stein, 1967; de Gier *et al.*, 1968; Kruyff *et al.*, 1973). Some folding or wrinkling has been observed after longer drying of cells and particularly with cell ghosts or various membranous vesicles such as sarcoplasmic reticulum.

It seems that a dry biological structure is more sensitive to the effects of gravity, changes in hydrostatic pressure (transfer from vacuum to the atmospheric pressure and back again), effects of shadowing, and electron beam irradiation. Our recent experience showed that vesicles of the reconstituted ATPase wrinkle to a certain extent even when fixed (stabilized) with UA or glutaraldehyde before freeze-drying. There is no doubt that it is much easier to maintain a spherical shape in a liquid medium than in a dry state. Shrinkage of tissue was also observed after CPD and was even greater than after freeze-drying in parallel experiments (Boyde, 1972).

In any case, the main domain of freeze-drying is smaller biological specimens such as macromolecules, viruses, and subcellular structures. The main advantage of the adsorption freeze-drying technique favored in our laboratory is its relative simplicity and versatility (possible combination with different techniques of visualization for electron microscopy). Freeze-drying of negatively stained specimens is a unique feature of our modification. The amount of specimen needed for freeze-drying is substantially smaller than for freeze-etching or spray freeze-drying.

The CPD technique has been successfully applied to the preparation of unicellular organisms or small pieces of tissues for scanning electron microscopy, but some experience with smaller specimens such as bacteria (Cohen *et al.*, 1968; Cohen, 1974; Bystricky *et al.*, 1972), viruses (de Harven *et al.*, 1973), and ribosomes (Spiess, 1973) has also been reported. Physical fixation (freezing) is replaced by a chemical one, and water is removed at room temperature without exposing the specimen to a high vacuum during the drying process. The purpose of fixation may be twofold: (1) to stabilize the biological structure particularly towards the organic solvents as used in the original version for dehydration; (2) to increase the permeability of cells for organic solvents and subsequently for carbon dioxide, Freon, or nitrous oxide.

However, none of the common fixatives prevents extraction of both proteins and lipids when tissue is exposed to organic solvents at room temperature. For example, glutaraldehyde fixes proteins and some lipids, while osmium tetroxide fixes predominantly lipids (Hayat, 1970). This factor is not significant at a level of cells and tissues at lower resolution of a scanning electron microscope, but it is quite important in the studies of enveloped viruses and membranes. In this case, stabilization with uranyl acetate is probably the best way to preserve at least an overall shape and structure (Nermut, 1972a; de Harven *et al.*, 1973; Hayat, 1975). Infiltration of pre-fixed tissue with nitrous oxide has been tried instead of organic solvents and carbon dioxide (Koller and Bernhard, 1964), but the whole process is fairly slow (overnight) and no further experience with this otherwise promising procedure has been reported. Fixation has generally been applied even when various fluorocarbons were used for critical-point drying (Fromme *et al.*, 1972; Spiess, 1973). In practical terms the critical-point drying cannot be used after negative staining, and for shadowing or replication the specimen has to be transferred to a suitable coating unit. This means that it is exposed first to air humidity, then to a vacuum.

In conclusion, freeze-drying is the technique of choice in the study of macromolecules, viruses, and small cell fragments, whereas critical-point drying is more practical in the preparation of tissues and whole cells for scanning electron microscopy (Boyde and Wood, 1969; Cohen, 1974; Hayat, 1977) as well as high voltage electron microscopy (Humphreys, 1976). One can therefore agree with Anderson (1956a) that both techniques "complement each other in a most desirable fashion."

CONCLUDING REMARKS

By eliminating the damaging effects of surface tension forces the freeze-drying technique provides more accurate information concerning the shape and size of biological structures, and its main field of application is in the study of macromolecules, viruses, and subcellular fractions. Surfaces of bacteria or eukaryotic cells can also be studied with freeze-drying, but only after replication which is accompanied by problems of mounting for transmission electron microscopy, particularly when large areas of cell monolayers are required for the study. Scanning electron microscopy of frozen or critical-point-dried tissues is more useful from this point of view. On the other hand, higher resolution of Pt-C shadowed carbon replicas enables detailed studies of small portions of cell surfaces.

The application of freeze-drying in virus research showed that the data concerning the size and shape of viruses obtained with conventional air-dry preparation techniques are inaccurate and unreliable. For example, after freeze-drying the diameter of the so-called pleomorphic viruses (e.g., influenza virus) is less by 25-30% than after air-drying, many "typical" structural features turn out to be artifacts, and surface structures are better revealed after freeze-drying and replication. It may well be that the present data on the size and shape of viruses (e.g., for purposes of classification) need to be re-evaluated.

New fields open also in macromolecular and membrane research. An appropriate illustration of these applications is provided by vesicles of reconstituted ATPase. The outer portion of the ATPase molecules is usually not seen at all after air-drying, but it is well revealed by freeze-drying, so that the number of ATPase molecules per square area can easily be established and correlated with chemical data, e.g., ratio of lipids and proteins and calcium transport.

A certain degree of shrinkage and wrinkling of vesicular structures, cell walls, and cell ghosts can certainly be regarded as artifacts of freeze-drying; on the other hand, it provides an interesting and useful information concerning the rigidity and behavior of such structures in the dry state. It is therefore expected that the freeze-drying technique would become a routine method in any advanced laboratory devoted to ultrastructural research.

Note: The manuscript of this chapter was submitted in April 1974 so that the references end with 1973 except for a few of them added before the text went into production.

I wish to thank many of my colleagues and friends who kindly supplied me with highly purified preparations of viruses, sarcoplasmic reticulum vesicles, bacteria, and tissue cultures for this study: Drs. N. M. Green, V. Mautner, D. Metz, W. C. Russell, G. Schild, and D. A. Thorley-Lawson from the National Institute for Medical Research, London, and Dr. J. T. Finch from the MRC Laboratory for Molecular Biology, Cambridge. I am grateful to Dr. H. Frank (Max-Planck-Institut für Virusforschung, Tübingen) for electron micrographs of murine leukemia viruses and bacteriophage T-5, the *Journal of Microscopy* for permission to reproduce Fig. 2.11B, Mrs. B. J. Ward and Mr. A. Morris for technical assistance, and Mrs. L. Lucas for secretarial assistance. I also appreciate helpful discussions with Dr. J. A. Armstrong concerning the manuscript.

Refèrences

Abermann, R., Salpeter, M. M., and Bachmann, L. (1972). High resolution shadowing. In: *Principles and Techniques of Electron Microscopy. Biological Applications*, Vol. 2 (Hayat, M. A., ed.), p. 197. Van Nostrand Reinhold Company, New York and London.

Anderson, T. F. (1951). Techniques for the preservation of three-dimensional structure in preparing specimens for the electron microscope. *Trans. N.Y. Acad. Sci.* **13,** 130.

Anderson, T. F. (1953a). The morphology and osmotic properties of bacteriophage systems. *Cold Spring Harbor Symp. Quant. Biol.* **18,** 197.

Anderson, T. F. (1953b). A method for eliminating gross artefacts in drying specimens. *C.R. Prem. Cong. Int. Micros. Electron* (Paris), p. 567. Revue d'Optique, Paris.

Anderson, T. F. (1954). Some fundamental limitations to the preparation of three-dimensional specimens for the electron microscope. *Trans. N.Y. Acad. Sci.* **16,** 242.

Anderson, T. F. (1956a). Electron microscopy of microorganisms. In: *Physical Technique and Biological Research*, Vol. 3 (Oster, G., and Polister, A. W., eds.), p. 177. Academic Press, Inc., New York.

Anderson, T. F. (1956b). Preservation of structure in dried specimens. *Proc. 3rd Int. Conf. Electron Micros.* (London, 1954), p. 122. Royal Micr. Society, London.

Bachmann, L., and Schmitt-Fumian, W. W. (1973). Spray-freeze-etching of dissolved macromolecules, emulsions and subcellular components. In: *Freeze-etching, Techniques and Applications* (Benedetti, E. L., and Favard, P., eds.), p. 63. Société Francaise de Microscopie Électronique, Paris.

Backus, R. C., and Williams, R. C. (1950). The use of spraying methods and of volatile suspending media in the preparation of specimens for electron microscopy. *J. Appl. Phys.* **21,** 11.

Bayer, M. E., and Remsen, C. C. (1970). Bacteriophage T_2 as seen with the freeze-etching technique. *Virology* **40,** 703.

Bayer, M. E., and Bocharov, A. F. (1973). The capsid structure of bacteriophage lambda. *Virology* **54,** 465.

Boyde, A., and Wood, C. (1969). Preparation of animal tissues for surface-scanning electron microscopy. *J. Microscopy* **90,** 221.

Boyde, A. (1972). Biological specimen preparation for the scanning electron microscope—An overview. *Scanning Electron Microscopy* (Part II). IIT Research Institute. (Johari, O. and Corvin, I., eds.) pp. 257–264, Chicago.

Brown, D. T., Waite, M. R. F., and Pfefferkorn, E. R. (1972). Morphology and morphogenesis of Sindbis virus as seen with freeze-etching techniques. *J. Virology* **10,** 524.

Bullivant, S. (1973). Freeze-etching and Freeze-fracturing. In: *Advanced Techniques in Biological Electron Microscopy* (Koehler, J. K., ed.), p. 67. Springer-Verlag, New York.

Bystricky, V., Fromme, H. G., Pfautsch, M., and Pfefferkorn, G. (1972). Scanning and transmission electron microscopy of an unusual soil bacterium: application of the critical point drying method. *Micron* **3,** 474.

Cohen, A. L. (1974). Critical point drying. In: *Principles and Techniques of Scanning Electron Microscopy: Biological Applications*, Vol. 1 (Hayat, M. A., ed.). Van Nostrand Reinhold, New York and London.

Cohen, A. L., Marlow, D. P., and Garner, G. E. (1968). A rapid critical point method using fluorocarbons "Freons" as intermediate and transitional fluids. *J. Microscopie* **7,** 331.

Dourmashkin, R. R., and Simons, P. J. (1961). The ultrastructure of Rous sarcoma virus. *J. Ultrastruct. Res.* **5,** 505.

Dubochet, J. (1973). Préparation de spécimens biologiques pour la microscopie électronique en fond noir. *Experientia* **29,** 770.

Dubochet, J., and Kellenberger, E. (1972). Selective adsorption of particles to the supporting film and its consequences on particle counts in electron microscopy. *Microscopica Acta* **72,** 119.

Dunlop, W. F., Parish, G. R., and Robards, A. W. (1972). Temperature, vacuum and fracturing in the freeze-etching technique. *Proc. 4th Eur. Cong. Electron Micros.* (Manchester), p. 248. The Institute of Physics, London and Bristol.

Fernández-Morán, H., van Bruggen, E. F. J., and Ohtsuki, M. (1966). Macromolecular organization of hemocyanins and apohemocyanins as revealed by electron microscopy. *J. Mol. Biol.* **16,** 191.

Fromme, H. G., Pfautsch, M., Pfefferkorn, G., and Bystricky, V. (1972). Die Kritische Punkt-Trocknung als Präparations-methode für die Raster-Elektronenmikroskopie. *Microscopica Acta* **73,** 29.

Gelderblom, H., Bauer, H., Bolognesi, D. P., and Frank, H. (1972). Morphogenese und Aufbau von RNS-Tumorviren: Elektronenoptische Untersuchungen an Virus-Partikeln vom C-Typ. *Zbl. Bakt. Hyg., I. Abt. Orig.* **A220,** 79.

Gier, J. de, Mandersloot, J. G., and Deenen, L. L. M. van (1968). Lipid composition and permeability of liposomes. *Biochim. Biophys. Acta* **150,** 666.

Glauert, A. M., and Lucy, J. A. (1969). Electron microscopy of lipids: effects of pH and fixatives on the appearance of a macromolecular assembly of lipid micells in negatively stained preparations. *J. Microscopy* **89,** 1.

Harven, E. de, and Friend, C. (1964). Structure of virus particles partially purified from the blood of leukemia mice. *Virology* **23,** 119.

Harven, E. de, Beju, D., Evenson, D. P., Basu, S., and Schidlovsky, G. (1973). Structure of critical point dried oncornaviruses. *Virology* **55,** 535.

Hayat, M. A. (1970). *Principles and Techniques of Electron Microscopy: Biological Applications*, Vol. 1. Van Nostrand Reinhold Company, New York and London.

Hayat, M. A. (1975). *Positive Staining for Electron Microscopy.* Van Nostrand Reinhold Company, New York and London.

Hayat, M. A. (1977). *Introduction to Biological Scanning Electron Microscopy*. Pergamon Press, London, New York.

Hayat, M. A., and Zirkin, B. R. (1973). Critical point drying method. In: *Principles and Techniques of Electron Microscopy: Biological Applications*, Vol. 3 (Hayat, M. A., ed.), p. 297. Van Nostrand Reinhold Company, New York and London.

Henderson, W. J., and Griffiths, K. (1972). Shadow casting and replication. In: *Principles and Techniques of Electron Microscopy: Biological Applications*, Vol. 2 (Hayat, M. A., ed.), p. 151. Van Nostrand Reinhold Company, New York and London.

Horne, R. W. (1965). The examination of small particles. In: *Techniques for Electron Microscopy* (Kay, D., ed.) (2nd edition), p. 311. Blackwell Scientific Publications, Oxford.

Horwitz, M. S., Maizel, J. V., and Scharff, M. D. (1970). Molecular weight of adenovirus type 2 hexon polypeptide. *J. Virology* **6,** 569.

Humphreys, C. J. (1976). High voltage electron microscopy. In: *Principles and Techniques of Electron Microscopy: Biological Applications*, Vol. 6 (Hayat, M. A., ed.). Van Nostrand Reinhold Company, New York and London.

Johnson, M. W., and Horne, R. W. (1970). Some observations on the relative dehydration rates of negative stains and biological objects. *J. Microscopy* **91,** 197.

Kleinschmidt, A. K., Lang, D., Jacherts, D., and Zahn, R. K. (1962). Darstellung und Längenmessungen des gesamten Desoxyribonucleinsäure-Inhaltes von T_2-Bakteriophagen. *Biochim. Biophys. Acta* **61,** 857.

Koehler, J. K. (1968). The technique and application of freeze-etching in ultrastructure research. *Adv. Biol. Med. Phys.* **12,** 1.

Koehler, J. K. (1972). The freeze-etching technique. In: *Principles and Techniques for Electron Microscopy: Biological Applications*, Vol. 2 (Hayat, M. A., ed.), p. 53. Van Nostrand Reinhold Company, New York and London.

Koller, T., and Bernhard, W. (1964). Séchage de tissus au protoxyde d'azote (N_2O) et coupe ultrafine sans matière d'inclusion. *J. Microscopie* **3,** 589.

Kruyff, B. de, Greef, W. J. de, Eyk, R. V. W. van, Demel, R. A., and Deenen, L. L. M. van (1973). The effect of different fatty acid and sterol composition on the erythritol flux through the cell membrane of Acheloplasma laidlawii. *Biochim. Biophys. Acta* **298,** 479.

Lange, J., Frank, H., Hunsmann, G., Moennig, V., Wollman, R., and Schäfer, W. (1973). Properties of mouse leukemia viruses. VI. The core of Friend virus; isolation and constituents. *Virology* **53,** 457.

Luftig, R. B., and Kilham, S. S. (1971). An electron microscope study of Rauscher leukemia virus. *Virology* **46,** 277.

Medzon, E. L., and Bauer, H. (1970). Structural features of vaccinia virus revealed by negative staining, sectioning and freeze-etching. *Virology* **40,** 860.

Meryman, H. T. (1960). Principles of freeze-drying. *Ann. N.Y. Acad. Sci.* **85,** 630.

Moor, H. (1970). Die Gefrierätzung. In: *Methodensammlung der Elektronenmikroskopie* (Schimmel, G., and Vogell, W., eds.). 2.4.2.2. Wiss. Verlagsgesellschaft MBH, Stuttgart.

Moor, H. (1973a). Cryotechnology for the structural analysis of biological material. In: *Freeze-etching, Techniques, and Applications* (Benedetti, E. L., and Favard, P., eds.), p. 11. Société Francaise de Microscopie Électronique, Paris.

Moor, H. (1973b). Etching and related problems. In: *Freeze-etching, Techniques, and Applications* (Benedetti, E. L., and Favard, P., eds.), p. 21. Société Francaise de Microscopie Électronique, Paris.

Mühlethaler, K. (1973). History of freeze-etching. In: *Freeze-etching, Techniques, and Applications* (Benedetti, E. L., and Favard, P., eds.), p. 1. Société Francais de Microscopie Électronique, Paris.

Munn, E. A. (1968). On the structure of mitochondria and the value of ammonium molybdate as a negative stain for osmotically sensitive structures. *J. Ultrastruct. Res.* **25,** 362.

Muscatello, U., and Horne, R. W. (1968). Effect of the tonicity of some negative-staining solutions on the elementary structure of membrane-bounded systems. *J. Ultrastruct. Res.* **25,** 73.

Nanninga, N. (1968). The conformation of the 50S ribosomal subunit of *Bacillus subtilis. Proc. Nat. Acad. Sci.* (U.S.A.) **61,** 614.

Nei, T. (1962). Freeze-drying in the electron microscopy of microorganisms. *J. Electron-microscopy* **11,** 185.
Nermut, M. V. (1972a). Negative staining of viruses. *J. Microscopy* **96,** 351.
Nermut, M. V. (1972b). Further investigation on the fine structure of influenza virus. *J. Gen. Virol.* **17,** 317.
Nermut, M. V. (1973a). Methoden der Negativkontrastierung. In: *Methodensammlung der Elektronenmikroskopie* (Schimmel, G., and Vogell, W., eds.). 3.1.2.3. Wiss. Verlagsgesellschaft MBH, Stuttgart.
Nermut, M. V. (1973b). Freeze-drying and freeze-etching of viruses. In: *Freeze-etching, Techniques and Applications* (Benedetti, E. L., and Favard, P., eds.), p. 135. Société Francaise de Microscopie Électronique, Paris.
Nermut, M. V. (1973c). Freeze-drying for electron microscopy. Balzers High Vacuum Report. VBP 2.
Nermut, M. V., and Frank, H. (1971). Fine structure of influenza A2 (Singapore) as revealed by negative staining, freeze-drying and freeze-etching. *J. Gen. Virol.* **10,** 37.
Nermut, M. V., Frank, H., and Schäfer, W. (1972). Properties of mouse leukemia viruses. III. Electron microscopic appearance as revealed after conventional preparation techniques as well as freeze-drying and freeze-etching. *Virology* **49,** 345.
Nermut, M. V. (1975). Fine structure of adenovirus Type 5. 1. Virus capsid. *Virology* **65,** 480.
Nermut, M. V., and Murray, R. G. E. (1967). Ultrastructure of the cell wall of *Bacillus polymyxa. J. Bact.* **93,** 1949.
Nermut, M. V., Schramek, S., and Brezina, R. (1970). Electron microscopy of Coxiella burneti phase I antigen. *Zbl. Bakt., I. Abt. Orig.* **214,** 236.
Nowinski, R. C., Old, L. J., Sarkar, N. H., and Moore, D. H. (1970). Common properties of the oncogenic RNA viruses (oncornaviruses). *Virology* **42,** 1152.
Pinto da Silva, P. (1973). Membrane intercalated particles in human erythrocyte ghosts: Sites of preferred passage of water molecules at low temperature. *Proc. Nat. Acad. Sci.* (U.S.A.) **70,** 1339.
Rebhun, L. (1972). Freeze-substitution and Freeze-drying. In: *Principles and Techniques of Electron Microscopy: Biological Applications*, Vol. 2 (Hayat, M. A., ed.), p. 3. Van Nostrand Reinhold Company, New York and London.
Rowe, T. W. G. (1960). The theory and practice of freeze-drying. *Ann. N.Y. Acad. Sci.* **85,** 641.
Russell, W. C., McIntosh, K., and Skehel, J. J. (1971). The preparation and properties of adenovirus cores. *J. Gen. Virol.* **11,** 35.
Spiess, E. (1973). Untersuchungen zur Struktur der Ribosomen. Präparation der 50-S Untereinheit von *Escherichia coli* nach der Kritische-Punkt-Methode. *Cytobiologie* **7,** 28.
Steere, R. L. (1973). Preparation of high-resolution freeze-etch, freeze-fracture, frozen-surface and freeze-dried replicas in a single freeze-etch module, and the use of stereo electron microscopy to obtain maximum information from them. In: *Freeze-etching, Techniques and Applications* (Benedetti, E. L., and Favard, P., eds.), p. 223. Société Francaise de Microscopie Électronique, Paris.
Steere, R. L., and Haga, J. Y. (1970). Pre-shadowed surface replicas of frozen biological specimens. *28th Ann. EMSA Meeting* (Arceneaux, C. J., ed.), p. 288. Claitor's Publishing Division, Baton Rouge, Louisiana.
Stein, W. D. (1967). *The Movement of Molecules Across Cell Membranes.* Academic Press, New York and London.
Stephenson, J. L. (1960). Fundamental physical problems in the freezing and drying of biological materials. In: *Recent Research in Freezing and Drying* (Parkes, A. S., and Smith, A. U. eds.), p. 122. Blackwell Scientific Publications, Oxford.

Valentine, R. C., and Pereira, H. G. (1965). Antigens and structure of the adenovirus. *J. Mol Biol.* **13,** 13.

Williams, R. C. (1952). Electron microscopy of sodium desoxyribonucleate by use of a new freeze-drying method. *Biochim. Biophys. Acta* **9,** 237.

Williams, R. C. (1953a). A method of freeze-drying for electron microscopy. *Exp. Cell Res.* **4,** 188.

Williams, R. C. (1953b). The shapes and sizes of purified viruses as determined by electron microscopy. *Cold Spring Harbor Symp. Quant. Biol.* **18,** 185.

Williams, R. C., and Fraser, S. (1953). Morphology of the seven T-bacteriophages. *J. Bact.* **66,** 458.

Witter, R., Frank, H., Moennig, V., Hunsmann, G., Lange, J., and Schäfer, W. (1973). Properties of mouse leukemia viruses. IV. Hemagglutination assay and characterization of hemagglutinating surface components. *Virology* **54,** 330.

Wrigley, N. G. (1968). The lattice spacing of crystalline catalase as an internal standard of length in electron microscopy. *J. Ultrastruct. Res.* **24,** 454.

Wyckoff, R. W. G. (1946). Frozen-dried preparations for the electron microscope. *Science* **104,** 36.

Zeigel, R. F., and Rauscher, F. J. (1964). Electron microscopic and bioassay studies on a murine leukemia virus (Rauscher). I. Effects of physico-chemical treatments on the morphology and biological activity of the virus. *J. Nat. Cancer Inst.* **32,** 1277.

3. IMAGE RECONSTRUCTION OF ELECTRON MICROGRAPHS BY USING EQUIDENSITE INTEGRATION ANALYSIS

Klaus-Rüdiger Peters

Heinrich-Pette-Institut für Experimentelle Virologie und Immunologie an der Universität Hamburg, Hamburg, Federal Republic of Germany

INTRODUCTION

A large number of methods for the evaluation of electron micrographs have been put forward during the last 10 to 15 years. The analysis of biological objects, the structural units of which possess a symmetrical arrangement, was of special interest. Mostly, however, these units do not show in high-resolution micrographs a uniform image because of numerous factors such as masking by uneven contrast, alterations during preparation, confusion by background information, superpositioning, radiation damage, and other electron-microscopic artifacts. According to the image quality of the periodic structures, direct or indirect methods of analysis may be subsequently applied (also see Ellison and Jones in this Volume).

Obviously when symmetrical substructures are not obscured, a direct comparison of the image with a model is possible with regard to the type of staining employed (Caspar, 1966). With "one-sided" stained objects extensive comparative experiments are necessary (Klug, 1965; Klug and Finch, 1965), but for the exact image analysis of "two-sided" stained objects even more involved studies are required (Finch *et al.*, 1970; Finch and Klug, 1966), which should advisably employ computer analysis like the Fourier synthesis (Crowther *et al.*, 1970).

Prior to computer analysis, methods were developed for the analysis of simple images. However, when the symmetrical substructures of an object appear in two-dimensional symmetry and are not clearly presented, indirect methods of image analysis may be used. In such a case the image content is first of all clarified by either optical methods or contrast analysis and then analyzed.

Optical methods include optical diffraction (Klug and Berger, 1964) and optical filtration (Klug and DeRosier, 1966; DeRosier and Klug, 1972) as well as self convolution (Elliott *et al.*, 1968; Fiskin and Beer, 1968), permitting analysis of the direction and frequency of periodic structures and an image reconstruction using optical filtration.

The contrast analysis of EM micrographs may be produced with the aid of indirect optical methods. According to a method by Galton (1878), contrast integration may be achieved by the superpositioning of various images with identical orientation. This "sandwich" procedure was also employed by Finch *et al.* (1966). Markham *et al.* (1963; 1964) and Gachet and Thiéry (1964) varied the Galton technique by using a single micrograph in such a way that the contrast of the periodic substructures was enhanced, resulting in an average image of the complete symmetrical arrangement. This stroboscopic integration technique (rotation technique, stroboscopic technique, harmonic technique) of Markham *et al.* may be easily utilized as it does not require special equipment. The disadvantage of this technique has often been discussed. Only a contrast filtration of the micrograph before the application of stroboscopic integration permits the method to be controlled, allowing the evaluation of the contribution of the information from the initial image to the end product. Crowther and Amos (1971) developed harmonic analysis in this connection, permitting an evaluation of EM micrographic contrasts by computer. After numerous procedures a rotationally integrated picture was produced with the strongest possible power spectrum, thus allowing the establishment of the periodicity.

Alternatively, defined contrast filtration is possible by employing a photographic copying process using equidensite negatives to control the integration of periodic structures as well as producing an average image (equidensite rotation method; Peters, 1974a). By means of such copying, a structural correlation of the contrast filtered image to the original is carried out, thus making possible the reconstruction of the image analogous to optical filtration.

BASIC METHODS

The equidensite integration method described here in more detail consists of two steps. By means of the equidensite technique, the contrast of an EM negative is analyzed photographically, and then with certain selected contrasts the integration technique according to Markham *et al.* is performed, followed by image reconstruction. The various techniques and the necessary materials must first be described and characterized.

Characterization of Photographic Material

The transfer characteristics of a photographic emulsion are represented by its characteristic *j*-curve (Fig. 3.1). It indicates the respective blackness (density *D*) of the emulsion after exposure (*E*) to a certain quantity of light. The density is expressed as a decadic logarithm of the reciprocal transparency. The exposure (brightness × time) is marked at the abscissa in logarithmic measure. The contrast transfer (*j*) of the emulsion is characterized by the gradient of the curve. In order to set up the characteristic curve a gray wedge can be used instead of varying the time of exposure. For instance, a density step of 0.3 *D* corresponds to a change in light intensity of a factor of 2, and a step of 0.15 *D* to a factor of 1.41. So for the demonstration of contrast reproduction a gray wedge (Agfa Gevaert AG., Repro-Graukeil) was sometimes copied, the densities of which varied by steps of 0.15 *D*, starting at white with 0.075 *D* (respective 0) and ending in black with 1.65 *D*.*

In order to change the contrast volume of a negative, it can be copied with an emulsion having a different contrast transfer ability. The steepness of the *j*-curve of the emulsion on which the copies are made may be varied by changing the time of development as well as by using other developers. Thus, the hard emulsion of Agfa ortho 25 professional may be developed at 20°C and unagitated with the balance developer Agfa Refinal for 5 to 8 min or with the hard X-ray developer Agfa G 150 (diluted 1 + 5) for 5 to 8 min.

Agfa pan 100 professional (21 DIN-100 ASA) may be used to produce halftone copies with development in Agfa Refinal at 20°C with interrupted agitation for 10 min. Hard copies are produced with the graphic emulsion Agfa Gevalith ortho developed for 2 min with constant agitation in a paper developer.

The special film Agfa contour professional is treated according to the advice of the manufacturer for 2 min with constant agitation in a special developer (Agfa-Gevaert, 1972, 1974). For yellow filtration of the light used for copying, a set of filters required for color work is used (Agfa Gevaert AG., Federal Republic of Germany).

All copies can be made easily with an enlarger in which a fixed frame may

*Unfortunately the reproduction of the gray wedge is falsified due to various nonlinear photographic transfers before printing.

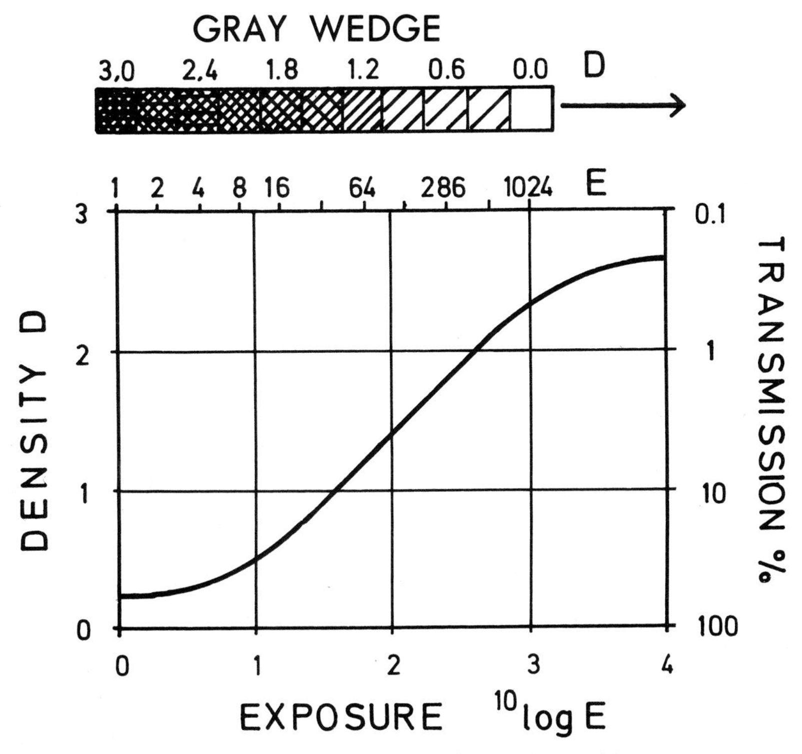

Fig. 3.1 Characteristic *j*-curve of a negative emulsion.

serve to orientate the direction of successive negatives for copying. The enlarger Fokomat II (Leitz GmbH, Federal Republic of Germany) was used in this case, being equipped for 1 : 1 reproduction and for color filtering. The exposure intensity may be adjusted by varying either the enlargement or the diaphragm. For the reproducible measurement of the light intensity, a good exposure meter is required, e.g., Luna six (Gossen AG., Federal Republic of Germany). The latter possesses a working scale, which may be transferred directly to lux values (or ft-c). The exposure of equidensite film is approximately 5 sec for light gray values and approximately 4 min for dark gray values with an illumination of 300 lux (28 ft-c).

Photographic Production of Equidensites

Equidensites represent the parts of a pattern in photography which have the same density, so that the equidensite negative shows areas of the object photographed possessing the same gray tone.

Photographically, equidensites can be obtained by the superpositioning of a negative and a positive of the same pattern. The gradation of the positive must

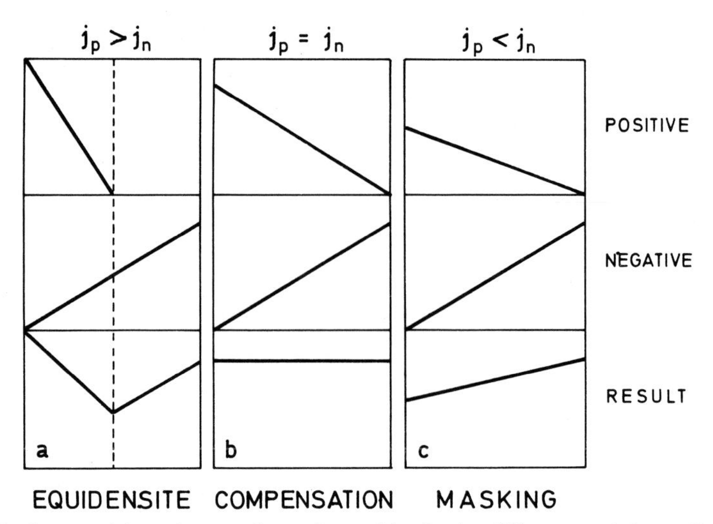

Fig. 3.2 Superposition of a negative and a positive having different gradations. Simplified j-curves.

be steeper than that of the negative (Fig. 3.2a); otherwise only a compensation or flattening of the negative gradation can take place (Fig. 3.2b and c). According to this sandwich procedure, a saddle develops in the resulting j-curve, the minimum of which represents a certain density (Fig. 3.3). With Agfa contour professional film the positive and the negative components are produced at the same time in the same emulsion layer during development. However, because of the fact that both components possess a different color sensitivity, the width of the equidensites, i.e., the contrast volume, which is to be demonstrated, may be influenced by yellow filtration.

Because of the higher light sensitivity of the positive component, the position of the equidensites may be altered by varying the time of exposure. For exam-

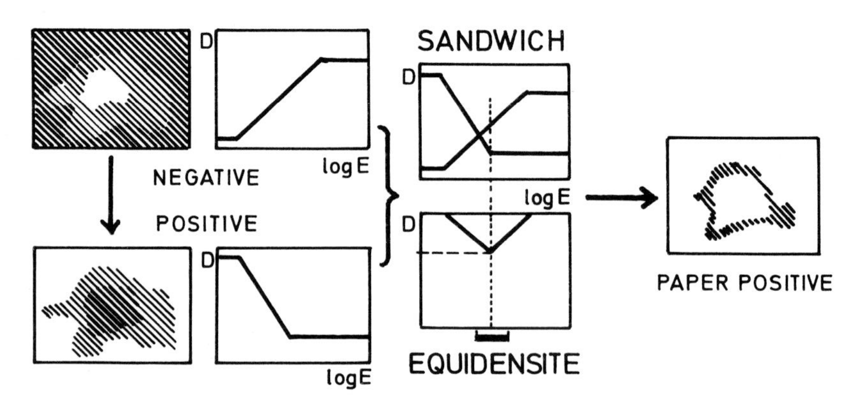

ple, in the case when a certain density D_1 is obtained with the time t_1, the density D_2 results from an exposure t_2 according to the equation

$$t_2 = t_1 \cdot 10^{D_2 - D_1}$$

In this way one can readily divide the pattern into single equidensites, thus forming a band of equidensites. Thus an adjacent density step in such a band may be obtained by prolonging the time of exposure with a factor in accordance to the width of the equidensites required:

width of equidensite	0.15	0.2	0.3	0.4
factor of exposure	1.4	1.6	2.0	2.5

The resulting equidensites form a pattern representing areas referred to as equidensites of the first order. When the latter are subsequently copied onto a contour film and exposed to produce a new equidensite for a middle gray tone, then very thin lines develop at the borders of the dense areas. It is only here that the corresponding density occurs and these lines are called equidensites of the second order.

Stroboscopic Integration Technique

According to Markham *et al.* (1963), rotationally symmetrical structures may be enhanced photographically. If one is dealing with an m-fold symmetry and a certain time of exposure is necessary to produce a photographic picture of the pattern, then one rotates the copy under the enlarger around the symmetrical

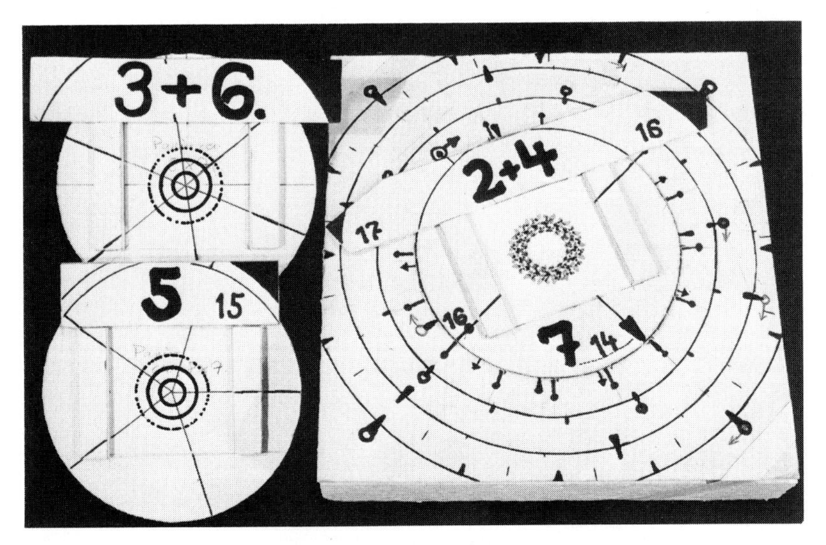

Fig. 3.4 Simple apparatus for rotational integration.

Fig. 3.5 Apparatus for linear integration.

axis making m exposures, each at a rotation of $360°/m$ and having an exposure time of t/m. From this stroboscopic compilation results a composite picture demonstrating symmetrical areas, composed from all individual areas. All substructures common to the single areas are represented in their correct density on the copy, and all others possess a weaker density, so that the common characteristics of the subunits are enhanced and thereby characterized by their density.

An apparatus for rotation may easily be constructed (Fig. 3.4). It consists of a calibrated plate, which can be rotated through 360° and on which photographic paper may be attached. A similar apparatus, which is just as simple, may be constructed for the linear integration technique. The photographic paper is attached to a movable plate, and linear shifting may be carried out in equal steps with an interchangeable scale. In addition, the table on the apparatus shown in Fig. 3.5 may be turned through certain angles in order to ascertain the direction of periodicities, which may have a sloping orientation to the main axis.

Micrograph Analysis for Image Reconstruction

The aim of every image analysis is to describe the relevant characteristics of an object, which means either the relationship of symmetry or the structural arrangement of its units. The prerequisite for such analysis is that certain ideas regarding the model of the symmetrical construction of the object must exist, and that this model must be proved or disregarded subject to a comparative description with the existing image contrast. In this way each analysis may be divided into the following sequence of single steps: object judgment and model conception; information filtration; model development; analysis and model construction; image reconstruction.

By equidensite integration analysis the information filtration is performed by the equidensite technique and the analysis by stroboscopic integration. The complete procedure is described below with reference to actual examples, and taking into account the most important steps.

CONTRAST ANALYSIS AND MODEL DEVELOPMENT

The first example will be taken from the electron-microscopic examination of Laelia red leafspot virus (Peters, 1976). In ultrathin sections these orchid viruses showed some deviating morphological characteristics. In some longitudinal sections of monomeric virus particles a cross striation of the inner component can be recognized. Because the periodicity appears to be almost twice as large as that of other similar rhabdoviruses (Knudson, 1973), it was interesting to analyze their fine structure. Several particle images (Fig. 3.6a) were therefore examined in this context.

The model concept refers to a helical construction of the inner component of rhabdoviruses and presumes a linear periodicity, which seems in this case twice as large as in other rhabdoviruses or in a combination of the substructures in pairs.

Contrast Filtration by Means of Equidensites

For a better interpretation of the micrographs, a contrast analysis is first of all performed. In order to reach a differentiated fission of the equidensites, the contrast of the original EM micrograph is spread as wide as possible, and at the same time the object picture is enlarged, so that the subsequent copying steps do not cause a substantial loss of detail. The easiest way is to work in a format of 6.5 $\times$ 10.5 cm.

The photographic contrast transfer (Fig. 3.6b) is characterized by a relationship of the j-values of the original emulsion (j_a) and of the copying emulsion (j_c, j_d). The original object volume of approximately seven gray steps is reduced at $j_a/j_c > 1$, as seen in Fig. 3.6c. In the case when j_a/j_d is < 1 a contrast increase results of approximately eight gray steps (Fig. 3.6d). This negative could have been developed even harder, in order to take full advantage of the contrast volume of the emulsion. Nevertheless this somewhat overexposed negative is used for the equidensite analysis.

When one is copying a pattern on the equidensite film, the desired width of the equidensite must first of all be estimated which is influenced by yellow filtration and to a lesser extent by the copying procedure.

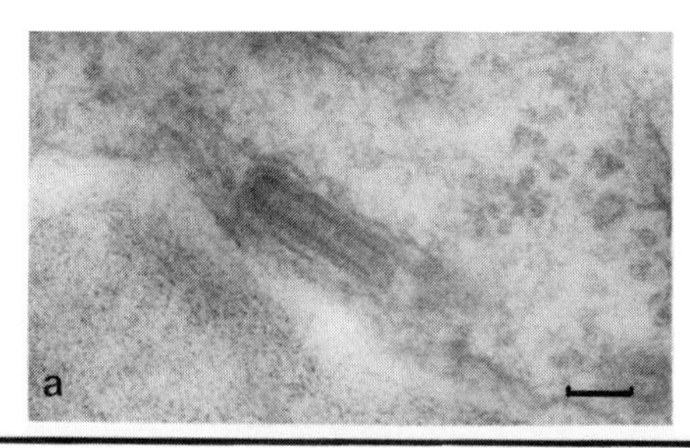

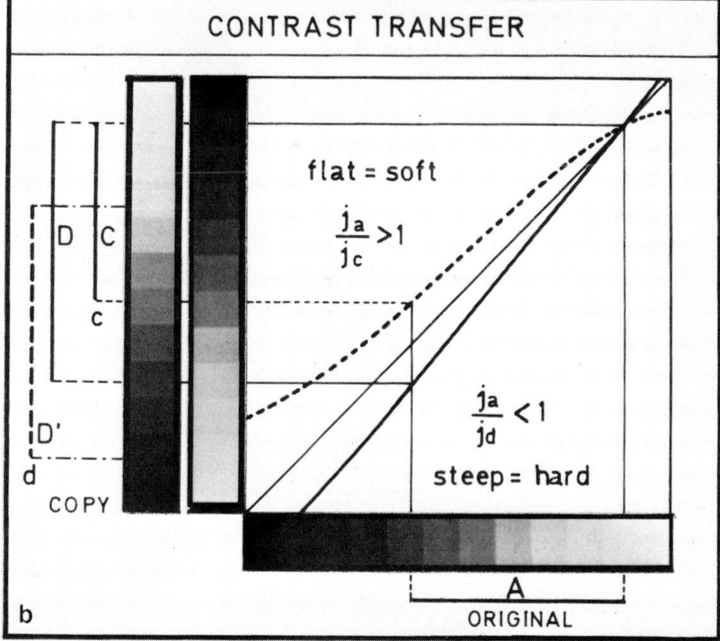

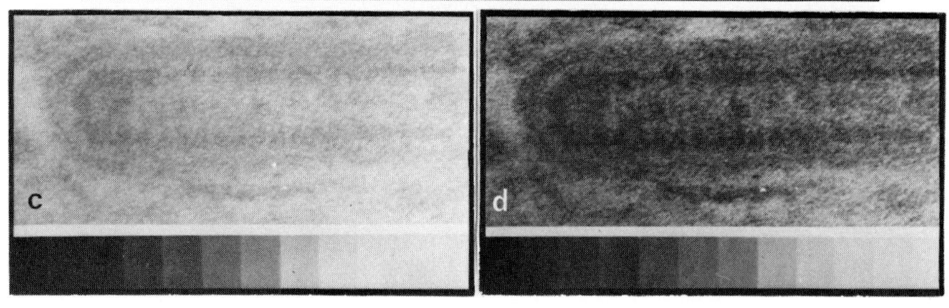

Fig. 3.6 Photographical increase of the contrast volume of the original EM micrograph. (a) Original EM micrograph of the Laelia red leafspot virus. ×60,000. Scale represents 100 nm. (b) Contrast transfer, with *A*, *C*, *D* and *D′* representing contrast volumes. (c) Copy of (a) with reduced contrast volume. ×210,000. (d) Copy of (a) with increased contrast volume. ×210,000.

Figure 3.7 demonstrates the alteration of the equidensite in relationship to yellow filtration. Because the same times of exposure are used, and because the yellow filter with increased color intensity absorbs more light, the equidensite will move towards the lighter gray values. However, with a corresponding increase in yellow filtration the width of the equidensite will decrease. The density reproduction of the gray wedge is shown schematically and refers to the contour negatives of Figs. 3.7a–e. The other way to influence the width of the equidensite is shown in Fig. 3.8. Figure 3.8a shows the contour negative of Fig. 3.7b. From the density characterization in Fig. 3.8b it is recognizable that in the case of hard copying, depending on the time of exposure, lighter gray steps are also includable. The result is that on the copies, equidensites with a width of two gray steps (Fig. 3.8c) can be extended to as much as four steps (Fig. 3.8e). Figure 3.8f demonstrates that in the case of overexposure the copy becomes overgrown.

To demonstrate the density distribution of the pattern, Fig. 3.6d was divided, considering these factors, into single gray steps of the gray wedge, of 0.15 *D*. Figure 3.9 shows the respective equidensites of the first order produced. The strongest density of the pattern is demonstrated in Fig. 3.9b analogous to the density step two (1.5 *D*); the lightest gray values appear in Fig. 3.9i, belonging to the wedge step eight (0.45 *D*). The division into single contrast steps can be observed clearly because of the fact that the areas of a certain density are missing in the other equidensite demonstrations. The equidensites of the second order of the negatives of Fig. 3.9 may be taken from Fig. 3.10. The broadening of the lines is partly due to the fact that the gray values resulting from the flat negative shoulder are not completely suppressed by the copying.

To demonstrate the contrast distribution of the pattern, it is possible to reassemble single equidensites of the second order, obtaining an equidensite band. Figures 3.11a, b comprise five corresponding density extracts representing a uniform variation of 0.15 *D*.

Model Development

The development of a model can now be performed by means of single density extractions. According to the model conception, periodically arranged contrasts in the region of the inner component of the virus particle are to be found in various equidensites, i.e., Figs. 3.7a, b, d, Figs. 3.8c, d, and Figs. 3.9c, d, e. Because of the uneven contrast of the particle, two density extracts are considered for the development of a model (Figs. 3.11c and d). The individually demonstrated density areas are then incorporated into the model, provided that they appear as isolated areas without open contact to neighboring areas. Since in both equidensite extracts the area representing the internal structure is clearly separated from the other features, these may be disregarded. Thus for the

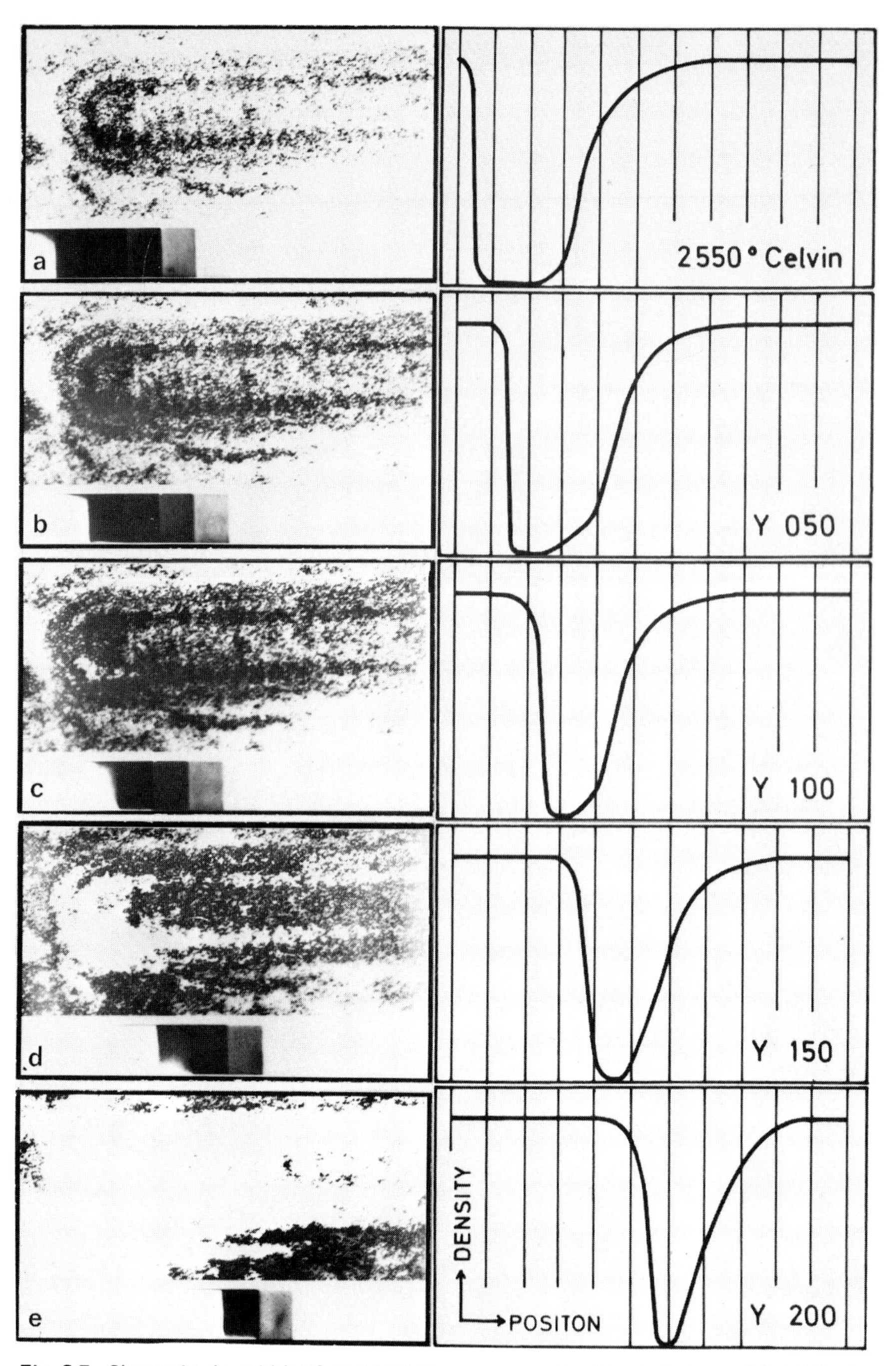

Fig. 3.7 Change in the width of equidensites by means of yellow filtration. (a)–(e) Equidensites of the negative from Fig. 3.6d. (a) Without yellow filtration. (b)–(e) With different yellow filtrations of increased color intensity using the same time of exposure. Next to the pictures the respective contrast reproduction of the gray wedge is demonstrated.

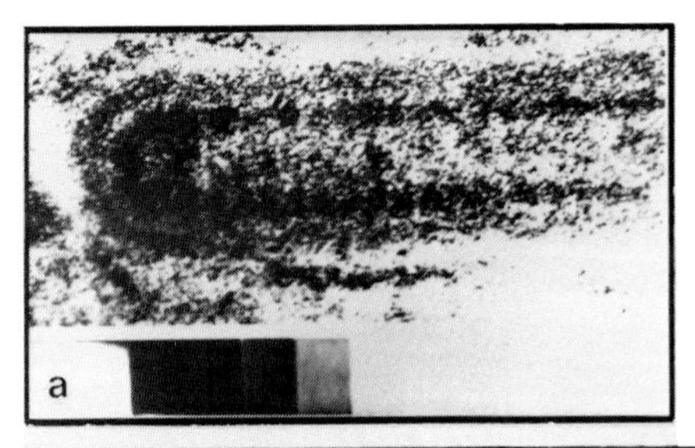

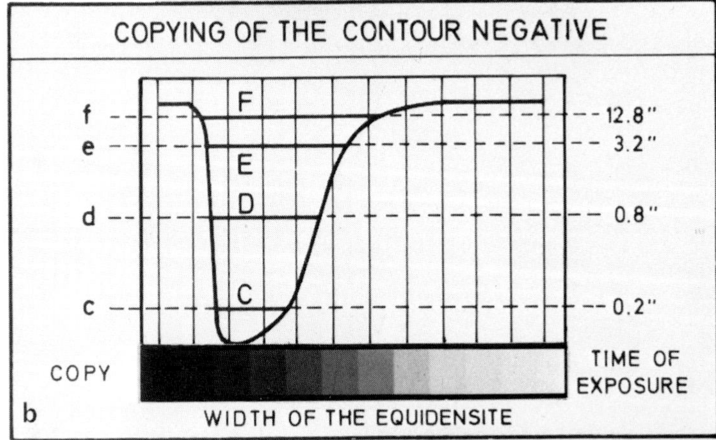

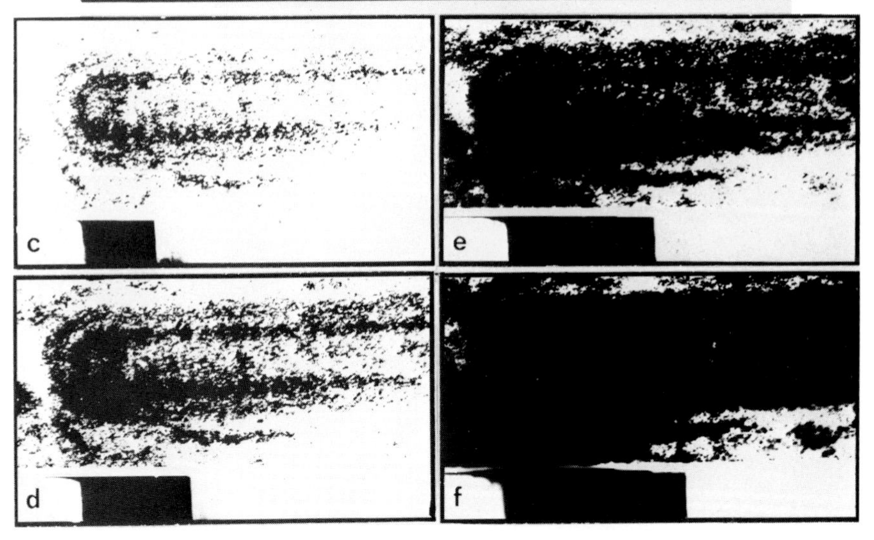

Fig. 3.8 Change of the width of equidensites by means of different copying. (a) Halftone copy of the contour negative. (b) Contrast transfer with marked copying volume for different times of exposure. (c)–(f) Copies according to (b).

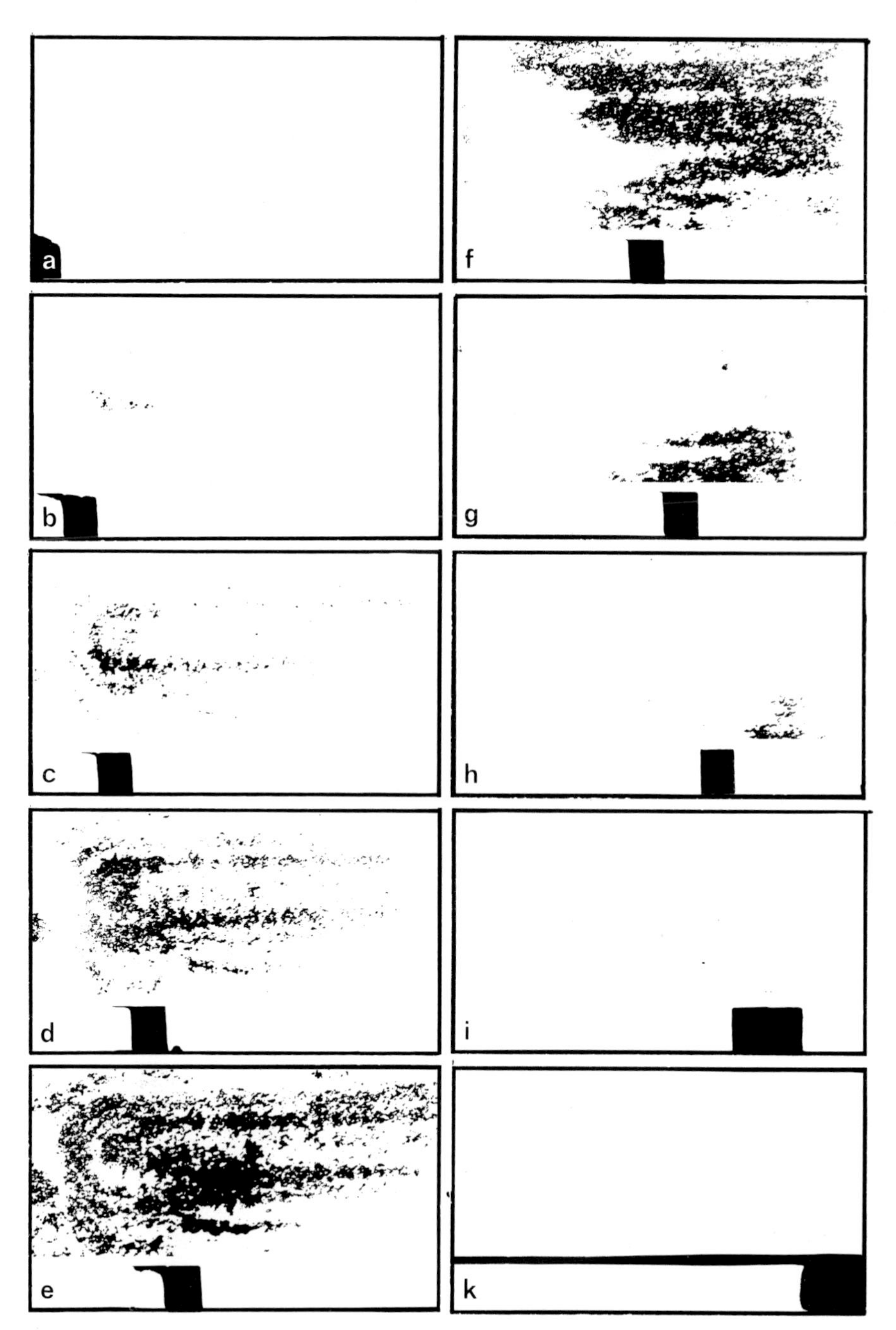

Fig. 3.9 Contrast excerpts of the negative from Fig. 3.6d with equidensites of the first order. (a)–(h) Width of equidensites according to steps of the gray wedge of 0.15 D. (i,k) Width of equidensites of 0.3 D.

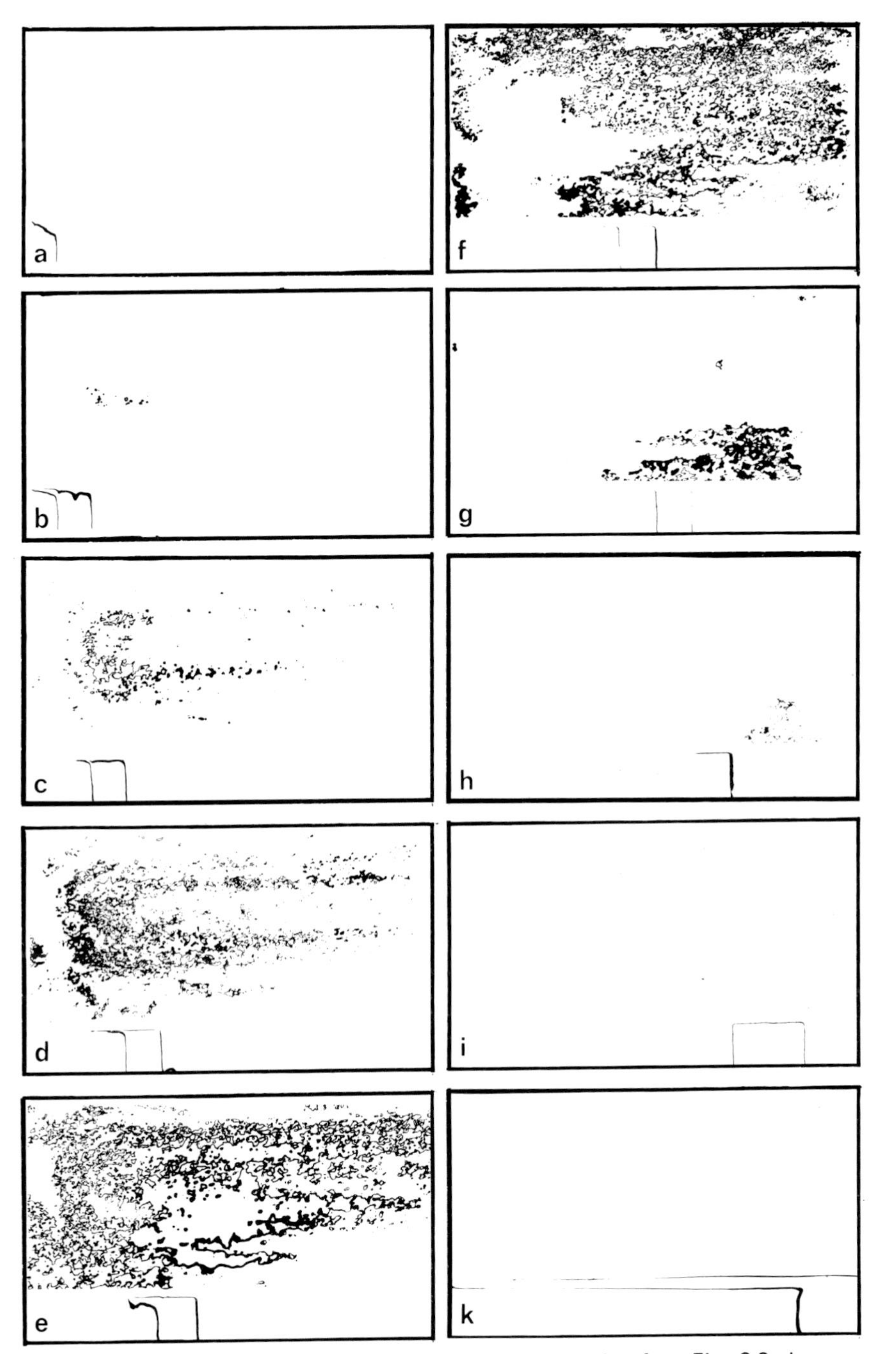

Fig. 3.10 Equidensites of the second order of the negatives from Figs. 3.9a–k.

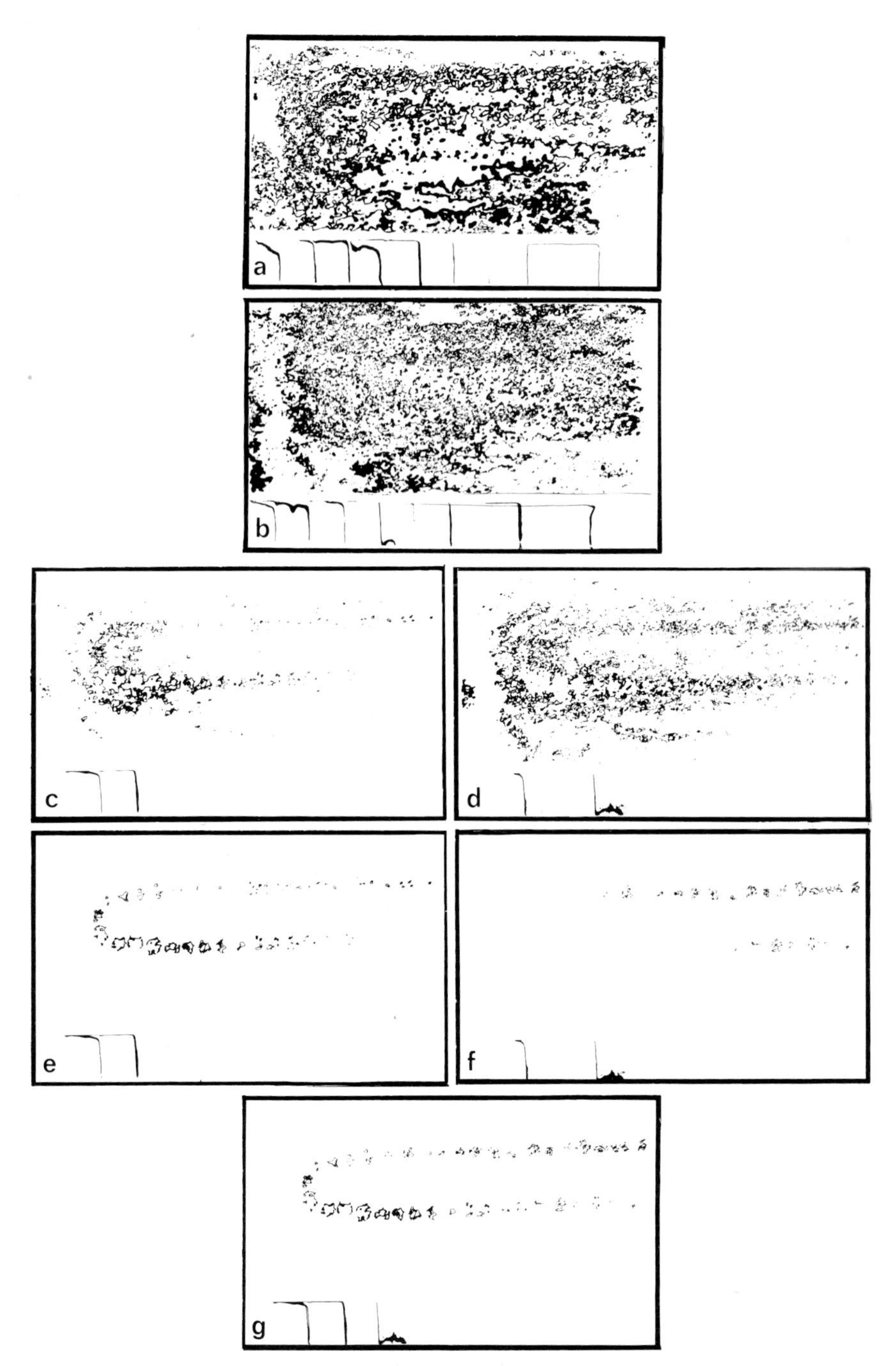

Fig. 3.11 Model construction from the contrast excerpts. (a,b) Band of equidensites with density differences of 0.15 *D*. (c,d) Selected equidensites, showing the isolated contrasts of the analyzed structures. (e,f) Excerpt of the representative contrasts. (g) Composition of pictures (e) and (f).

selection of contrasts at the half round closed end of the inner component, the equidensites of higher density must also be taken into consideration (Figs. 3.10b and c).

The same holds true for the representation of the open side of the inner component. Here, for areas representing the upper structure, the corresponding gray areas from Fig. 3.11d were attached to the points at Fig. 3.11c, which surround the latter. For the lower area of the structure the largest, collective individual areas were used. The manually filtered contrast in this complex case is represented in Figs. 3.11e, f, and are collected in Fig. 3.11g. The development of the model shows that the strongest contrasts in the various picture areas, analogous to the picture of the equidensite excerpt (Fig. 3.12a) almost correspond to the model conception of a periodic arrangement along the section line of the inner component.

Final Steps for Image Reconstruction

By means of the stroboscopic integration technique, the possibility of the periodic arrangement of the single contrast areas may be checked. For that purpose the linearly preceding area of the contrast excerpt (Fig. 3.12b) is analyzed. Of the linear shifting frequencies tried analogous to 9.5, 12, and 14.5 nm, only the shifting frequency according to 12 nm showed a resolution (Fig. 3.12c). All filtered density areas are located in single, separate integration areas. This proves the model conception of a given periodicity. The isolated integration areas are shown roughly circular. They can reflect the average form of the single periodic elements. This integration analysis now permits a direct model construction. Because the density areas at the semicircular closed end of the inner structure show almost the same periodicity and extension as the integration areas, a model can be constructed according to Fig. 3.12d.

In order to allow the image reconstruction, the model must now be adjusted according to the actual contrasts, from which it was built (Fig. 3.12e). For this reason, the model is transferred (manually or photographically) onto a transparent foil according to the proportions ascertained and is then adjusted to the negative of Fig. 3.12b. In this position it is then fixed. For the purpose of reconstruction the model is transferred onto the pattern (Fig. 3.12f).

From this analysis using the amount of image contrast selected it is shown that the inner component of the Laelia red leafspot virus may show a periodicity of approximately 12 nm in ultrathin sections and that the periodic units may appear individually and not in pairs.

MODEL CONSTRUCTION AND IMAGE RECONSTRUCTION

The development of a model of the fine structure of the nitrobacter particle Nb_1 (Bock *et al.*, 1974) could be executed step by step in the same way as previously

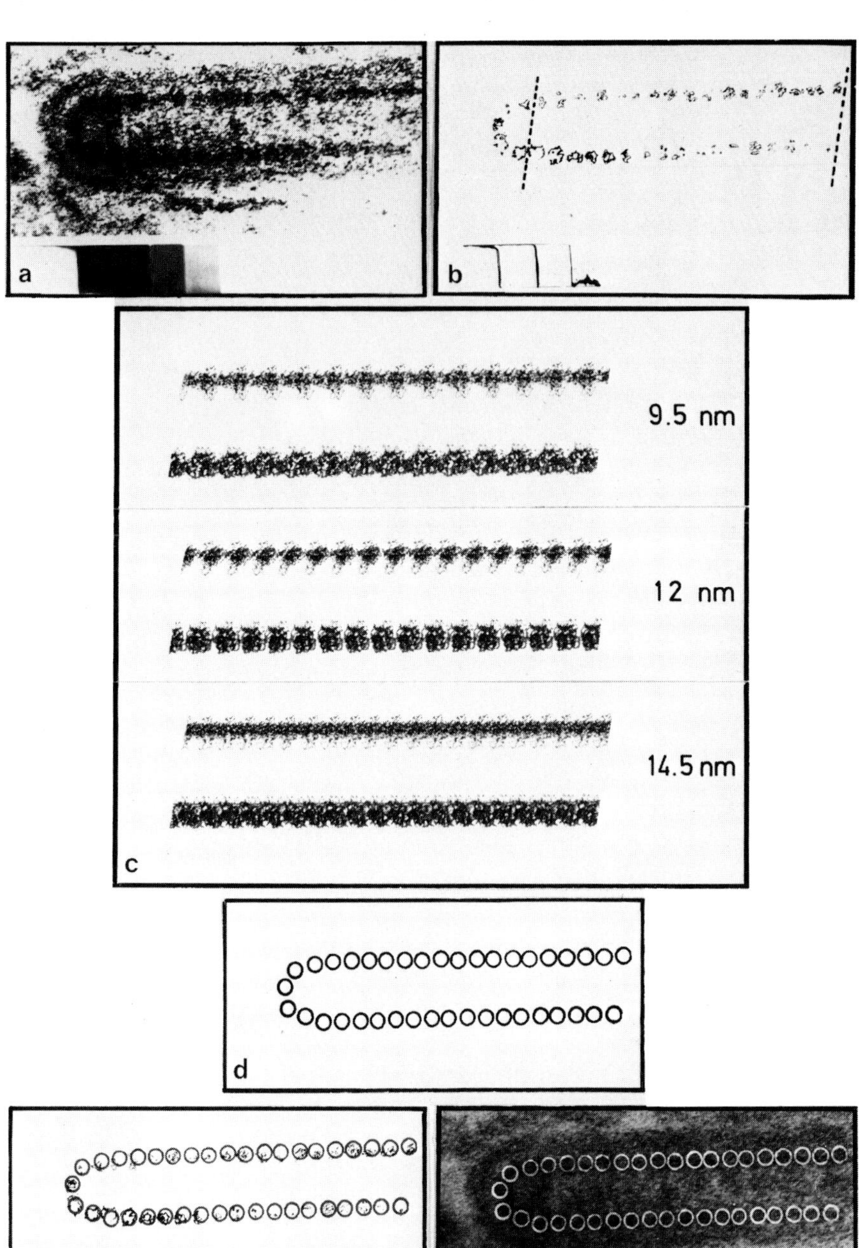
a
b
9.5 nm
12 nm
14.5 nm
c
d

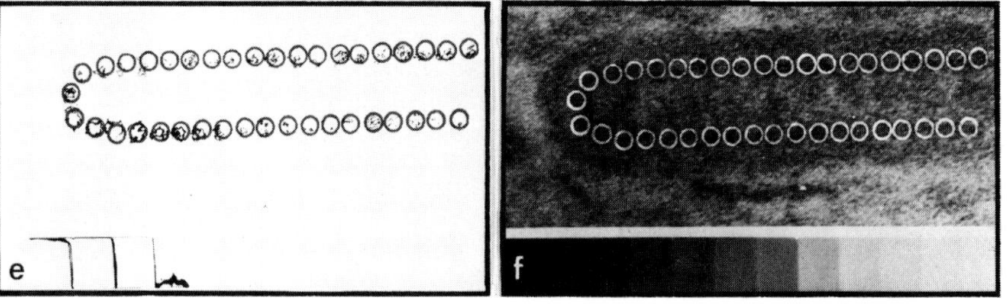
e
f

described. The model concept was based on the assumption that the particle might be composed of an icosahedral capsid. To substantiate the model series of Caspar and Klug (1962) an analysis of the substructure of the particle envelope was necessary and will here be partly demonstrated.

First Steps for Model Development

The particles showed a granular substructure in their envelopes (Fig. 3.13a). To produce a one-sided contrast a special technique using negative staining was employed (Peters, 1974b). Figure 3.13b shows such an image with reversed contrast because of the copying of the original micrograph. In order to develop a model, a contrast filtration with equidensites was again executed. The equidensites of the first and second orders were filtered from the hard printed copy of Fig. 3.13c. Figures 3.13d and g show light gray values produced through a great width of equidensites so that higher densities appear white. In this granular pattern, some substructures were evident and subsequently filtered by a narrow width of equidensites excluding the middle gray tones (Figs. 3.13e and h). In the upper right part of the particle image (*circle*) a group of six contrast-rich areas which seem to be arranged rotationally symmetrical may be easily observed.

However, as can be seen from the equidensite of the second order, single areas are partly in open contact with those adjacent, which means they are not to be considered as representing individually situated substructures at this stage. However, after a slight variation in yellow filtration a further contrast excerpt revealed all six structure areas to be isolated in their equidensite of the first order (Fig. 3.13f). In this case during the copying of the equidensite of the second order, the upper structural unit became again merged to its surroundings (Fig. 3.13i). Nevertheless, it could be concluded that we might be dealing here with a centrally located unit symmetrically surrounded by five further units (Fig. 3.13k).

Contrast Analysis and Model Construction

As a prerequisite for stroboscopic integration analysis these contrasts had to be evaluated according to their possible symmetry. At the same time for rotation

---

Fig. 3.12 Image reconstruction of the inner virus component, using the linear stroboscopic integration technique. (a) Excerpts from equidensites of the first order with filtered relevant contrasts for the model construction. (b) Isolated, characteristic contrasts according to the model conception. (c) Linear stroboscopic integration analysis of the area indicated in (b). The shifting frequency according to 12 nm is resolved; the shifting frequencies according to 9.5 nm (*above*) and 14.5 nm (*below*) are unresolved. (d) Model constructed in accordance with (c). (e) Control of the model with the contrast excerpt of (b). (f) Reconstruction of the analyzed structure areas in the original picture.

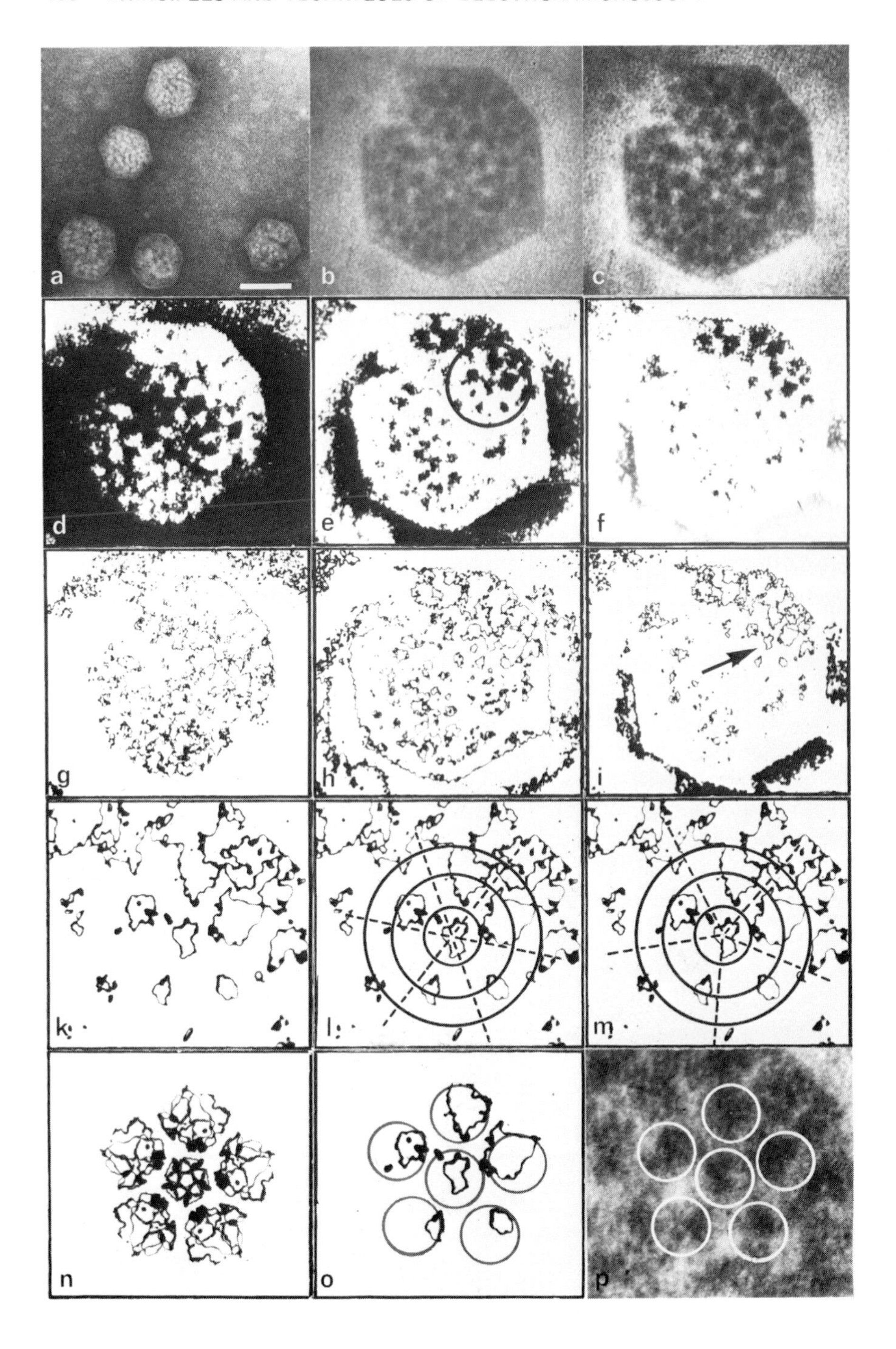
a
b
c
d
e
f
g
h
i
k
l
m
n
o
p

analysis the center axis of rotation had to be established. Both steps are closely connected. In order to obtain the highest possible correlation of the given contrast, its contribution to the rotation picture had to be evaluated. To get the strongest power spectrum a moveable transparent shablone (pattern) showing radially symmetrical rays and centric rings may be used. The axis of rotation can be fixed more accurately in this way rather than by visual alignment. The optimal point is found as soon as the given single contrasts are located in the middle of their respective sectors and situated at an averaged distance from the center.

From Figs. 3.13l and m an evaluation of the structures which are to be analyzed with the power spectrum pattern according to a sixfold and fivefold rotational symmetry may be made. This evaluation can easily be performed under the enlarger by the insertion of the pattern into the rotation apparatus. Figure 3.13m shows the best agreement with the fivefold symmetry pattern, presenting a stronger power spectrum than Fig. 3.13l, where the five contrast-rich areas have to be distributed in six integration areas. The rotational integration with a frequency of 5 is shown in Fig. 3.13n. All single contrasts are located in common integration areas and do not overlap, so that this symmetry is considered as being confirmed. In accordance with the model conception, structure units of an identical size are suspected, so that a model in accordance with Fig. 3.13n may be constructed.

Image Reconstruction

Using equidensites for the rotation analysis, the prints of the elements which are copied one upon the other can be easily recognized in the integrated picture. In the rotation picture (Fig. 3.13n) the contours of the single areas may be at least partially seen at the borders. In this way it is possible to obtain the exact location of the individual areas in the integrated picture and thus to orientate the model according to these contrasts. Because of the fact that the contrasts

---

Fig. 3.13 Image reconstruction of a capsid substructure by means of the rotation stroboscopic integration technique. (a) Image of the phagelike nitrobacter particle Nb_1, ×72,500. Scale represents 100 nm. (b) Enlarged demonstration of a particle with one-sided contrast. ×262,000. (c) Figure (b) with enlarged contrast volume serving as pattern for the contrast analysis. (d)–(f) Equidensites of the first order for middle and high densities. (g)–(i) Equidensites of the second order of (d), (e), and (f). (k) Enlarged area from (i) (*arrow*), serving as the pattern for the model construction. (l, m) Evaluation of contrast distribution in view of the rotationally symmetrical distribution by means of a simple transparent shablone, with (l) for a sixfold symmetry, and (m) for a fivefold symmetry. (n) Rotation analysis for a frequency of 5. Isolated integration areas demonstrate resolution. (o) Control of the model constructed from (n), on the basis of analyzed contrasts. (p) Reconstruction of the analyzed structure area.

represent the analyzed image, the model can be incorporated precisely into the original (Fig. 3.13p). The result of this integration analysis can thus be demonstrated graphically. At the same time the reconstruction may be critically viewed.

DISCUSSION

The Markham method may be criticized for its deficiency in allowing the control of the contribution of the contrast elements (Agrawal, 1964) because all contrasts present in the image contributed to the enhancement of the symmetrical units. By rotation those disordered "noise" signals are filtered which have an identical periodicity (Norman, 1966). Especially in the case of a distorted object the unspecific contrast contribution is increased (Crowther and Amos, 1971). Deficient contrast processing may also result in pseudosymmetries, which holds true for the inegration of too many small single units (Agrawal *et al.*, 1965). The correct selection of the rotation axis is of great importance in order to present contrast intensifications, which may obscure the object's symmetry.

Limitations of Integration Techniques in Image Analysis

To demonstrate the limits of these techniques, the analysis of ultrathin sections of tobacco mosaic virus (TMV) will be undertaken. Tobacco mosaic virus shows per turn of the basic helix 16 $\frac{1}{3}$ subunits (Klug and Caspar, 1960). In a section of a cytoplasmic inclusion showing TMV (Fig. 3.14a), one observes that the virus particles show a certain periodicity (Fig. 3.14b). One of these cross sections was

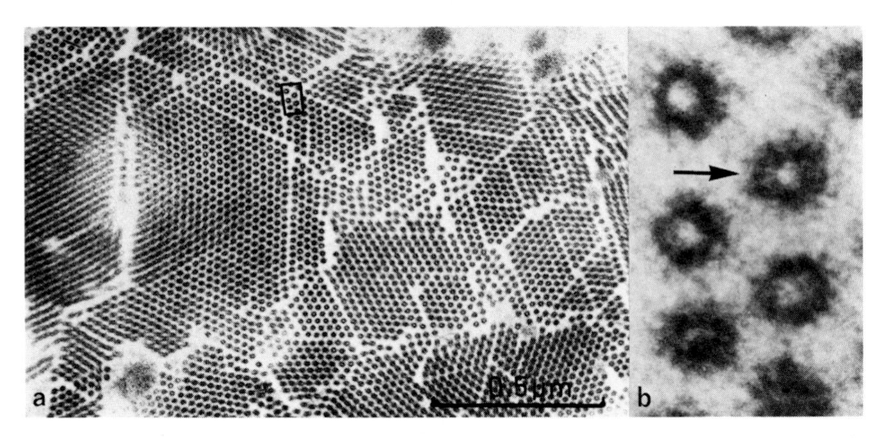

Fig. 3.14 Ultrathin section through a cytoplasmic inclusion showing tobacco mosaic virus. (a) Survey. ×67,000. Scale represents 0.5 μm. (b) Excerpt from (a)-insert. ×950,000.

analyzed in detail (Fig. 3.15a). The negative reveals limited contrast variations. The contrast-rich ringlike structure is surrounded by a weakly contrasted area shown in the equidensite of Fig. 3.15d. The image reveals only thin areas of middle contrast (Fig. 3.15b) and uneven areas of high contrast, being only partly separated (Fig. 3.15c).

Rotation analysis with equidensites of Figs. 3.15b and c with a frequency of 8 permits the observation of an apparent symmetry in Figs. 3.15e and f. It is also obvious in the half tone integration pictures taken from the original micrograph at a frequency of 16 (Fig. 3.15h) and seems weaker in the inner area of the rotation figure at a frequency of 15 (Fig. 3.15g). The stroboscopic integration of the contrast-rich parts of the image using the equidensite of Fig. 3.15c, demonstrates apparent resolution at a frequency of 15 (Fig. 3.15l).

To interpret these apparent symmetries, the basic idea of the integration technique must be considered. Through the special copying technique the single symmetrical units are given such an emphasis that in the integrated picture only "true symmetrical" features are enhanced. The rotation pictures shown in Figs. 3.15e, f, k, l, and m are given an increased contrast, allowed in order to recognize the single units. Overlapping is seldom seen in the apparently integrated areas at frequencies of 15 and 16 (Figs. 3.15l and m), so that the contrast distribution for both cases does not appear significant (Figs. 3.15n and o). The contrasts are distributed in such an uneven way that a reconstruction is impossible, and the model in Fig. 3.15p is not permissible.

Friedman (1970) as well as DeZoeten (1969) suggested contrast analysis in the integrated pictures, in order to improve the recognition of frequencies resolved, but such an analysis does not speak for the quality of integration. The pseudosymmetries in this case are based on an uncritical contrast evaluation before integration and wrong contrast enhancement during integration. These are considered the most frequent mistakes, leading to false results.

Comparison with Other Methods for Image Analysis

The contrast filtration is theoretically based on the fact that the objects are analyzed in the scattered contrast images, eliminating weaker phase contrast and unspecific noises. The filtration may be executed electronically as with the harmonious analysis as well as photographically in the same way. The electronic arithmetical evaluation of the most favorable rotation frequency and rotation axis may be compared with the evaluation by means of a simple power spectrum pattern. They are easily applied, and the results are sufficient. Both methods have the same mutual problem, namely the decision, at which stage the rotation method may not usefully be further applied. It ought to be possible to estimate a minimal contrast threshold at filtration and a minimal contrast cover of the resulting integrated areas. Both parameters could be controlled by means of the equidensite technique.

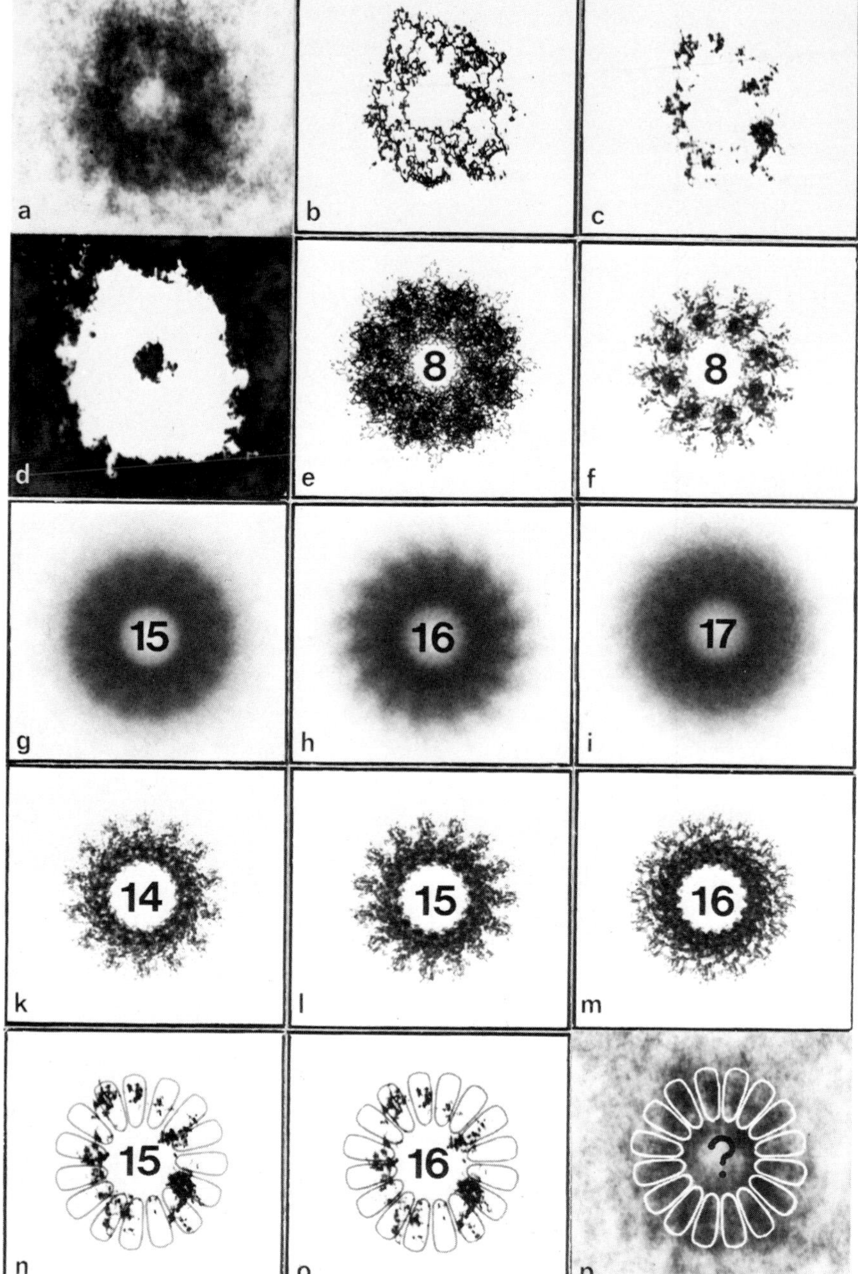
a
b
c
d
8
e
8
f
15
g
16
h
17
i
14
k
15
l
16
m
15
n
16
o
?
p

The advantage of the equidensite method may be seen not only in its simplicity but also in the possibility of picture reconstruction. The latter is also possible by means of an optical filtration method, in which case a filtration is also executed first, while certain diffraction points are selected for the optical reconstruction of the image according to a model concept. The only point is that no rotationally symmetrical objects can be analyzed. Subjectivity in the evaluation of the filtration and the resulting misinterpretations may be found in all three methods. Therefore, the respective techniques must be critically applied and resulting models must be proved by other independent means.

CONCLUDING REMARKS

The subjective interpretation of periodic structures of electron-micrographic images can be made considerably more objective by means of a very simple photographic method using equidensite negatives. In comparison with the use of an expended light-optical apparatus or the application of computer programming, considerably less effort and time are required, as only normal darkroom equipment and techniques are necessary. The method provides for the quick recognition of periodicity, the simple production of reconstruction models, and a direct, critical judgement of their agreement with the original image investigated.

The electron microscopy was carried out in the laboratory of Professor Doctor D. Düvel, Institut für Angewandte Botanik der Universität Hamburg, to whom I am most grateful.

References

Agfa-Gevaert (1972). *Agfacontour Professional in der Photographik.* Afga-Gevaert AG, Leverkusen.

Agfa-Gevaert (1974). *Agfacontour Professional in Wissenschaft und Technik.* Agfa-Gevaert AG, Leverkusen.

Agrawal, H. O. (1964). Electron microscope observations on the structure of the protein shell of turnip yellow mosaic virus. *Neth. J. Plant Path.* **70,** 195.

Fig. 3.15 Image analysis demonstrating the limits of the application of integration techniques. (a) Enlarged cross section of a virus from Fig. 3.14b. ×2,500,000. (b)–(d) Contrast excerpts by means of equidensites of the first order for different densities. (e,f) Rotation pictures of the equidensites (b) and (c) with a frequency of 8. A model construction does not seem possible. (g)–(i) Half tone rotation analysis with frequencies 15, 16, 17. Part (h) shows a contrast intensification for a frequency of 16. (k)–(m) Equidensite rotation using (c) for frequencies of 14, 15, 16. The apparent resolution of the frequency is not significant, because of insufficient contrast superpositioning. (n,o) Control of model conception of 15 (n) or 16 (o) subunits on the basis of analyzed contrasts. No sufficient contrast distribution. (p) Image reconstruction is not significant.

Agrawal, H. O., Kent, J. W., and MacKay, D. M. (1965). Rotation technique in electron microscopy of viruses. *Science* **148,** 638.

Bock, E., Düvel, D., and Peters, K.-R. (1974). Charakterisierung eines phagenähnlichen Partikels aus Zellen von Nitrobacter. I. Wirts-Partikelbeziehung und Isolierung. *Arch. Microbiol.* **97,** 115.

Caspar, D. L. D. (1966). An analog for negative staining. *J. Mol. Biol.* **15,** 365.

Caspar, D. L. D., and Klug, A. (1962). Physical principles in the construction of regular viruses. *Cold Spring Harbor Symp. Quant. Biol.* **27,** 1.

Crowther, R. A., and Amos, L. A. (1971). Harmonic analysis of electron microscope images with rotational symmetry. *J. Mol. Biol.* **60,** 123.

Crowther, R. A., Amos, L. A., Finch, J. T., DeRosier, D. J., and Klug, A. (1970). Three dimensional reconstructions of spherical viruses by Fourier synthesis from electron micrographs. *Nature* **266,** 421.

DeRosier, D. J., and Klug, A. (1972). Structure of the tubular variants of the head of bacteriophage T4 (polyheads). *J. Mol. Biol.* **65,** 469.

DeZoeten, G. A. (1969). A technique for interpretation of high resolution electron micrographs. *Neth. J. Plant Path.* **75,** 19.

Elliott, A., Lowy, J., and Squire, J. M. (1968). Convolution camera to reveal periodicities in electron micrographs. *Nature* **219,** 1224.

Finch, J. T., and Klug, A. (1966). Arrangement of protein subunits and the distribution of nucleic acid in turnip yellow mosaic virus. II. Electron microscopic studies. *J. Mol. Biol.* **15,** 344.

Finch, J. T., Leberman, R., Chang, Y.-S., and Klug, A. (1966). Rotational symmetry of the two turn disk aggregate of tobacco mosaic virus protein. *Nature* **212,** 349.

Finch, J. T., Klug, A., and Leberman, R. (1970). The structures of turnip crinkle and tomato bushy stunt viruses. II. The surface structure: Dimer clustering patterns. *J. Mol. Biol.* **50,** 215.

Friedman, M. H. (1970). A re-evaluation of the Markham rotation technique using model systems. *J. Ultrastruct. Res.* **32,** 226.

Fiskin, A. M., and Beer, M. (1968). Autocorrelation functions of noisy electron micrographs of stained polynucleotide chains. *Science* **159,** 1111.

Gachet, J., and Thiéry, J.-P. (1964). Application de la méthode de tirage photographique avec rotations ou translations à l'étude de macromolecules et de structures biologiques. *J. Microscopie* **3,** 253.

Galton, F. (1878). Composite portraits. *Nature* **18,** 97.

Klug, A. (1965). Structure of viruses of the papilloma-polyoma type. II. Comments on other work. *J. Mol. Biol.* **11,** 424.

Klug, A., and Berger, J. E. (1964). An optical method for the analysis of periodicities in electron micrographs, and some observations on the mechanism of negative staining. *J. Mol. Biol.* **10,** 565.

Klug, A., and Caspar, D. L. D. (1960). The structure of small viruses. *Adv. Virus Res.* **7,** 225.

Klug, A., and DeRosier, J. (1966). Optical filtering of electron micrographs: Reconstruction of one-sided images. *Nature* **212,** 29.

Klug, A., and Finch, T. J. (1965). Structure of viruses of the papilloma-polyoma type. I. Human art virus. *J. Mol. Biol.* **11,** 403.

Knudson, D. L. (1973). Rhabdoviruses. *J. Gen. Virol.* **20,** 105.

Markham, R., Frey, S., and Hills, G. J. (1963). Methods for the enhancement of image detail and accentuation of structure in electron microscopy. *Virology* **20,** 88.

Markham, R., Hitchborn, J. H., Hills, G. J., and Frey, S. (1964). The anatomy of the tobacco mosaic virus. *Virology* **22,** 342.

Norman, R. S. (1966). Rotation technique in radially symmetric electron micrographs: Mathematical analysis. *Science* **152,** 1238.

Peters, K.-R. (1974a). Rekonstruktion von Capsidstrukturen isometrischer Viren mit einer Äquidensiten-Rotationsmethode. *Mikroskopie* **30,** 270.

Peters, K.-R. (1974b). Charakterisierung eines phagenähnlichen Partikels aus Zellen von Nitrobacter. II. Struktur und Größe. *Arch. Microbiol.* **97,** 129.

Peters, K.-R. (1976). Orchid viruses: A new rhabdovirus in Laelia red leafspots (in press).

4. G-BANDING OF CHROMOSOMES

F. Ruzicka

Histologisch-Embryologisches Institut der Universität, Wien, Austria

INTRODUCTION

Several methods which produce banding patterns within chromosomes are in use. G-banding enables chromosomes (including those of the human C-, D-, F-, and G-group chromosomes) to be differentiated. Caspersson *et al.* (1970) demonstrated the differential staining of chromosomes by quinacrine fluorescence analysis. Only those methods involving procedures carried out on chromosome preparations and subsequent staining with Giemsa (G-banding) will be described here. The banding technique is derived from *in situ* hybridization methods. Single strands of DNA hybridize with DNA of a complementary base sequence. After dissociation of double-stranded DNA (e.g., with NaOH treatment) the preparations are incubated with tritium-labeled single-stranded DNA. After hybridization the labeled nucleic acid molecules can be localized by autoradiography. This method is not very sensitive, and only repetitive DNA sequences can be studied. Arrighi and Hsu (1971) were the first to use NaOH-treatment followed by 2 × SSC (0.3 M NaCl and 0.03 M trisodium citrate, pH 7.0) and Giemsa staining to produce bands.

Banding patterns of chromosomes can also be demonstrated with HCl (Greilhuber, 1973), enzyme treatment, and Giemsa staining (Dutrillaux *et al.*, 1971). Pretreatments with potassium permanganate (Utakoji, 1972), sodium dodecyl sulfate (Yosida and Sagai, 1972), strong bases, alkaline salt solution (Kato and Moriwaki, 1972), or surface active agents also produce banding patterns in chromosomes. A method involving the electron microscope for the study of chromosome preparation has been developed in our laboratory (Ruzicka, 1971). Using this method one is able to study the same preparation with the light microscope, and the conventional and scanning electron microscopes. Further studies have revealed that the various G-banding methods yield similar results with some differences in the quality of the ultrastructure of the banding patterns (Ruzicka, 1973; Ruzicka and Schwarzacher, 1974; Burkholder in this volume).

Special methods could be developed to demonstrate premature chromosome condensation. The G1-, S-, and G2-chromosomes show differentially stained regions along the entire length of the chromatids after the G-banding procedure. It should be pointed out that on these elongated chromosomes, unlike on metaphase chromosomes, the differently stained regions can be clearly observed.

SPECIMEN PREPARATION

Preparation of Chromosomes

Culture. Peripheral blood cells of normal male and female individuals were cultured according to the method of Moorhead *et al.* (1960). Fibroblast cultures were obtained from fascia biopsy material of people with normal chromosomes. The medium consisted of 20% human umbilical cord serum, 50% Medium 199 and 30% Gey's solution with embryonic extract and antibiotics. HeLa-cells were cultured and synchronized (Rao and Johnson, 1972), and cell fusion and premature chromosome condensation with Sendai virus was performed (Harris *et al.*, 1966; Okada, 1969). Chromosome preparations were made after colcemide treatment (0.04% for 3 hr) using the standard air-drying method.

Treatment of Chromosome Spreads

Incubation in Buffer Only (Sumner et al., *1971).* Conventionally air dried chromosome preparations are incubated in 2 × SSC (0.3 M NaCl and 0.03 M trisodium citrate, pH 7.0) for 1 hr at 60°C. The slides are then rinsed in demineralized water and stained in Gurr's Giemsa R 60 solution (1 ml Giemsa stock solution and 50 ml buffer made with Gurr's buffer tablets, pH 6.8) for 60–90 min. A modified ASG-technique is used in our laboratory. Air-dried chromosome preparations are incubated in SSC for 36–72 hr at 60°C. This step is followed by Giemsa staining for 20–30 min. The Giemsa stain is prepared by mixing 1 part Merck Giemsa solution and 9 parts phosphate buffer (1.14 gm $Na_2HPO_4 \cdot 2H_2O$, 0.49 gm KH_2PO_4 in 1 liter distilled water, pH 7.2).

Sodium Hydroxide Treatment Followed by Incubation in Buffer (Schnedl, 1971). Colchicinized metaphase cells are incubated for 10 min in a 4 × diluted Gey's solution for hypotonic treatment, fixed with a mixture of methanol

(3 parts) and glacial acetic acid (1 part), dropped onto cooled glass slides, and air-dried. The chromosome preparations are treated with 0.002 N NaOH for 90 sec. The preparations are transferred in turn to three 70% ethanol solutions, one 96% ethanol solution, absolute ethanol, and air-dried. They are then incubated in phosphate buffer (pH 6.8) at 59°C for at least 15 to 24 hr. The chromosomes are stained for 20 min in Giemsa solution (1 part Merck Giemsa solution and 9 parts $\frac{1}{15}$ M phosphate buffer, pH 6.8), washed in water, and air-dried.

The Giemsa-9-technique of Patil et al. *(1971).* Cells treated hypotonically with 0.075 M KCl are fixed, spread on cooled wet slides, and dried on a warmed plate at 60°C for 90–120 sec. The chromosomes are stained with Giemsa solution (2 ml Harleco Giemsa blood stain original Azure blend type, 2 ml 0.14 M Na_2HPO_4, and 96 ml distilled water or Gurr's Giemsa R 66 at pH 9.0). The duration of staining is critical, and an optimal duration between one and 30 min should be selected and used. After rinsing in distilled water the preparations are air-dried.

The Reverse Band Method of Dutrillaux and Lejeune (1971). The cells receive hypotonic treatment (15 parts of a serum diluted 1:6 with distilled water, 1 part 3.39% $MgCl_2$ with a final hyaluronidase concentration of 2.5 U/ml) for 15 min, after which they are fixed, first for 35 min in a solution containing 3 parts chloroform, 1 part glacial acetic acid and 6 parts 100% ethanol and then for 15 min in an ethanol/glacial acetic acid (3:1) solution, and air-dried. The chromosome preparations are incubated at 87°C in 20 mM phosphate buffer (pH 6.5) for 10–12 min and rinsed in tap water for a few seconds. They are stained in Giemsa (4 parts Merck Giemsa stock solution, 4 parts phosphate buffer at pH 6.7, and 92 parts distilled water) for 10 min and rinsed in distilled water.

Method of Yunis et al. *(1971).* Conventionally air-dried chromosomes are incubated in $\frac{1}{15}$ M phosphate buffer (pH 6.8) at 85°–100°C for 5–10 min. The slides are then transferred to $\frac{1}{15}$ M phosphate buffer (pH 6.8) at 65°C and incubated for at least 10 hr. After cooling, staining of the slides with the normal Giemsa solution is performed. It is possible to demonstrate centromeric heterochromatin (C-bands) with this method.

Method of Gagné et al. *(1971).* Air dried 0.014 N chromosome preparations are treated for 15–20 sec with 0.014 N NaOH, transferred in turn to three 70% ethanol solutions, and one 95% ethanol solution, and air-dried again. The specimens are incubated in 6 × SSC at 66°C for 18 hr, washed twice in 6 × SSC, transferred in turn to 70% and 95% ethanol solutions, and air-dried. Finally, Giemsa staining is performed (5 ml Harleco Giemsa solution, 1.5 ml citric acid at pH 6.9, 1.5 ml methanol, and 50 ml distilled water) for 30 min followed by rinsing in distilled water and air-drying.

Method of Drets and Shaw (1971). Since 1% sodium citrate is used as the hypotonic solution, these authors use flame-dried chromosome preparations. The chromosome preparations are immersed in a solution of 0.07 M NaOH in 0.112 M NaCl for 30 sec, and dipped for 10 min each time in three solutions containing 12 × SSC (pH 7.0). They are incubated in 12 × SSC at 65°C for 60–72 hr, after which they are rinsed three times in 70% ethanol solution for a total of 10 min. After rinsing in 95% ethanol the slides are air-dried. Finally, the slides are stained in Giemsa for 5 min (10 ml W. H. Curtin Giemsa stock solution and 100 ml 0.01′ M phosphate buffer, pH 7.0). Modifications of the NaOH treatment have been developed to demonstrate the secondary constriction of chromosome No. 9.

Method of Bobrow et al. *(1972).* Three- to four-day-old chromosome preparations are stained in freshly prepared Giemsa solution (2% Gurr Giemsa solution adjusted to pH 11 with NaOH). The slides are rinsed in distilled water and air-dried.

Method of Gagné and Laberge (1972). After fixation in methanol/glacial acetic acid mixture (3 : 1), the cells are suspended in 45% acetic acid for 3–4 min, dropped onto cooled wet slides, and air-dried. The chromosome preparations are stained for 5 min in a 2% Giemsa solution (Harleco azure blend type) in 0.1% Na_2HPO_4 adjusted to pH 11.6 with NaOH). The slides are washed in water and air-dried.

Enzymatic Treatment Followed by Giemsa Staining (Dutrillaux et al., *1971).* Chromosomes are prepared as for the reverse band method (Dutrillaux and Lejeune, 1971). The preparations are treated with a 0.005% pronase solution at 37°C for 3–6 min. The slides are stained with the normal Giemsa solution.

Method of Seabright (1971). The chromosome preparations are air-dried, as flame-drying is less suitable. They are treated with 0.25% trypsin (Difco) in an isotonic salt solution at 37°C for 10–60 sec, followed by staining for 3–4 min with Leishman stain (Gurr) diluted 5 times with phosphate buffer (pH 6.8).

Method of Wang and Fedoroff (1972). Chromosomes are prepared with the flame-drying technique. The slides are treated with 0.025–0.05% trypsin in Ca- and Mg-ion free-balanced salt solution (pH 7.0) at 25°–30°C for 10–15 min. The chromosome preparations are successively rinsed in 70%, 96%, and absolute ethanol and dried, after which they are stained with Giemsa (10% Fischer

Scientific Giemsa solution in 0.01 M phosphate buffer, pH 7.0), rinsed twice in distilled water, and air-dried.

Combined Treatment Followed by Giemsa Staining (Arrighi and Hsu, 1971). Centromeric heterochromatin (C-bands) can be demonstrated with this method. Flame- or air-dried chromosome preparations are treated with pancreatic RNase (100 μg/ml in 2 × SSC) at 37°C in a damp chamber for 60 min. The solution should be heated in a boiling water bath for 5–10 min before use to inactivate the DNase present in the commercially available RNase. The preparations are rinsed three times in 2 × SSC, once in 70% ethanol, and once in 95% ethanol, and air-dried. The chromosomes are then treated with 0.07 N NaOH for 2 min, washed three times in 70% ethanol and once in 95% ethanol, and air-dried. They are incubated in 2 × SSC or 6 × SSC at 65°C for 12–24 hr. After rinsing in 70% and 95% ethanol the chromosomes are stained in Giemsa solution (10% W. H. Curtin Giemsa solution in 0.01 M phosphate buffer, pH 7.0) for 15–30 min.

In addition to the methods presented above, several modifications have been published involving the use of salt solutions with alkaline pH, strong bases, surface active agents (Kato and Moriwaki, 1972), sodium dodecyl sulfate solution (Yosida and Sagai, 1972), or potassium permanganate (Utakoji, 1972). All these techniques reveal banding patterns, but their actual mechanisms are not known.

HCl Treatment Followed by Aceto-carmine Staining (Greilhuber, 1973). This method allows a differential heterochromatin staining of plant chromosomes. The preparations are fixed in ethanol/acetic acid (3:1) for 15 min to 24 hr and rinsed in distilled water for 5 min. They are then incubated in 0.2 N HCl at 60°C for 30 min (or 70°C for 10 min or 80°C for 5 min), dipped in water, and stained with cool aceto-carmine for several minutes to many hours. There is no danger of overstaining.

MICROSCOPY

Whole Mount Conventional and Scanning Transmission Electron Microscopy

1. The chromosome preparations pretreated with one of the already mentioned banding techniques are rinsed twice in a chloroform/ether (1:1) mixture and air-dried.
2. The slides are coated with carbon having a thickness of 30 nm. The specimen should be rotated in a vacuum which is higher than 10^{-5} torr, and a "Meisner" trap cooled with nitrogen must be used.
3. With the aid of a light microscope, suitable mitotic figures and interphase nuclei are selected and marked with a diamond marker (Reichert, Vienna).

4. The carbon films supporting the chromosomes are floated off the slide with 0.3 N hydroflouric acid.
5. With a glass rod the carbon films are transferred onto a clean water surface.
6. The round carbon films are fixed onto single-hole grids with holes of 0.6 mm.

These preparations can be studied with both conventional and scanning transmission electron microscopes.

Whole Mount Scanning Electron Microscopy

For scanning electron microscopy the surface of the glass slides are scored with a diamond glass cutter before being broken into small pieces. The slides are placed onto specimen mounts using colloidal graphite or silver as conducting cement. One can use any of the previously described banding techniques for pretreatment.

Perspective Representation (Pseudo-three-dimensional)

To get a perspective representation similar to that obtained with the scanning electron microscope, the specimen is tilted at an angle of 30°–60° and photographed with a conventional transmission electron microscope. The negative is copied onto a film, and the picture obtained is enlarged. This picture is a negative of that seen on the flourescent screen. This method yields a similar result to that obtained with the scanning electron microscope.

EQUIDENSITOMETRY

Equidensitometry allows the densitometric interpretation of pictures. A yellow filter (No. 99 Agfa enlarger Varioscope 60) is used. The light is directed at the plane of the photo at 100 lux, and four different exposures, for 5.5, 14, 32, and 75 sec, are made. In our laboratory Agfa contour film is used and developed as suggested by Agfa. These photographs are rated as first-order equidensities. In order to get all four equidensity photographs onto one picture the single black-and-white photographs are rephotographed together, using one colored positive of Agfa-positive-M film. Enlarged color prints from this negative are made.

OBSERVATIONS

Ultrastructure of Chromosomes after Hypotonic Treatment and Fixation in Methanol (Ethanol) and Glacial Acetic Acid

It is often possible, especially after Giemsa staining, to observe the so-called primary or major coils under the light microscope. Using the electron microscope, these primary coils appear as electron-dense structures. The coiled chromatids of such metaphase chromosomes enable the characterization of the chromosomes by establishing a mean relative pitch and a possibility for counting the coils. Ohnuki (1965) published a method making primary coils visible by the light microscope. In a subsequent publication (Ohnuki, 1968) he found that prophase chromosomes have more coils than metaphase chromosomes. Primary coils can easily be counted with the electron microscope (Fig. 4.1).

Moreover, it is possible to work out a mean relative pitch. This can be calculated in the same way as the mean relative length of a chromosome. It is expressed as a percentage of the sum of the lengths of the 22 autosomes (and the X-chromosome). The number of coils in prophase chromosomes is significantly higher than in metaphase chromosomes, and values in prophase of up to twice those in metaphase have been observed. The number of coils was found to be inversely proportional to the diameter of the chromatid. Measurements of the pitch of a group of 10 metaphase chromosomes gave values of 0.36–0.58 μm. No significant difference could be demonstrated between these findings and those from prophase chromosomes (Student t-test). Since pitch is influenced by the method of preparation, it is useful to calculate a mean relative pitch.

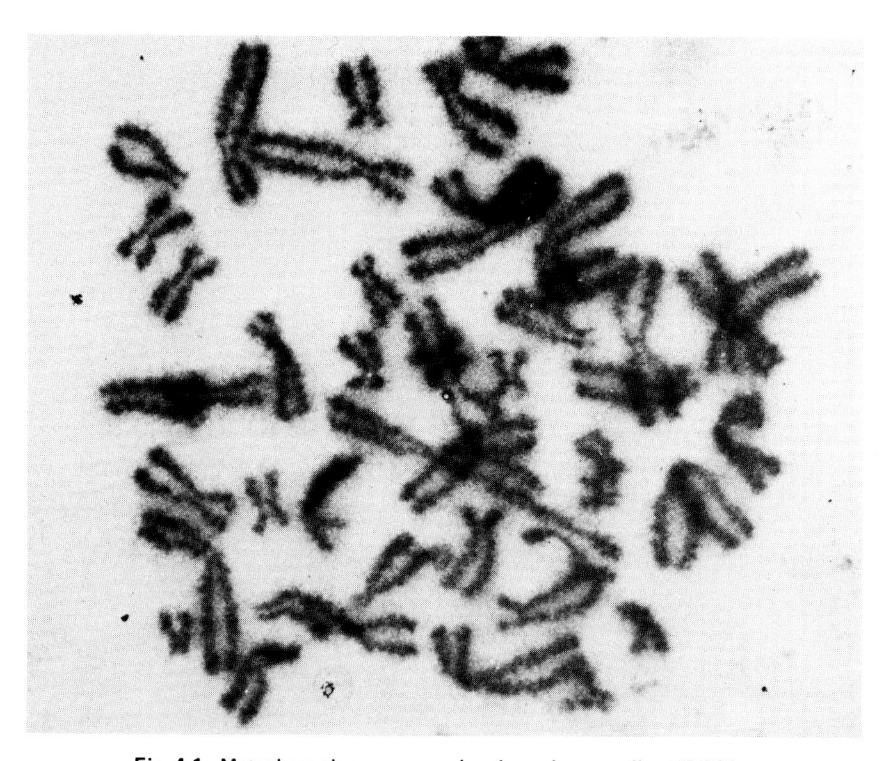

Fig. 4.1 Metaphase chromosomes showing primary coils. ×2,500.

Two further observations should be mentioned here. After culturing cells with the alkylating and antimetabolic cytostatic drug peptichemio, an unwinding of the primary coils revealed the presence of secondary coils. It is interesting to note that Actinomycin D produces banding patterns identical to those produced by postfixation techniques (Shafer, 1973). This finding lends support to the special chromosome theory that chromosomes are composed of coils with differing proportions and distributions of adenine/thymine and guanine/cytosine base pairs. The second observation was that wet preparations of Giemsa-stained chromosomes, enclosed between two layers of carbon, have a different ultrastructure under the electron beam than dry preparations have, in that thick fibrils of mean diameter of 100 nm are present. At higher magnifications it appeared that 25-nm fibrils are arranged parallel within 100-nm units.

Ultrastructure of Chromosomes after Incubation in Buffer

A swelling process seems to take place in the chromosomal structures. Chromatids become fused and coils can no longer be observed. After Giemsa staining, banding patterns can be seen, the dark areas of which are composed of closely

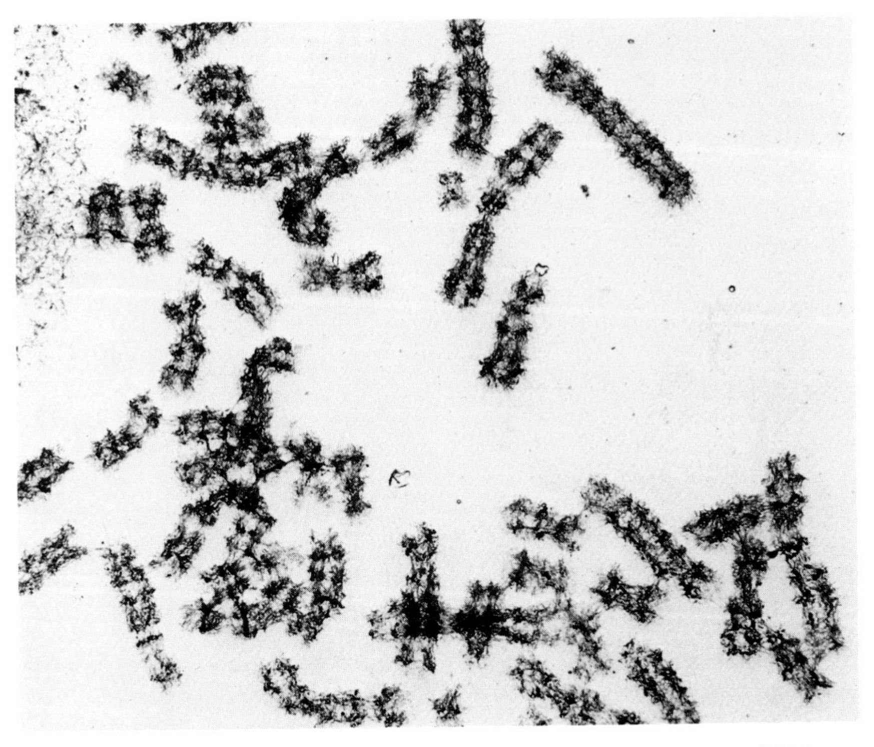

Fig. 4.2 Metaphase chromosomes prepared by using the ASG-technique. ×2,500.

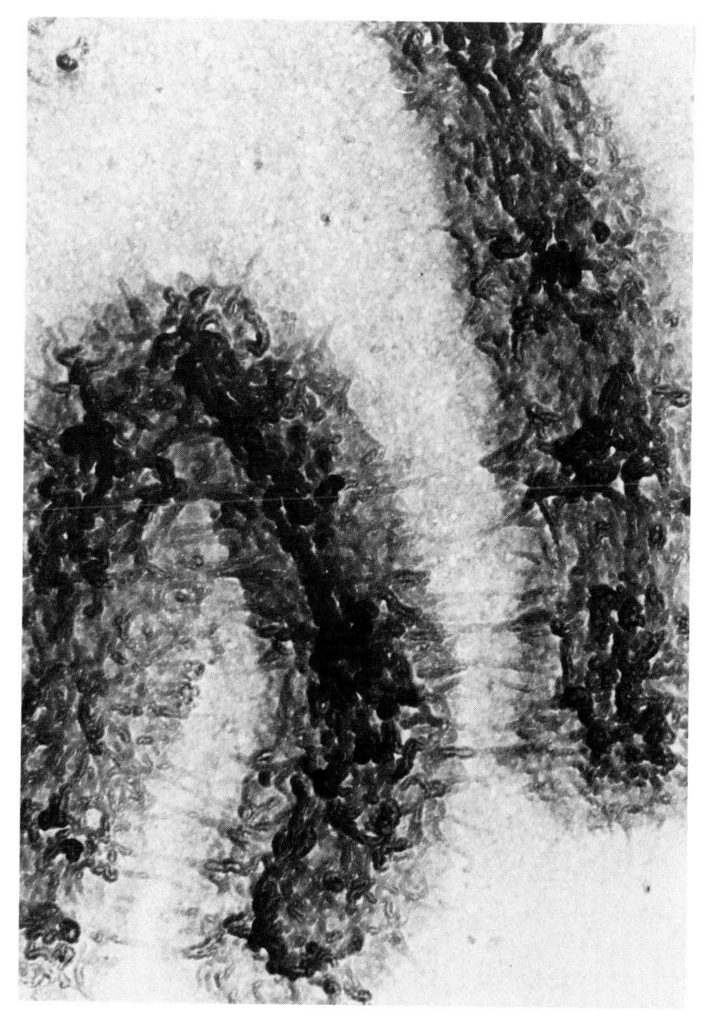

Fig. 4.3 Metaphase chromosomes prepared by using the method of Schnedl (1971). ×15,000.

folded fibrils. In unstained regions similar fibrils can be seen, but not so tightly folded as in the G-band regions (Figs. 4.2 and 4.3).

Ultrastucture of Chromosomes after NaOH Treatment

The ultrastructure of chromosomes is disturbed after treatment with NaOH. No fibrils and coils can be seen. If the NaOH treatment is not followed by an incubation in heated buffer solution and Giemsa staining, no banding pattern can be observed.

Ultrastructure of Chromosomes after Enzymatic Treatment

Using the light microscope, chromosomes treated with pure trypsin without NaOH often appear swollen with indistinct bands. When they are examined by light and electron microscopy, bands can be demonstrated at identical places. As proteolytic enzymes probably dissolve structural proteins, thereby altering the fine structure of the chromosome, it is difficult to achieve reproducible results. Thus, only relative positions of the bands are indicated. With the electron microscope, electron-dense areas are only partially visible in the region of the G-bands.

Difficulties in reproducibility have led to a series of modifications of the original technique. Cervenka *et al.* (1973) have examined trypsin-treated chromosomes with Normarski contrast microscopy and SEM. After long periods of treatment with trypsin, they found that G-bands dissolve, whereas C-bands are resistant. Pawlowitzki *et al.* (1968) described a method using proteolytic enzymes. In the published pictures primary coils and fibrilar structures could be seen, whereas banding patterns were absent.

Comparison of Light and Electron Microscope Studies

We were interested in examining chromosomes after G-band staining with a combination of both light and electron microscope techniques. The G-banding technique of Schnedl (1971) was applied, and a clear band structure could be observed with both the light and electron microscopes (Fig. 4.4). At a higher magnification, fibrils with a mean diameter of 30 nm were apparent. These fibrils form loops and are folded. The diameter of thick fibrils can be up to 100 nm. With perspective representation one can see these structures easily (Fig. 4.5).

Ultrastructure of Chromosomes after HCl Treatment

After hydrolysis in HCl, coils can be observed. Satisfactory results depend upon the concentration and the duration and temperature of incubation. An incubation of 5 min in 1 N HCl at 60°C is satisfactory; however after 10 min incubation in these conditions, the coils show signs of breaking. After 15 min no coils can be seen, and a granular structure is observed. The same results can be obtained using Orcein staining with HCl.

Interphase Nuclei

In interphase, nuclear fibrils of the same dimensions and fine structure as those of metaphase chromosomes are seen (Fig. 4.6). They are arranged in irregular loops and folds. There are regions in which the fibrils appear to be more folded, more tightly packed, and thicker. Such regions correspond to the G-bands of

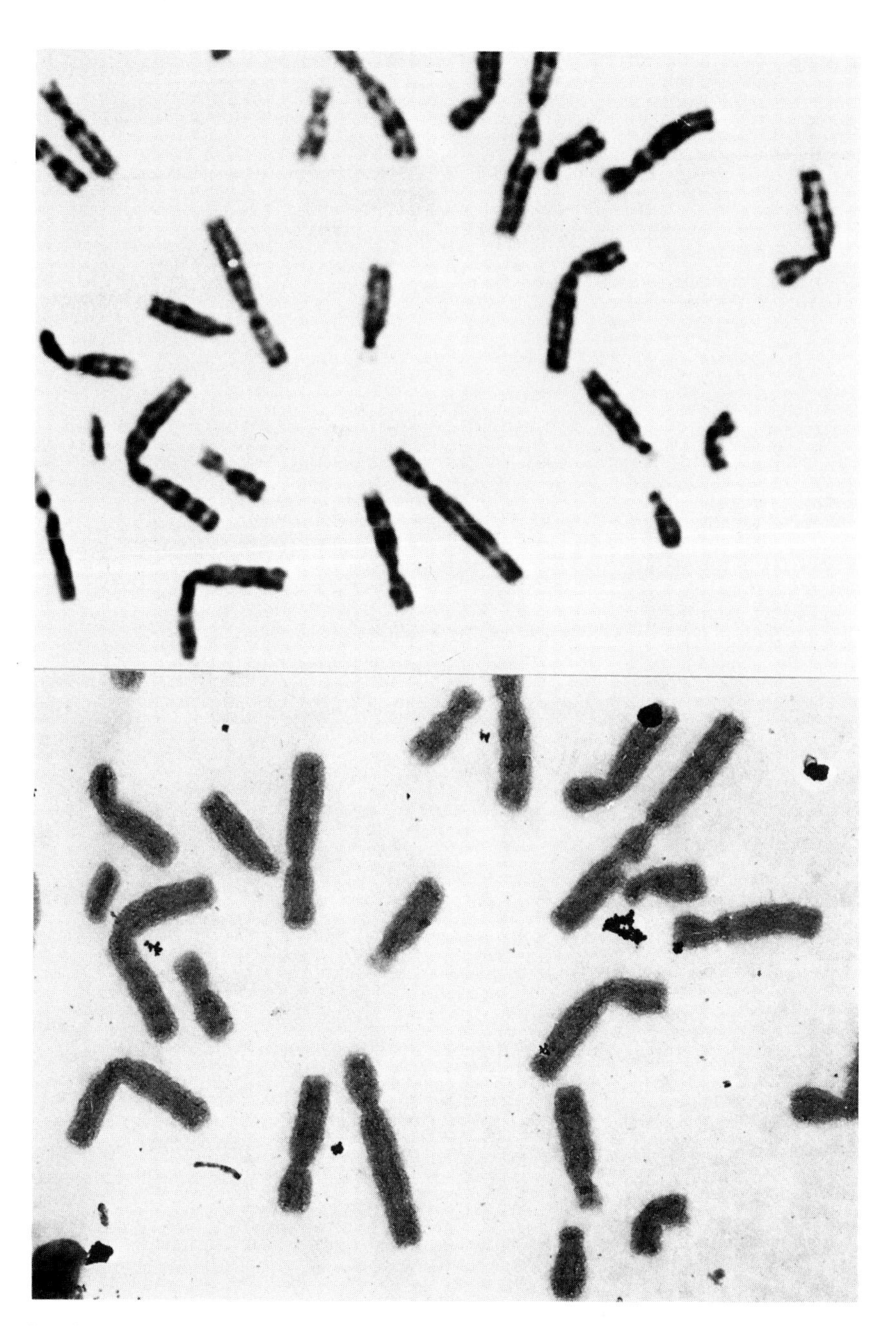

Fig. 4.4 Metaphase chromosomes prepared by using the method of Schnedl (1971). ×2,000. (a) Photomicrograph with the light microscope. (b) Micrograph with the electron microscope.

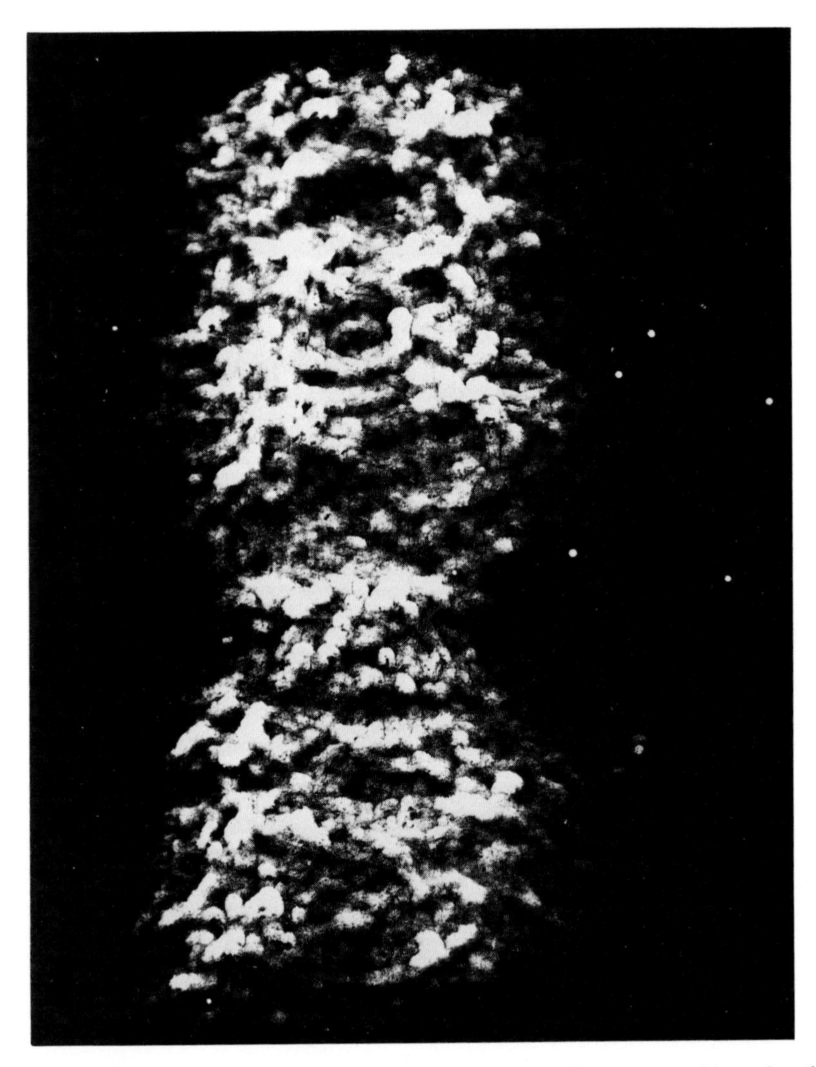

Fig. 4.5 Perspective representation of chromosome preparations prepared by using the method of Schnedl (1971). ×20,000.

mitotic chromosomes. Fibrils seem to stick through the border of the nucleus in the same way as those of metaphase chromosomes. Higher magnifications clearly show that these fibrils are loops. Bridges of fibrils are frequently seen between neighboring nuclei. They appear very similar to the connecting fibrils between adjacently-lying chromosomes or between sister chromatids (Fig. 4.7). These findings clearly indicate that such interconnecting fibrils are artifacts. It seems that the fibrils are loosened during the preparation, probably by the hypotonic treatment, forming loops which project from the nucleus or from the

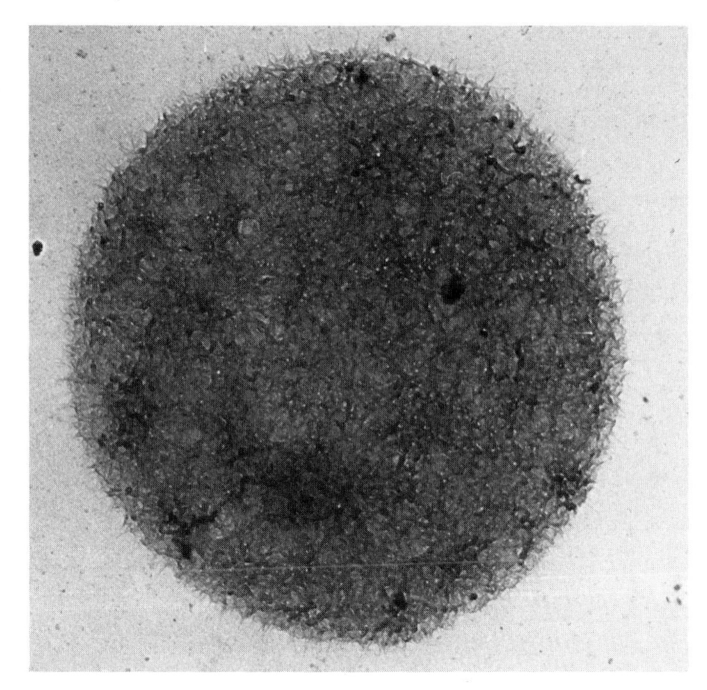

Fig. 4.6 Interphase nuclei prepared by using the method of Schnedl (1971). ×4,000.

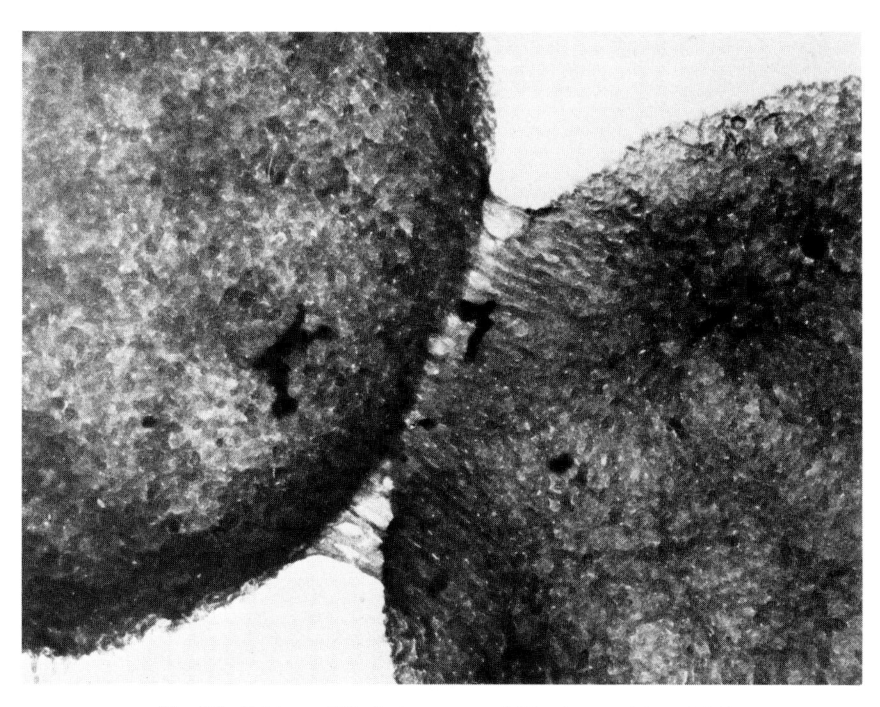

Fig. 4.7 Bridges of fibrils between neighboring nuclei. ×6,000.

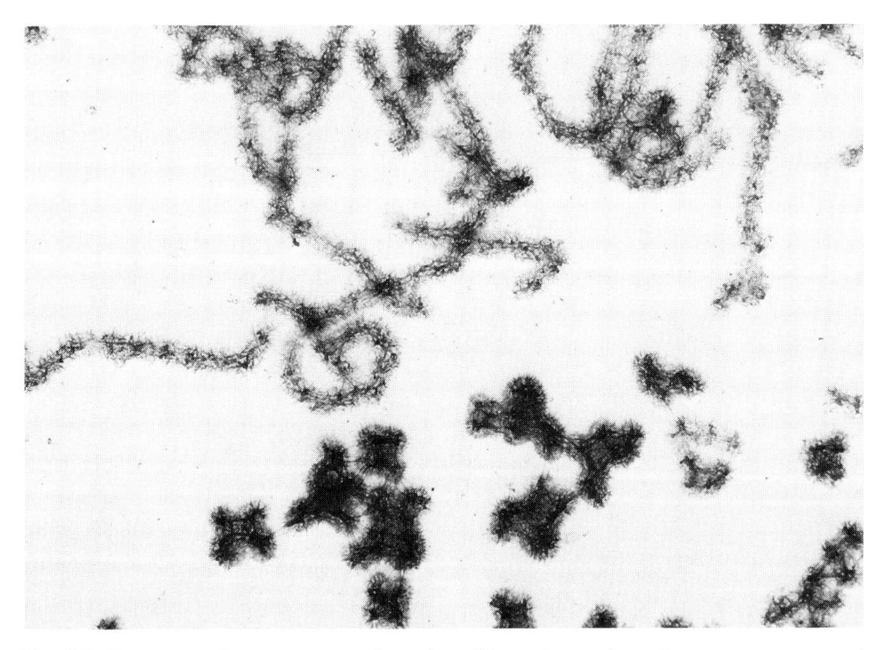

Fig. 4.8 Premature chromosome condensation; G1- and metaphase chromosomes prepared by using the method of Schnedl (1971). ×3,000.

chromatids. When such loops are sufficiently close, they bind together forming interconnecting bridges.

Premature Chromosome Condensation

The mechanism of premature chromosome condensation is well documented (Rao and Johnson, 1970, 1971, 1972). After G-band staining, G1-, S-, and G2-chromosomes show fibrils of the same dimensions and fine structure as those of metaphase chromosomes (Figs. 4.8–4.10). They are arranged in coils and folds quite similar to the arrangement seen in prophase chromosomes (Fig. 4.11). G1-chromosomes have one and G2-chromosomes have two chromatids composed of regions in which fibrils are close together, and others in which the fibrils are more dispersed representing the G-band and interband areas, respectively. S-chromosomes are particularly interesting because instead of the expected regions of DNA replication with the electron microscope being observed, folded fibrils with the same fine structure as in metaphase chromosomes were seen.

CONCLUDING REMARKS

As already described previously (Ruzicka, 1973), fibrils of 30–100 nm in diameter become visible after treatment of the chromosomes with the G-banding

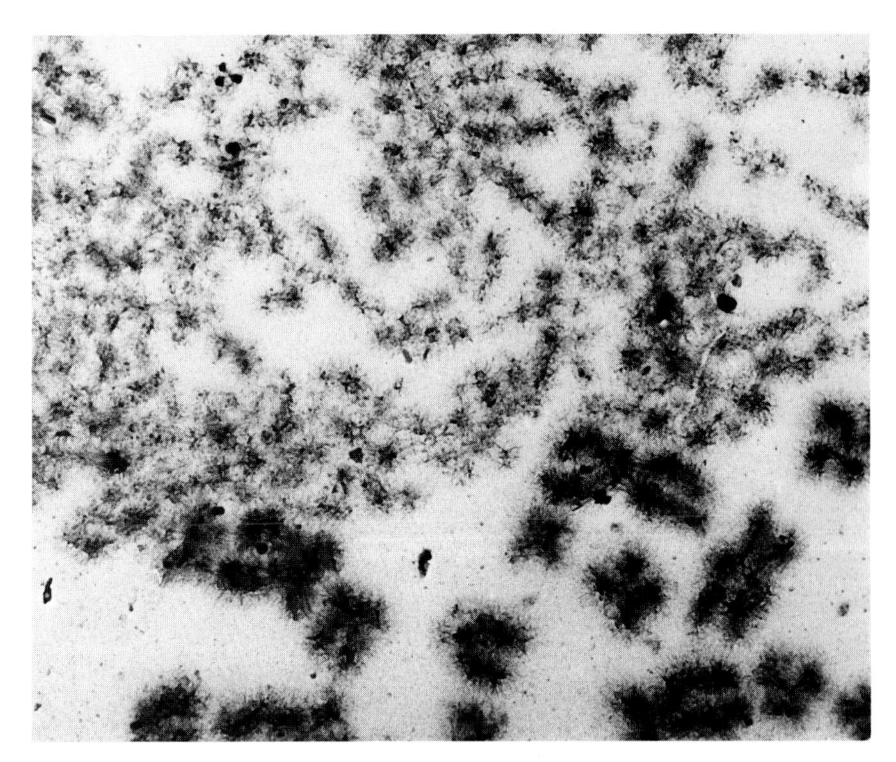

Fig. 4.9 Premature chromosome condensation showing S- and metaphase chromosomes. The specimens were prepared according to the method of Schnedl (1971). ×3,000.

technique. Similar observations were published by Comings *et al.* (1973). Fibrils have been found as main constituents of chromosomes with a variety of preparations, although the diameter of the fibrils reported varies considerably (Kaufmann *et al.*, 1960; Ris, 1961; Wolfe, 1965; Wettstein and Sotelo, 1965; Gall, 1966; Schwarzacher and Schnedl, 1967; Lampert and Lampert, 1970; Du Praw, 1970; Ruzicka, 1971; Comings and Okada, 1970).

A comparison between the light and electron microscope pictures of G-band stained chromosome spreads reveals two main differences. With the electron microscope, bands are smaller than with the light microscope, and they appear to have a halo of fibrils. The larger G-bands observed with the light microscope are caused by a diffraction phenomenon, and the halo of fibrils observed with the electron microscope is caused by a spreading effect.

Very often a preparation artifact makes the center of the fibrils appear less electron-dense. If one uses carbon shadowing, more carbon is condensed on steep surfaces than one would expect because of the lightness of the carbon atoms and their dispersion in the remaining gas atmosphere.

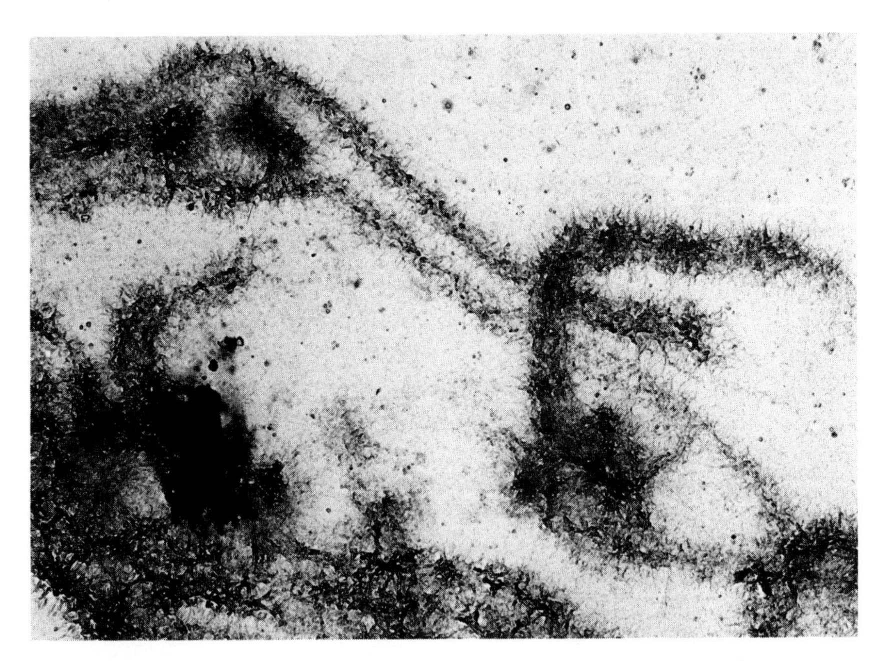

Fig. 4.10 Premature chromosome condensation showing G2- and metaphase chromosomes. The specimens were prepared according to the method of Schnedl (1971). ×3,000.

With the electron microscope the G-band regions are darker than the interband regions. The closer coiling of the fibrils and the thickness of the chromatids is responsible for this effect. Thickenings of the chromatids were also reported by Bahr (1972) with electron microscope preparations of water surface spread and dried chromosomes. The electron microscope reveals other bands which are finer (EM-bands) and seem to correspond to primary coils (Ruzicka, 1973). Evans (1973) also described fibrils and thickenings of chromatids in G-band regions with electron microscope preparations of chromosome replicas. A similar surface topography after ASG banding was reported by Gormley and Ross (1972). Light microscopy findings also show that the G-bands reflect a stronger chromatin condensation (Ross and Gormley, 1973; McKay, 1973; Yunis and Sanchez, 1973).

McKay (1973) put forward the hypothesis that G-bands may develop from a differential rearrangement of the chromosomal structure after disruption by the special pretreatment which removes divalent cations. The rearrangement of the chromatin structure could be induced by the Giemsa stain acting in a manner similar to divalent cations. The G-bands are considered to be regions of higher chromatin density since they are more resistant to these treatments. The micrographs published by Gormley and Ross (1972), Evans (1973), Bahr (1972), and

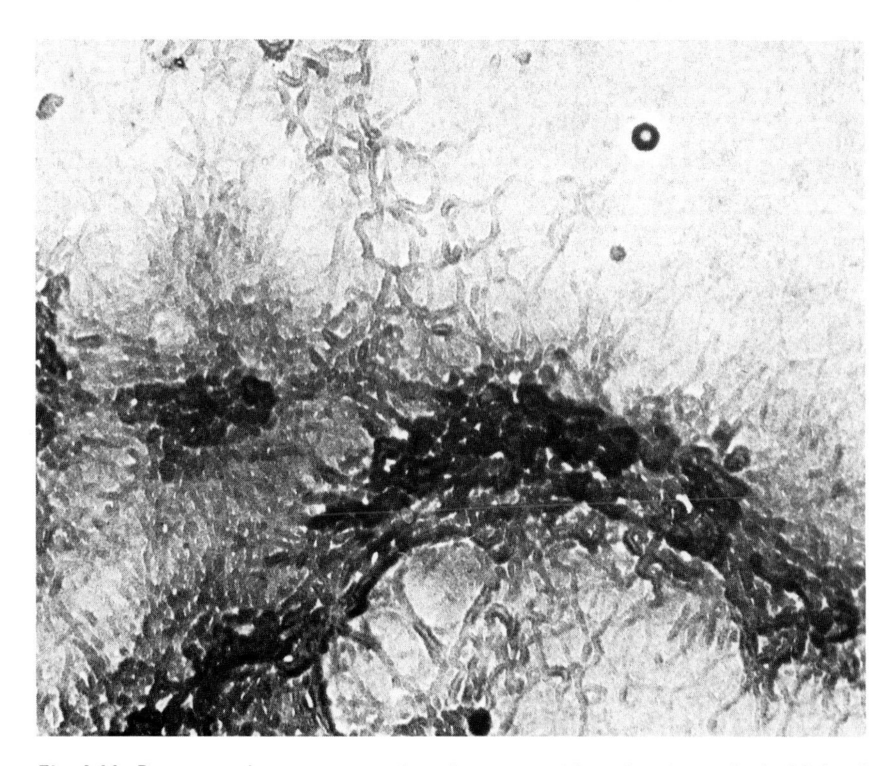

Fig. 4.11 Premature chromosome condensation prepared by using the method of Schnedl (1971). ×15,000.

Comings *et al.* (1973), as well as our own clearly indicate that the fibrillar folds accumulate in the G-band regions.

References

Arrighi, F. E., and Hsu, T. C. (1971). Localization of heterochromatin in human chromosomes. *Cytogenetics* **10**, 81.

Bahr, G. F. (1972). The ultrastructural basis of chromosomal banding patterns observed after renaturation and Giemsa staining. *10th Ann. Somat. Cell Genet. Conf.* (Aspen, Colorado).

Bobrow, M., Madan, K., and Pearson, P. L. (1972). Staining of some specific regions of human chromosomes, particularly the secondary constriction of No. 9. *Nature, New Biol.* **238**, 122.

Caspersson, T., Zech, L., Johannson, C., and Modest, E. J. (1970). Identification of human chromosomes by DNA-binding fluorescent agents. *Chromosoma* **30**, 215.

Cervenka, J., Thorn Hattie, L., and Gorlin, R. J. (1973). A structural basis of banding pattern of human chromosomes. *Cytogenet. Cell Genet.* **12**, 81.

Comings, D. E., Avelino, E., Okada, T. A., and Wyandt, H. E. (1973). The mechanism of C- and G-banding of chromosomes. *Exp. Cell Res.* **77**, 469.

Comings, D. E., and Okada, T. A. (1970). Electron microscopy of chromosomes. In:

Perspectives in Cytogenetics (Wright, S. W., Chandall, B. F., and Boyer, L., eds.), p. 223. Charles C. Thomas, Springfield, Illinois.

Drets, M. E., and Shaw, M. W. (1971). Specific banding patterns of human chromosomes. *Proc. Nat. Acad. Sci.* **68,** 2073.

Du Praw, E. J. (1970). *DNA and Chromosomes.* Holt, Rinehart and Winston, Inc., New York.

Dutrillaux, B., and Lejeune, J. (1971). Sur une nouvelle technique d'analyse du caryotype humain. *C.R. Acad. Sci. Paris, Série D* **272,** 2638.

Dutrillaux, B., de Grouchy, J., Finaz, C., and Lejeune, J. (1971). Mix en evidence de la structure fine des chromosomes humains per digestion enzymatique (pronase en particulier). *C.R. Acad. Sci. Paris, Série D* **273,** 587.

Evans, H. J. (1973). Molecular architecture of human chromosomes. *Br. Med. Bull.* **20,** 196–203.

Gagné, R., and Laberge, C. (1972). Specific cytological recognition of the heterochromatic segment of number 9 chromosome in man. *Exp. Cell Res.* **73,** 239.

Gagné, R., Tangnay, R., and Laberge, C. (1971). Differential staining patterns of heterochromatin in man. *Nature, New Biol.* **232,** 29.

Gall, J. G. (1966). Chromosome fibers studied by a spreading technique. *Chromosoma* (Berl.) **20,** 221.

Gormley, I. P., and Ross, A. (1972). Surface topography of human chromosomes examined at each stage during ASG banding procedure. *Exp. Cell Res.* **74,** 585.

Greilhuber, J. (1973). Differential staining of plant chromosomes after hydrochloric acid treatments (Hy-Bands). *Osterr. Bot. Z.* **122,** 333.

Harris, H., Watkins, J. F., Ford, C. E., and Schoefl, G. I. (1966). Artificial heterokaryons of animal cells from different species. *J. Cell Sci.* **1,** 1.

Kato, H., and Moriwaki, K. (1972). Factors involved in the production of banded structures in mammalian chromosomes. *Chromosoma* (Berl.) **38,** 105.

Kaufmann, B. P., Gay, H., and McDonald, M. R. (1960). Organizational patterns within chromosomes. *Int. Rev. Cytol.* **9,** 77.

Lampert, F., and Lampert, P. (1970). Ultrastructure of the human chromosome fiber. *Humangenetik* **11,** 9.

McKay, R. D. G. (1973). The mechanism of G and C banding in mammalian metaphase chromosomes. *Chromosoma* (Berl.) **44,** 1.

Moorhead, P. S., Nowell, P. C., Mellmann, W. J., Bettips, D. M., and Hungerford, D. A. (1960). Chromosome preparations of leucocytes cultured from human peripheral blood. *Exp. Cell Res.* **20,** 613.

Ohnuki, Y. (1965). Demonstration of the spiral structure of human chromosomes. *Nature* **203,** 916.

Ohnuki, Y. (1968). Structure of chromosomes. I. Morphological studies of the spiral structures of human somatic chromosomes. *Chromosoma* **25,** 402.

Okada, Y. (1969). Factors in fusion of cells by HVJ. In: *Current Topics in Microbiology and Immunology* **48,** 102. Springer-Verlag, Berlin and New York.

Patil, S. R., Merrick, S., and Lubs, H. A. (1971). Identification of each human chromosome with a modified Giemsa stain. *Science* **173,** 821.

Patil, S. R., Rao, P., and Lubs, H. A. (1972). Banding patterns in G1 and G2 chromosomes. *Mammal. Chromosome Newsletter* **13,** 91.

Pawlowitzki, J. H., Blaschke, R., and Christenhuss, R. (1968). Darstellung von Chromosomen im Raster-Elektronenmikroskop nach Enzymbehandlung. *Naturwissenschaften* **55,** 63.

Rao, P. N., and Johnson, R. T. (1970). Mammalian cell fusion: Studies on the regulation of DNA synthesis and mitosis. *Nature* **225,** 159.

Rao, P. N., and Johnson, R. T. (1971). Mammalian cell fusion. IV. Regulation of chromosome formation from interphase nuclei by various chemical compounds. *J. Cell Physiol.* **78,** 217.

Rao, P. N., and Johnson, R. T. (1972). Cell fusion and its application to studies on the regulation of the cell cycle. In: *Methods in Cell Physiology*, Vol. V (Prescott, D. M., ed.), 75-126. Academic Press, New York.

Ris, H. (1961). Ultrastructure and molecular organization of genetic systems. *Canad. J. Genet. Cytol.* **3,** 95.

Ross, A., and Gormley, I. P. (1973). Examination of surface topography of Giemsa-banded human chromosomes by light and electron microscopic techniques. *Exp. Cell Res.* **81,** 79.

Ruzicka, R. (1971). Ein einfaches Verfahren zur Darstellung von humanen Metaphasenplatten für Elektronenmikroskop. *Humangenetik* **13,** 199.

Ruzicka, F. (1973). Uber die Primärwindungen menschlicher Chromosomen. *Humangenetik* **20,** 335.

Ruzicka, F., and Schwarzacher, H. G. (1974). The ultrastructure of human mitotic chromosomes and interphase nuclei treated by Giemsa banding techniques. *Chromosoma* (Berl.) **46,** 443.

Schnedl, W. (1971). Banding pattern of human chromosomes. *Nature, New Biol.* **233,** 93.

Schwarzacher, H. G., and Schnedl, W. (1967). Elektronen-mikroskopische Untersuchung menschilicher Metaphasen-Chromosomen. *Humangenetik* **4,** 153.

Seabright, M. (1971). A rapid banding technique for human chromosomes. *Lancet* **II,** 971.

Shafer, D. A. (1973). Banding human chromosomes in culture with Actinomycin D. *Lancet* 828.

Sumner, A., Evans, T., and Buckland, R. A. (1971). New technique for distinguishing between human chromosomes. *Nature, New Biol.* **232,** 31.

Utakoji, T. (1972). Differential staining patterns of human chromosomes treated with potassium permanganate. *Nature* **239,** 168.

Wang, H. C., and Fedoroff, S. (1972). Banding in human chromosomes treated with trypsin. *Nature, New Biol.* **235,** 52.

Wettstein, R., and Sotelo, J. R. (1965). Fine structure of meiotic chromosomes. The elementary components of metaphase chromosomes of *Gryllus argentinus. J. Ultrastruct. Res.* **13,** 367.

Wolfe, S. L. (1965). The fine structure of isolated metaphase chromosomes. *Exp. Cell Res.* **37,** 45.

Yosida, T. H., and Sagai, T. (1972). Banding pattern analysis of polymorphic karyotypes in the black rat by a new differential staining technique. *Chromosoma* **37,** 387.

Yunis, J. J., Roldau, L., Yasmineh, W. G., and Lee, J. C. (1971). Staining of satellite DNA in metaphase chromosomes. *Nature* **231,** 532.

Yunis, J. J., and Sanchez, O. (1973). G-banding and chromosome structure. *Chromosoma* (Berl.) **44,** 15.

5. AUTORADIOGRAPHIC LOCALIZATION OF DNA, IN NONMETABOLIC CONDITIONS

M. Geuskens

Laboratoire de Cytologie et d'Embryologie moléculaires, Université libre de Bruxelles, Belgium

INTRODUCTION

Rapid progress in the development of preparatory procedures for preserving the fine structure has led to the development of the field of electron microscope cytochemistry. In connection with the detection of nucleoproteins, considerable effort has been made to develop electron microscopy techniques which may provide information comparable to that obtained with the Feulgen staining of DNA. Similarly, attempts have been made to match Unna (methyl green-pyronine) staining combined with enzymatic extraction with ribonuclease (proposed by Brachet) for revealing RNA-containing structures at the light microscope level.

Recently, reviews on the cytochemical techniques employed for detecting proteins and nucleic acids at the subcellular level have been published (Hayat, 1975; Bouteille *et al.*, 1975). Several methods especially involving the Feulgen-type reactions for detecting DNA have been proposed. These methods although quite complex, do provide reasonable information with regard to specificity, contrast, resolution, and reproducibility. Nevertheless, several techniques have to be combined in order to characterize with certainty the DNA nature of some structures.

Autoradiography is a very useful method for studying DNA metabolism and for localizing newly synthesized molecules inside the cells which have incorporated ^{3}H-thymidine. If standardized technical conditions are used, semiquantitative results may be derived from experiments of labeled precursor incorporation. The different steps involved in the autoradiographic process and the problems encountered at the electron microscope level have been reviewed and discussed by Salpeter and Bachmann and Jacob and Budd in Volumes 2 and 6, respectively, of this series. More information on the autoradiographic techniques and references concerning the results obtained with that method at the ultrastructural level can be found in the reviews by Bouteille *et al.* (1974) and Jacob (1971).

Recently, a method allowing autoradiographic detection of DNA in ultrathin sections of cells not having previously incorporated tritiated thymidine has been developed by Fakan and Modak (1973). They have used an enzyme, terminal deoxynucleotidyl transferase, to add ^{3}H-dATP to free 3′-OH ends of DNA and oligodeoxynucleotide molecules located at the surface of ultrathin sections of glycol methacrylate (GMA)- or Epon-embedded cells.

In the present chapter we shall describe two methods for autoradiographic detection of DNA, which, like the one mentioned above, were first developed at the light microscope level. They are carried out under nonmetabolic conditions, as they are performed on fixed material without enzyme intervention, and are based only upon specific affinities and interactions between molecules. As with the other autoradiographic methods for the detection of DNA referred to above, these two methods do not reveal all cellular DNA located in ultrathin sections and do not yield the same information. The first method allows the localization of double-stranded DNA accessible to Actinomycin, while the second allows the detection of specific repetitive DNA sequences in cells.

LOCALIZATION OF DOUBLE-STRANDED DNA BY ^{3}H-ACTINOMYCIN D BINDING

This method, first developed at the light microscope level by Brachet and Ficq (1965), is based on the fact that Actinomycin D binds to double-stranded DNA with a deoxyguanosine dependency and therefore inhibits RNA synthesis. Many reviews concerning the action of Actinomycin and the way it binds to DNA are available (Reich and Goldberg, 1964; Goldberg and Friedman, 1971; Sobell, 1973, 1974).

The incubation of fixed or unfixed cytological preparations as well as paraffin-embedded tissue sections in a radioactive Actinomycin D solution has allowed the detection of small amounts of DNA in cells (references in Geuskens, 1974).

The first reports regarding the intracellular detection of the labeled antibiotic binding sites at the ultrastructural level concern experiments performed on living cells (Simard, 1967; de Harven, 1968). The silver grains were seen to be more concentrated over the condensed masses of heterochromatin, and the labeling was proportional to the duration of uptake. These observations led de Harven (1968) to propose the use of ^{3}H-Actinomycin D binding as a specific method to detect DNA at the ultrastructural level using a nonmetabolic type of autoradiography.

Fixed Cells

Steinert and Van Gansen (1971) were the first to succeed in improving the resolution of the technique using fixed material incubated with the labeled antibiotic before embedding. They detected the ^{3}H-Actinomycin D binding sites in *Xenopus laevis* oocytes fixed in glutaraldehyde or methanol–acetic acid. Although their statistical results are in favor of a specific ^{3}H-Actinomycin D binding to yolk platelet periphery, it is especially the labeling observed over the nuclei of the follicular cells surrounding the oocytes which is convincing and shows that the technique can be used under their experimental conditions (Fig. 5.1). Aldehyde fixation does not prevent but probably impairs the binding of Actinomycin D to its receptor sites, as a heavier labeling is observed when methanol–acetic acid fixed oocytes are incubated with the labeled antibiotic. However, the cell fine structure preservation is of course better after aldehyde fixation.

The technique used by these investigators was as follows. Mature amphibian oocytes were fixed in 3% glutaraldehyde in the Millonig buffer (pH 7.4) for 30 min at 20°C. The oocytes could then be cut into halves for better penetration of labeled Actinomycin D, before washing them in the Millonig buffer. The specimens were incubated in the Millonig buffer containing ^{3}H-Actinomycin D (40 μg/ml; 100 μCi/ml) (Schwarz) of high specific activity (3.38 Ci/mmole) for 1 hr at room temperature. The oocytes were then washed in a concentrated solution of unlabeled Actinomycin D (500 μg/ml) in the Millonig buffer for 1 hr at 4°C. This step was followed by postfixation in osmium tetroxide, dehydration in alcohol, and embedding in Araldite. Gold-colored sections were subjected to the autoradiographic process. For comparison, oocytes fixed in methanol–acetic acid mixture (6 vol/4 vol) for 30 min at 20°C were used.

Ultrathin Frozen Sections

Bernier *et al.* (1972) have directly incubated ultrathin frozen tissue sections in a ^{3}H-Actinomycin D solution by floating them on the surface of the solution. Extraction and enzymatic digestion tests have shown that the labeled antibiotic molecules bind only to DNA under these experimental conditions, confirming

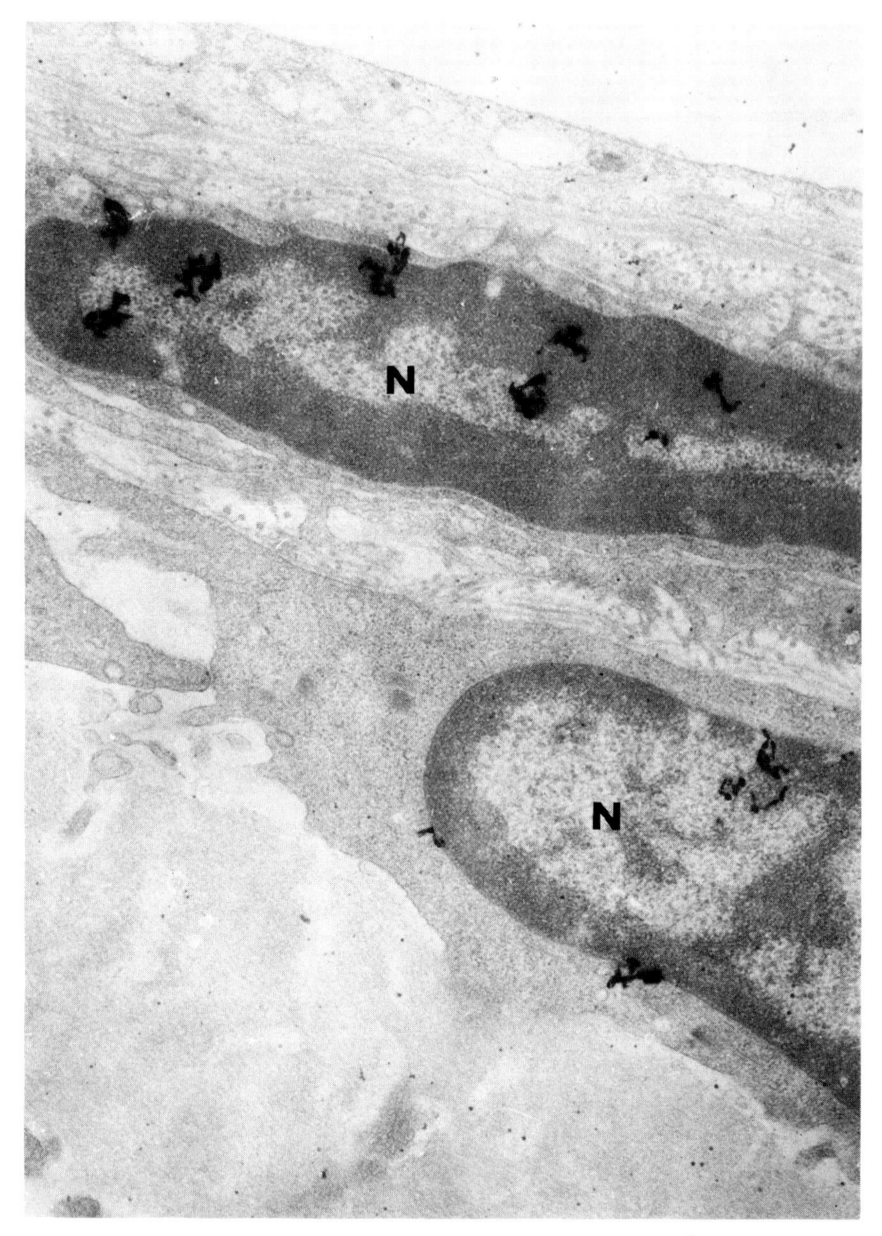

Fig. 5.1 Follicular cells surrounding a *Xenopus* oocyte treated with ^{3}H-Actinomycin D after fixation with glutaraldehyde and subsequently embedded in araldite. N: nuclei of follicular cells X27,000. (*Courtesy of G. Steinert.*)

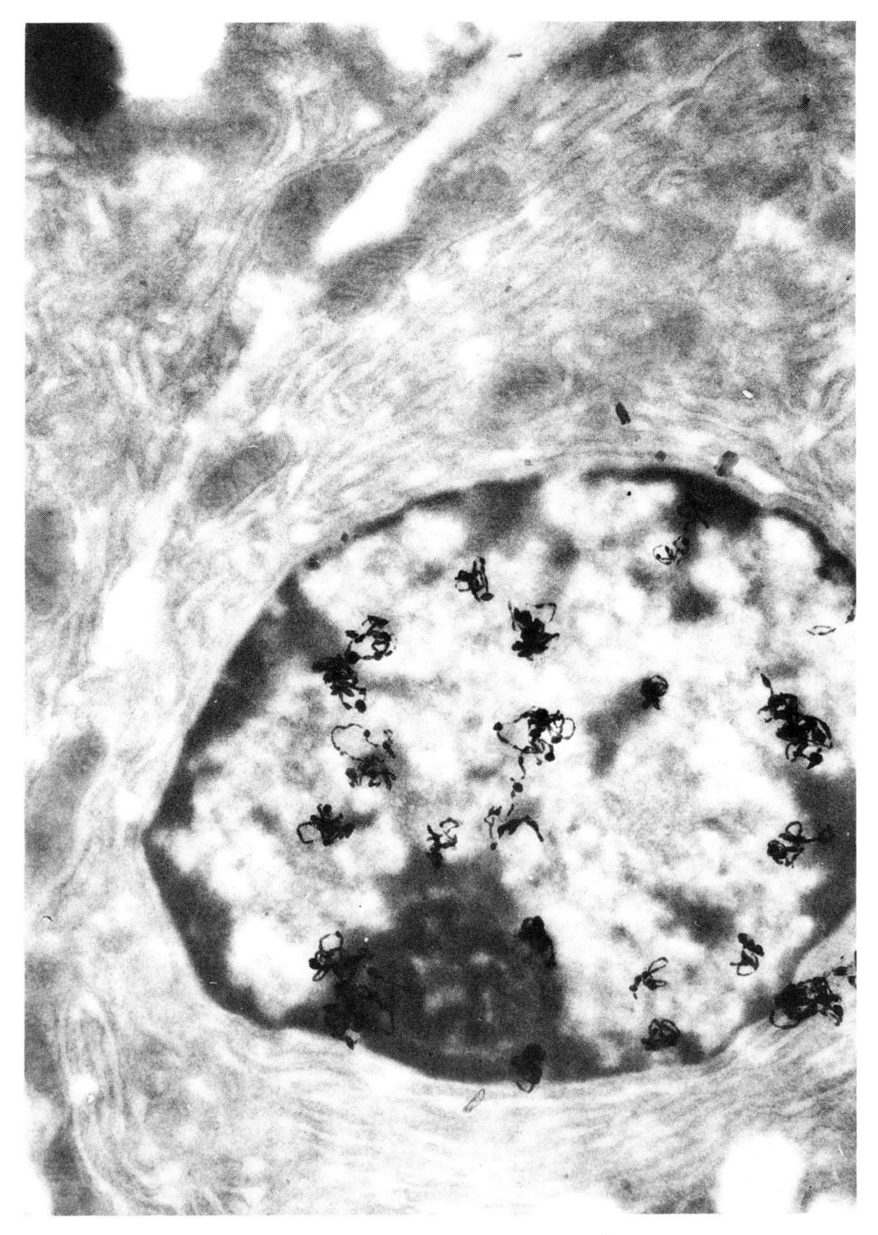

Fig. 5.2 Ultrathin frozen section of rat pancreas floated on ^{3}H-Actinomycin D for 1 hr. The silver grains are localized over the nucleus of this exocrine cell. X18,000.

the conclusions of specificity tests performed on squashes or paraffin sections at the light microscope level (Brachet and Ficq, 1965; Camargo and Plaut, 1967; Ebstein, 1967). Adult rat liver was used by these investigators; the method is given below.

Small tissue fragments (1 mm^3) were fixed in 2.5% glutaraldehyde in 0.2 M cacodylate buffer (pH 7.2) for 15 min or for 1 hr at 4°C. The specimens were washed overnight in the same buffer and soaked in a 25% glycerol solution for 10 min at room temperature. Methyl cellulose was used as a supporting medium before rapid freezing of tissue fragments in liquid nitrogen.

Ultrathin frozen sections were obtained on 50% dimethyl sulfoxide (DMSO), collected with plastic rings according to the Marinozzi technique (1964), and floated on decreasing concentrations of DMSO before storage on distilled water in a watch glass until use. The sections were floated on the ^{3}H-Actinomycin D solution (8.4 Ci/mmole; 100 μCi/ml) for 1 to 3 hr at 4°C.

The sections were rinsed in a solution of unlabeled Actinomycin D (1 mg/ml) for 1 hr at 4°C. According to Bernier *et al.* (1972), this step is important for removing by isotopic exchange much of the labeled molecules which adhere to sections. The sections were then floated on 20 successive distilled water baths and finally subjected to the autoradiographic process. The results of these investigators have confirmed the value of the ^{3}H-Actinomycin D binding technique for the direct detection of DNA in ultrathin frozen sections.

Our own preliminary results with ultrathin frozen rat pancreas sections (Geuskens, 1972) have shown that it is possible to obtain satisfactorily low nonspecific binding when sections, after floating for 1 hr on a ^{3}H-Actinomycin D solution in the dark, are floated on an unlabeled Actinomycin D solution (200 μg/ml) for 1 hr, briefly rinsed in two distilled water baths, floated for 20 min on a protease solution (Sigma, type VI; 0.01% in phosphate buffer) at 20°C, and finally rinsed in distilled water before being harvested on carbon-Formvar-coated grids and submitted to the autoradiographic process (Fig. 5.2).

Ultrathin Plastic Sections

Steinert and Van Gansen (1971) have tried to treat directly Araldite-embedded oocyte sections with the ^{3}H-Actinomycin D. Ultrathin sections, transferred as usual with plastic rings, were floated on a ^{3}H-Actinomycin D solution (100 μCi/ml; 40 μg/ml; 3.38 Ci/mmole) for 1 hr. The sections were then floated for 1 hr on an unlabeled Actinomycin D solution (500 μg/ml), and transferred through 20 distilled water baths, for 20 sec in each bath. Then the sections were deposited on collodion-coated slides which were subsequently dipped into emulsion.

This technique did not give satisfactory results because a strong unspecific label covered the whole sections. The overall labeling resulted from the presence of labeled molecules adhering to the sections even after many washes. Therefore, we have tried to improve the washing step of the technique by using organic solvents such as pyridine and alcohol, which are better solvents than water for the antibiotic molecules. The following technique has led to a successful specific labeling of DNA located in ultrathin sections of GMA-embedded tissues (Geuskens, 1974) (Fig. 5.3).

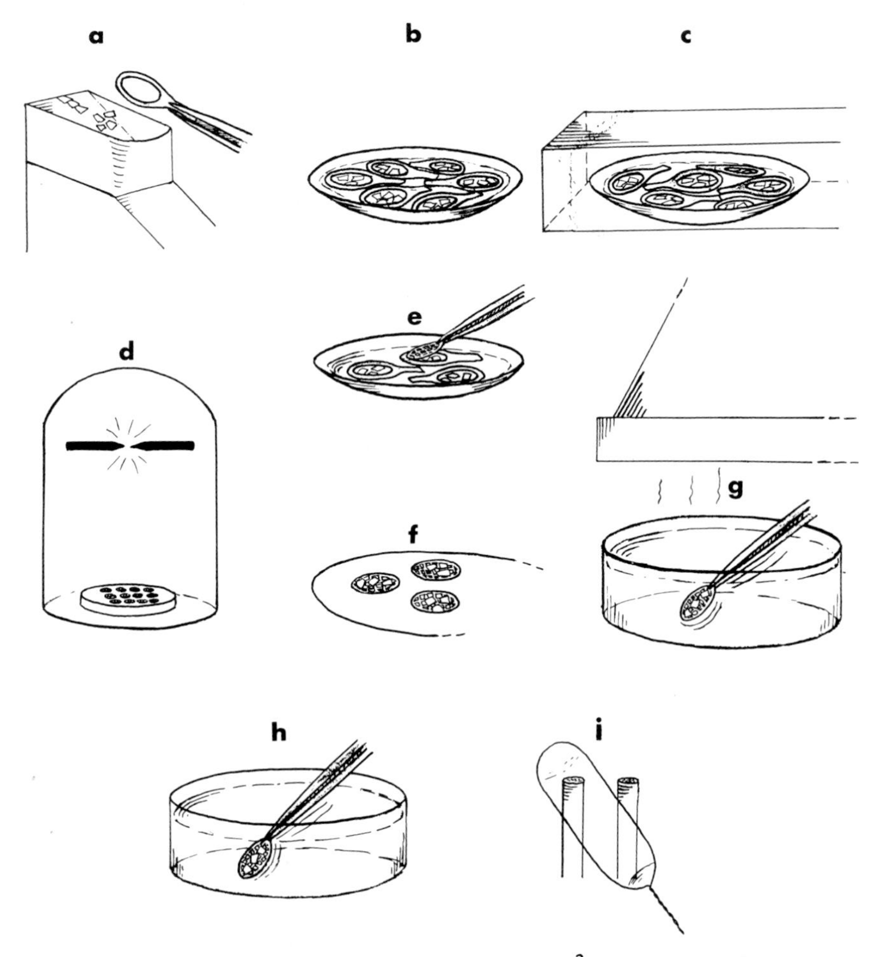

Fig. 5.3 Diagrammatic representation of the technique of ^{3}H-Actinomycin D binding to DNA localized in ultrathin sections of GMA-embedded tissues: (a) harvest of sections with plastic rings; (b) the plastic rings are stored on distilled water until use; (c) the sections are floated on the ^{3}H-Actinomycin D solution under a dark cover; (d) a thick carbon film is deposited on both sides of Formvar-coated grids; (e) these grids are used for harvesting the sections floating on water; (f) the grids are dried on a filter paper; (g) the grids are agitated in a pyridine-methanol solution under a hood; (h) the grids are then agitated in two baths of methanol; (i) finally, the grids are covered with emulsion using a loop.

Tissue fragments were fixed in 1.6% glutaraldehyde in 0.1 M Sörensen phosphate buffer (pH 7.4) for 15 min to 1 hr at 4°C. They were embedded in GMA according to Leduc and Bernhard (1967). Ultrathin sections were floated for 1 hr on a ^{3}H-Actinomycin D solution (100 μCi/ml; 6.1 Ci/mmole) in the dark. The sections were transferred to two distilled water baths for 5 min in each and harvested onto Formvar-coated grids which had been previously covered with a thick layer of carbon on both sides.

The next step of the technique may be performed on the next day. The grids, taken with tweezers, were individually agitated in a 50% solution of pyridine in methanol for 5 to 10 min under a suction hood and subsequently in two successive baths of methanol for 2 min in each bath. The grids were dried by touching their edge to a filter paper. Ilford L 4 emulsion was applied on the grids using a loop. After development, the grids were stained for 10 min with uranyl acetate followed by lead citrate. Between the two staining steps, the grids were dried for 30 min in an oven at 40°C. As shown in Figs. 5.4 and 5.5, the label is localized over the nuclei and more concentrated over the condensed chromatin.

Discussion

The main advantage of the technique of ^{3}H-Actinomycin D binding to DNA located in plastic-embedded tissue ultrathin sections is its simplicity once the autoradiographic technique itself is controlled. Some technical points will be briefly discussed here.

Labeled or unlabeled Actinomycin baths are made in small watch glasses (30 mm in diameter). Bernier *et al.* (1972) recommend that the watch glasses be washed in an aqueous solution of cold Actinomycin D before use in order to saturate their surface, as the antibiotic molecules bind strongly to glass (Ebstein, 1967). This precaution is valid when paraffin sections are treated with ^{3}H-Actinomycin D on glass slides which are subsequently dipped into emulsion, but it is unnecessary here.

It is advisable to incubate the sections on Actinomycin solutions under a light-tight cover, as the antibiotic is sensitive to light. Only small volumes of solution are necessary for the incubations, and the labeled solution, replaced in a dark bottle, can be used several times, at least as long as it appears clean and keeps its original color.

When working on nonembedded material, elimination of nonspecifically bound ^{3}H-Actinomycin D molecules is improved by washing in alcohol, since the antibiotic is more soluble in alcohol than in water. This has already been observed as a way to lower the background of preparations in light microscope autoradiography. Dehydration of fixed tissue fragments previously treated with ^{3}H-Actinomycin D in ethanol contributes to the elimination of nonspecific label in such a way that satisfactory nuclear localization of the silver grains can be observed (Fig. 5.1). The use of cold alcohol solution should further improve the

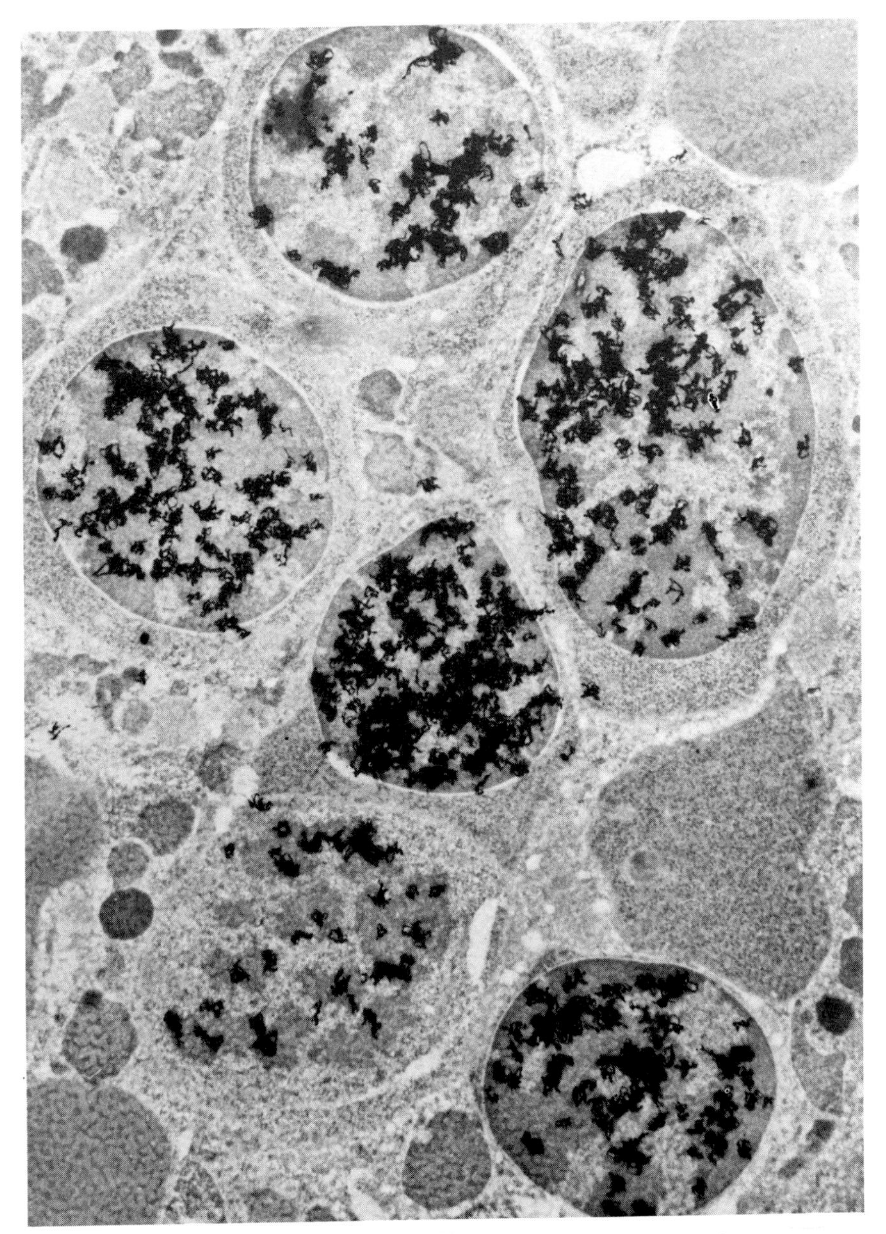

Fig. 5.4 Ultrathin section of GMA-embedded rat lymph node floated on ^{3}H-Actinomycin D for 1 hr and subsequently rinsed in pyridine-methanol. Kodak D 19 developer. ×6,600.

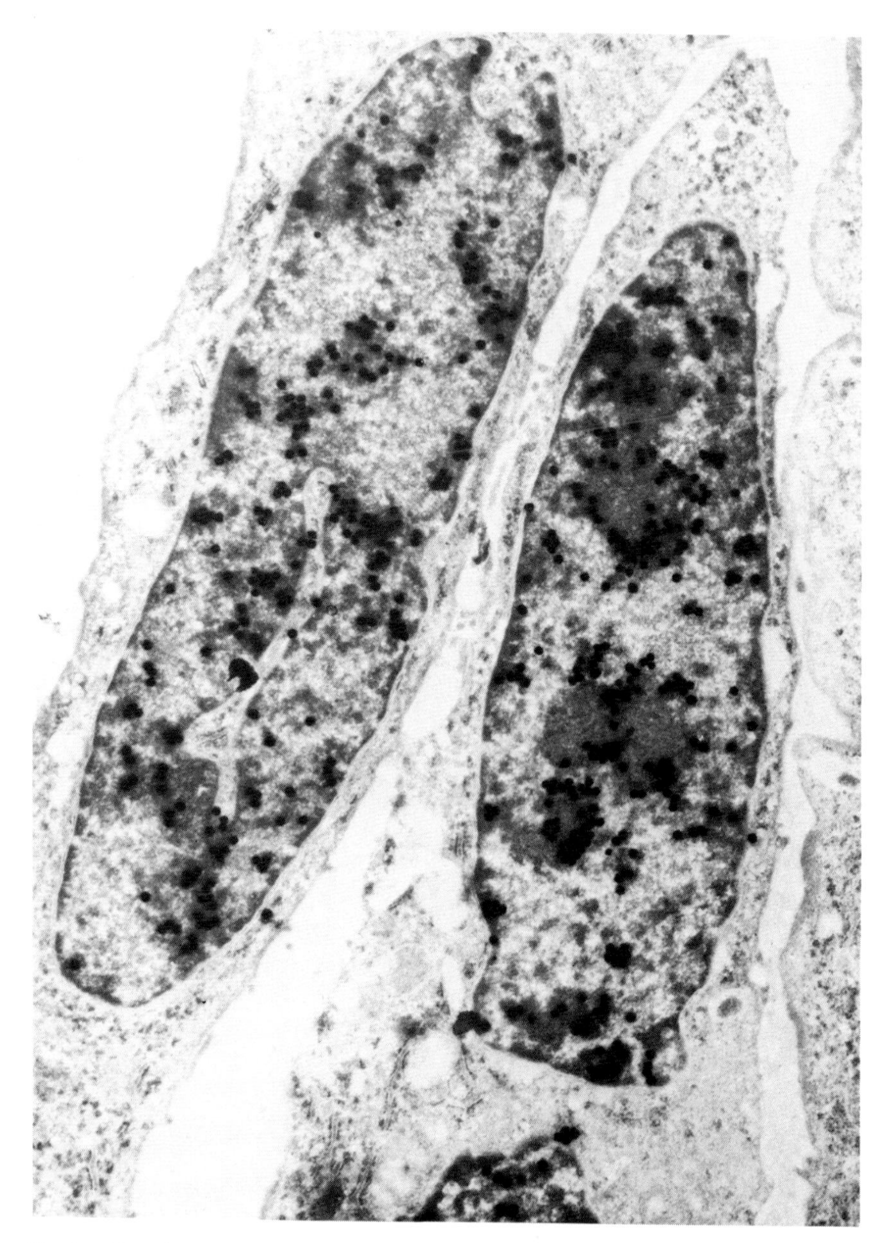

Fig. 5.5 Ultrathin section of GMA-embedded monkey kidney cells in culture floated on ^{3}H-Actinomycin D and subsequently rinsed in pyridine-methanol. Phenidone developer. ×10,800.

washing efficiency as Actinomycin has a negative temperature coefficient of solubility (Gellert *et al.*, 1965).

Plastic rings do not float on concentrated alcohol solutions. Therefore, sections must be harvested on grids after the rough rinse in water which follows floating on the labeled antibiotic solution. As pyridine rapidly dissolves plastic, the harvesting of sections on grids is a requisite for rinsing them in this solvent. For the same reason, it is necessary to use Formvar-coated grids previously protected on both sides with a carbon layer.

We have not tried to carry out the pyridine-methanol rinsing technique after floating epoxy resin-embedded tissue sections on the ^{3}H-Actinomycin D solution, but attempts by Steinert and Van Gansen (1971) seem to indicate that a specific binding can also occur to Araldite-embedded tissue sections. The nonspecific labeling of these sections can probably be eliminated or drastically reduced by rinsing them in a pyridine-methanol solution.

Sometimes the support film becomes detached from the grid during the shaking in the organic solvent solution. Therefore, it is necessary to provide for ~10% fit-for-use grid loss at this step of the technique by treating an adequate number of sections of each sample. The rinsing time in the pyridine solution can probably be adapted for each material between 1 and 10 min. The washing-solutions renewing frequency depends upon the volume used and the number of treated grids.

As very heavy labeling can be observed, at least when GMA-embedded specimens are used, it is advisable to use developers producing small silver grains during the photographic process [gold latensification–elon ascorbic acid or Phenidone developers; see the review by Salpeter and Bachmann in Volume 2 of this treatise or that by Bouteille *et al.* (1975)] in such a way that the underlying structures are not hidden. Concerning the labeling intensity, let us remark that the results of Bernier *et al.* (1972) show that, for ultrathin frozen sections at least, there is an increase in ^{3}H-Actinomycin D binding as a function of the treatment time with the labeled antibiotic. Moreover, experiments performed at the light (Camargo and Plaut, 1967) and electron microscope (Steinert and Van Gansen, 1971) levels point out the fact that the nature of the fixative used has an effect on the capacity of DNA to bind Actinomycin D. A lower ^{3}H-Actinomycin D binding capacity of cells fixed in glutaraldehyde can result from the preservation of DNA-associated proteins as well as increased DNA molecule cross-linking, both reducing DNA accessibility to antibiotic molecules.

^{3}H-Actinomycin D binding to DNA as a function of chromatin condensation and DNA-proteins association have been discussed elsewhere (Geuskens, 1974). Our results with GMA-embedded tissue sections have shown that the labeling is always heavier over the condensed chromatin, and the ^{3}H-Actinomycin D binding seems to occur in correlation with DNA concentration inside the nuclei. However, when ultrathin frozen sections are used, this preferential localization is not observed according to Bernier *et al.* (1972), in partial contradiction with our own observations (Geuskens, 1972) (Fig. 5.2). As several studies have shown that Actinomycin D binds to DNA in places where the genome is not repressed by associated proteins (references in Geuskens, 1974), it is likely that the antibiotic molecules bind only to "free" DNA, where proteins do not interfere with their intercalation, under our experimental conditions too.

Actinomycin D binds to double-stranded DNA with a deoxyguanosine dependency. Therefore, one may hope for a preferential intracellular localization of G-C rich DNA satellite–containing structures using the ^{3}H-Actinomycin D binding technique (Geuskens, 1974). However, such localization would have to be confirmed by *in situ* molecular hybridization.

In conclusion, the ^{3}H-Actinomycin D binding technique combined with autoradiography can be very useful for detecting small amounts of double-stranded DNA at the ultrastructural level. Carrying out the technique on ultrathin sections eliminates the problems of irregular penetration of the antibiotic molecules into tissues. The use of plastic-embedded tissue ultrathin sections is favored, as these sections can be obtained with a better yield than frozen ultrathin sections, and previously embedded materials can be used. However, the use of plastic sections makes it difficult to perform enzymatic controls on these preparations.

LOCALIZATION OF SPECIFIC DNA SEQUENCES BY *IN SITU* MOLECULAR HYBRIDIZATION

In situ molecular hybridization between radioactive polynucleotidic sequences in solution and complementary DNA of cytological preparations allows localization of specific DNA sequences with autoradiography. This method has only recently and tentatively been carried out at the electron microscope level, as an extension of work which succeeded in extending to cytological preparations a technique developed by biochemists. Initially this method was used by biochemists with a view to checking the degree of homology of two nucleic acid molecule populations and thus the similarity of the genetic information encoded in their base sequences.

The *in situ* molecular hybridization method is based upon the technique used for studying *in vitro* pairing reactions between radioactive RNA or DNA molecules in solution and complementary DNA also in solution (Marmur and Doty, 1961; Hall and Spiegelman, 1961; Yankovsky and Spiegelman, 1963) or bound to nitrocellulose filter (Nygaard and Hall, 1963; Gillepsie and Spiegelman, 1965; McCarthy and Church, 1970). In this case, hybrid molecules are detected by using a scintillation counter. Under the cytological molecular hybridization conditions, hybrid molecules are detected by autoradiography. As DNA is maintained in its original cellular location, it is theoretically possible to localize sequences complementary to the labeled probe used. Deoxyribonucleic acid/DNA and DNA/RNA hybrid formation reactions are basically similar and may be considered together.

The principle of the molecular hybridization method, which is simply explained in the book by Davis *et al.* (1968), for instance, is as follows: the two strands of native DNA molecules are held together by hydrogen bonds of nucleotide pairs (and other noncovalent bonds between bases). Disruption of the hydrogen bonds causes the separation of the two strands and denaturation of DNA. This can be brought about by increasing the temperature, raising the pH of the medium, or using substances such as formamide ($NCONH_2$) which compete with the DNA bases for formation of hydrogen bonds.

Renaturation of DNA molecules denatured at high temperature is obtained by slow cooling, and is maximal in annealing conditions, ~25° below their melting temperature (midpoint of thermal denaturation) (Marmur and Doty, 1961; Marmur *et al.*, 1963), at high ionic strength for decreasing electrostatic repulsion between molecules. Addition of formamide to the medium allows denaturation of DNA at a lower temperature (Marmur and Ts'o, 1961). Association between complementary RNA and DNA molecules can be obtained under similar conditions to those used for DNA renaturation, but difficulties arise because of the fact that RNA/DNA hybrid formation must compete with the reassociation of the DNA strands.

After hybridization for several hours in annealing conditions, noncomplexed or incompletely complexed nucleic acid molecules are removed by enzymatic digestion with heated pancreatic ribonuclease if radioactive RNA was present in the hybridization medium or a nuclease specific for single-stranded DNA molecules if previously denaturated labeled DNA was used for complementation studies.

When the method is performed on cytological preparations (squashes, smears, or paraffin-embedded tissue sections) of DNA which has been previously denatured in its natural cellular location, it is hybridized with labeled RNA or DNA molecules in solution. After the enzymatic step for removal of the uncomplexed radioactive sequences and elimination with dilute cold trichloracetic acid of the degradation products resulting from that treatment, the cytological preparations are subjected to the autoradiographic process. After an adequate duration of exposure, autoradiograms are developed, and silver grains are observed over the hybridization sites.

However, as we have been reminded by Hennig (1973) in a recent review on the *in situ* molecular hybridization method, interpretation of such experiments is still open to discussion. The specificity of the method and the cytological results so far obtained at the light microscope level are discussed in detail in this review. Therefore, we shall only consider here the technical modifications resulting from the extension of the method at the ultrastructural level.

Publications from three laboratories have reported results obtained by carrying out the method at the electron microscope level. These results show that this extension is realizable and that specific DNA sequences can be localized in cells at the ultrastructural level, at least in some cases. To date no standard technical procedure can be recommended, as these works are the very first ones, and the three research groups used different technical conditions.

According to the type of specimen used, some of the technical steps used by these investigators will be preferred over those by others. These technical procedures can still be subjected to modifications in the future. Presently these publications show how the difficulties raised by the use of the method at the ultrastructural level have been overcome in the three laboratories and furnish useful indications for starting new research in that field in other laboratories. Some modifications with a view to improve the efficiency of the method are already pointed out in the discussion.

Ultrathin Plastic Sections

Jacob *et al.* (1971) first succeeded in carrying out the method at the ultrastructural level using the following procedure for hybridizing tritiated ribosomal RNA to complementary DNA in *Xenopus* oocytes.

Ovaries of metamorphosing tadpoles were fixed in 2.5% glutaraldehyde in 0.05 M phosphate buffer (pH 7.2) for 20 min at 4°C. Small blocks of tissue were transferred through increasing concentrations of GMA in water and embedded in this plastic using ultraviolet light for polymerization at a low temperature.

Sections with a gold interference color were harvested on carbon-Formvar-coated gold grids. The grids were floated on a 0.2% bovine pancreatic ribonuclease solution in 2 X SSC (SSC:0.15 M NaCl; 0.015 M Na citrate) for 1 hr at 37°C in order to remove the RNA from the tissue sections. After washing in large volumes of distilled water, the grids were floated on 0.01% protease (from *Streptomyces griseus*, type VI, Sigma Chemical Co.) in phosphate buffer for 30 min at 37°C. After thorough washing in distilled water, the grids were dried and stored.

Deoxyribonucleic acid contained in the sections was denatured by floating the sections on 0.1 N NaOH for 1 to 2 hr at room temperature. The grids were washed in 2 X SSC and floated on ^{3}H-28S RNA contained in small depressions in glass slides (sealed with a coverslip). Incubation was carried out for 5 hr at 70°C.

The tritiated RNA was extracted from cultures of *Xenopus* kidney cell line incubated in the presence of ^{3}H-uridine. The molecular weight of the RNA was reduced to ~2.10^{4} daltons (3.5 S) by treatment with NaOH for 30 min at 20°C. Its specific activity was 1.5×10^{6} dpm/μg.

After the hybridization period, the grids were washed in 2 X SSC and treated with ribonuclease (0.05% in 2 X SSC) for 15 min at 22°C. After thorough washing in 2 X SSC, the grids were stained and covered with a thin carbon layer and finally with Ilford L4 emulsion with the aid of a loop.

According to our own attempts, three grids can be incubated in a 20 μl drop of

^{3}H-RNA. Indeed, trying to apply the technique described by Jacob *et al.* (1971) for detecting SV 40 DNA in lytically infected cells by hybridization with tritiated complementary RNA (cRNA), we put a 20-μl drop of the labeled probe dissolved in 2 X SSC in the center of a histologic slide previously dipped into a dilute silicone solution or covered with a 2% parlodion film. Three grids, carefully held by their edges with tweezers, were successively placed onto the drop. The slide was then turned down on a thick slide with a central cavity 2 mm in depth, whose edges had been coated with a silicone grease using a toothpick. The extremities of the two slides were then firmly clamped together with Aclé grippers as shown in Fig. 5.6. The grids were thus incubated on a hanging drop. There was no problem of evaporation inside the cavity even after overnight incubation at 65°C. However, we did not obtain convincing reproducible results, at least with our specimens.

Jacob *et al.* (1974) have subsequently carried out the technique for localizing satellite DNA in mouse L cells in culture (Fig. 5.7). Complementary ^{3}H-RNA was transcribed from purified mouse liver satellite DNA using *Escherichia coli* RNA polymerase. The four tritiated nucleoside triphosphates were present in the synthesis medium in approximately equimolar amounts. The specific activity of the ^{3}H-cRNA obtained was estimated to be $\sim 1.2 \times 10^8$ dpm/μg (3.3×10^4 or 9.3×10^4 dpm/μl were used in the hybridization medium).

Rae and Franke (1972) have also studied the distribution of satellite DNA-containing heterochromatin by *in situ* molecular hybridization in mouse interphase nuclei. Mouse liver and testis smears were treated according to the incubation conditions in current use for *in situ* hybridization at the light microscope level, and were subsequently embedded in Epon. Semithin sections of this material were used for autoradiography in order to obtain a better picture of the location of satellite DNA-containing structures. Simultaneously, blocks of mouse liver were fixed in glutaraldehyde followed by osmium tetroxide and embedded in Epon. Ultrathin sections of this material were used for comparison in the electron microscope.

This was a first step in the direction of examination by high-resolution autoradiography of the localization of hybrid molecules in cells which were embedded in plastic after the completion of the hybridization reaction. However, the usual ethanol–acetic acid fixation was used. The next step consisted of using fixation and hybridization conditions which would allow satisfactory preservation of the fine structure so that labeled hybrid molecules could be directly localized under the electron microscope. This was carried out by Croissant *et al.* (1972).

Thick Nonembedded Tissue Sections or Fragments

Orth *et al.* (1971) carried out the usual *in situ* molecular hybridization method for studying the vegetative replication of viral DNA in tumors induced by the

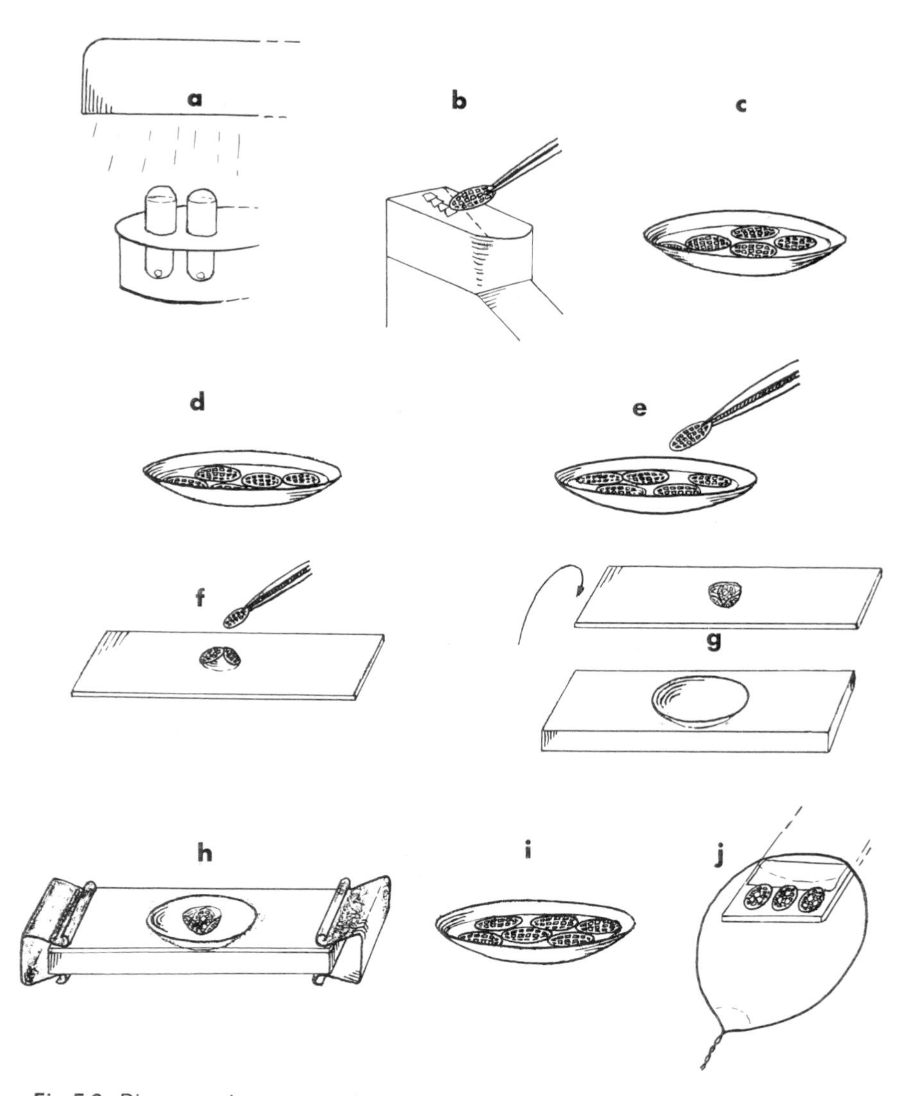

Fig. 5.6 Diagrammatic representation of the technique of Jacob *et al.* (1971) for *in situ* molecular hybridization on GMA ultrathin sections, under our technical conditions for incubation of the grids: (a) polymerization of the blocks under UV light; (b) harvest of sections on carbon-Formvar-coated gold grids; (c) the grids are floated, sections down, on a ribonuclease solution; (d) the grids are then floated on a protease solution; (e) the DNA contained in the sections is denatured by floating the grids on NaOH; (f) three grids are deposited on a 20-μl ^{3}H-cRNA solution in the center of an object slide; (g) the slide is turned down on a thick slide with a central cavity whose edges are greased; (h) the two slides are firmly held together with grippers during incubation in the oven; (i) the grids are floated on a ribonuclease solution; (j) the grids, fixed on slides, are finally covered with emulsion.

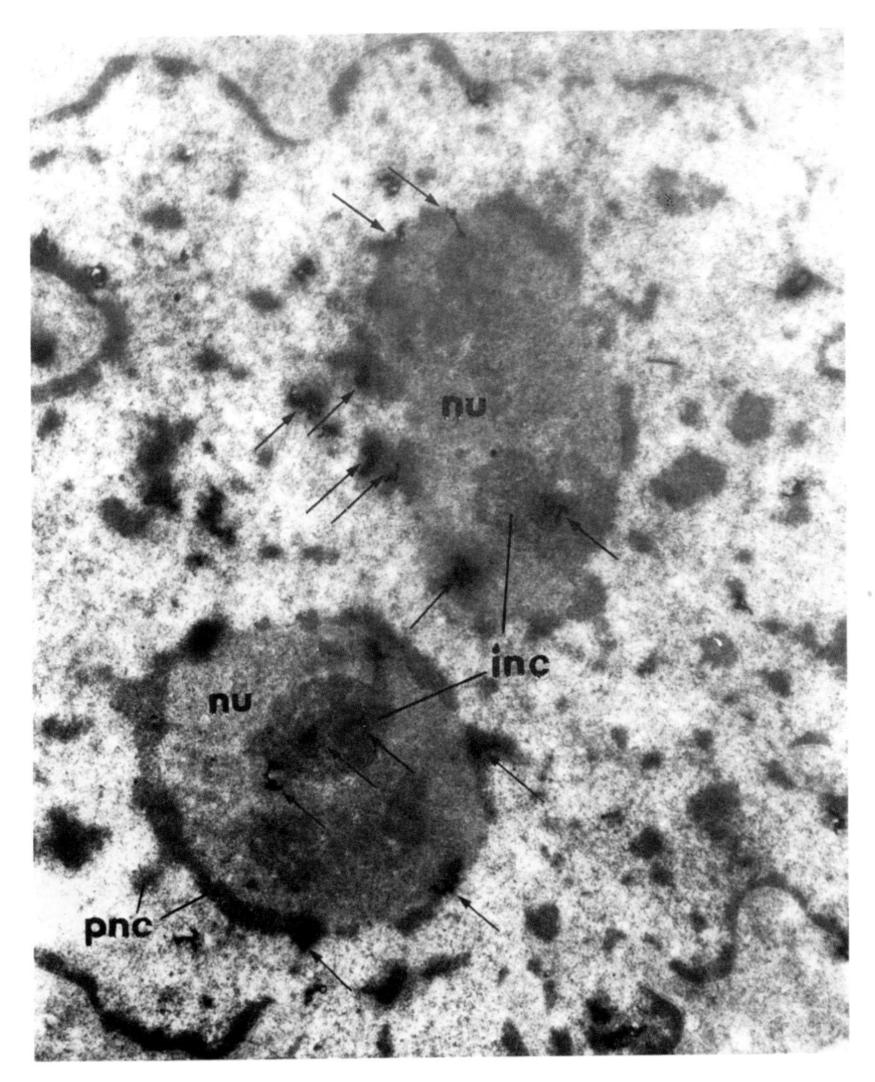

Fig. 5.7 *In situ* molecular hybridization of mouse satellite DNA-complementary RNA onto ultrathin section of mouse L cell. In this preparation, the distribution of silver grains over nucleolus-associated chromatin is seen especially well (*arrows*). Both intranucleolar chromatin (inc) and perinucleolar chromatin (pnc) are labeled. nu, nucleolus. ×8,000 (From Jacob *et al.*, 1974)

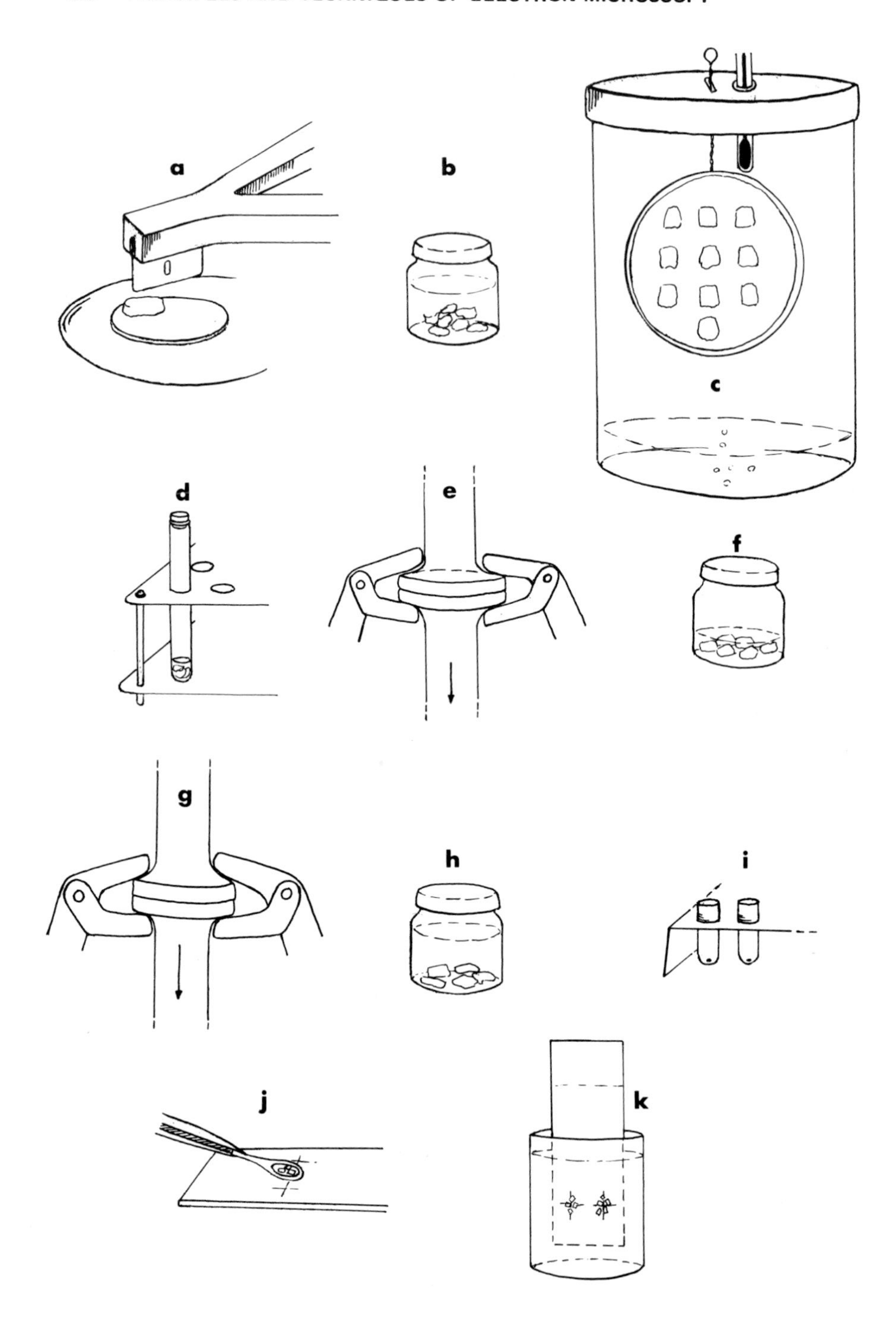

a
b
c
d
e
f
g
h
i
j
k

Shope papilloma virus in cottontail rabbits. Subsequently, they extended their observations to the ultrastructural level, using the following procedures (Croissant *et al.*, 1972) (Fig. 5.8):

One hundred to 150 μm thick sections of unfixed papilloma of a cottontail rabbit were sectioned with a TC 2 Sorvall tissue sectioner. The sections were fixed in 0.5% glutaraldehyde in 0.1 M phosphate buffer (pH 7.4) for 10 min. They were washed in the same buffer containing 0.22 M glucose for 6 hr and harvested onto a millipore filter (GSWPO 47 SO). The filter with the sections was maintained for 10 min in water vapor at 100°C and then rapidly immersed into 0.1 X SSC at 0°C. Afterwards, the filter was dipped into 70% and 95% ethanol for 2 min, and finally plunged into 6 X SSC. The sections, detached from the filter, were then incubated for 16 hr at 66°C in a small tube (Kahn tube) containing 0.3 ml of tritiated RNA (0.5 μg/ml) (specific activity: 1.7 X 10^7 cpm/μg). The labeled RNA was transcribed *in vitro* from purified viral DNA using *E. coli* RNA polymerase.

After the hybridization period, the sections were washed by filtration with 200 ml of 6 X SSC at 66°C and then with 200 ml of 2 X SSC at 20°C. Subsequently, the sections were treated for 1 hr at 37°C with pancreatic ribonuclease (10 μg/ml in 2 X SSC) and washed by filtration with 6 X SSC at 66°C, with 6 X SSC at 20°C, and finally with 0.1 X SSC at 20°C. The sections were fixed successively in 1% glutaraldehyde and 1% osmium tetroxide before dehydration in alcohol. Small section fragments (~2 mm^2) were embedded in Araldite/Epon. Ultrathin sections of this material were subjected to the autoradiographic process using the dipping technique and Ilford L 4 emulsion. Replication of viral DNA was localized in cells in the process of keratinization of the cutaneous papilloma (Figs. 5.9 and 5.10).

Based partly upon the two techniques mentioned above, we have obtained, in collaboration with de Recondo and Chevaillier, preliminary results concerning the ultrastructural localization of crab poly d(A-T) satellite after ^{3}H-DNA/DNA hybridization (Geuskens *et al.*, 1974): small fragments (1 mm^3) of crab intestine were fixed in cold 0.5% glutaraldehyde in 0.1 M phosphate buffer (pH 7.2) for 10 min.

After thorough washes in the same buffer for 4 hr, tissue specimens were

---

Fig. 5.8 Diagrammatic representation of the technique used by Croissant *et al.* (1972) for *in situ* molecular hybridization on thick tissue sections: (a) thick sections are made with a "Sorvall tissue sectioner"; (b) these sections are briefly fixed in 0.5% glutaraldehyde; (c) the DNA contained in the sections adsorbed on a millipore filter, placed itself on a membrane glued to a plastic ring, is denatured in water vapor at 100° C; (d) the sections are incubated in 0.3 ml of ^{3}H-cRNA in a small tube; (e) the sections deposited on a millipore filter are washed by filtration; (f) they are treated with ribonuclease; (g) the sections are washed again by filtration; (h) they are fixed in glutaraldehyde and osmium tetroxide; (i) small fragments of the sections are embedded in Epon; (j) ultrathin sections are deposited on slides for autoradiography; (k) the slides are dipped into emulsion.

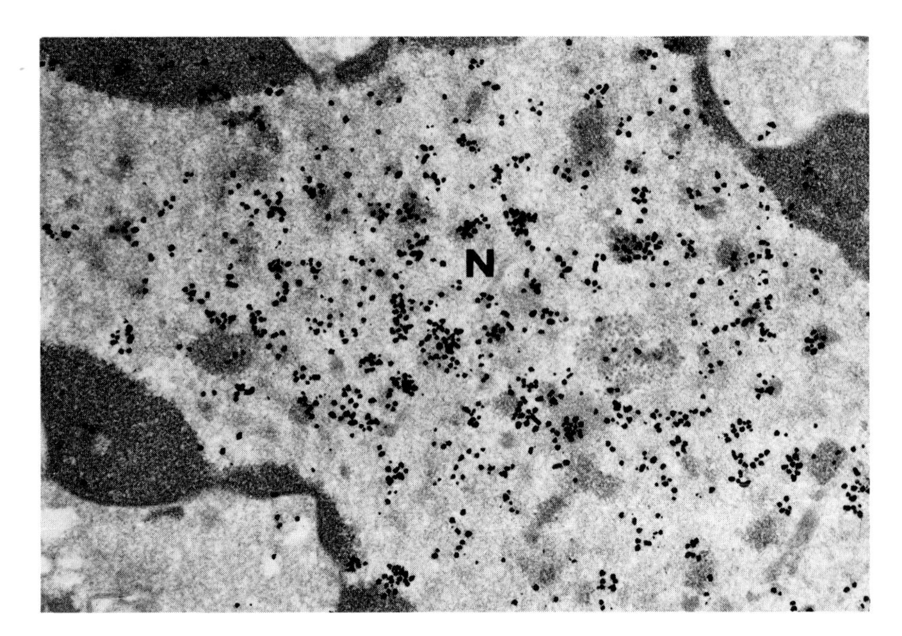

Fig. 5.9 Keratinizing cell of the upper side of the granular layer. Many silver grains are dispersed over the nucleoplasm (N) (250 grains/10 μ^2), pointing to viral DNA accumulation sites. No virus particle detectable.

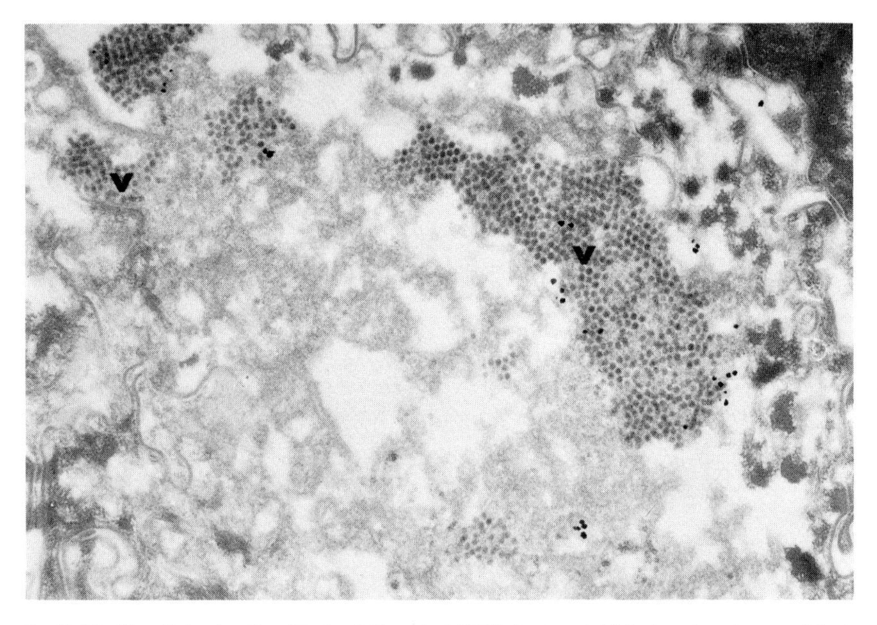

Fig. 5.10 Keratinized cell. Most of the viral DNA is encapsidated and not accessible to ^{3}H-cRNA. Only a very small quantity of viral DNA is detectable. v: virus particles.

treated with preheated ribonuclease (0.02% in 2 × SSC) for 30 min at 37°C and washed twice in 2 × SSC. Then, they were treated with a protease according to Jacob *et al.* (1971) for 20 min at 37°C and rinsed twice with buffer. The DNA contained in the tissue was denatured by immersion into 0.1 N NaOH for 10 min. The tissue specimens were washed twice in 0.1 × SSC at 0°C, in 70% and 95% alcohol for 2 min in each, and in 6 × SSC immediately before use for hybridization.

^{3}H-poly d(A-T) was transcribed *in vitro* with *E. coli* DNA polymerase, using the four deoxynucleoside triphosphates, dATP and TTP being labeled with tritium. The product had a specific activity of 3×10^6 dpm/μg. This DNA was denatured by heat, in boiling water, for 10 min, and the tube rapidly immersed in water at 0°C before transferring to the hybridization medium. Hybridization was carried out in 6 × SSC at 54°C. Crab poly d(A-T) containing only 2–3% G-C is indeed denatured at lower temperatures than the main DNA. Consequently, the annealing temperature is also lower. The DNA concentration used was 7.8 μg in 0.3 ml and the incubation period was 16 hr.

After the hybridization period, the tissue fragments were washed three times in 6 × SSC at 54°C, once in 6 × SSC at 20°C, and three times in a 0.02 M acetate buffer (pH 4.5). They were incubated in a solution of *Aspergillus oryzae* nuclease specific for single-stranded DNA in the acetate buffer (Vogt, 1973) for 1 hr at 45°C. After three washes in the same buffer and two washes in phosphate Sörensen buffer, the tissue specimens were successively fixed in glutaraldehyde and osmium tetroxide as usual and embedded in Epon. Semithin sections of this material were covered with Ilford L 4 emulsion and autoradiograms developed after four weeks exposure. Silver grains were seen concentrated over the nuclei, particularly at the periphery of the tissue sections (Fig. 5.11a).

Ultrathin sections of the same material were mounted on carbon-Formvar-coated grids and covered with the same emulsion using a loop. After three or four months of exposure, a few silver grains were observed over the nuclei, without particular localization (Fig. 5.11b), confirming results obtained at the light microscope level (Chevaillier *et al.*, 1974).

Cells in Culture

As the use of thick tissue sections or fragments may involve an irregular penetration of the labeled nucleic acid molecules through many cell layers, it seemed advisable to try carrying out the *in situ* molecular hybridization method on cell monolayers, using cells in culture. The search for a biological system allowing

---

Figs. 5.9 and 5.10 Cells of a cutaneous papilloma induced by the Shope virus in a cottontail rabbit. *In situ* molecular hybridization with ^{3}H-RNA complementary to viral DNA. Three months exposure. Gold latensification–elon ascorbic acid developer. ×25,000. (*Courtesy of O. Croissant.*)

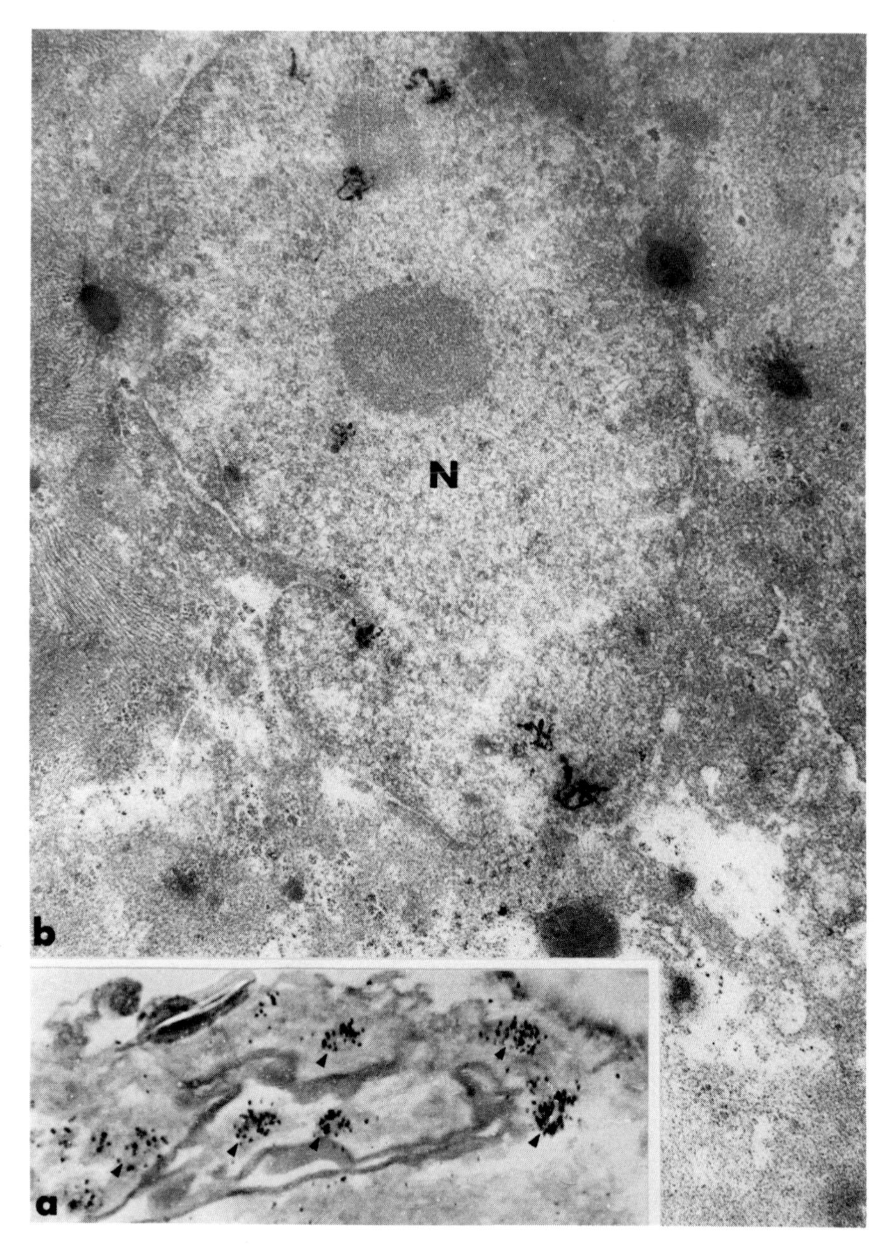

Fig. 5.11 *Cancer pagurus* intestine fragment incubated with ^{3}H-poly d(A-T) and subsequently embedded in Epon. (a) Semi-thin section; $2\frac{1}{2}$ months exposure. Arrow heads show labeled nuclei. X1,000. (b) Ultrathin section of the same material. Silver grains are seen localized over the nucleus (N); six months exposure. X1,800.

frequent collisions between ^{3}H-cRNA in solution and complementary DNA sequences led us to the use of cells lytically infected with virus, containing many newly synthesized viral genomes. ^{3}H-RNA complementary to purified SV 40 viral DNA was synthesized *in vitro* with *E. coli* RNA polymerase, using the four tritiated nucleoside triphosphates in order to obtain a very high specific activity. The technical procedure used for this work (Geuskens and May, 1974) is given below (Fig. 5.12).

Monkey kidney cell monolayers grown in wells of Microtest II tissue culture plates (Falcon plastics) were infected with crude SV 40 viral lysate. Control and infected cultures were fixed *in situ* for 17 hr at 4°C in a 4% formaldehyde solution freshly prepared from paraformaldehyde in 0.1 M Sörensen buffer (pH 7.4). The cells were washed for 48 hr in the same buffer at 4°C. Deoxyribonucleic acid was denatured by treating the cells with a 50% formamide solution

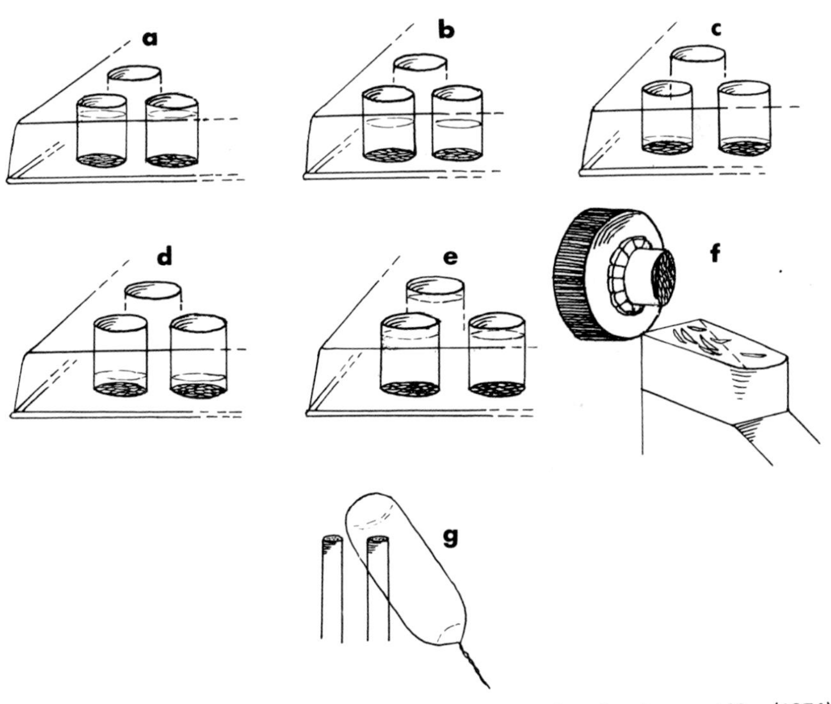

Fig. 5.12 Diagrammatic representation of the technique used by Geuskens and May (1974) for *in situ* molecular hybridization on cell monolayers: (a) cells grown in wells of tissue culture plate are fixed *in situ*; (b) DNA contained in this material is denatured by incubation in 50% formamide at 70°C; (c) hybridization with ^{3}H-cRNA is carried out in 50% formamide at 37°C; (d) the cells are treated with ribonuclease; (e) the cells are embedded in Epon *in situ*; (f) nontrimmed blocks are sectioned; (g) the sections, harvested on grids, are covered with emulsion using a loop. (*Redrawn from Geuskens and May, 1974*.)

in 0.1 × SSC at 70°C for 15 min. The cells were washed twice in 0.1 × SSC at 0°C and twice in 2 × SSC at 4°C before using them for hybridization.

Fifty microliters of ^{3}H-cRNA (0.8 μg/ml) in 50% formamide made in buffered 2 × SSC were placed in each well. These were sealed with paraffin-sheet and the culture plates incubated in a moist atmosphere for 45 hr at 37°C. ^{3}H-cRNA synthesized *in vitro* on superhelical viral DNA had a specific activity of 1.6 × 10^8 dpm/μg and sedimented with a peak at 4.5 S in sucrose density gradient. After hybridization, the cells were washed four times in 2 × SSC at 37°C and treated with both pancreatic and T1 ribonucleases in 2 × SSC for 1 hr at 37°C.

After three washes in 2 × SSC and three washes in 5% trichloracetic acid at 4°C, the cells were washed again in 0.1 M Sörensen before postfixation in glutaraldehyde and osmium tetroxide as usual. Finally, they were embedded in Epon *in situ*. After polymerization, the Epon blocks can be separated from their plastic mold by snapping the plate around them with cutting pliers and then tightening the blocks themselves in grippers until the plastic cracks.

The side where the cells are more concentrated can be chosen by placing the cell-containing face of the block onto an object-slide with an immersion oil droplet between the two and observing it under an inverted microscope. Non-trimmed blocks were cut and the ultrathin sections harvested onto carbon-Formvar-coated grids. These were covered with an emulsion monogranular layer using a loop. Silver grains were seen localized over nuclei, particularly over nucleoli of infected cells, suggesting an important role of this organelle during the lytic infection with SV 40. (Figs. 5.13-5.15).

Discussion

The experimental procedures used in the above studies vary widely. Indeed, these are derived from the procedures previously used for light microscope cytological hybridization, which are themselves derived from those used for molecular hybridization on filters; biochemists have developed many different denaturation and hybridization procedures. The details concerning the specificity of the pairing reactions between nucleic acid molecules using DNA immobilized on a membrane can be found, for instance, in the studies by Nygaard and Hall (1963), Gillepsie and Spiegelman (1965), and McCarthy and Church (1970).

Various aspects of the *in situ* molecular hybridization method using labeled RNA or DNA, as well as the cytological results obtained until 1972, have been discussed by Hennig (1973). A detailed description of the technique used by Gall and Pardue in their pioneering work in this field is published (1971). For additional references concerning the detection of viral genomes in cells by *in situ* molecular hybridization, the reader is referred to Geuskens and May (1974).

In order to obtain reproducible results when studying the association of polynucleotidic sequences on nitrocellulose filter, temperature, saline concentration, fragment size, and concentration of molecules must be controlled. As pointed

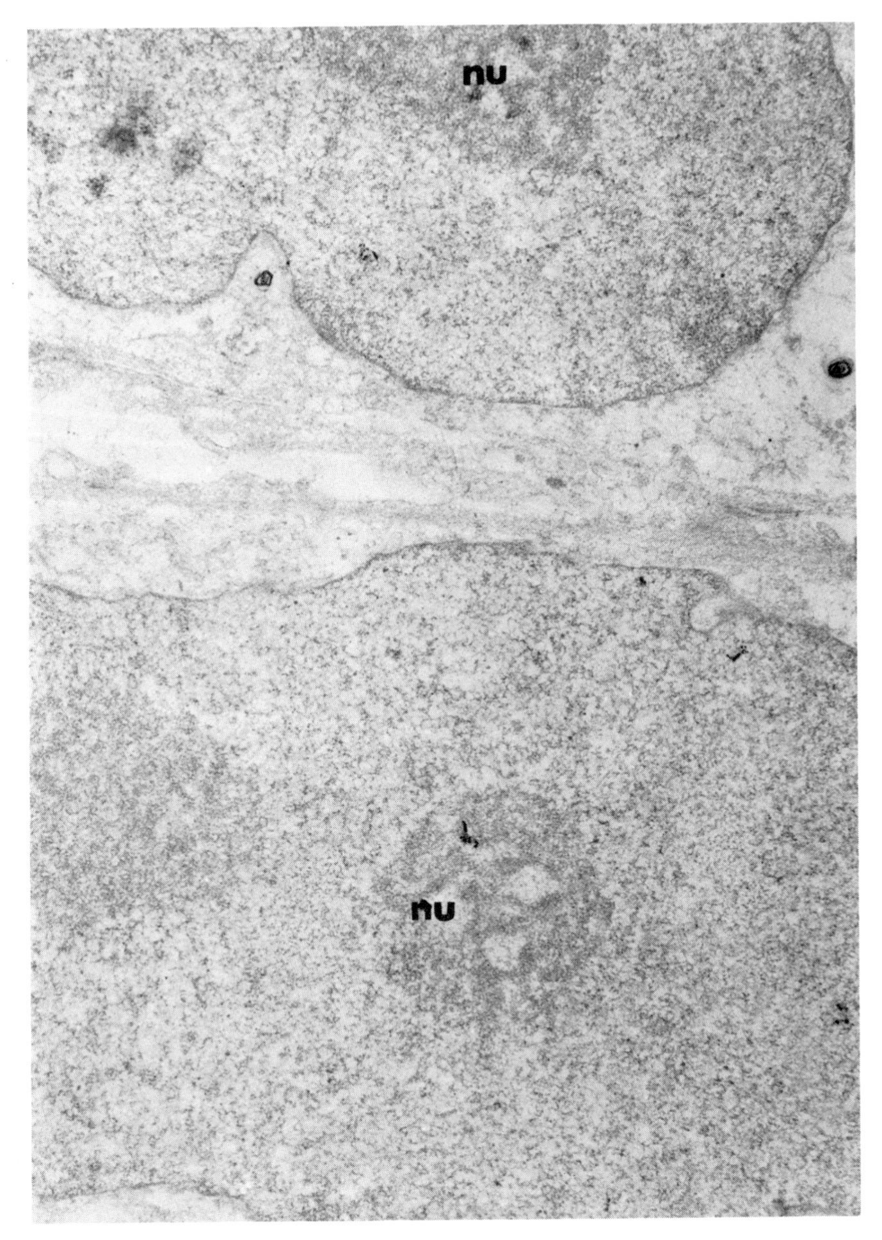

Fig. 5.13 Control CV1 cells: *In situ* molecular hybridization with ^{3}H-RNA complementary to SV 40 viral DNA before embedding. N: nucleus; nu: nucleolus. Autoradiogram exposed for two months. ×11,200.

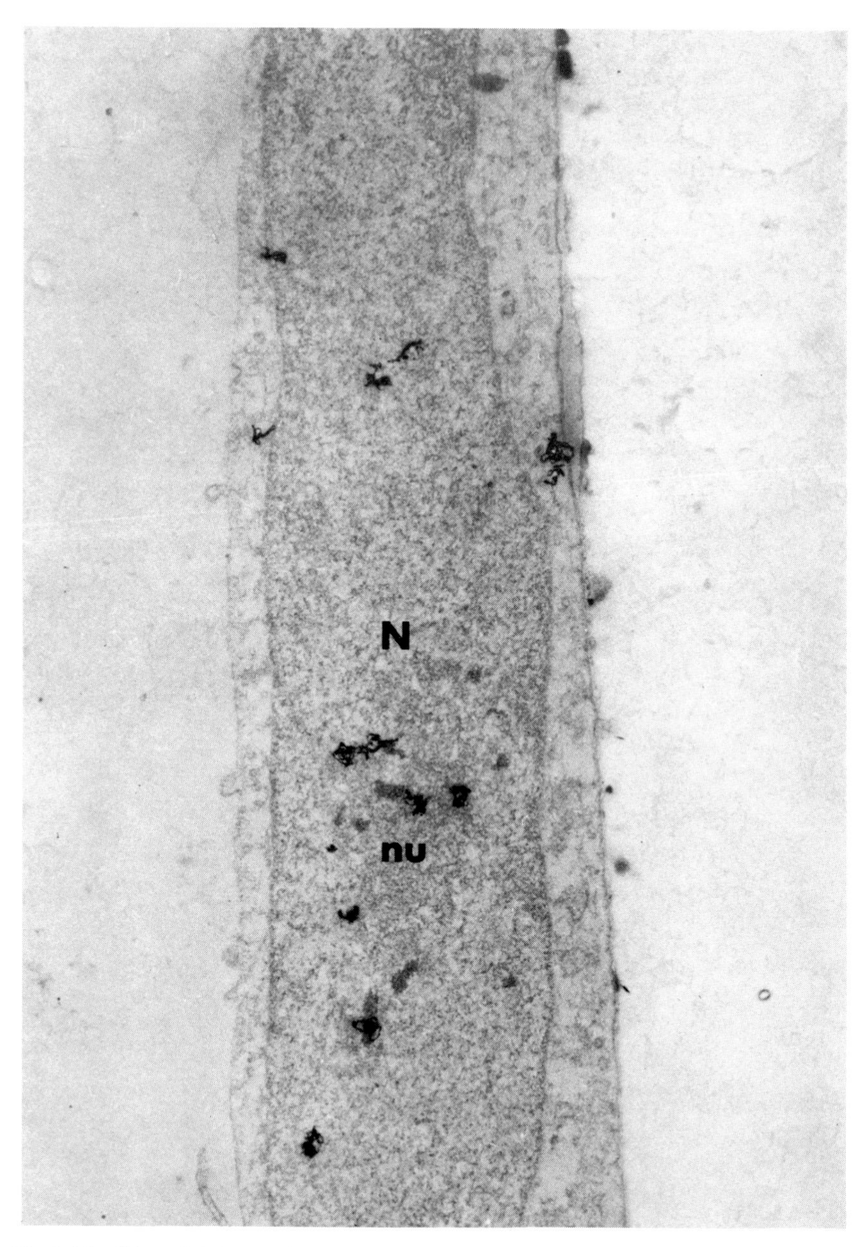

Fig. 5.14 CV1 cell infected for 20 hr with SV 40 virus: *in situ* molecular hybridization with ^{3}H-RNA complementary to SV 40 viral DNA before embedding. The labeling is localized over the nucleus. nu: nucleolus. The cell is on the edge of the section (plastic on the left and Formvar film alone on the right side of the cell). Autoradiogram exposed for two months. ×10,500.

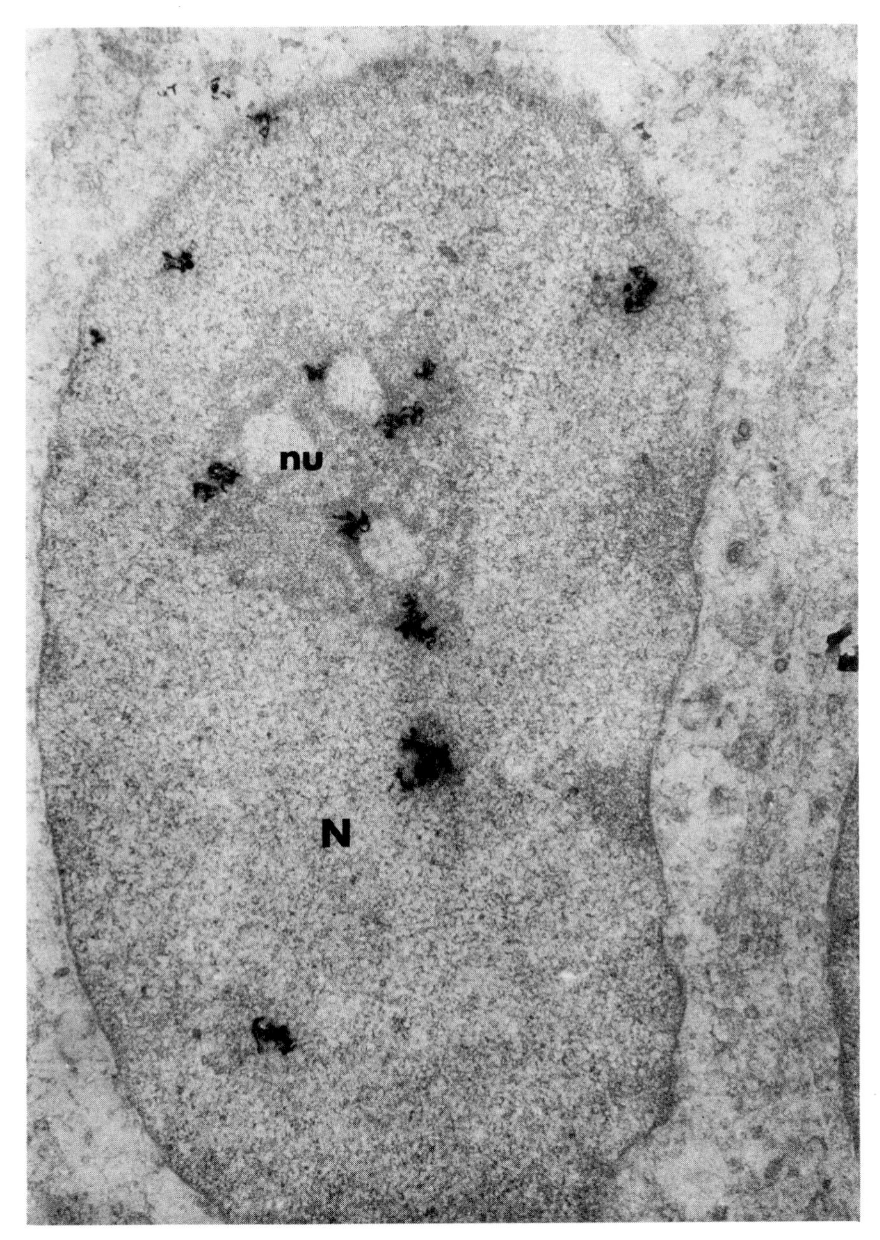

Fig. 5.15 CV1 cell infected for 30 hr with SV 40 virus: *in situ* molecular hybridization with ^{3}H-RNA complementary to SV 40 viral DNA before embedding. The silver grains are seen localized over the nucleus (N) and more particularly over the nucleolus (nu). Autoradiogram exposed for two months. ×14,500.

out by Hennig (1973), the physico-chemical parameters of the *in situ* reactions are not known. However, evidence supporting the validity of the method is provided by the results of several biological tests including competition experiments with unlabeled RNA or hybridization with labeled heterologous RNA. Moreover, a preferential cytological localization of the labeling has been observed in many cases as reported in the review by Hennig (1973). The interpretation of the results is sometimes controversial as unknown factors are involved in the reactions. Moreover, no quantitative interpretation can be inferred from the results.

Efficiency

The efficiency of the *in situ* molecular hybridization method has been estimated to be 1 to 10% according to various investigators (Hennig, 1973). The *in situ* molecular hybridization method performed at the electron microscope level compromises between hybridization efficiency and the preservation of cell fine structure. The majority of the studies carried out at the light microscope level deal with specimens fixed in ethanol–acetic acid mixture. This fixative cannot be used for electron microscopy because the fine structure of the cells, especially their nuclei, is poorly preserved. Fixatives currently used for electron microscope work could interfere with nucleic acid hybridization if DNA cross-linking affects hybridization. Moreover, cross-linking of other cell chemical constituents can prevent penetration of the labeled molecules and their access to DNA. However, a brief fixation in glutaraldehyde, while possibly reducing the efficiency of the molecular hybridization reaction, does not prevent the hybridization (Croissant *et al.*, 1972).

Formaldehyde seems at first sight to be inadequate as a fixative because of its cross-linking action on DNA. Nevertheless, we have chosen to fix cells overnight in 4% formaldehyde at 4°C because these parameters yield a satisfactory preservation of nuclear structures, and according to Kuhlman (1973), this fixation does not prevent penetration of as large molecules as peroxidase-labeled antibodies into cells when immunocytochemical methods are used for detecting antigens at the ultrastructural level. Probably plasma and nuclear membrane alterations also allow penetration of low molecular weight RNA molecules. It is possible that part of the action of formaldehyde on DNA is reversible during subsequent incubation conditions because molecular hybridization appears possible (Geuskens and May, 1974).

The fact that molecular hybridization can still occur between ^{3}H-RNA and DNA located at the surface of ultrathin sections of glutaraldehyde-fixed and plastic-embedded tissues is more surprising. However, under these conditions, the sensitivity of the method is probably poor.

The results obtained by Croissant *et al.* (1972 and personal communication) have shown that regular penetration of the labeled probe in 100–150-μm-thick papilloma sections is possible after brief fixation. The use of frozen 10–15-μm-thick sections cut in a cryostat for incubation in the presence of the labeled nucleic acid, followed by fixation and embedding in epoxy resins, could furnish still better conditions for obtaining regular intracellular penetration when soft specimens are used. After fixation in ethanol–acetic acid, a treatment with 45% acetic acid or hydrochloric acid is generally carried out for removing histones which could interfere with the nucleic acid association.

Attempts by Das and Alfert (1969) with a view to hybridizing ^{3}H-RNA to complementary DNA located in fixed animal and plant cells had led them to conclude that RNA binds to nuclear and cytoplasmic basic proteins and that this bound RNA is relatively resistant to ribonuclease digestion. This difficulty, which is also encountered when insufficiently deproteinized DNA is used for hybridization on filters, has not been reported by other investigators as a cause of nonspecific binding in cytological hybridization studies. However, it is generally considered that the hybridization level is lower when a treatment for removing basic proteins has not been carried out. Enzymatic digestion treatment with pronase or another protease of broad action specificity is preferable to acid treatment because for electron microscopy the cell morphology will be better preserved. A treatment with pancreatic ribonuclease for removing messenger RNAs which could form hybrid molecules with the labeled probe is generally also carried out before DNA denaturation. A thorough washing of the specimen is then necessary for removing residual enzyme molecules which could hydrolyze ^{3}H-cRNA subsequently used.

The denaturation step which is essential for obtaining hybrid molecules can be carried out in different ways. The use of alkaline pH is the most effective and frequently used treatment because the reaction does not depend upon DNA or saline concentrations. As pointed out by Hennig (1973), treatment with NaOH has the additional advantage of hydrolyzing cellular RNA which could interfere with hybridization; previous treatment with ribonuclease is then unnecessary.

Denaturation by NaOH is probably not the most favorable treatment for cell fine structure preservation when nonembedded specimens are used, but is particularly convenient when plastic-embedded tissue ultrathin sections are used. Treatment of up to two hours does not cause additional damage in that case (Jacob *et al.*, 1971).

Denaturation by heat has been carried out by Croissant *et al.*, (1972). It must be remembered that the cytological conditions are different from those used for hybridization on nitrocellulose filters. Higher temperatures must be used, especially if double-stranded DNA is still stabilized by associated proteins. If DNA denaturation is not complete, parts of the double-stranded molecules will not be open to complementary RNA, and the level of hybridization will be reduced. On the contrary, treatment at a high temperature necessary for disruption of G-C bonds, which account on an average for 41% of the bonds in DNA, can damage briefly fixed tissues. Addition of formamide to the denaturation medium allows this step to be carried out at a lower temperature. This procedure has been used for cytological hybridization by Alonso (1973) and Alonso *et al.* (1974). The cell fine structure is probably less damaged by this treatment. However, if membranes are not locally disrupted by heat treatment, ^{3}H-RNA penetration can be impaired.

One cause for low efficiency of molecular hybridization is supposed to be the rapid reassociation of complementary DNA strands. Indeed, under the cytological conditions, the complementary DNA strands are maintained in close association. The usual way for avoiding DNA/DNA reassociation is to rapidly immerse preparations into an ice-cold solution of low ionic strength (0.1 × SSC) and eventually into 70% and 95% ethanol (Croissant *et al.*, 1972). Drying the preparation at this stage, as usually done with material used for light microscopy, must be avoided in order to preserve cell fine structure. If the preparations have been stored in alcohol after denaturation, they must be bathed again in ice-cold 0.1 × SSC before the incubation of the material in the presence of the labeled nucleic acid.

Two recent publications have contributed to our understanding of the modifications induced in nuclei (Franke *et al.*, 1973) and chromosomes (Comings *et al.*, 1973) by the procedures carried out during the *in situ* hybridization and denaturation-renaturation techniques. Losses of DNA are dramatic when the standard denaturation treatments are performed on isolated nuclei (Franke *et al.*, 1973), and this is a supplementary argument against quantitative interpretation of *in situ* hybridization results. Up to 80% of fixed chromosomal DNA is removed by NaOH treatment (Comings *et al.*, 1973). Losses of proteins and RNA are also important.

The work by Comings *et al.* (1973) has also emphasized the fact that repetitious DNA renatures in seconds while nonrepetitious DNA renatures in minutes, confirming the difficulty of competition of DNA/DNA reassociation. Therefore, these investigators suggest a denaturation of the chromosomal DNA in formamide-saline solution at 100°C for 2 min, and then bringing the solution to 60° or 65°C for hybridization. Denaturation of cellular DNA for 30 min at 65°C, in the presence of formamide and ^{3}H-RNA, followed by progressive lowering of temperature to 37°C, a temperature at which hybridization was allowed to take place for 18 hr, was carried out by Jones and Robertson (1970). Wimber *et al.* (1974) have also used such conditions in their *in situ* hybridization experiments with ^{125}I-5S RNA, the slides being kept in a 50% formamide-salt solution for 1 hr at 60°C and then transferred to 40°C for 3 hr.

Denaturation and hybridization in the presence of 50% formamide in 2 × SSC at the respective temperatures of 70° and 37°C have constituted favorable conditions for ultrastructural localization of SV 40 viral DNA by hybridization with ^{3}H-cRNA (Geuskens and May, 1974). In these experiments, the only difference between denaturation and hybridization medium was the addition of buffer to the hybridization medium in order to avoid pH modifications during the long incubation period. However, the omission of the rinse between these two technical steps would probably have improved the efficiency of the hybridization reaction. Indeed ^{3}H-RNA could advantageously be added directly to the denaturation medium. After denaturation at 70°C, the temperature of the oven containing the Microtest plates could be progressively lowered to 37°C for hybridization. The ^{3}H-RNA already present during the denaturation step could more efficiently compete with DNA/DNA reassociation under the annealing conditions. This technical improvement is of course still more valid for ^{3}H-DNA/DNA hybridization experiments as DNA in solution itself will rapidly reassociate.

The main point which restricts the use of ^{3}H-DNA for *in situ* hybridization experiments is that reassociation occurs between the molecules in solution, reducing the efficiency of the hybridization reaction (Hennig, 1973). However, this difficulty is avoided if the two DNA strands can be separated by centrifugation in CsCl gradient. Moreover, it is advisable to shear the DNA molecules into fragments of ~1,000 nucleotides as recommended by biochemists for experiments performed on filters. This will also improve the penetration of molecules into cells.

The so-called hybridization conditions are the same as those used for light microscope experiments. A temperature of 65°C (or 37°C if formamide is included in the medium) and dissolution of the labeled probe in 2 X SSC are generally used.

Durations of incubation varying from 5 (Jacob *et al.*, 1971), to 16 (Croissant *et al.*, 1972), to 45 hr (Geuskens and May, 1974) have been used so far for electron microscope experiments. However, it seems that hybrid formation, as DNA renaturation, is very rapid. Therefore, incubations longer than a few hours are probably useless if the hybridization reaction is carried out at 65°C. However, if formamide is present in the incubation medium, the reaction rate could be slower (Hennig, 1973) and hybrid molecule formation can proceed for several hours (Bonner *et al.*, 1967; Alonso *et al.*, 1974). Addition of formamide to the hybridization medium is recommended by Alonso *et al.* (1974) because it improves the efficiency of the reaction. The preparations may be incubated at a lower temperature if formamide is added to the medium (Bonner *et al.*, 1967), which probably allows better preservation of the cell fine structure.

After hybridization, the preparations are extensively washed with SSC at the same temperature and concentration as in the incubation medium and treated afterwards with pancreatic ribonuclease for elimination of nonhybridized labeled RNA molecules and unpaired loops, RNA complexed with DNA being resistant to the enzyme (in 2 X SSC, at 37°C). Takadiastase T1 ribonuclease can be used in addition for removing unpaired loops consisting of purine bases. After ^{3}H-DNA/DNA hybridization performed on *Xenopus* oocyte squashes, Pardue and Gall (1969) thoroughly washed their preparations with hot 2 X SSC to eliminate nonhybridized molecules. A supplementary treatment with a nuclease specific for single-stranded DNA, followed by rinses in cold dilute trichloracetic acid is however advisable for eliminating this cause of unspecific labeling. An endonuclease from *Neurospora crassa* (Linn and Lehman, 1965a,b) or a nuclease from *Aspergillus oryzae* (Vogt, 1973) may be used for this purpose.

Sensitivity

The low sensitivity of the method is the main restriction on its use at the electron microscope level, although it may allow the detection of smaller quantities of viral DNA than other biochemical methods, for example, a Hirt extraction. There is only a very small amount of material which can hybridize with the labeled probe at a point of a nucleus, for instance. This amount is still more greatly reduced in quantity in electron microscope preparations by the necessity to use ultrathin sections. This requires the use of ^{3}H-labeled nucleic acid molecules of very high specific activity in order to yield a sufficient number of disintegrations at the site of hybridization, taking into account the efficiency of the autoradiographic technique itself. Gall and Pardue (1971) consider 10^5 cpm/μg of RNA as the useful lower limit for cytological hybridization experiments performed at the light microscope level. For electron microscope work, it is necessary to raise the specific activity of the labeled nucleic acids used with an order of magnitude of 100 to 1,000.

In vitro enzymatic transcription of DNA with a bacterial DNA-dependent RNA polymerase in the presence of the four tritiated nucleoside triphosphates imparts the nucleic acid a high specific activity (up to 10^8 dpm/μg). However, the labeled polynucleotides are obtained in mixed sizes and proportions. *In vitro* copying of isolated RNA molecules by reverse transcriptase can also be used for obtaining DNA molecules of high specific activity.

A new approach to *in situ* molecular hybridization method at the ultrastructural level seems attractive for future use. Nucleic acid molecules which can theoretically reach 10^9 dpm/μg in specific activity can be obtained by introducing ^{125}I into their molecular structure after the completion of biological synthesis. This ^{125}I labeling of the nucleic acid molecules does not noticeably alter their hybridization capacity. Single-stranded DNA or RNA can be labeled in this way with 5-iodocytosine formation (Commerford, 1971; Getz *et al.*, 1972; Altenburg *et al.*, 1973; Tereba and McCarthy, 1973; Scherberg and Refetoff, 1974). The most frequently used iodination procedure is described by Prensky *et al.* (1973).

Low energy electrons of ^{125}I can be used in both light (Appelgren *et al.*, 1963) and electron microscope (Kayes *et al.*, 1962) autoradiography. The Auger electrons resulting from ^{125}I decay will produce autoradiograms with greater efficiency than ^{3}H (Prensky *et al.*, 1973; Fertuck and Salpeter, 1974). However, the resolution will be somewhat poorer. Elimination of protein contamination from the nucleic acid molecules is essential before iodination. This *in vitro* labeling method of nucleic acid molecules with ^{125}I allows obtainment of higher specific activities than those by *in vitro* labeling with tritium. Moreover, only very low quantities of nucleic acid (less than 1 μg) are necessary for the reaction.

In vitro transcription of DNA corresponding to a single structural gene is not possible, since isolation of a single gene DNA population has not been technically possible as yet. On the other hand, isolation of a single RNA population (mRNA for histones, hemoglobin, L and H chains of immunoglobulins, silkworm fibroin) has already been realized. The *in vitro* labeling of these mRNAs with ^{125}I at very high specific activity is thus theoretically possible for localizing the corresponding structural genes by cytological molecular hybridization. However, this theoretical possibility is limited by the fact that the usual conditions for *in situ* molecular hybridization allow in practice detection of only repeated DNA sequences or highly amplified genes in the genome.

Giant dipterian chromosomes, however, are a favorable material where localization of less repeated sequences has been possible (Hennig, 1973). Iodine-labeled ribosomal RNAs (18S + 28S, and 5S) have already been successfully used in cytological hybridization experiments with dipterian salivary gland chromosomes (Prensky *et al.*, 1973; Wen *et al.*, 1974). A satisfactory labeling resolution was obtained. However, no clear chromosomal localization was observed with ^{125}I-4S RNA. More recently, 5S RNA genes have been localized in *Zea mays* by *in situ* RNA-DNA hybridization using iodinated RNA (Wimber *et al.*, 1974).

For detection of unique DNA sequences, high concentrations of the labeled probe and long durations of incubation are necessary, at least when molecular hybridization is carried out on filters. In practice, excess ^{3}H-RNA is always furnished in cytological experiments. Incubations up to 45 hr in 50% formamide at 37° seem to do no harm to nonembedded cell fine structure (Geuskens and May, 1974). Anyway, the localization of single DNA sequences at the electron microscope level is seriously limited by the impossibility of observing whole nuclear or chromosomal preparations. The necessity to section the material renders the observation of localized silver grain concentration very hazardous at the electron microscope level.

The use of N,N-dimethyl formamide as denaturing agent has been recommended by Scherberg and Refetoff (1973) as a way to improve the efficiency of the hybridization reaction. According to these workers, the agent allows one to carry out hybridization reaction at a low temperature and with higher efficiency than with formamide. A very low background is obtained under filter conditions. The use of this denaturing agent is unusual for experiments performed on filter because it causes shrinkage of the filter and DNA losses. However, this drawback does not exist in cytological studies.

CONCLUDING REMARKS

The method of *in situ* molecular hybridization performed at the electron microscope level is still in its infancy. While the first paper on this subject was published in 1971, to our knowledge only three other publications reporting results obtained at the ultrastructural level have appeared so far; two were contributed by other laboratories. Confirmation of the reproducibility of these results in the case of other biological systems studied in other laboratories is necessary before it is possible to recommend any one technical procedure. However, as mentioned above, the technical conditions must be chosen according to the material used.

Molecular hybridization carried out directly on ultrathin sections is an attractive technique as it eliminates problems of labeled probe penetration into cells. Moreover, plastic-embedded material is more resistant to alterations due to the denaturation and hybridization conditions. However, the efficiency of this technique seems to be low and the results are not reproducible according to our own experience. Moreover, factors difficult to control are involved in the plastic polymerization reaction.

The use of briefly fixed material embedded in plastic after completion of the hybridization reaction has the advantage of allowing simultaneous use of semithin sections for light microscope autoradiography. As the autoradiographic technique performed at the electron microscope level requires months of exposure and is thus very time-consuming, the significance of obtaining rapid results with the light microscope cannot be underestimated. Moreover, if silver grains are seen concentrated over some structure under the light microscope, a few grains observed over the same structure in the electron microscope will appear more significant and convincing. Indeed, as the expected labeling is low, especially if the DNA sequences to localize are not highly reiterated and clustered, the radioactivity resulting from true molecular hybrids will be difficult to distinguish from the labeling resulting from other types of unspecific labeling and the background resulting from the autoradiographic process itself.

Problems of labeled probe penetration may be overcome by using 10- or 15-μm-thick frozen sections or monolayers of cells in culture for incubation. Also, shearing of the nucleic acid molecules into fragments of ~1,000 nucleotides or less is advisable to improve this penetration. Theoretically, the specificity of the pairing reaction is lowered in such conditions, but with RNA fragments of a transfer RNA size, unspecific matching will be minimal (McCarthy and Church, 1970).

Denaturation and hybridization in the presence of formamide or dimethyl formamide and the labeled nucleic acid, the temperature being progressively lowered between the two steps, are recommended in order to improve the efficiency of the method. Labeling of nucleic acid molecules to a very high specific activity with ^{125}I is advisable for improving the sensitivity of the method.

However, the extension of the method at the ultrastructural level will always be restricted. To avoid making useless attempts, it is advisable to carry it out only after obtaining convincing results with the same material using standard light microscope hybridization procedure.

Unless one uses a large number of serial sections of the same cell, the probability of detecting clustered hybrid molecules is very low. Therefore, to avoid losing much time, only particular problems should be examined with this technique. Only localization of highly reiterated DNA whose rough localization is already known from the results of other experiments will be reasonably approached with this method. Up to now, the search for viral genomes in virus-producing cells, where a great number of the same DNA sequences are dispersed in the nuclei, has furnished the most favorable biological test system to carry out the technique. However, even when a convincing number of grains are observed over a structure, care must be taken in the interpretation of the results, as labeling can be due to nonspecific binding or trapping.

ADDENDUM

Since writing this chapter, the presence of ribosomal DNA in the Feulgen-positive bodies (which appear during maturation in the cytoplasm of *Xenopus laevis* oocytes) has been demonstrated (Steinert *et al.*, 1976). This was accomplished by studying *in situ* hybridization of ^{125}I-labeled ribosomal RNA. The procedure used was that presented by Jacob *et al.* (1971) in that GMA-tissue sections were incubated with ^{125}I-cRNA, under the technical conditions described in Fig. 5.6.

A new method for gene mapping at the chromosome level using *in situ* molecular hybridization and scanning electron microscopy has recently been developed (Manning *et al.*, 1975). According to this method, biotin-labeled rRNA is hybridized to denatured DNA in a chromosome squash. Upon incubation with a solution of polymethacrylate spheres attached to the protein avidin, some of the biotin sites become labeled with spheres because of the strong non-covalent interaction between biotin and avidin. The chromosome squash is then examined in the scanning electron microscope and polymer spheres which label the nucleolus can be observed.

A recent publication by Gall and Pardue (1975) describes nucleic acid hybridization to the DNA in cytological preparations. This publication presents the procedures used in their laboratory. Also, Altenburg *et al.* (1975) and Scherberg and Refetoff (1975) studied the radioiodination of nucleic acids and the application of these molecules to biology, especially for *in situ* hybridization experiments. A comprehensive review on the ultrastructural localization of DNA in ultrathin tissue sections, including the binding of organic compounds revealed by autoradiography, has recently been published (Gautier, 1976).

The author is grateful to Professor J. Brachet and Doctor W. Bernhard, who have maintained a continuous interest in this work. He is also indebted to Drs. O. Croissant, A. M. de Recondo, and P. May who have critically read this review and to Dr. Ch. Thomas for helpful suggestions. During the development of these techniques in the laboratory of Dr. Bernhard (Institute for Scientific Research on Cancer, Villejuif, France), the excellent technical assistance of M. J. Burglen has been greatly appreciated. The assistance of Dr. A. Morris and Dr. M. A. Hayat for improvement of the English version of the manuscript is acknowledged with gratitude. The author is "Maître de recherches" of the Belgian National Fund for Scientific Research.

References

Alonso, C. (1973). Improved conditions for *in situ* RNA/DNA hybridization. *FEBS Letters* **31,** 85.

Alonso, C., Helmsing, P. J., and Berendes, H. D. (1974). A comparative study of *in situ* hybridization procedures using cRNA applied to *Drosophila hydei* salivary gland chromosomes. *Exp. Cell Res.* **85,** 383.

Altenburg, L. C., Getz, M. J., Crain, W. R., Saunders, G. F., and Shaw, M. W. (1973). ^{125}I-labelled DNA-RNA hybrids in cytological preparations. *Proc. Nat. Acad. Sci.* **70,** 1536.

Altenburg, L. C., Getz, M. J., and Saunders, G. F. (1975). ^{125}I in molecular hybridization experiments. In: *Methods in Cell Biology*, Vol. 10 (Prescott, D. M., ed.), p. 325. Academic Press, New York.

Appelgren, L. E., Söremark, R., and Ullberg, S. (1963). Improved resolution in autoradiography with radioiodine using the extranuclear electron radiation from ^{125}I. *Biochim. Biophys. Acta* **66,** 144.

Bernier, R., Iglesias, R., and Simard, R. (1972). Detection of DNA by tritiated Actinomycin D on ultrathin frozen sections. *J. Cell Biol.* **53,** 798.

Bonner, J., Kung, G., and Bekhor, I. (1967). A method for the hybridization of nucleic acid molecules at low temperature. *Biochemistry* **6,** 3650.

Bouteille, M., Dupuy-Coin, A. M., and Moyne, G. (1975). Localization of proteins and nucleoproteins in the cell nucleus by high resolution autoradiography and cytochemistry. Methods in enzymology **40,** Part E, 3.

Bouteille, M., Laval, M., and Dupuy-Coin, A. M. (1974). Localization of nuclear function as revealed by ultrastructural autoradiography and cytochemistry. In: *The Cell Nucleus*, Vol. I (Busch, H., ed.), p. 3. Academic Press, New York.

Brachet, J., and Ficq, A. (1965). Binding sites of ^{14}C-Actinomycin in amphibian ovocytes and an autoradiography technique for the detection of cytoplasmic DNA. *Exp. Cell Res.* **38,** 153.

Camargo, E. P., and Plaut, W. (1967). The radioautographic detection of DNA with tritiated Actinomycin D. *J. Cell Biol.* **35,** 713.

Chevaillier, P., de Recondo, A. M., and Geuskens, M. (1974). Deoxyribonucleic acid of the crab *Cancer pagurus*. III. Intracellular localization of poly d(A-T) of *Cancer pagurus* by hybridization *in situ*. *Exp. Cell Res.* **86,** 183.

Comings, D. E., Avelino, E., Okada, T. A., and Wyandt, H. E. (1973). The mechanism of C- and G-banding of chromosomes. *Exp. Cell Res.* **77,** 469.

Commerford, S. L. (1971). Iodination of nucleic acids *in vitro*. *Biochemistry* **10,** 1993.

Croissant, O., Dauguet, C., Jeanteur, P., and Orth, G. (1972). Application de la technique d'hybridation moléculaire *in situ* à la mise en évidence au microscope électronique de la réplication végétative de l'ADN viral dans les papillomes provoqués par le virus de Shope chez le lapin cottontail. *C. R. Acad. Sci.* **274,** 614.

Das, N. K., and Alfert, M. (1969). Binding of labelled ribonucleic acid to basic proteins, a major difficulty in ribonucleic acid–deoxyribonucleic acid hybridization in fixed cells *in situ*. *J. Histochem. Cytochem.* **17,** 418.

Davis, B. D., Dulbecco, R., Eisen, H. N., Ginsberg, H. S., and Wood, W. B. (1968). *Microbiology*. Hoeber Medical Division, Harper and Row, New York.

Ebstein, B. S. (1967). Tritiated Actinomycin as a cytochemical label for small amounts of DNA. *J. Cell Biol.* **35,** 709.

Fakan, S., and Modak, S. P. (1973). Localization of DNA in ultrathin sections incubated with terminal deoxynucleotidyl transferase, as visualized by electron microscope autoradiography. *Exp. Cell Res.* **77,** 95.

Fertuck, H. C., and Salpeter, M. M. (1974). Sensitivity in electron microscope autoradiography for ^{125}I. *J. Histochem. Cytochem.* **22,** 80.

Franke, W. W., Deumling, B., and Zentgraf, H. (1973). Losses of material during cytological preparation of nuclei and chromosomes. *Exp. Cell Res.* **80,** 445.

Gall, J. C., and Pardue, M. L. (1971). Nucleic acid hybridization in cytological preparations. In: *Methods in Enzymology,* XXI D, 470. Academic Press, New York.

Gautier, A. (1976). Ultrastructural localization of DNA in ultrathin tissue sections. *Int. Rev. Cytol.* **44,** 113.

Gellert, M., Smith, C. E., Neville, D., and Felsenfeld, G. (1965). Actinomycin binding to DNA: mechanism and specificity. *J. Mol. Biol.* **11,** 445.

Getz, M. J., Altenburg, L. C., and Saunders, G. (1972). The use of RNA labelled *in vitro* with iodine-125 in molecular hybridization experiments. *Biochim. Biophys. Acta* **287,** 485.

Geuskens, M. (1972). Fixation d'Actinomycine D tritiée sur la chromatine de coupes à congélation. *J. Microscopie* **13,** 153.

Geuskens, M. (1974). ^{3}H-Actinomycin D binding to DNA localized in ultrathin sections of plastic-embedded biological materials. *J. Ultrastruct. Res.* **47,** 179.

Geuskens, M., and May, E. (1974). Ultrastructural localization of SV 40 viral DNA in cells, during lytic infection, by *in situ* molecular hybridization. *Exp. Cell Res.* **87,** 175.

Geuskens, M., de Recondo, A. M., and Chevaillier, P. (1974). Localization of DNA in fixed tissue by radioautography. 2. DNA specific sequences. *Histochem. J.* **6,** 69.

Gillepsie, D., and Spiegelman, S. (1965). A quantitative assay for DNA-RNA hybrids with DNA immobilized on a membrane. *J. Mol. Biol.* **12,** 829.

Goldberg, J. H., and Friedman, P. A. (1971). Antibiotics and nucleic acids. *Ann. Rev. Biochem.* **40,** 775.

Hall, B. D., and Spiegelman, S. (1961). Sequence complementarity of T2-specific RNA. *Proc. Nat. Acad. Sci.* **47,** 137.

Harven, E. de. (1968). Electron-microscope autoradiography. In: *Radioisotopes in Medicine:* in vitro *Studies,* Proc. Symp. Oak Ridge Associated Univ., Nov. 1967 (Hayes, R., Goswitz, F. A., and Murphy, B. E. P., eds.), p. 661. U.S. Atomic Energy Comm., Div. Tech. Inform., Oak Ridge, Tennessee, U.S.A.

Hayat, M. A. (1975). *Positive Staining for Electron Microscopy.* Van Nostrand Reinhold Company, New York and London.

Hennig, W. (1973). Molecular hybridization of DNA and RNA *in situ. Int. Rev. Cytol.* **36,** 1.

Jacob, J. (1971). The practice and application of electron microscope autoradiography. *Int. Rev. Cytol.* **30,** 91.

Jacob, J., Gillies, K., MacLeod, D., and Jones, K. W. (1974). Molecular hybridization of mouse satellite DNA-complementary RNA in ultrathin sections prepared for electron microscopy. *J. Cell Sci.* **14,** 253.

Jacob, J., Todd, K., Birnstiel, M. L., and Bird, A. (1971). Molecular hybridization of (^{3}H)ribosomal RNA with DNA in ultrathin sections prepared for electron microscopy. *Biochim. Biophys. Acta* **228,** 761.

Jones, K. W., and Robertson, F. W. (1970). Localization of reiterated nucleotide sequences in *Drosophila* and mouse by *in situ* hybridization of complementary RNA. *Chromosoma* **31,** 331.

Kayes, J., Maunsbach, A. B., and Ullberg, S. (1962). Electron microscopic autoradiography of radioiodine in the thyroid using the extranuclear electrons of ^{125}I. *J. Ultrastruct. Res.* **7**, 339.

Kuhlman, W. D. (1973). Ultrastructural localization of antigens by peroxidase-labelled antibodies. In: *Electron Microscopy and Cytochemistry* (Wisse, E., Daems, W. T., Molenaar, I., and van Duijn, P., eds.), p. 155. North-Holland Publ. Co. Amsterdam, The Netherlands.

Leduc, E., and Bernhard, W. (1967). Recent modifications in the glycol methacrylate embedding procedure. *J. Ultrastruct. Res.* **19,** 196.

Linn, S., and Lehman, I. R. (1965a). An endonuclease from *Neurospora crassa* specific for polynucleotides lacking an ordered structure. I. Purification and properties of the enzyme. *J. Biol. Chem.* **240,** 1287.

Linn, S., and Lehman, I. R. (1965b). An endonuclease from *Neurospora crassa* specific for polynucleotides lacking an ordered structure. II. Studies of enzyme specificity. *J. Biol. Chem.* **240,** 1294.

Manning, J. E., Hershey, N. D., Broker, T. R., Pellegrini, M., Mitchell, H. K., and Davidson, N. (1975). A new method of *in situ* hybridization. *Chromosoma* **53,** 107.

Marinozzi, V. (1964). Cytochimie ultrastructurale du nucléole. RNA et protéines intranucléolaires. *J. Ultrastruct. Res.* **10,** 433.

Marmur, J., and Doty, P. (1961). Thermal renaturation of deoxyribonucleic acids. *J. Mol. Biol.* **3,** 585.

Marmur, J., Rownd, R., and Schildkraut, C. L. (1963). Denaturation and renaturation of deoxyribonucleic acid. *Progr. Nucl. Acid Res.* **1,** 231.

Marmur, J., and Ts'o, P. O. (1961). Denaturation of deoxyribonucleic acid by formamide. *Biochim. Biophys. Acta* **51,** 32.

McCarthy, B. J., and Church, R. B. (1970). The specificity of molecular hybridization reactions. *Ann. Rev. Biochem.* **39,** 131.

Nygaard, A. P., and Hall, B. D. (1963). A method for detection of RNA-DNA complexes. *Biochem. Biophys. Res. Commun.* **12,** 98.

Orth, G., Jeanteur, P., and Croissant, O. (1971). Evidence for and localization of vegetative viral DNA replication by autoradiographic detection of RNA-DNA hybrids in sections of tumors induced by the Shope papilloma virus. *Proc. Nat. Acad. Sci.* **68,** 1876.

Pardue, M. L., and Gall, J. G. (1969). Molecular hybridization of radioactive DNA to the DNA of cytological preparations. *Proc. Nat. Acad. Sci.* **64,** 600.

Pardue, M. L., and Gall, J. G. (1975). Nucleic acid hybridization to the DNA of cytological preparations. In: *Methods in Cell Biology*, Vol. 10 (Prescott, D. M., ed.), p. 1. Academic Press, New York.

Prensky, W., Steffensen, D. M., and Hughes, W. L. (1973). The use of iodinated RNA for gene localization. *Proc. Nat. Acad. Sci.* **70,** 1860.

Rae, P. M., and Franke, W. W. (1972). The interphase distribution of satellite DNA-containing heterochromatin in mouse nuclei. *Chromosoma* **39,** 443.

Reich, E., and Goldberg, I. H. (1964). Actinomycin and nucleic acid function. *Progr. Nucl. Acid Res. Mol. Biol.* **3,** 183.

Scherberg, N. H., and Refetoff, S. (1973). Hybridization of RNA labelled with ^{125}I to high specific activity. *Nature, New. Biol.* **242,** 142.

Scherberg, N. H., and Refetoff, S. (1974). The radioiodination of ribopolymers for use in hybridizational and molecular analyses. *J. Biol. Chem.* **249,** 2143.

Scherberg, N. H., and Refetoff, S. (1975). Radioiodine labeling of ribopolymers for special applications in biology. In: *Methods in Cell Biology*, Vol. 10 (Prescott, D. M., ed.), p. 343. Academic Press, New York.

Simard, R. (1967). The binding of Actinomycin D-^{3}H to heterochromatin as studied by quantitative high-resolution radioautography. *J. Cell Biol.* **35,** 716.

Sobell, H. M. (1973). The stereochemistry of Actinomycin binding to DNA and its implications in molecular biology. *Progr. Nucl. Acid Res. Mol. Biol.* **13,** 153.

Sobell, H. M. (1974). How Actinomycin binds to DNA. *Scient. American* **231,** 82.

Steinert, G., and Van Gansen, P. (1971). Binding of ^{3}H-Actinomycin to vitelline platelets of amphibian oocytes. A high resolution autoradiographic investigation. *Exp. Cell Res.* **64,** 355.

Steinert, G., Thomas, C., and Brachet, J. (1976). Localization by *in situ* hybridization of amplified ribosomal DNA during *Xenopus laevis* oocyte maturation (a light and electron microscopy study). *Proc. Nat. Acad. Sci.* **73,** 833.

Tereba, A., and McCarthy, B. J. (1973). Hybridization of ^{125}I-labelled ribonucleic acid. *Biochemistry* **12,** 4675.

Vogt, V. M. (1973). Purification and further properties of single-strand-specific nuclease from *Aspergillus oryzae*. *Eur. J. Biochem.* **33,** 192.

Wen, W. N., Leon, P. E., and Hague, D. R. (1974). Multiple gene sites for 5 S and 18 + 28 S RNA on chromosomes of *Glyptotendipes barbipes* (Staeger). *J. Cell Biol.* **62,** 132.

Wimber, D. E., Duffey, P. A., Steffensen, D. M., and Prensky, W. (1974). Localization of the 5 S RNA genes in *Zea mays* by RNA-DNA hybridization *in situ*. *Chromosoma* **47,** 353.

Yankofski, S. A., and Spiegelman, S. (1963). Distinct cistrons for the two ribosomal RNA components. *Proc. Nat. Acad. Sci.* **49,** 538.

6. OPTICAL ANALYSIS AND RECONSTRUCTION OF IMAGES

A. J. Gibbs and A. J. Rowe

Department of Biochemistry and Electron Microscope Section, School of Biological Sciences, University of Leicester, Leicester, England

INTRODUCTION

During recent years the resolution obtainable by transmission electron microscopes has approached close to atomic dimensions. It is inevitable however that with most biological specimens the resolution obtainable is not limited by the imaging system itself, but by the method used to make the specimen visible. The resolution that can be obtained from thin sectioned material is limited by chromatic aberration (Hayat, 1970), while for specimens that have been shadowed with heavy metals the resolution is limited to ~20 Å because of the size of the metal grain (Rowe, 1974). Positive and negative contrasting techniques which use heavy metal salts to visualize the specimen are limited in resolution by the size of the metal ions, and also by the extent of specimen penetration. For a detailed discussion of staining with heavy metals, the reader is referred to Hayat (1975). The quality of the micrograph produced by the electron microscope is further reduced by radiation damage produced by the electron beam and by the quantized nature of the beam itself. Along with operator error these effects distort the contrast of the image as it is recorded in the micrograph, and reduce the likelihood of the structure's being represented correctly. Isaacson has presented a comprehensive discussion of radiation damage of biological specimens in this volume.

The structural detail that is recorded in a micrograph can be enhanced in the microscope itself or by optical methods after it has been processed. Dark field imaging and in-focus phase contrast techniques are examples of the methods which operate in the microscope itself, and their principles and practices have been presented by Dubochet (1973) and Johnson (1973).

In this chapter we consider an alternative to in-microscope detail enhancement techniques; that is, the optical analysis of the micrograph for structural detail and its optical reconstruction with that detail enhanced. The methods described can equally well be applied to any form of micrograph or any photographically presented information.

One widely used method of optical image processing that can be used to enhance the contrast of a repeating structure in a micrograph is photographic superimposition (Markham *et al.*, 1963). The micrograph is photographed, then moved (rotationally or linearly) until the structure has been displaced by the distance of one structural repeat, whereupon it is rephotographed. This process is continued until the maximum possible number of structural repeats have been photographed and an averaged image has been produced. Great care must be taken in determining the correct displacement between photographs if artifacts are to be avoided. When a large number of structural repeats are recorded on one micrograph, this method of image enhancement can give good results (e.g., for large virus crystals); but offers little improvement for certain micrographs, and suffers from the disadvantage that the size of the repeat interval must be known (or strongly suspected) in advance. Artifactual "repeats" are of necessity generated in all multiple superimpositions (Fig. 6.1), and hence the application of this procedure to genuinely "unknown" specimens is hazardous in the extreme.

An alternative approach to image analysis and enhancement rests upon the Fourier analysis of the micrograph, i.e., separating the structure present into its linear or spatial frequency components. Such a Fourier analysis of the micrograph could be carried out on a digital computer. However, the information content of a typical electron micrograph poses problems for the storage and speed of even present-day machines. Fortunately, Fourier analysis of a micrograph can be carried out directly with the aid of the analogue optical computing device known as an optical diffractometer, which performs the Fourier analysis of the intensity distribution recorded on a photographic plate or film by producing its Fraunhofer diffraction pattern.

The development of optical diffractometry was carried out by the team of X-ray crystallographers at Manchester (Lipson and Taylor, 1951; Taylor and Lipson, 1951), who used it to produce the Fraunhofer diffraction patterns of proposed crystal structures for comparison with X-ray diffraction patterns. These early optical diffractometers used filtered arc lamps as their light sources, and as a result of the low level of illumination obtainable, the photographic recording of the diffraction patterns produced lasted many hours. In the mid 1960s came the development of continuous wave gas lasers which provided parallel beams of coherent monochromatic light, intense enough to reduce ex-

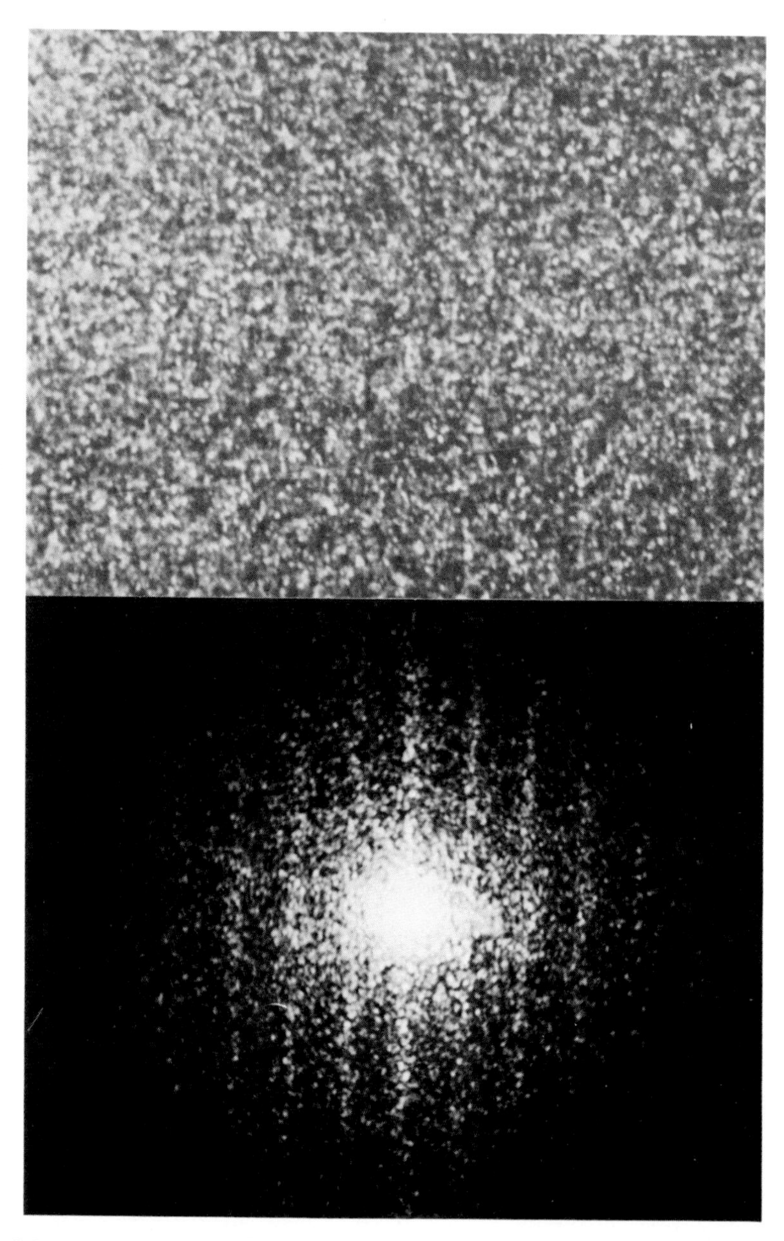

Fig. 6.1 An image of noise (above), and the optical transform (below) of the same after eight superimpositions with constant linear displacement between each (Gibbs and Rowe, unpublished results). The "layer lines" arise from the displacement, and are not related to any feature in the original image.

posure times to a few seconds and to produce diffraction patterns from objects possessing less structural order than crystallographic models. The availability of cheap helium-neon lasers has turned optical diffraction into a widely applicable research tool that has produced many important developments in the field of electron microscopy, and commercial instruments are now available (see Appendix to this chapter).

One of the first biological uses of optical diffraction was by Klug and Berger (1964), who were seeking a method of micrograph analysis that was both more sensitive and more unequivocally interpretable than the various methods of photographic averaging referred to above. It is a direct result of the work by Klug and Berger that optical diffraction has become a successful analytical tool for electron microscropists.

SIMPLE DIFFRACTION THEORY

The principles of optical, X-ray, and electron diffraction are identical; this connection has undoubtedly contributed towards the development of optical diffraction and the interpretation of the results it produces (Klug and Berger, 1964). The greater similarity exists between optical and electron diffraction because these radiations, unlike X rays, can be focused by lenses.

When a propagating wavefront passes normally through an object with periodic structure (i.e., a periodic amplitude or phase modulation), it is diffracted: the angle between the first diffraction maxima and the undiffracted beam is given by Bragg's equation (Fig. 6.2):

$$n\lambda = d \sin \theta \tag{6.1}$$

The factors in Eq. (6.1) are as shown in Fig. 6.2; n is an integer called the diffraction order. Thus it can be seen that the distribution of light in a plane (P)

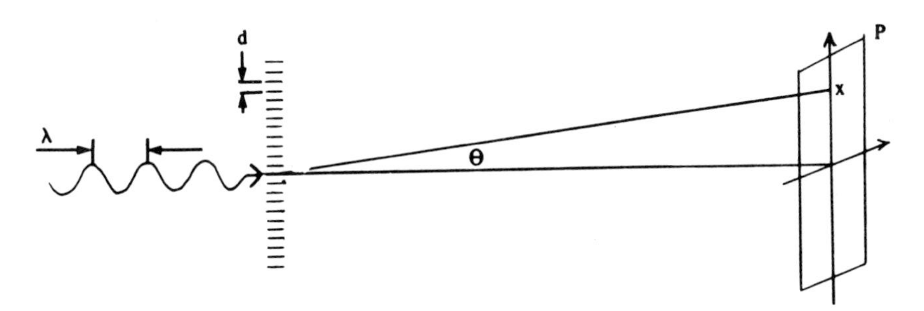

Fig. 6.2 Diagram illustrating the generation of a distribution of light intensity in a plane P, arising from the normal incidence of a wavefront on an object with periodic structure. At point x in plane P the condition for a maximum in light intensity is given in Eq. (6.1) (Bragg's Law).

intersecting the diffracted beam normally, is a reciprocal function of the periodicities producing the diffraction.

It can be shown that the introduction of a converging lens into the light beam on either side of the diffracting object will produce a Fourier transform of the object in the back focal plane of the lens. The back focal plane of a lens may be defined as the plane onto which a parallel beam of light will be focused by the lens. The mathematical proof of the Fourier transforming properties of lenses can be found in most optical textbooks (Lipson and Lipson, 1969).

Fourier transforms are complex functions, consisting of both amplitude and phase information. If an attempt is made to record the Fourier transform photographically, or with any other photodetector, the phase information will be lost, and the amplitude will be recorded as intensity (amplitude squared), which can only assume positive values. This distribution of intensity in the back focal plane of a lens is called the optical transform or optical diffraction pattern.

As a simple example, Fig. 6.3 shows results of an optical diffraction analysis. Electron microscope specimens are in general supported on some type of film, with the result that the micrograph recorded shows the structure of the specimen superimposed upon the pseudorandom structure of the support film. Optical diffraction of such a micrograph yields two components (Fig. 6.3), a pattern of relatively sharp diffraction spots produced by the specimen itself and a random background resulting from the support film. The symmetry and size of the background noise pattern can yield valuable information concerning the accuracy of focusing of the microscope and the degree of any astigmatism present. The use of an optical diffractometer for such an assessment of a microscope's performance can be found in Vol. 3 of this series (page 131).

We have already shown that Fourier-analyzing a structure splits it up into its harmonically related spatial frequencies, and that an optical system can do this in two dimensions at once. The mathematical inverse of Fourier analysis is called Fourier synthesis, which is the summation of a number of complex, harmonically related sinusoidal functions to form a new function. For example a square wave of amplitude A can be Fourier-analyzed into a series of harmonically related sine waves, all in phase and with amplitudes (a) given by the following equation:

$$a_n = \frac{A}{n\pi} \sin \frac{n\pi}{2} \tag{6.2}$$

where n is an integer representing the position in the harmonic series, i.e., $a_0 = \frac{1}{2}$, $a_1 = 1/\pi$, $a_2 = 0$, etc. Fourier synthesis of this example would consist of summation of these components to yield the original square wave.

In the above example the Fourier analysis contained only harmonically related components, showing that the original function contained no noise; if noise had been present, the analysis would have also contained a number of other com-

Fig. 6.3 Electron micrograph (*above*) of an A-segment of rabbit psoas muscle, and the optical transform (*below*) of the micrograph. The regular features of the transform, which arise from the highly ordered structure of this organelle are superimposed as a background noise spectrum. The latter arises partly from the support film used, and partly from actual degradation of the specimen during preparation for examination. (*From results by J. A. Trinick and A. J. Rowe.*)

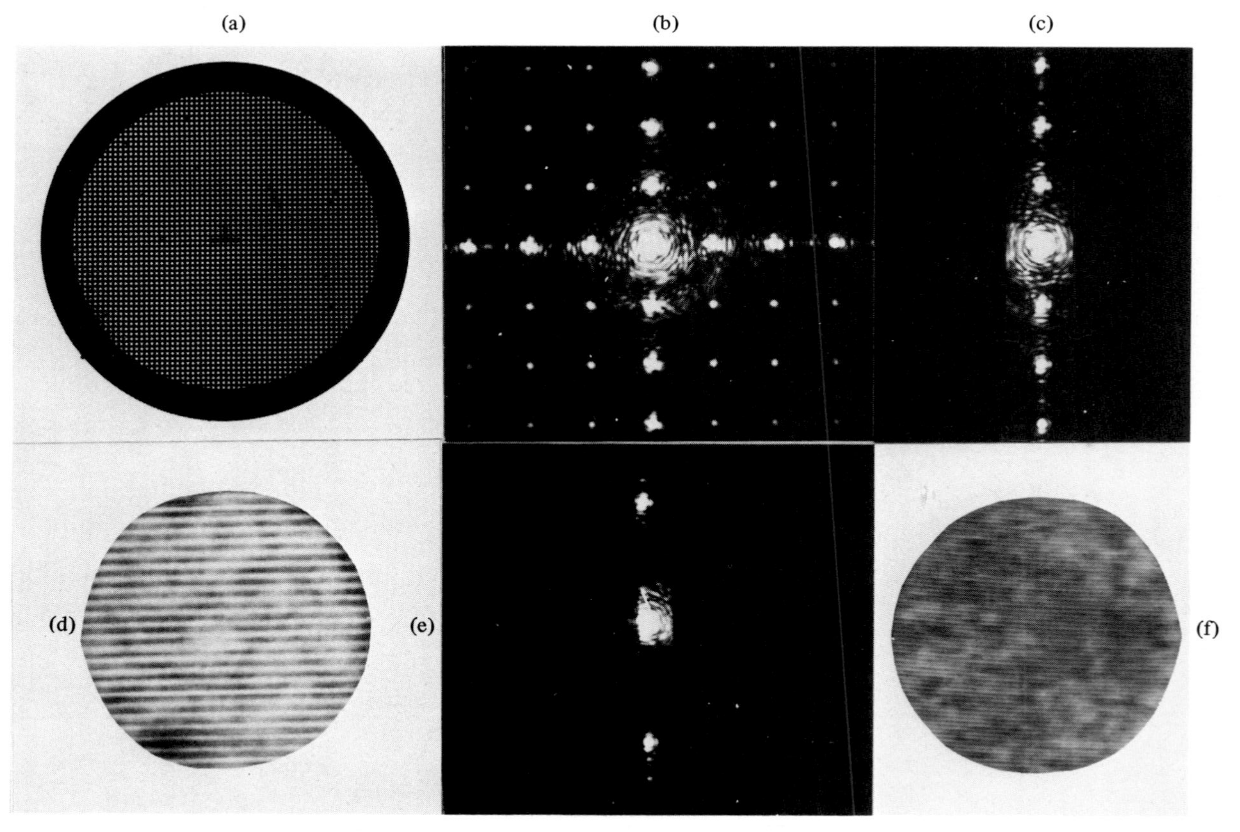

Fig. 6.4 A single optical diffraction analysis of a photograph of an electron microscope grid. (a) Original specimen, ×20. (b) Optical transform of the specimen. (c) Filtered optical transform; only the vertical frequency components are being transmitted. (d) Reconstructed image, ×40; the horizontal lines have the same periodicity as the original specimen. (e) Filtered optical transform, passing only the second and higher orders. (f) Reconstructed image, ×40; the periodicity is now doubled as compared to the original specimen.

ponents. By omitting the nonharmonic components from the Fourier synthesis the square wave would be reformed in a "cleaned up" or spatially filtered form.

There are strong mathematical similarities between Fourier analysis and Fourier synthesis; it can be shown that the Fourier transform of a Fourier transform of a function is the original function itself multiplied by 2π (e.g., Lipson and Lipson (1969), *Optical Physics*, p. 56). The process of Fourier synthesis (sometimes called second transforming) can be carried out optically by the addition of another lens to the diffractometer. The resulting instrument will be able to Fourier-analyze, spatially filter, and Fourier-resynthesize the structure in a micrograph.

Experiments in optical image processing of this form were first carried out by Abbé (1873) in an attempt to support the theory of the operation of optical microscopes. Similar experiments were undertaken by Porter in 1906, with similar results. Both workers used a single lens to carry out the diffraction and the reconstruction of the spatial frequencies present in an illuminated grid. Figure 6.4 illustrates an experiment following their ideas and demonstrates some of the elementary properties of optical transforms and spatial filtering. Figure 6.4a shows the grid itself while Fig. 6.4b is its optical transform. By limiting the parts of the transform which are used to reconstruct the image, the different spatial frequency components of the grid are recorded. Figure 6.4c shows the vertical frequency components of the transform being transmitted by the spatial filter to recombine, thus yielding the horizontal periodicity of the grid (Fig. 6.4d). If the spatial filter permits only the even orders of the vertical diffraction components to pass, the spatial frequency reconstructed is doubled (Fig. 6.4e).

The undiffracted beam of light passing through the spatial filter can be reduced in amplitude and will result in an increase in the contrast of the reconstructed image. If this beam is blocked, a contrast reversal and doubling of spatial frequencies is seen in the reconstructed image. By removing one-half of the diffraction pattern, the reconstructed image converts phase changes in the specimen into intensity changes; such an arrangement is called a Schlieren system. Fritz Zernike's phase contrast microscope operates on a spatial filtering technique and, like Schlieren optics, renders phase changes in a transparent specimen visible.

CONSTRUCTION OF OPTICAL DIFFRACTOMETER FOR IMAGE ANALYSIS AND RECONSTRUCTION

Certain criteria must be considered when one is designing an optical diffractometer to produce diffraction patterns or spatially filtered images from electron micrographs. The factor which exerts the most influence over the cost of an optical diffractometer is the size of the periodicities it is to examine. By using the Bragg equation (6.1) for a known diffraction camera length (L; see Fig. 6.5),

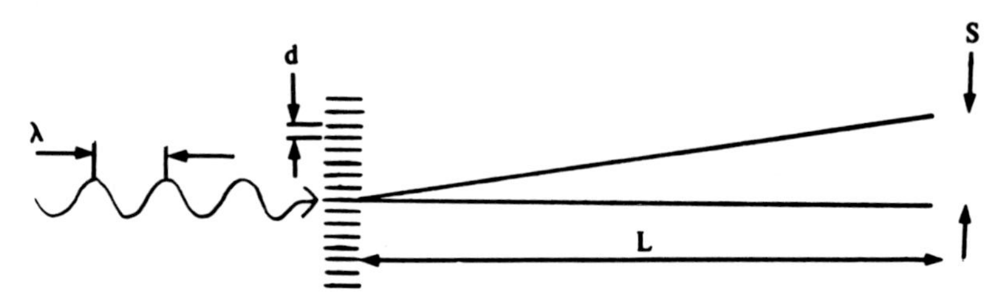

Fig. 6.5 Diagram to illustrate the relationship between camera length (L) and the spacing (S) of diffraction maxima [see Eq. (1.3)].

we can calculate the distance (s) between the zero order and first order maxima of a diffraction pattern:

$$\frac{d}{L} = \frac{\lambda}{S} \tag{6.3}$$

The highest spatial frequencies that can be examined on a micrograph are of course limited by the size of the photographic grain; there is no point in making a diffractometer capable of resolving structure smaller than this. We shall soon see that the lowest spatial frequency recorded by a diffractometer is limited by the light source.

The authors' diffractomer (manufactured commercially by Rank Taylor Hobson, and illustrated in Fig. 6.6) has a camera length of 0.5 meter and produces useful results from specimens with structure in the range 0.01 to 4 mm. The resulting diffraction patterns have first order maxima between 0.2 and 65 mm apart. Some workers (Horne and Markham, 1972) prefer larger diffraction patterns and use camera lengths up to 5 meters. While a longer camera length will produce a larger transform, it does not automatically lead to resolution of longer periodicities, and the increased cost of such a large structure and the need for a special room to contain it are very serious drawbacks.

There are three basic arrangements of a micrograph with respect to a convex lens that can be used to produce its diffraction pattern (Fig. 6.7). The lens can be placed before the micrograph (Fig. 6.7a), just after the micrograph (Fig. 6.7b) or well after the micrograph (Fig. 6.7c). In each case the camera length is the distance between the micrograph and the back focal plane of the lens. These methods of producing the optical diffraction pattern all project it onto a curved surface, not onto a plane; i.e., they introduce a quadratic phase factor into the Fourier transform. For long focal length lenses this phase deviation is small and is not important if the diffraction pattern is to be recorded photographically. However, if the system is to be used for image reconstruction, the spatial filters should really be curved to fit the back focal surface correctly.

One optical configuration which produces a flat back focal plane is shown in Fig. 6.7d. This is a modification of Fig. 6.7c, in which the micrograph has been moved to the front focal plane of the lens. In this case the Fourier transform of

Fig. 6.6 The optical diffractometer in use in the authors' laboratory, showing a diffraction pattern displayed on a TV monitor. Permanent records of transforms are obtained from either the 5″ X 4″ plate camera back (center of the instrument) at first transform plane magnification, or from the 35 mm camera (lower center) at the rather lower magnification of the second transform plane. Reconstructed images are photographed on the 35 mm camera back.

the micrograph is faithfully reproduced in the back focal plane of the lens, assuming that all the diffraction orders fall within its aperture. Thus to produce a given space-bandwidth product* the configuration shown in Fig. 6.7d needs a very large and hence expensive lens. The authors' system uses the arrangement shown in Fig. 6.7b: if the same space-bandwidth product were to be obtained from the last arrangement, a 500 mm F1 lens would be needed! Other optical systems have been developed which produce the Fourier transform on a flat plane. One such system developed by Blandford (1970) consists of a four-element lens of simple design, and, while being more complex than the arrangements hitherto mentioned, results in a very high space-bandwidth product and a small physical size. Future optical diffractometers may well be built on this system.

The optical systems just described are all suitable (within their individual space-bandwidth limitations) for obtaining diffraction patterns from electron micrographs. If, however, the system is to be used to spatially filter and reconstruct an image, other lenses are required.

*Space-bandwidth product may be defined as the product of the area of a micrograph to be examined and the highest spatial frequency to be recorded.

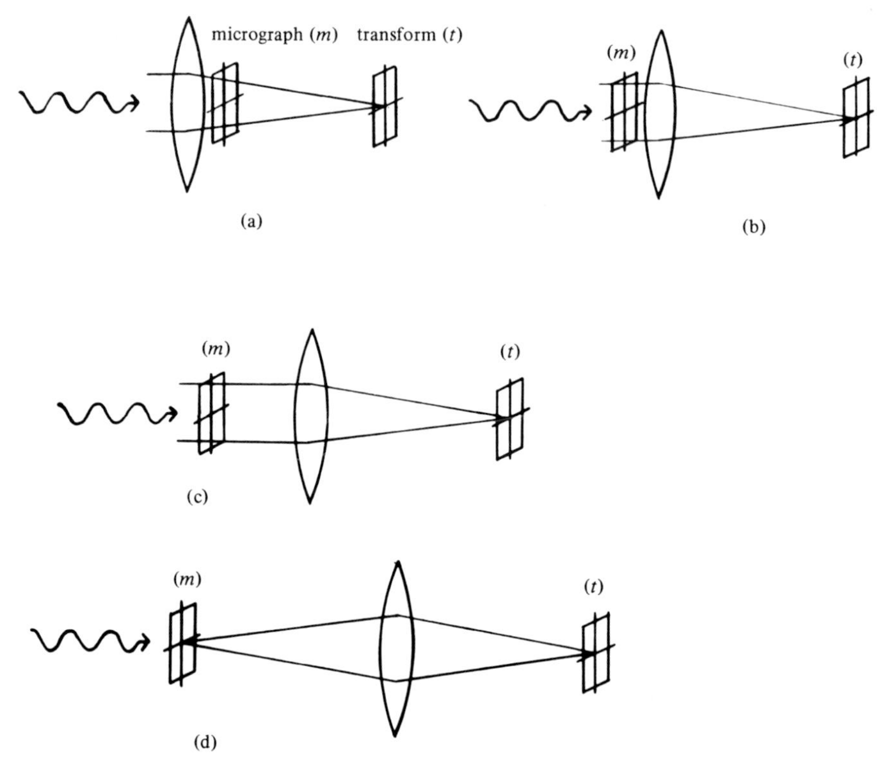

Fig. 6.7 (a)–(d) Various lens-specimen arrangements for producing diffraction patterns in a plane t from a micrograph m (see text for details).

We have already mentioned the experiments by Abbé and Porter, in which they demonstrated that diffraction and reconstruction can be carried out with one lens. A system which would do this is shown in Fig. 6.8a. The input plane of the system (the micrograph) is P_1, the transform is obtained at plane P_2 and the reconstructed image at plane P_3. The large distance between plane P_1 and the lens results in the system's having an extremely low space-bandwidth product, rendering it of little use for analyzing micrographs. The system shown in Fig. 6.7d can be "doubled up" to reconstruct an image and carry out spatial filtering; such an arrangement is shown in Fig. 6.8b. As has been pointed out, this system needs large lenses if it is to be of any practical use; the cost of two such lenses makes its use prohibitively expensive.

The flat Fourier transform plane of the two lens system shown in Fig. 6.8b enables holographic filters to be used so that micrographs can be corrected for deviations in the transfer function of the electron microscope, thus resulting in increased resolution. The subject of holographic filtering is beyond the scope of this monograph, but interested readers are referred to the work by Stroke (1970).

The optical system proposed by Maréchal (1953) (Fig. 6.8c) maintains a high space-bandwidth product at the expense of holographic filtering. In this configuration the transforming lens is placed close to the micrograph (P_1) so that the widest possible range of diffraction orders is presented to the lens. The imaging or reconstruction lens is placed close to the transform plane (P_2), making the transforming lens (L_1) into a field lens. By using suitable focal length lenses the image magnification can be varied or, more conveniently, made unity. The close proximity of the reconstruction lens and the transform plane maintains the high space-bandwidth product of the system, permitting very high quality reconstructed images to be obtained from modest sized lenses.

A modification of Maréchal's arrangement is shown in Fig. 6.8d; a third lens (L_3) has been introduced to relay the transform onto the same plane at which the reconstructed image is formed. The authors have used a commercially available instrument based upon this system, in which the third lens (L_3) is fixed to a motor-driven slide so that the image and diffraction pattern can be recorded on the same film in rapid succession. We have found that this system considerably

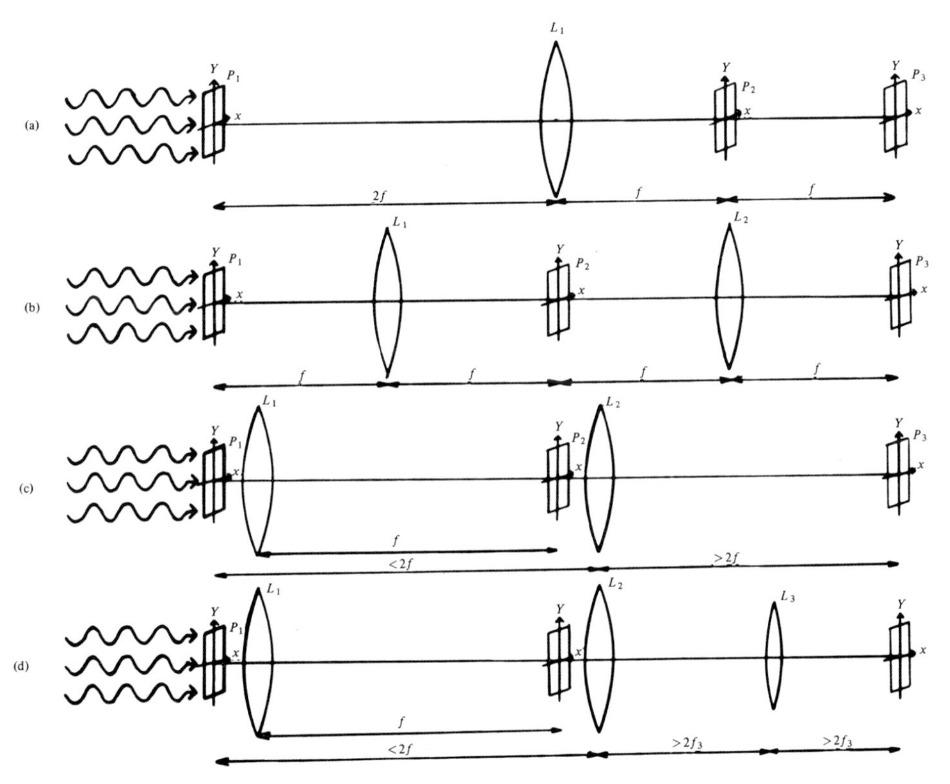

Fig. 6.8 (a)–(d) Optical arrangements for image reconstruction. P_1, P_2, and P_3 are the specimen, transform, and reconstructed image planes, respectively (see text for details).

speeds up the use of the instrument by simplifying the alignment of spatial filters. This is further helped by a closed circuit television system, the surface of the camera videcon tube being placed at plane P_3. With this system every small change in the position of a spatial filter can be observed at considerable magnification and its effect on the reconstructed image assessed. The addition of a television system also makes the diffractometer a useful teaching aid.

LIGHT SOURCES

The impact which cheap helium-neon lasers have had upon optical diffraction has already been mentioned. Today even the lowest power He-Ne lasers offers advantages so great (e.g., light intensity, coherence, monochromicity) when compared with arc lamps, that their use becomes obligatory.

Most He-Ne lasers produce a beam diameter of 1 to 2 mm, which is far too small to produce useful diffraction patterns from electron micrographs. A diffracting beam diameter of 20 mm is most useful for the optical arrangement described above; such a beam can be produced by simple methods. A concave lens placed in the laser beam will make it divert, and a suitably positioned convex lens can be used to realign the beam (Fig. 6.9a).

This method of beam expansion permits any stray reflections produced in the laser to emerge into the diffracting system and produce multiple imaging. An alternative system, which cuts out such reflections, can be made with two convex lenses. The lenses are placed with their back and front focal planes together (Fig. 6.9b). The ratio of focal lengths gives the magnification of the beam. The distribution of light in the back focal plane of lens L_1 is the Fourier transform of the incoming laser beam. If the laser beam is perfectly parallel, its Fourier transform will be an Airy disc, the size of which is determined by the beam diameter and the lens focal length (from Bragg's equation). Any stray reflections will show up in the plane, and can be removed with a carefully positioned pinhole (Fig. 6.9c). The pinhole should not touch the central peak of the Airy disk, or subsidiary rings will be formed, defeating the whole aim of the pinhole. The Airy disk is imaged on to the transform plane of the diffractometer with its size increased by the ratio of the focal length of the transforming lens and the focal length of lens L_2 in the beam expander. It is this projection of the Airy disc that limits the lowest spatial frequencies that can be resolved.

The diameter of the beam reaching the micrograph can be changed in a number of ways; a new beam expander could be placed in the beam each time that a different-sized portion of micrograph is to be examined. Such a system, while being optically ideal, would be intolerably slow to use in practice. By placing a variable-diameter iris in the expanded beam just in front of the micrograph different-sized areas can readily be examined. However, this system has one disadvantage, in that the detail in the micrograph is convoluted with the iris. Thus, if a small round iris is used, every spot in the transform produced by the

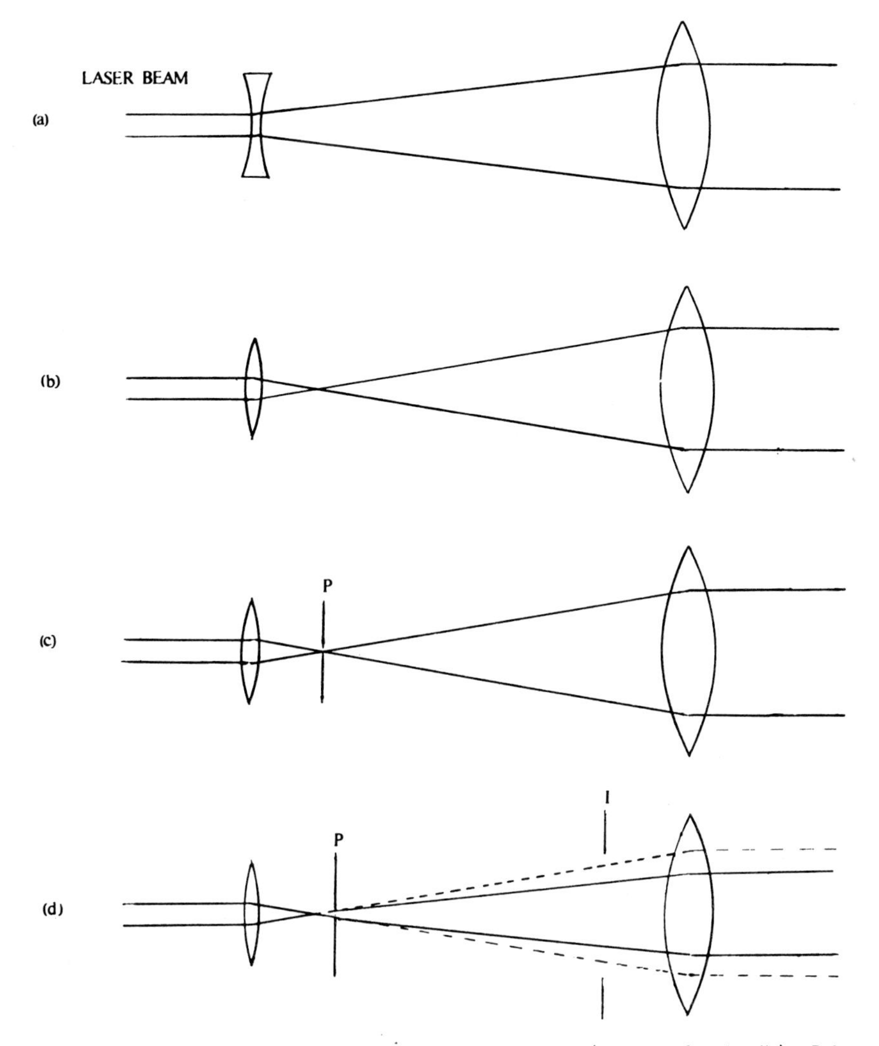

Fig. 6.9 (a)–(d) Optical arrangements for beam expansion (see text for details). P is a pinhole and I an iris diaphragm.

micrograph will be surrounded with small rings. By keeping the iris as near the micrograph as possible, the convolution effect is kept to a minimum. The convolution effect can be avoided completely if the beam reaching the micrograph has a Gaussian intensity distribution. Such a beam can be obtained by moving the pinhole in the beam expander away from the focal points of the two lenses, so that it produces a set of rings, the inner one having the desired intensity distribution. The rings themselves are cut off with a variable iris (*I* in Fig. 6.9d). By moving the pinhole and iris together one can obtain different beam diameters. Unfortunately, the use of a Gaussian beam introduces problems in interpreting the optical transforms produced, because any detail in the micrograph will produce different diffraction spot intensities when in different parts of the beam.

To focus the beam expander the beam should be projected onto a distant wall, where its diameter can be measured and compared with that at the exit of the expander. By this method the beam can be set parallel to better than 0.1 mrads (.1 mm/10 m). If the system is to be used for image reconstruction and not just diffraction, then great care must be taken to align the beam parallel if good image quality is to be obtained.

OTHER OPTICAL COMPONENTS

The monochromatic and coherent nature of the light used in an optical diffractometer demands that the lenses be of the highest quality; any blemishes, bubbles, or abberations can produce disastrous results. If the maximum resolution is to be obtained, the lenses should be designed for the particular job they are to carry out (as in the Rank Taylor-Hobson instrument). A number of workers have used astronomical telescope objectives for the long focal length lenses, with good results (see Lipson and Lipson (1969), *Optical Physics*). At least one commercially made optical diffractometer (see Appendix to this chapter) uses camera optics for the transforming lens. Insufficient results are available to enable a judgment to be made on the adequacy of such lenses in a diffraction mode.

The structure upon which the optics are mounted depends mainly upon the focal lengths of the lenses used. While an optical bench is the logical base for the instrument, it would be expensive if focal lengths over 0.5 m are to be used. A number of workers have used steel beams with success (Horne and Markham, 1972). By folding the system, its overall size can be kept small enough to put on a laboratory bench.

PRACTICAL USE OF OPTICAL DIFFRACTOMETER FOR IMAGE ANALYSIS AND RECONSTRUCTION

The size of periodicities which can be examined in an optical diffractometer has already been discussed and shown to be a few mm at most. With many specimens the magnification in the microscope is set to give an image of the periodic structure equal to or greater than this. Micrographs taken at such a magnification are not suitable for analyzing in the diffractometer directly, but must be photographically reduced first.

Many electron micrographs show a small specimen and a relatively large area of support film, often of opposite contrast. The contrast of the plate or film to be examined in the optical diffractometer should be such that the detail to be observed has a high transmission factor while the support film is low. In this way the highest signal-to-noise ratio is obtained in the optical transform, which in turn eases the problem of making spatial filters to clean up the image.

The detail contained in the optical transform of a micrograph is complex; i.e., it consists of both amplitude and phase information derived from the transmission and phase structure in the plate or film being examined. The phase struc-

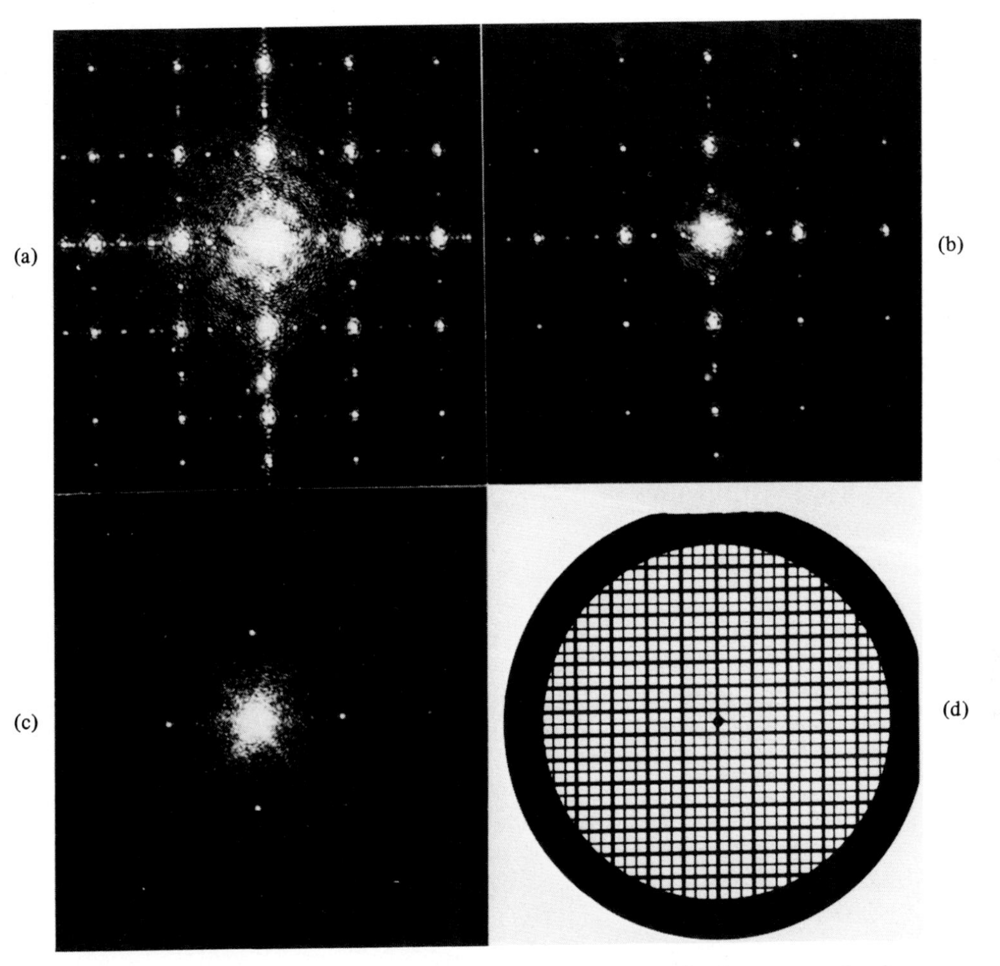

Fig. 6.10 Diffraction patterns arising from the phase and amplitude components of a specimen. (a) Total optical transform of the specimen, a photograph on glass plate of a grid, shown in (d). (b) Optical transform derived from the amplitude components; phase components have been eliminated by cementing optical flats over immersion oil onto the emulsion of the specimen. (c) Optical transform of the phase components only, produced by chemically bleaching the plate until no visible structure remained. (d) Original specimen, $\times 20$.

ture of photographic materials can be divided into three types. The film or plate onto which the emulsion is fixed may not be optically flat, resulting in errors in the transform. These phase changes normally have a long periodicity compared with the structure being examined; hence they have little detrimental effect.

Gelatin emulsions often suffer from large numbers of small surface defects which all introduce noise into the diffraction pattern, and hence artifacts in the reconstructed image. A third phase component is caused by the surface structural changes produced by the exposure and development of the emulsion itself. Exposure-related phase structure in an emulsion can be seen in Fig. 6.10, which compares the transforms produced by the intensity and phase structure (Fig. 6.10a), and the phase structure left after the image had been completely bleached out (Fig. 6.10c). It can be seen that the phase structure is similar to, but not identical to, the intensity structure (Fig. 6.10b); the original structure is shown in Fig. 6.10d.

The effect of phase noise in the micrograph can be removed by placing the film or plate in a suitable immersion oil between two optical flats. This method produces good results but is slow and inconvenient to use. The authors have found that an alternative method is to fix a microscope-slide coverslip over the emulsion using Canada balsam. This method has been found to give extremely consistent results, the plate does not need cleaning after analyzing, and the coverslip affords it some degree of permanent protection against physical damage. While this system works well with plates, care must be taken if it is used with film.

For routine examination, the micrographs or the reduced copies are placed directly in the diffractometer; areas that yield information can thus be rapidly located. To fix the coverslip, a small blob of balsam in xylene is placed over the structure to be examined, and the coverslip is lowered onto it. The coverslip flattens out the balsam under surface tension, whereupon the micrograph is placed back in the input plane of the diffractometer.

The type of stage upon which the film or plate is mounted depends on the overall structure of the instrument; most optical bench systems have suitable *X-Y* stages. A rotating stage greatly simplifies the process of aligning spatial filters for image reconstruction. A degree of movement along the axis (*Z*-axis) of the system enables the optimum focus to be obtained for all thicknesses of plates or film.

FOCUSING

The reconstructed image produced by the diffractometer will only be interpretable if it is correctly focused. A method of focusing an optical diffractomer has been proposed by Horne and Markham (1972). In this method a pair of small holes in a filter is placed in the transform plane, and the main diffraction lens is replaced by one of smaller focal length. The image reconstructed by this

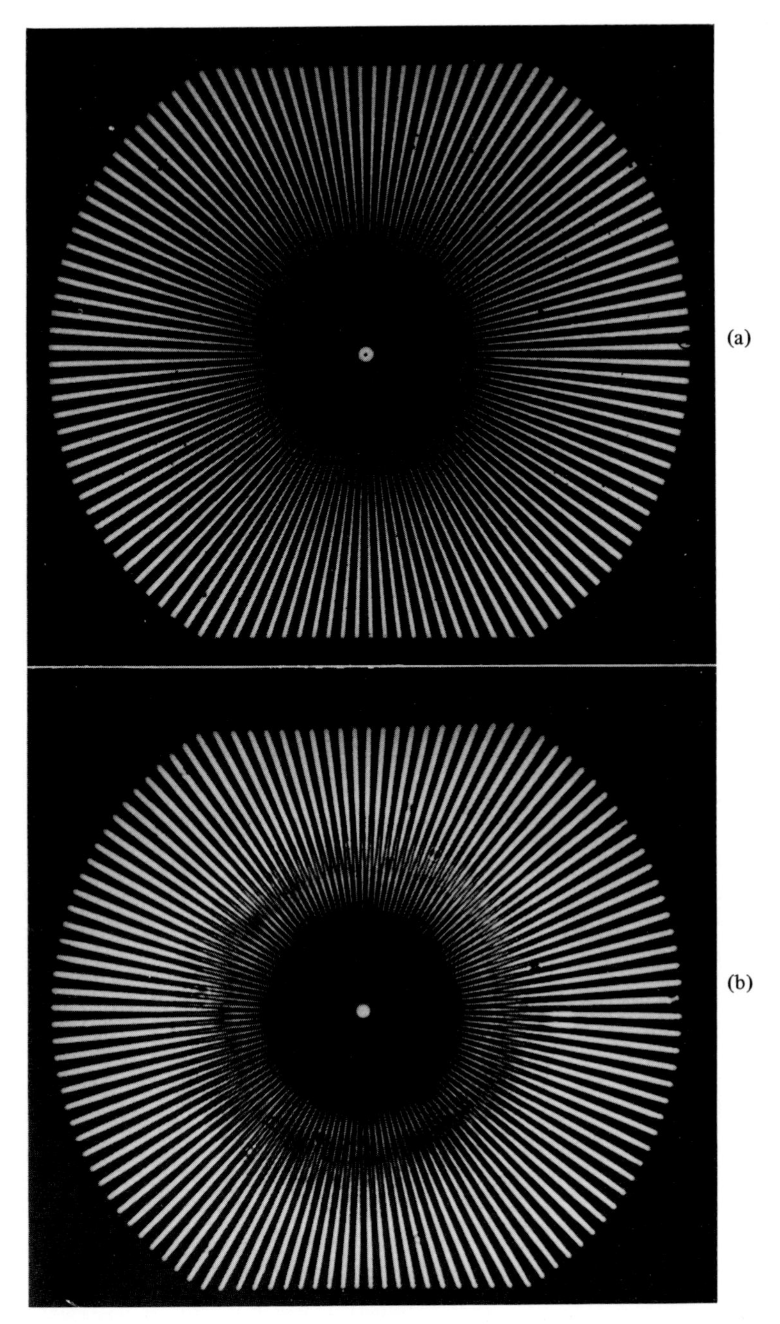

Fig. 6.11 A spoked-wheel chart used for focusing. (a) Correctly focused. (b) Incorrectly focused, showing a ring of contrast reversal at a moderately high spatial frequency.

system is an Airy disc crossed by interference fringes: the image position giving the sharpest fringes is found, and used as the focal plane.

It is questionable as to whether this procedure is valid. Only one spatial frequency is used to produce the "reconstructed image," and as it can be shown that the transfer function will oscillate with a number of positive maxima, for varying degrees of defocus (Goodman, 1968), it is hard to see how a unique focal plane can be established with certainty.

We employ a more conventional optical procedure, using a spoked-wheel chart (Fig. 6.11a) to generate a wide range of spatial frequencies. If the aberrations (including defocus) of the system are small, the optical transfer function remains positive up to high spatial frequencies, and hence these are recorded with positive contrast. A significant focusing error will cause the transfer function to become negative at a spatial frequency within the range of the instrument's resolution (Fig. 6.11b). The transition from positive to negative contrast is clearly seen as a ring of zero contrast. As the position of true focus is more closely approximated, by axial translation of the specimen, this ring moves towards the center of the reconstructed image of the chart, and finally disappears when true focus (within the resolution limits of the system) is achieved.

PRACTICAL METHODS FOR PREPARING SPATIAL FILTERS

Either physical or photographic techniques may be used to provide a filter to place in the transform plane to remove selected spatial frequencies from the reconstructed image. Physical methods involve cutting holes in some suitable opaque material, such as card, metal foil, or densely exposed sheet film. A range of mounted needles and small scalpels is useful here. For choice all material to be removed should be completely cut out: holes which are made by pressure application invariably have jagged and potentially reflecting edges. Where a 5″ × 4″ back can be inserted into the transform plane, then a Polaroid print of the transform provides a simple filter material on which the areas to be removed can be delineated using a sharp pencil, and then cut out. Care must be taken however in mounting the cut-out print so that it lies flat in the transform plane; and as the cut edge of photographic paper is often rather "hairy," it can be desirable to paste the Polaroid print by its perimeter onto a material more suited to producing a clean cut, and then cutting through both materials.

Physical filters of this type are successful, but very difficult to make where the transform is small. In these cases it is much easier to make a photographic filter. An image of the transform on plate or sheet film is placed in the carriage of a photographic enlarger equipped with baseboard illumination facilities. At a high magnification (e.g., ~×15) the image of the transform is projected onto a sheet of white board. The parts of the transform which are to be removed are then "blocked out" using a black, felt-tip pen. The enlarger light-source is then turned off, an unexposed negative is placed in the carriage, and an exposure is given using the baseboard illumination system. With a high contrast film developer combination, a little experimentation soon establishes conditions under which very precise, clearly defined filters can be produced.

An obvious disadvantage of photographically produced filters is the potential introduction of uncontrolled phase modulation in the transform plane, in addition to the desired amplitude modulation. The filters should therefore be mounted between optical flats in a medium of suitable refractive index. The method suggested earlier for specimens, using Canada balsam, is equally applicable here.

INTERPRETATION OF RECONSTRUCTED IMAGES AFTER SPATIAL FILTERING

Unless all lenses and components of a diffractometer are made and located to within tolerances of a fraction of a wavelength, which would be prohibitively expensive for most purposes, it is inevitable that the quality of the reconstructed image obtained with coherent illumination will be less good than is normal in a high-quality enlarger or other copying device employing noncoherent illumination. Figure 6.12 illustrates this point. The *resolution* of a high contrast specimen is maintained very well indeed in a modern instrument; but on close examination the background of the reconstructed image is found to show multiple interference effects. These lead to an overall loss of contrast, which could be significant for a specimen of low initial contrast. The practice of using a high-contrast copy of the original specimen for optical diffraction analysis is a sound one, provided that care is taken to ensure that significant detail is recorded within the response curve of the copy material. The small loss in optical transfer function entailed in the copy process will be more than compensated for by the contrast gain, which can be quite large, especially if a size reduction is involved.

It having been established then that a specimen of reasonable contrast level suffers little loss in reconstruction, to what extent may spatial filters be used to produce an actual gain in discernible information content in the reconstructed image? A variety of possible approaches can be considered:

(1) Spectral Termination. This procedure entails progressive deletion of the higher frequency components, and is a test procedure rather than one which would generally be used to produce image enhancement. Figure 6.12 shows that progressive removal of the high frequency components results in a loss of information in the final image at the predicted level, without the introduction of artifactual information. Where the higher frequency components arise predominantly from background noise, this procedure could be used to produce a "cleaned-up" image. A noise-contaminated specimen containing a sinusoidally varying signal would be an ideal system for processing in this way.

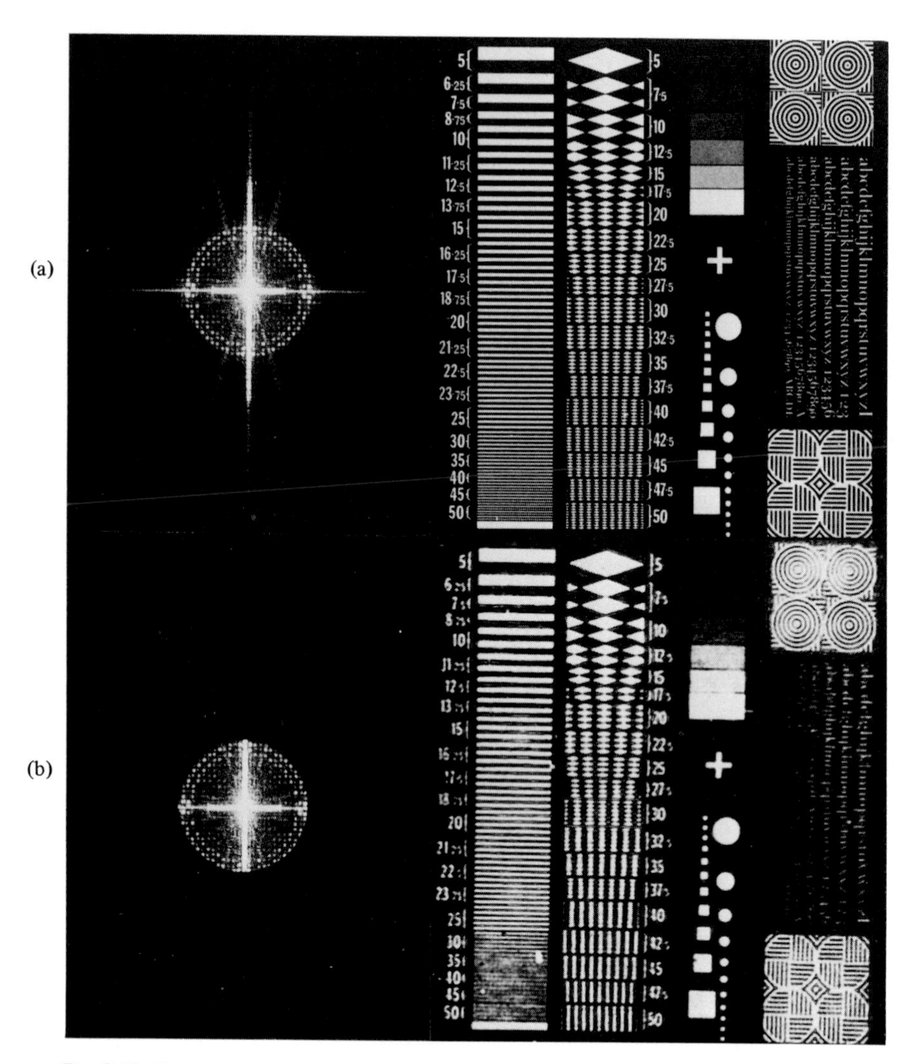

Fig. 6.12 The reconstruction of a resolution test chart with varying degrees of frequency cut-off in the spatial filter plane. (a) The entire transform (*left*) and reconstructed image ×16 (*right*) of the test chart. The small amount of degradation of the background caused by the use of coherent light is just visible on the original but could not be reproduced in this picture. (b) – (e) Increasing degrees of spatial filtering, terminating at defined frequencies by means of circular apertures in the spatial filter plane (*left*), resulting in reconstructed images (*right*) from which detail has been subtracted at the appropriate level. Interference effects caused by use of coherent light beam are more noticeable at extreme spatial filtering. (f) Enlarged section ×50 of the reconstruction of the original specimen (a), showing the high resolution of which the optical arrangement is capable. (*Resolution test chart reproduced by kind permission of Paterson Products Ltd, London.*)

(c)

(d)

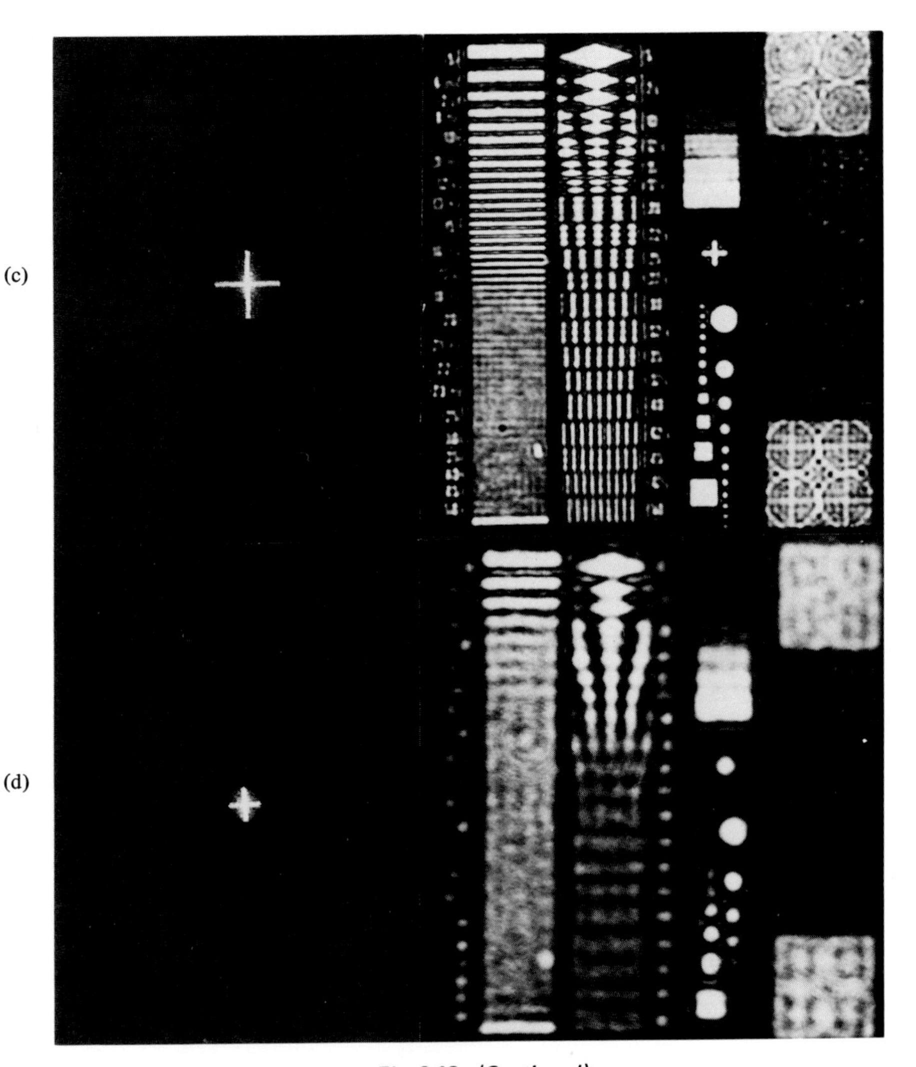

Fig. 6.12 *(Continued)*

(2) Histogram Correction or Equalization. Elegant digital procedures have been devised (Andrews, 1972), enabling the amplitudes of the frequencies in the power spectrum of an image to be modified, generally in a manner designed to produce reinforcement of the high frequency components (fine detail) at the expense of low frequency information. It should in principle be possible to make a spatial filter of radially varying optical density to achieve the same effect in an optical diffractometer. In practice we have found that satisfactory filters

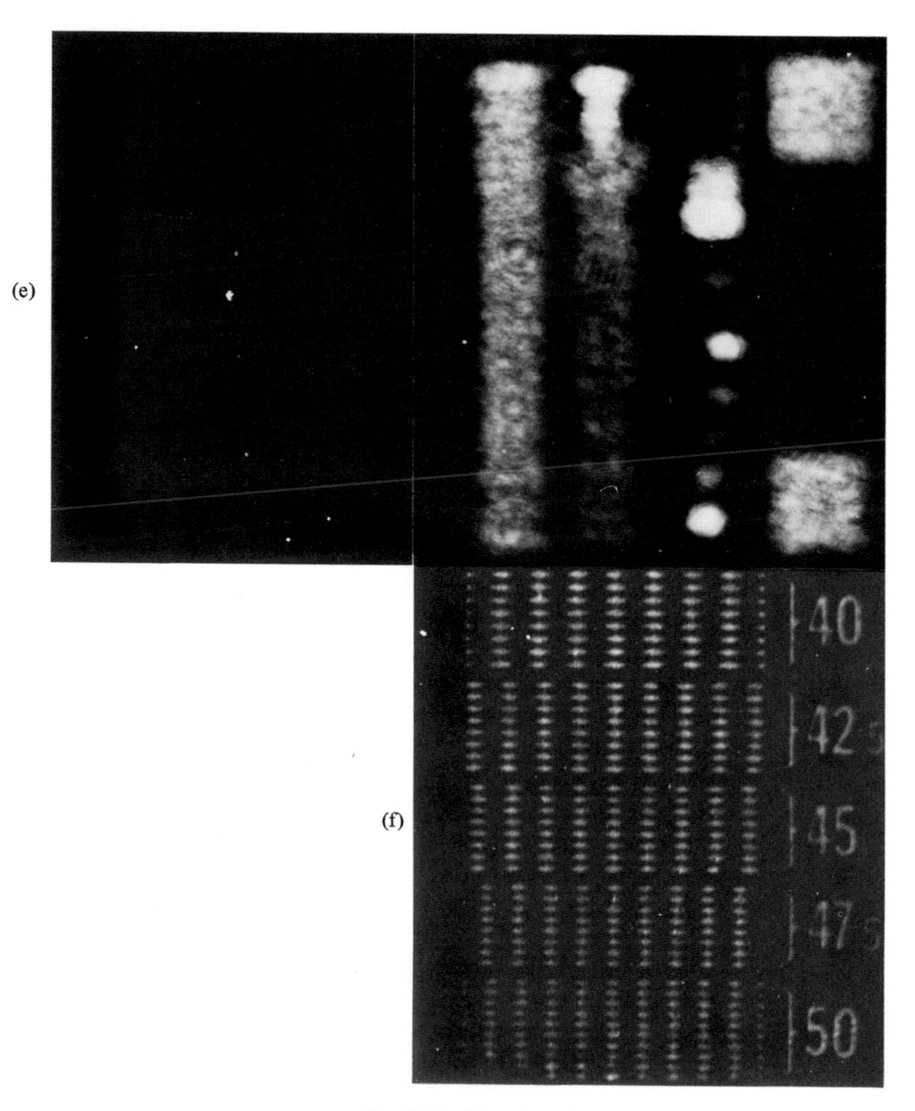

Fig. 6.12 *(Continued)*

are not easy to make, but current investigations are proceeding on this topic (Gibbs and Rowe, unpublished results).

(3) Noise Removal by Transform Filtering. When the noise spectrum overlaps the specimen transform it should be possible to "clean up" the reconstructed image by allowing through the transform plane only the reference beam and those frequencies corresponding to identified spots in the diffraction pattern.

The method works best where there is external evidence, for example from X-ray diffraction data, to indicate which are the genuine spots which are to be allowed through. Figure 6.13 shows an early and successful example of a filtered reconstructed image, of the I-filament protein from muscle, F-actin. In this case the X-ray diffraction pattern is well known. Horne and his co-workers have suggested that for protein crystals the electron diffraction pattern, obtained in the electron microscope, could be used in a similar way.

It is important however that one should not use a spatial filter which is an accurate complement of the expected structure. It has been shown (Gibbs and Rowe, 1973) that such a filter may reconstruct an image irrespective of whether it was present in the specimen or not. Irregular areas should be cut out, decidedly larger than the apparent size of the spots, in order to allow through satellite maxima. It is usually desirable also to allow through at least one diffraction order higher than the observed ones, to allow for faint orders. A recently devised procedure (Gibbs and Rowe, 1974) enables weak orders to be spotted with greater certainty.

The use of reconstructed, filtered images is twofold. Firstly, since phase information is conserved, they actually contain more information than do the corresponding diffraction patterns. To take a simple example: in an assay of arrowhead patterns, the information concerning the direction of the arrowheads will be lost in the recorded diffraction pattern. Secondly, they enable information concerning a structure, deduced from its diffraction pattern, to be pre-

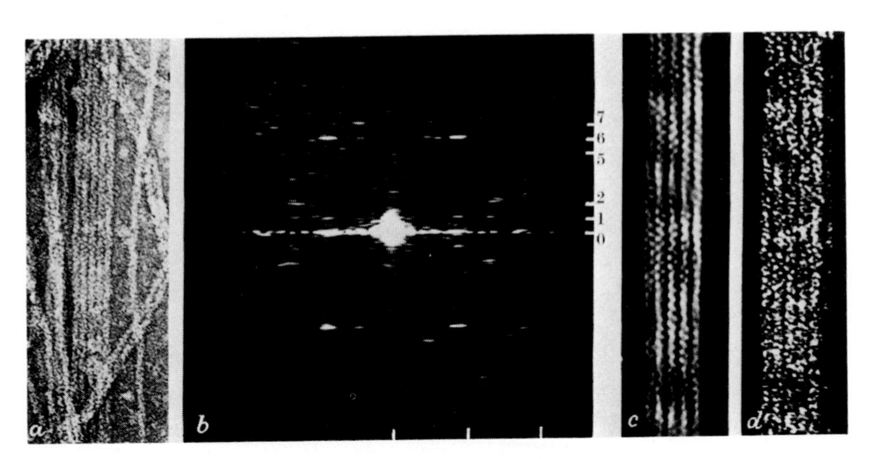

Fig. 6.13 The reconstruction of a spatially filtered image of a paracrystal of impure F-actin, negatively contrasted in uranyl acetate. (a) Original micrograph. (b) The optical transform. (c) Reconstructed, filtered image, formed from the primary beam plus the strong diffracted rays. (d) The reconstructed, unfiltered image for comparison. The complex helical structure of the strands is obvious in (c), but virtually invisible in (d). (*Reproduced by permission of the Royal Society*.)

sented in a concise manner, intelligible to those having no crystallographic knowledge. The importance of this latter point should not be underestimated.

(4) Unscrambling Complex Images. This is the most hazardous procedure of all. It involves selective removal not merely of the noise spectrum but of spatial frequencies and diffraction spots corresponding to parts of the specimen, with a view to presenting a filtered, reconstructed image of one or more identifiable components of the specimen: for example, of one layer in a membrane, or of one side of a tube. Since the transforms of the different periodic components of the specimen are convoluted, and not merely added, what is being attempted is an analogue deconvolution.

It is a commonplace observation that arbitrary removal of parts of the transform of a periodic structure can result in the formation of reconstructed images bearing little obvious visual relation to the original structure (Horne and Markham, 1972). We ourselves have produced many examples of this phenomenon. But for certain types of complex pattern, in particular rotational Moiré patterns caused by the superimposition of two patterns of hexagonal symmetry, and the pattern produced by superimposition of the two sides of a helix, the transform does indeed approximate the additive sum of the components. Indexing and removal by spatial filtering of one component may then enable an interpretable, reconstructed image of the other component to be produced (Fig. 6.14).

The method is a perfectly legitimate one, and though it is obvious that care must be taken in focusing, we do not consider that this causes insuperable difficulties. The onus, however, is on the experimenter to show that his procedure is valid, and it currently seems that the two types of complex patterns quoted above are the only ones likely to be met in practice where a deconvolution is possible by the simple analogue technique of spatial filtering.

CONCLUDING REMARKS

High-quality, high-resolution reconstructed images can be produced in a modern optical diffractometer. The effect of modifying the amplitudes of components of the spatial spectrum can be tested with much greater ease than if the same procedure were performed using digital techniques. It seems likely that the difficulties with weak orders, for which digital procedures have been recommended (Amos, 1974), have been overcome by the use of fully computed diffraction optics and by the use of processing techniques on the transform (Gibbs and Rowe, 1974). The use of complex filters to modify phase as well as amplitude would enable all types of spatial filtering to be performed on an analogue basis. While this has been achieved (e.g., Redman, 1973), much further research is required before it becomes an economic and practical possibility in the ordinary laboratory. It does not seem at all unlikely, however, that the difficulties here will be overcome in the course of the next few years.

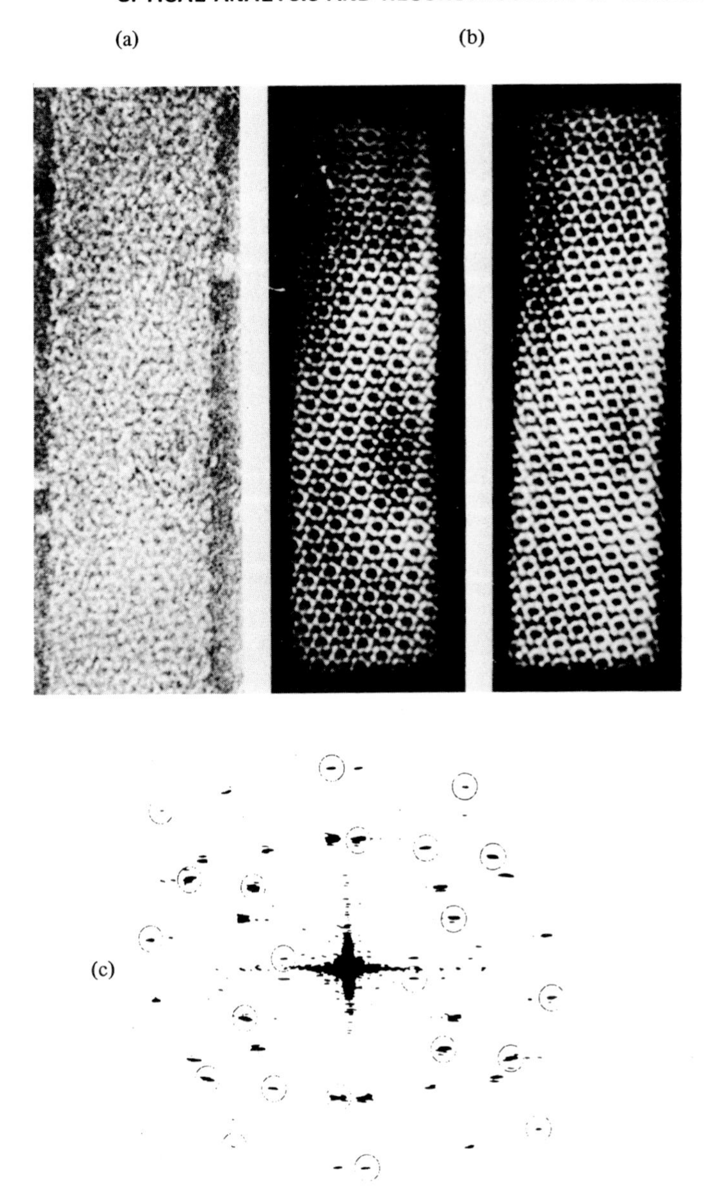

Fig. 6.14 (a) Electron micrograph of a negatively contrasted tubular structure formed from the head protein of T4 bacteriophage. (b) Optical transform of (a). The spots clearly lie on a pair of hexagonal lattices, arising from the upper and lower sides of the flattened structure. (c) The reconstructed image formed by allowing rays from one hexagonal lattice only to pass through a spatial filter. The noise spectrum also is attenuated, and the reconstructed image shows details of the packing of individual subunits. *(From Yanagida, De Rosier, and Klug, 1972; reproduced by permission of Academic Press.)*

APPENDIX

Commercial Optical Diffractometers

The principles which have been outlined above should enable the reader to construct for himself an optical diffractometer which is adequate at least so far as obtaining diffraction patterns is concerned. The building of an instrument which will perform reconstructions is not so simple, and it is our own view that unless the reader has considerable skill and experience in practical optics, the results are liable to be of a very inferior quality.

Three commercial instruments are available, and their purchase would generally seem to be justified in laboratories where a significant amount of optical diffraction analysis is to be performed.

The Polaron Diffractometer. This is a straightforward optical bench device incorporating facilities for both transforms and reconstructed images. The overall length of the bench is 3 m, and the instrument must be used in a fully darkened room. The alignment of the lenses is by manual adjustment. The transform lens is located immediately in front of the specimen, giving a good space-bandwidth product, and the results obtained from this instrument, which has been on the market for some years, appear to be of a high quality provided that the alignment is correctly maintained.

The Rank Taylor Hobson Image Analyzer. A brief description of this instrument was given earlier in the text. A folded optical bench is used, the whole of the optics being enclosed in a light-tight box of approximate dimensions 1 m × 60 cm × 56 cm. All the optics are prealigned and fixed except for the positioning of the accessory reconstruction lens and mirror for photograph, which is performed by means of motorized drives. The final transform or reconstructed image is formed on the front face of a TV camera and from thence displayed on a TV monitor at a rather large magnification.

The quality of transforms and reconstructed images obtained with this instrument are of a very high order, and it is particularly convenient and simple to use. As an accessory device a "Contourographic" accessory is available which displays transforms in simulated three-dimensional space, enabling the intensity of development spots to be measured as well as their co-ordinates.

The coherent optical transform function of the author's instrument is shown in Fig. 6.15, and yields valuable information concerning the resolution in different parts of the image.

The Talbot Bench Top Optical Diffractometer; Talbot Research. This is a folded optical bench instrument using high-quality camera lenses. Transforms, reconstructed images, and special filtering are all within the capability of this

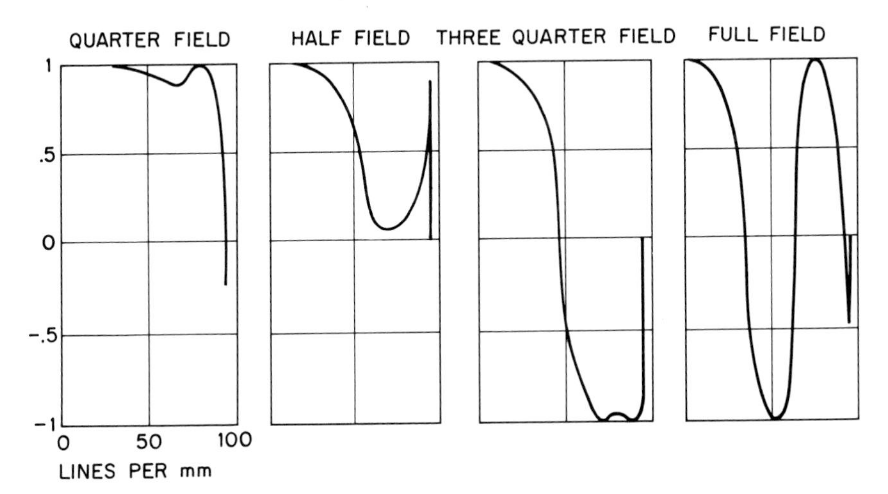

Fig. 6.15 Coherent optical transfer function-tangential section.

machine, but insufficient published results are available to enable any worthwhile comment to be made on the quality of the final images.

Suppliers

Polaron Equipment Ltd.,
60, Greenhill Crescent,
Holywell Estate,
Watford, England

Rank Taylor Hobson,
East Park Road
Leicester, England

Talbot Research,
P.O. Box 56200,
Pinegowrie
TVL.,
South Africa

Hogstow Place
Minsterley,
SHREWSBURY,
Salop, SY5 OHU,
England

References

Abbé, E. (1873). "Beiträge zur theorie des mikroskops und der mikroskopischen wahrsemmung." *Arch. Mikrosk. Anat.* **9**, 413.

Amos, L. A. (1974). Image analysis of macromolecular structures. *J. Microscopy* **100**, 143.

Andrews, H. C. (1972). Digital computers and image processing. *Endeavour* **31**, 88.

Blandford, B. A. F. (1970). A new lens system for use in optical data processing. In: *Conference on Optical Instruments and Techniques*. Oriel Press, New Castle upon Tyne.

Dubochet, J. (1973). High resolution dark-field electron microscopy. In: *Principles and*

Techniques of Electron Microscopy: Biological Applications, Vol. 3 (Hayat, M. A., ed.). Van Nostrand Reinhold Company, New York and London.

Gibbs, A. J., and Rowe, A. J. (1973). Reconstruction of images from transform by an optical method. *Nature* **246**, 509.

Gibbs, A. J., and Rowe, A. J. (1974). Processed optical transforms: An approach to the detection of periodic structure in noisy images. *I.C.R.S. (Biochemistry)* **2**, 1092.

Goodman, J. W. (1968). *Introduction to Fourier Optics*. McGraw Hill Book Company, London.

Hayat, M. A. (1970). *Principles and Techniques of Electron Microscopy: Biological Applications*, Vol. 1. Van Nostrand Reinhold Company, New York and London.

Hayat, M. A. (1975). *Positive Staining for Electron Microscopy*. Van Nostrand Reinhold Company, New York and London.

Horne, R., and Markham, R. (1972). *Practical Methods in Electron Microscopy*, Vol. 1. North Holland Publishing Company, London.

Johnson, H. M. (1973). In-focus phase contrast electron microscopy. In: *Principles and Techniques of Electron Microscopy: Biological Applications*, Vol. 3 (Hayat, M. A., ed.). Van Nostrand Reinhold Company, New York and London.

Klug, A., and Berger, J. E. (1964). An optical method for the analysis of electron microscope plates and some observations on the mechanism of negative staining. *J. Mol. Biol.* **10**, 565.

Lipson, H., and Taylor, C. A. (1951). Optical methods in X-ray analysis. 2. Fourier transforms and crystal structural determination. *Acta Cryst.* **4**, 458.

Lipson, S. G., and Lipson M. (1969). *Optical Physics*, Cambridge University Press.

Maréchal, A. (1953). Un filtre de frequences spatiales pour l'amelioration du contraste des images optiques. *C.R. Acad. Sci., Paris* **237**, 607.

Markham, R., Frey, S., and Hills, G. (1963). Methods for the enhancement of image detail and accentuation of structure in electron microscopy. *Virology* **20**, 88.

O'Brien, E. J., Bennett, P. M., and Hanson, J. (1971). Optical diffraction studies of myofibrillar structure. *Proc. Roy. Soc.* **B261**, 201.

Porter, A. B. (1906). On the diffraction theory of microscope vision. *Phil Mag* **11**, 154.

Redman, J. D. (1973). *Suppression of Coherent Optical Noise*. United Kingdom Atomic Energy Authority, AWRE, Aldermaston, England.

Rowe, A. J. (1974). Replication techniques. In: *Practical Methods in Electron Microscopy* (Glauert, A., ed.). North Holland Book Company, London.

Stroke, G. W. (1970). *An Introduction to Coherent Optics and Holography*. Academic Press, London.

Taylor, C. A., and Lipson, H. (1951). Optical methods in crystal structure determination. *Nature* **167**, 809.

Yanagida, M., De Rosier, D. J., and Klug, A. (1972). The structure of tubular variants of the head of bacteriophage T4 (Polyheads). *J. Mol. Biol.* **65**, 489.

7. MIRROR ELECTRON MICROSCOPY

R. S. Gvosdover and B. Ya. Zel'dovich*

Electron Optics Laboratory, Physics Department, Moscow State University, Moscow, U.S.S.R.

INTRODUCTION

The mirror electron microscope (MEM) is an instrument for investigating topographical inhomogeneities, potential distributions, and magnetic structures on the surface of bulk specimens. The main imaging element of the MEM is the electron mirror (Henneberg and Recknagel, 1935), in which the primary electron beam reflected at the equipotentials very close to the surface of the specimen reverses its direction. By examining experimentally the properties of electron mirrors, Hottenroth (1937) first showed that electrons reflected very close to the surface of the specimen make it possible to obtain the image of topographical inhomogeneities on the surface of a reflecting electrode. These observations led subsequently to the development of mirror electron microscopy.

In the MEM the specimen is maintained at a slightly negative bias voltage with respect to the filament of the electron gun. As a result, the electrons are reflected at some distance from the surface under investigation, and the specimen is not affected by the electron beam. The MEM is a sensitive instrument, allowing the depiction of surface inhomogeneities and electrical and magnetic microfields. The high sensitivity of the instrument is a result of the fact that the electron beam is slowed down close to the point of reversal, and therefore it interacts with perturbations for a sufficient long period of time.

*Permanent address: Quantum Radiophysics Laboratory, P. N. Lebedev Physical Institute, Academy of Sciences of the U.S.S.R., Moscow.

Both high sensitivity and the possibility of surface investigations without electron bombardment are the main advantages of the MEM. Because of these characteristics the MEM is capable of competing in some fields of applications with the electron microscopes of other types (transmission, scanning, emission, and reflection electron microscopes). The MEM, for instance, is a unique instrument for investigating magnetic microfields on the surface of bulk specimens (Spivak *et al.*, 1955) and for examining potential profiles on the surface (Orthuber, 1948).

Presently, many publications on mirror electron microscopy, such as the reviews by Mayer (1961) and Oman (1969), which are concerned with the investigation of magnetic structures, are available. Of particular interest are publications by Bok (1968) and Bok *et al.* (1971), in which the authors concentrate their attention on the development, design, and investigation of the MEM with focused images. The review by Lukianov *et al.* (1973) contains detailed descriptions of the construction of various types of MEM and a discussion of the image contrast formation and applications of the MEM in physical research. This review also contains a rather complete list of references dealing with the MEM.

The present chapter deals with the physical principles of the mirror electron microscopy. The section on principles contains a brief description of the MEM design and its modes of operation. The section on image contrast formation presents a detailed treatment of the classical theory of image contrast formation and displays the experimental results obtained in the shadow projection imaging mode of the MEM. The remaining sections deal with the theoretical and experimental results on the measurements of microfields by the MEM, the quantum effects in the MEM, and the applications of the MEM.

PRINCIPLES OF MEM

Construction and Modes of Operation

In the MEM the electrons emitted from a distant source and formed in a narrow beam refract in the field of the immersion objective and slow down in the uniform retarding field, reversing at zero-equipotential surface over the specimen mirror. The electrons again refract in the immersion objective and by falling on the viewing screen produce the image. There are two types of MEM—the so-called straight design and a separated beam system. Figure 7.1 shows a schematic scale diagram of different constructions of the MEM. The axes of the illumination and reflection beams coincide in the case of a "straight" design (Fig. 7.1a).

In the separated beam system (Fig. 7.1b), the illumination beam and the beam carrying the information on the surface of the specimen are separated at a small angle by the magnetic field which is perpendicular to the axis. The "straight" design of the MEM is simpler because it makes no use of a deflecting magnetic

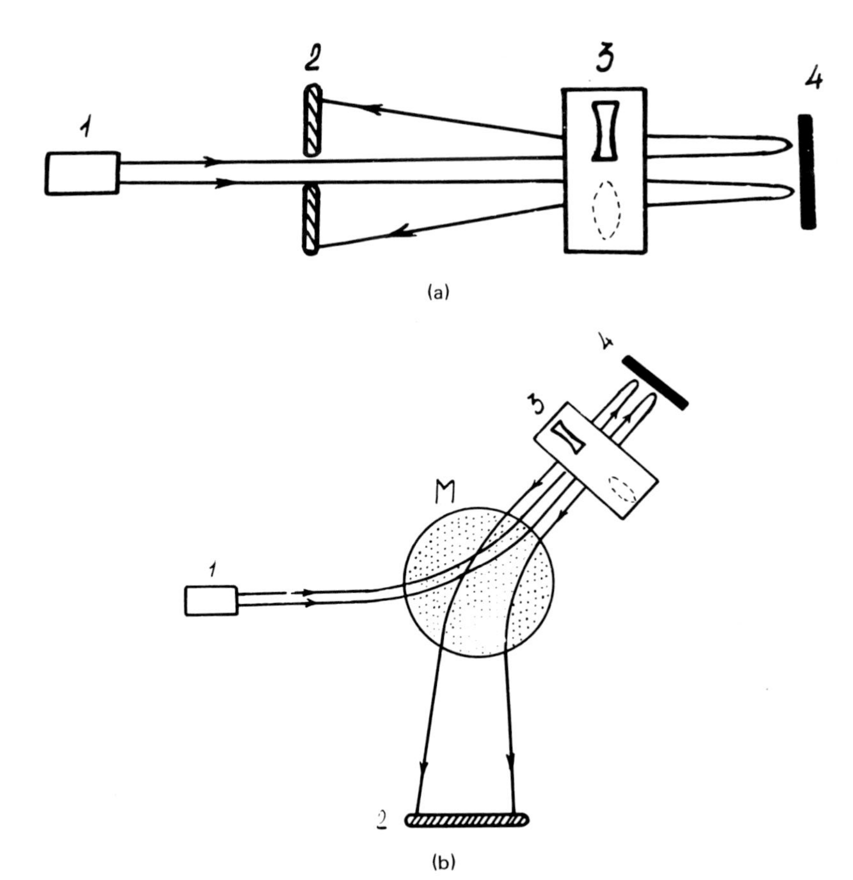

Fig. 7.1 Schematic diagram of various types of MEM. (a) "Straight" design; (b) separated beam system. 1, illumination source; 2, viewing screen; 3, immersion objective; 4, specimen; M, separating magnetic field.

field; such a field generally produces an additional astigmatism. The design of a separated beam system is more complicated. However, in the latter case the illumination beam and the beam carrying the image can be influenced independently.

The first models of the separated beam system of the MEM were presented by Hottenroth (1937) and Orthuber (1948). Subsequently, the MEM of that type was constructed by Bartz *et al.* (1956). In this instrument, a two-electrode objective system was used, and a resolving power of the order of 0.1 μm was reported. An MEM of the "straight" design with geometrical resolution of ~0.35 μm was first designed by Mayer (1955). Glass devices followed by an improved instrument were constructed by Spivak *et al.* (1961a) (Fig. 7.2).

Mirror electron microscopes for different fields of application have been constructed (Bethge *et al.*, 1960; Forst and Wende, 1964; Igras and Warminski,

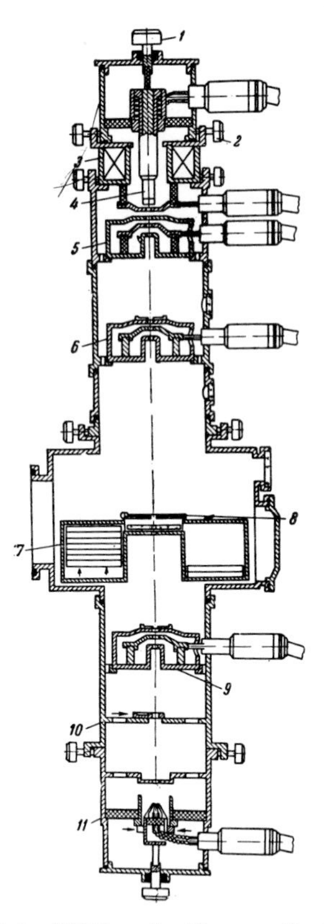

Fig. 7.2 Schematic section of the MEM at the Electron Optics Laboratory, Moscow State University (Spivak *et al.*, 1961a). 1, adjustment of the specimen in a vertical plane; 2, adjustment of the specimen in a horizontal plane; 3, magnetic coil and magnetic shielding; 4, specimen; 5, five-electrode immersion objective; 6, projection lens; 7, camera; 8, fluorescent screen for visual observation; 9, condenser lens; 10, gun aperture stage; 11, electron gun. (*Courtesy of G. V. Spivak.*)

1965; Shwartze, 1966; Ivanov and Abalmazova, 1966, Barnett and Nixon, 1967a; Bok *et al.*, 1968a; Someya and Watanabe, 1968). Artamonov (1968) has devised a simple spherical mirror electron microscope.

Progress in instrumentation techniques and the feasibility of MEM applications promoted the industrial production of the conventional MEM, JEMM-I by JEOL in 1968 and an instrument with separated beams by Igras and Warminski (1965–1968). One should also note the combined instruments such as an emission-mirror microscope (Delong and Drahos, 1964), scanning mirror microscopes (Vertzner and Chentsov, 1963; Garood and Nixon, 1968; Cline *et al.*, 1969; Ogilvie *et al.*, 1969), and related instruments (Kasper and Wilska, 1967; Vassoille *et al.*, 1970).

From the point of view of optics, the general modes of operation of the MEM are the shadow projection imaging mode (MEMSI) and the focused imaging mode (MEMFI). The first mode is realized when the electron rays originating from a virtual point source produce a shadow image of the specimen. In other words, the contrast is produced the same way as in a shadow optical or shadow transmission electron microscope (Leisegang, 1956). In the shadow projection imaging mode of the MEM, a unique ray of the illumination beam corresponds to each point on the surface of the specimen. Microirregularities on the surface of the sample lead to velocity modulation of the electron beam. During further motion of the electron beam from the specimen to the screen, a redistribution of current density occurs; i.e., contrast is produced.

In the focused imaging mode, the electron beam after reflection is focused in the immersion objective and each point of the specimen is transformed to the conjugate point of the screen by means of projection lenses. When the focused mode is realized, a contrast aperture is mounted in the cross-over of the objective lens focal plane. A part of the deflected electrons is cut off by the edge of the contrast aperture, resulting in a decrease of intensity at a corresponding point of the image; thus, the contrast appears. The focused imaging mode was experimentally realized in the instrument with separated beams by Shwartze (1967) and Bok *et al.*, (1968b). Reversibility of electron passes makes the focused image unobtainable in an instrument of "straight" design with electrostatic lenses. Shwartze (1967) also suggested the use of magnetic lenses for obtaining focused images in the instrument of "straight" design.

The MEM with focused images is characterized by a large angle of the illumination beam ($\sim 10^{-2}$–10^{-3} radian), and consequently by a high intensity of the image on the viewing screen. At the same time shadow projection imaging is often used in the instruments with separated beams, since the sensitivity to the depiction of both the electric and magnetic inhomogeneities is substantially higher in the shadow projection mode. Possibilities of shadow projection and of the focusing modes were compared by Bok *et al.* (1971).

Until recently MEMs were used only for the observation of static electric and magnetic microfields on the surface of specimens under investigation. The application of stroboscopic principles in the field of electron microscopy led to the construction of a stroboscopic mirror electron microscope (Spivak and Lukianov, 1966a; Spivak *et al.*, 1968a). A schematic diagram of the stroboscopic electron mirror microscope is shown in Fig. 7.3. The electron gun of the instrument is normally cut off by the negative bias voltage applied to the modulator of the electron gun; other kinds of realizations of the stroboscopic principle are also possible (Spivak *et al.*, 1968b). The electron gun is unlocked at definite times by the positive strob-pulses from the pulse generator. The latter has the tunable delay line and is synchronized with the a.c. generator which supplies voltage to the specimen.

The observation of images in the stroboscopic mode is made only at the time

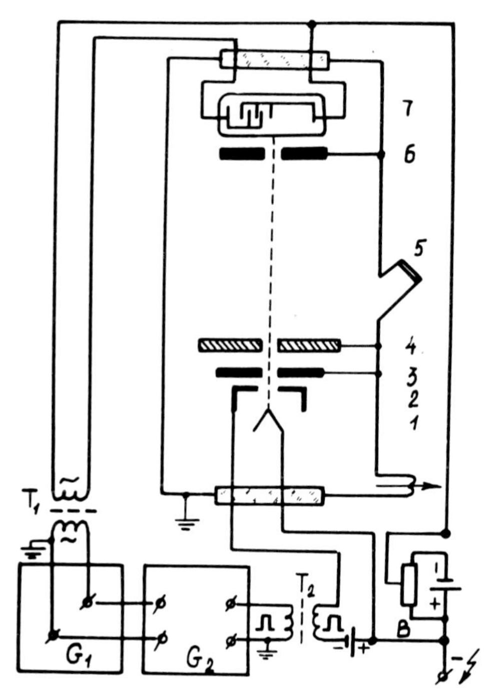

Fig. 7.3 Schematic diagram of a stroboscopic mirror electron microscope. 1–3, electrodes of the electron gun; 4, the fluorescent screen; 5, window for observation; 6, anode aperture; 7, specimen; T_1 and T_2, transformers; B, bias battery; G_1, a.c. generator; G_2, generator of strob-pulses.

when the electron gun is unlocked by positive strob-pulses. The development of stroboscopic MEM extended essentially the fields of application of the instrument. A stroboscopic MEM for examining signals at frequencies up to 100 MHz was constructed by Szentesi (1972). A more detailed description of the construction and modes of operation of the MEM is given in a review by Lukianov *et al.* (1973).

Estimation of Sensitivity and Resolving Power in Different Modes of Operation

The quantitative estimation of the main phenomena which accompany the image formation in the MEM is presented below. A detailed theory of the MEM operation will be discussed later.

The unperturbed electron trajectory in the uniform retarding field with field strength E_0 is:

$$r(t) = r_0 + tm^{-1} p_0, \qquad z(t) = z_0 + eE_0 t^2/2m$$

where an electron reaches the reversal point z_0 at $t = 0$, $\boldsymbol{p}_0 = (p_{0x}, p_{0y})$ is the unperturbed transversal momentum, $\boldsymbol{r}_0 = (x_0, y_0)$ is the coordinate on the surface of the specimen.

In the MEMSI the current density distribution of electrons $j(\boldsymbol{r})$ in some "defocusing plane" at the distance $\tilde{l}$ (typically the value of $\tilde{l} \sim 0.5 \div 3$ cm) from the specimen is imaged on the screen. The image contrast formation in the MEMSI is due to the bunching of electrons with modulated transversal velocity during the transit time of electrons $t_{\text{eff}} = (2m\tilde{l}/eE_0)^{1/2}$ from the specimen to the "defocusing plane." Because of the action of microfields the electrons obtain an additional momentum $\Delta \boldsymbol{p}$ and the electron trajectory is displaced to the amount $s = t_{\text{eff}} \, \Delta p/m$ in that plane. In that case the image contrast $K(\boldsymbol{r}) = [j(\boldsymbol{r}) - j_0]/j_0$ approximately equals $K \sim ds/dx \sim t_{\text{eff}} \, \Delta p/(m\delta x) \sim \Delta p \cdot (\tilde{l}/meE_0)^{1/2} \cdot (\delta x)^{-1}$ where δx is the transverse size of perturbation.

Contrast formation in the MEMFI is due to the cut-off of the deflected electrons by the contrast aperture. The magnitude of the contrast may be estimated as $K \sim \Delta p/\delta p_0$ where δp_0 is the extension of momentum distribution transmitted by the aperture. Usually δp_0 coincides with the width of momentum distribution in the illuminating beam. The order of magnitude of δp_0 may be estimated by the size r_a of the cross-over in aperture plane

$$\delta p_0 \sim r_a f^{-1} (2meU_0)^{1/2},$$

where f is the focal length of the objective, and U_0 is the accelerating voltage.

Microfields over the specimen satisfy Laplace's equation:

$$(\nabla^2)\varphi_1(x, y, z) = 0, \quad (\nabla^2)\boldsymbol{B}_1(x, y, z) = 0$$

where φ_1 is the potential of the electric microfield, and $\boldsymbol{B}_1$ is the strength of the magnetic microfield. One of the possible solutions of Laplace's equation which decreases at $z \to +\infty$ is:

$$\varphi_1(\boldsymbol{r}, z) = \varphi_1 \cos(\boldsymbol{k}\boldsymbol{r}) \exp\{-|\boldsymbol{k}|z\} \tag{7.1}$$

where $\boldsymbol{k} = (k_x, k_y)$ is the "wave vector" of the microfield. This solution illustrates the general property of microfields which is of essential importance for the estimation of the sensitivity and resolving power of the MEM: the vertical extension of microfields δz coincides with the characteristic size if inhomogeneities in transverse direction, $\delta z \sim |\boldsymbol{k}|^{-1} \sim \delta x$.

For vertically incident electrons (i.e., at $\boldsymbol{p}_0 = 0$) the height of reflection z_0 is determined by the energy of electrons ϵ and the bias voltage U_B referred to the filament of the electron gun: $z_0 = (eU_B - \epsilon)/eE_0$. Comparison of this relation with Eq. (7.1) shows that the energy spread of electrons $\delta\epsilon$ leads to a strong damping of the influence of the microfields with extension δx on the electron trajectory if

$$\delta x \lesssim \delta z_0 = \delta\epsilon/eE_0 \tag{7.2}$$

This relation determines the limit of resolution associated with the energy spread of the illuminating electron beam for any mode of operation of the MEM. This point is quite specific for the MEM: the energy spread is already important at the stage of interaction of electrons with the microfields on the surface of the specimen. (Typically, $\delta\epsilon \sim 1$ eV, $E_0 \sim 5.10^6$ V/m and $\delta x \sim 0.2$ μm).

In the case of monoenergetic illumination the classical estimations of the limit of resolution for both MEMSI and MEMFI are associated with the spread δp_0 of transverse momenta. For the MEMFI the limit of resolution is related again to the very process of interaction of electrons with microfields. The transverse displacement obtained by the electron during the transit time $t_{tr} \approx (m\delta z/eE_0)^{1/2} \approx (m\delta x/eE_0)^{1/2}$ of the electron passing the microfield with the size δx should be smaller than the size of the inhomogeneity: $p_0\, t_{tr}\, m^{-1} \leqslant \delta x$. Then it follows that:

$$\delta x \gtrsim \delta x_{\mathrm{FI}} = \frac{(\delta p_{0\mathrm{FI}})^2}{meE_0} \sim \frac{r_a^2\, U_0}{f^2\, E_0} \tag{7.3}$$

With monoenergetic illumination the limit of resolution for the MEMSI is related to the propagation of the beam from the specimen to the "defocusing plane." The spread of momenta $\delta p_{0\mathrm{SI}}$ leads to the spread of coordinates $m^{-1}\, t_{\mathrm{eff}}\, \delta p_{0\mathrm{SI}}$ and therefore,

$$\delta x \gtrsim \delta x_{\mathrm{SI}} \approx m^{-1}\, t_{\mathrm{eff}}\, \delta p_{0\mathrm{SI}} \approx \delta p_{0\mathrm{SI}}\, (\tilde{l}/meE_0)^{1/2} \tag{7.4}$$

The value of the momenta spread δp_0 is determined by the size of the illumination source. Usually the size of the source for the MEMSI is considerably smaller than that for the MEMFI; therefore $\delta p_{0\mathrm{SI}} \ll \delta p_{0\mathrm{FI}}$ and the intensity on the viewing screen of the MEMFI is considerably higher than that in the MEMSI.

For the estimation of the contrast it should be considered that the microfields over the specimen modify the transverse component of the momentum (velocity); this primary modulation of the electron beam occurs identically in both the MEMSI and MEMFI. In the case of electrical microfield with strength $E_1 \sim \varphi_1/\delta x$, the change of momentum Δp equals:

$$\Delta p \approx eE_1\, t_{tr} \approx E_1 \cdot (em\delta x/E_0)^{1/2}$$

The effect of a magnetic microfield $\boldsymbol{B}_1$ on the trajectory may be described as a result of rotation of the vector $\boldsymbol{p}_0$ because of Larmor precession with the angular velocity $\omega \sim eB_1/m$. During the transit time t_{tr} vector $\boldsymbol{p}_0$ rotates at an angle $\gamma \approx \omega t_{tr}$ so that:

$$\Delta p \approx \gamma p_0 \approx B_1\, (e\delta x/mE_0)^{1/2}\, p_0$$

The relations obtained allow the estimation of the image contrast of both electrical and magnetic microfields for two imaging modes of the MEM. For magnetic microfields it should be taken into account that the value of the average momentum $\overline{p_0}$ in the illumination beam of the MEMSI is proportional to the distance r_0 on the specimen from the electrical center of the image: $\overline{p_0} = mr_0/2t_{\text{eff}}$ (see the section on "Paraxial Geometrical Optics"). This corresponds to the fact that in the MEMSI the projection of the specimen to the screen occurs from the point illuminating source located at a distance $\sim\tilde{l}$ from the specimen. As a result the following estimations for the contrast are obtained:

$$(\text{SI}, E_1) \quad K \approx \frac{E_1}{E_0} \cdot \left(\frac{\tilde{l}}{\delta x}\right)^{1/2} \tag{7.5}$$

$$(\text{FI}, E_1) \quad K \approx \frac{E_1}{\delta p_{0\,\text{FI}}} \cdot \left(\frac{em\delta x}{E_0}\right)^{1/2} \approx \frac{E_1}{E_0} \cdot \left(\frac{\delta x}{\delta x_{\text{FI}}}\right)^{1/2} \tag{7.6}$$

$$(\text{SI}, B_1) \quad K \approx B_1 r_0 \left(\frac{e}{mE_0 \delta x}\right)^{1/2} \approx \frac{1}{2}\gamma \cdot \left(\frac{r_0}{\delta x}\right) \tag{7.7}$$

$$(\text{FI}, B_1) \quad K \approx \frac{\overline{p_0}}{\delta p_{0\text{FI}}} \cdot B_1 \left(\frac{e\delta x}{mE_0}\right)^{1/2} \approx \gamma \cdot \frac{\overline{p_0}}{\delta p_{0\text{FI}}} \tag{7.8}$$

These estimations show a high sensitivity of the MEM for the depiction of microfields.

For electrical microfields the contrast $K \sim 1$ may be achieved because of the action of the microfield with strength E_1 much smaller than that of the uniform retarding field E_0. The electrical contrast K in the MEMSI is of the order of the small dimensionless parameter (E_1/E_0) multiplied by the "gain factor" $(\tilde{l}/\delta x)^{1/2}$. Typically $\tilde{l} \sim 1$ cm, $\delta x \sim 10^{-2}$ to 10^{-4} cm, and $(\tilde{l}/\delta x)^{1/2} \sim$ 10 to 100; note that the "gain factor" of the contrast increases when δx decreases. For the MEMFI the "gain factor" equals $(E_0\, me\delta x/\delta p_{0\,\text{FI}})^{1/2} \approx (\delta x/\delta x_{\text{FI}})^{1/2}$; usually this value is of the order of 50 to 1 and tends to diminish when δx decreases.

Thus, both imaging modes possess considerably high sensitivity to the electrical microfields. The shadow imaging mode is more sensitive to the depiction of rather small inhomogeneities δx; the focusing mode is more sensitive when δx is sufficiently large.* The boundary depends on the value of t_{eff} (for the MEMSI) and δp_0 (for the MEMFI); this boundary lies at $\delta x \sim |k_B|^{-1} \sim 80\ \mu$m for the instrument described in the review by Bok *et al.* (1971).

In principle, the sensitivity of the MEMFI may be increased by the reduction of δp_0, followed by a decrease of intensity on the viewing screen of the instrument (high intensity being one of the advantages of the focusing mode). However, the manufacturing of small-size apertures and the alignment of the aperture in the cross-over may present technical difficulties.

*Here we define the sensitivity through the value of contrast. For the MEMSI it is also possible to register the action of large-scale microfields by, for example, measuring the displacement of the image of the mesh deposited on the surface of the specimen; this fact enlarges the possible applications for the MEMSI.

The image contrast of magnetic microfields in the MEMSI rises from the center to the periphery of the image. The magnetic contrast K in the MEMSI is of the order of the small dimensionless parameter γ (the deflection angle) multiplied by the "gain factor" $(r_0/\delta x)$; usually $(r_0/\delta x) \sim 1$ to 100.

In the MEMFI usually vertical illumination is used; therefore, one can estimate the "gain factor" $(\overline{p_0}/\delta p_{0\mathrm{FI}})$ as 1 or even less than 1. This is probably the reason why the magnetic contrast has not yet been revealed in the MEMFI. To achieve a considerable gain $(\overline{p_0}/\delta p_{0\mathrm{FI}})$ in the MEMFI, certain special conditions are necessary (for example, a tilt of the specimen, tilted illumination, gun with the cathode in a magnetic field, and placing the specimen in a magnetic field).

In conclusion, let us estimate the wave mechanical limit of resolution. The classical estimations [Eqs. (7.3) and (7.4)] are improved with the decrease of momentum spread δp_0 of the monoenergetic illumination beam. Substitution of δp_0 from the Heisenberg uncertainty principle $\delta p_0 \gtrsim \hbar/\delta x$ into Eqs. (7.3) and (7.4) yields the following quantum limits of resolution:

$$\text{(Quant., SI)} \quad \delta x \gtrsim (\hbar t_{\text{eff}}/m)^{1/2} \sim (\tilde{l})^{1/2}\,\lambda_0^{1/2}, \tag{7.9}$$

$$\text{(Quant., FI)} \quad \delta x \gtrsim (\hbar^2/2meE_0)^{1/3} \sim (\tilde{l})^{1/3}\,\lambda_0^{2/3} \tag{7.10}$$

where $\lambda_0 = 2\pi\hbar \cdot (2meE_0\tilde{l})^{-1/2}$ is the wavelength of the electron corresponding to the accelerating voltage $U_0 = E_0\tilde{l}$. Besides, for the MEMFI the following inequality should hold (Bok *et al.*, 1971):

$$\text{(Quant., FI)} \quad \delta x \gtrsim \frac{\hbar}{\delta p_a} \sim \frac{\lambda_0 f}{r_a}. \tag{7.11}$$

Usually $(\tilde{l})^{1/2} \cdot \lambda_0^{1/2} \sim 100$ nm, $(\tilde{l})^{1/3}\,\lambda_0^{2/3} \sim 1.5$ nm, $\lambda_0 f r_a^{-1} \sim 1.5$ to 15 nm, respectively. Apparently, practically achievable resolution seems to be limited by the energy spread of electrons and is of 200 to 100 nm of the order of magnitude. This limit is analogous to the resolution of the emission electron microscope with the same values of $\delta\epsilon$ and E_0.

IMAGE CONTRAST FORMATION IN MEM (THEORY AND PRACTICE)

Now we shall consider the classical theory of image contrast formation in mirror electron microscopy. The experimental data included concern shadow projection imaging in the MEM, which is characterized by high sensitivity to microfields. Readers interested in the experimental results obtained with the MEM with focused images are referred to the review by Bok *et al.* (1971).

The mechanism of image contrast formation in the MEM is presented here. Because of the action of perturbations caused by topographical inhomogeneities and electrical or magnetic microfields on the surface of the specimen under investigation, the electron obtains an additional transverse momentum. This leads to a displacement of the electron trajectory compared with that in the absence of perturbations. Below, the process of modification of transverse momenta will be referred to as the process of primary modulation. The latter is identical for both the shadow and focused imaging modes of operation of the MEM. The electron beam traveling from the specimen to the screen undergoes a process during which primary velocity modulation transforms into density modulation, which is termed demodulation. As a result, an image contrast is produced on the viewing screen. The process of demodulation is determined by the electron optical properties of the system and is essentially different for shadow projection and for focused modes of operation of the MEM.

The classical theory described in the present section contains the quantitative consideration of the electron optics of the instrument, the modulation and the demodulation processes, image contrast formation of electric microfields, topographical inhomogeneities and magnetic microfields.

Paraxial Geometrical Optics

In the majority of the published papers the geometrical optics of the MEM has been considered in terms of the geometrical rays, focal length, principle planes, etc. In contrast, we shall consider the electron trajectory introducing time as a parameter. This treatment seems to be more adequate for the specific features of the MEM, namely, for the existence of the reversal point of the electrons close to the surface of the specimen. Indeed, in the coordinate treatment there are two beams (the incident and the reflected) in the same cross section z = const, whereas in the temporal treatment the electron beam at each moment of time travels in a definite direction. Furthermore, the trajectories of electrons $\boldsymbol{r}(z)$ close to the point of reversal $z = z_0$ have a singularity $\boldsymbol{r}(z) \backsim \pm\sqrt{z - z_0}$. Contrary to that, the dependence of the coordinate $\boldsymbol{r}(t)$ and of the momentum $\boldsymbol{p}(t)$ on time t is quite regular. Besides, the introduction of the temporal dependence simplifies the treatment of the modulation of the beam by microfields both in classical and wave-mechanical approaches.

In the paraxial approximation, the differential equation for the electron trajectory $\boldsymbol{r}(t)$ in the presence of a magnetic field is (Glaser, 1952):

$$\frac{d^2\boldsymbol{r}}{dt^2} = -\left[\frac{e}{2m}\frac{\partial^2 U(z(t), 0)}{\partial z^2} + \frac{e^2}{4m^2} B_z^2(z(t), 0)\right]\boldsymbol{r}(t) \tag{7.12}$$

where $U(z, |\boldsymbol{r}|)$ is the potential of the electric field and $B_z(z, |\boldsymbol{r}|)$ is the z-component of the magnetic field. The dependence of z on the time is determined by the law of conservation of energy.

It is assumed that retarding electric field E_0 is homogeneous and that the magnetic fields are absent close to the surface of the specimen. Let the moment $t = 0$ correspond to the reversal of electrons over the surface of the specimen.

Then, in the absence of perturbing microfields the electron trajectory $\boldsymbol{r}(t)$ may be written as:

$$\boldsymbol{r}(t) = \boldsymbol{r}_0 + m^{-1}t\boldsymbol{p}_0, \quad z(t) = z_0(\epsilon, \boldsymbol{p}_0) + eE_0 t^2/2m \tag{7.13}$$

where $\boldsymbol{r}_0$ is the coordinate of the point reversal over the specimen, and $\boldsymbol{p}_0 = m\dot{\boldsymbol{r}}$ is the transverse component of momentum with which an electron approaches the specimen. Equation (7.13) indicates that close to the specimen the trajectories of the electrons are parabolas.

The influence of the perturbations on the electron trajectory is strongly dependent on the height of reflection of the electrons over the specimen. In a general case, $z_0(\epsilon, \boldsymbol{p}_0)$ is described by:

$$z_0(\epsilon, \boldsymbol{p}_0) = \left(eU_B + \frac{\boldsymbol{p}_0^2}{2m} - \epsilon\right)\bigg/ eE_0 \tag{7.14}$$

where U_B is the value of a small negative voltage applied to the specimen with respect to the filament of the electron gun, and ϵ is the total kinetic energy with which the electron leaves the cathode.

For the general description of the propagation of the electron beam from the specimen to the screen it is convenient to introduce two fundamental solutions $g_1(t)$ and $g_2(t)$ of the linear differential equation (7.12) which satisfy the following conditions:

$$g_1(0) = 1, \quad \dot{g}_1(0) = 0; \qquad g_2(0) = 0, \quad \dot{g}_2(0) = 1 \tag{7.15}$$

where the point corresponds to the differentiation over the time. Then the solution of the Eq. (7.12) corresponding to the unperturbed trajectory may be represented as:

$$\boldsymbol{r}(t) = g_1(t)\boldsymbol{r}_0 + m^{-1}g_2(t)\boldsymbol{p}_0; \quad \boldsymbol{p}(t) = m\dot{g}_1(t)\boldsymbol{r}_0 + \dot{g}_2(t)\boldsymbol{p}_0 \tag{7.16}$$

The influence of the perturbations on the electron trajectory may be described by means of the magnitudes of change of the transverse momentum and coordinate $\Delta\boldsymbol{p}_0$ and $\Delta\boldsymbol{r}_0$, respectively. Indeed, the sizes of perturbations are supposed to be small compared with the distance from the specimen to the closest optical elements. Therefore, it is possible to attribute these changes of the values of the $\boldsymbol{p}$ and $\boldsymbol{r}$ to the moment $t = 0$, i.e., to the point close to the surface of the specimen. Thus, the changes may be said to occur as if instantaneously. Therefore, in the presence of perturbations, the electron trajectory is determined by the solution of differential equation (7.12) with the following initial conditions:

$$\boldsymbol{r}(t=0) = \boldsymbol{r}_0 + \Delta\boldsymbol{r}_0(\boldsymbol{r}_0, \boldsymbol{p}_0, \epsilon); \quad \boldsymbol{p}(t=0) = \boldsymbol{p}_0 + \Delta\boldsymbol{p}_0(\boldsymbol{r}_0, \boldsymbol{p}_0, \epsilon) \tag{7.17}$$

In a final form at $t > 0$ the trajectory of the electron is:

$$\boldsymbol{r}(t) = (\boldsymbol{r}_0 + \Delta\boldsymbol{r}_0)g_1(t) + (\boldsymbol{p}_0 + \Delta\boldsymbol{p}_0)m^{-1}g_2(t) \tag{7.18}$$

It is convenient to consider the process of imaging of the beam from the specimen to the screen separately for the MEMSI and for the MEMFI.

Shadow Imaging Mode. For the MEMSI with a point illumination source the illumination beam is homocentrical. Therefore, the values of parameters $\boldsymbol{r}_0$ and $\boldsymbol{p}_0$ are subject to a linear relation:

$$\boldsymbol{p}_0 = mD_{\mathrm{sp}}\,\boldsymbol{r}_0. \tag{7.19}$$

The constant D_{sp} in Eq. (7.19) has the dimension of $(\mathrm{seconds})^{-1}$. The relation of the type of Eq. (7.19) between $\boldsymbol{p}_0$ and $\boldsymbol{r}_0$ in the illuminating beam of a transmission electron microscope (compare with Fig. 7.5) would characterize the spherical divergence of the beam with the distance from the source $R = v_z/D_{\mathrm{sp}}$, where v_z is the velocity along the axis. However, $v_z = 0$ in the MEM at the surface of reflection, and therefore the value R has no meaning here. Contrary to that, the value of D_{sp} remains finite.

Note that the height of the reversal in the MEMSI increases with the distance $|\boldsymbol{r}_0|$ from the electrical center of the image. Substitution of Eq. (7.19) into Eq. (7.14) yields the so called "parabola of heights of reflection" $z_0(\boldsymbol{r}_0)$:

$$z_0(\boldsymbol{r}_0) = \frac{U_B}{E_0} + \frac{mD_{\mathrm{sp}}^2}{2eE_0}\boldsymbol{r}_0^2 - \frac{\epsilon}{eE_0} \tag{7.20}$$

This effect leads to the decrease of the contrast of small-scale structures at the periphery of the image (see below Fig. 7.6).

For a further consideration of the homocentrical beam in the MEMSI, it is convenient to introduce a normalized unperturbed trajectory $g_0(t)$. Substitution of Eq. (7.19) into Eq. (7.16) yields:

$$\boldsymbol{r}(t) = g_0(t)\boldsymbol{r}_0;\quad g_0(t) = g_1(t) + D_{\mathrm{sp}}g_2(t) \tag{7.21a}$$

Thus, $g_0(t)$ is the solution of Eq. (7.12) satisfying the initial conditions:

$$g_0(0) = 1,\quad \dot{g}_0(0) = D_{\mathrm{sp}} \tag{7.21b}$$

For the illustration of general concepts we shall give a detailed consideration of the MEMSI with a two-electrode objective system, the latter consisting of the specimen and the aperture at the distance 1 from it (Fig. 7.4). It is well known (Zworykin *et al.*, 1945; Kelman and Yavor, 1968) that such an aperture in assumption that its diameter d is small compared with $l (d \ll l)$ acts as a thin diverging lens with a focal length $f = 4l$. However, we shall not use this relation and shall obtain the functions $g_1(t)$ and $g_2(t)$ directly from Eq. (7.12). The strength of the retarding field between the specimen and the aperture is $E_0 = U_0/l$, where U_0 is the accelerating voltage, and $v_0 = \sqrt{2eU_0/m}$ is the velocity of the electron.

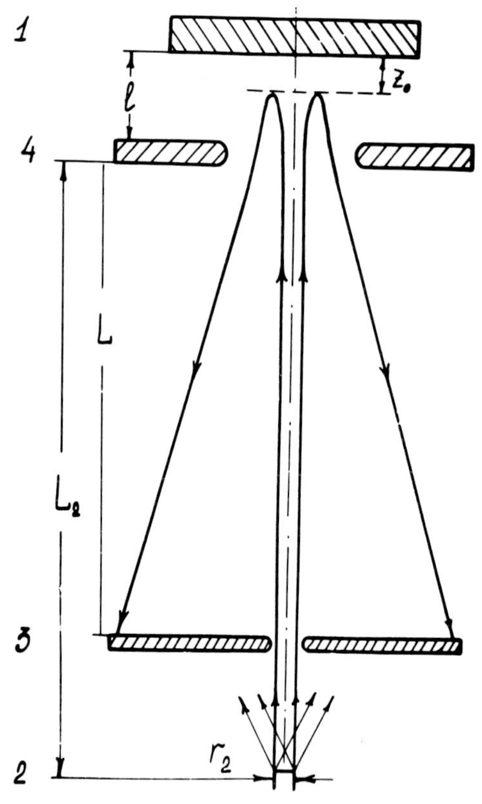

Fig. 7.4 Schematic drawing of MEM with two-electrode objective system (shadow projection imaging). 1, specimen; 2, electron source; 3, viewing screen; 4, anode aperture; r_2 denotes the size of the source.

The transit time from the aperture to the screen is $\tau_l = 2l/v_0$. In this case, the equation of motion (7.12) becomes:

$$\frac{d^2r}{dt^2} = \frac{1}{2\tau_l}\,[\delta(t+\tau_l) + \delta(t-\tau_l)]\; r(t) \tag{7.22}$$

where $\delta(t)$ is the Dirac δ-function. Taking Eq. (7.15) into account we obtain from Eq. (7.22):

$$|t| \leqslant \tau_l: \;\; g_1(t) \equiv 1, \;\; g_2(t) = t, \;\; z(t) = z_0 + eE_0 t^2/2m;$$

$$|t| \geqslant \tau_l: \;\; g_1(t) = \frac{1}{2} + \frac{|t|}{2\tau_l}, \;\; g_2(t) = \left(\frac{3}{2}|t| - \frac{\tau_l}{2}\right) \cdot \operatorname{sgn}(t), \;\; z(t) = l + v_0(|t| - \tau_l)$$

(7.23)

(here we neglect the weak dependence of τ_l on z_0). The assumption about the point illuminating source situated at an infinite distance from the objective system leads to the following expression for D_{sp}:

$$D_{sp} = \frac{v_0}{6l} = \frac{1}{3\tau_l}. \tag{7.24}$$

In the case of the MEMSI with an arbitrary objective system we shall observe that the trajectory of an electron at $t > 0$ with the perturbations taken into account is:

$$\boldsymbol{r}(t) = g_0(t)\boldsymbol{r}_0 + g_1(t)\Delta \boldsymbol{r}_0 + g_2(t)m^{-1}\Delta \boldsymbol{p}_0 \tag{7.25}$$

The coordinate of point $\boldsymbol{R}$ at which the electron hits the screen is calculated from Eq. (7.25) at $t = T_{scr}$, where T_{scr} is the moment of intersection of the electron trajectory with the screen plane. Introducing the magnification M of the MEMSI by the relation:

$$M = g_0(T_{scr}) \tag{7.26a}$$

and considering the coordinates of the screen related to the specimen: $\boldsymbol{r}' = \boldsymbol{R}/M$ (compare with the Fig. 7.5), one obtains from Eq. (7.25) that:

$$\boldsymbol{r}' = \boldsymbol{R}/M = \boldsymbol{r}_0 + \boldsymbol{s} \tag{7.26b}$$

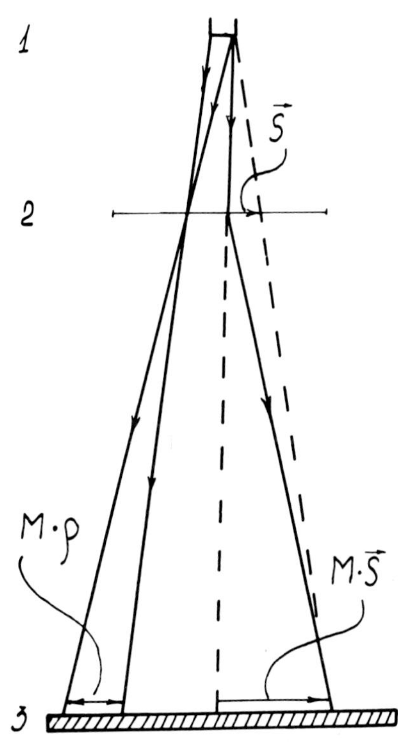

Fig. 7.5 Schematic representation of the projection from equivalent source in the shadow projection imaging mode. 1, image of the source; 2, specimen plane; 3, screen; $\boldsymbol{s}$, related to the specimen displacement vector due to the action of microfields; ρ, related to the specimen size of a source; M, magnification.

The vector s in Eq. (7.26b) describes the related to the specimen displacement of the electron trajectory which occurs because of the action of perturbations. From Eq. (7.19) and Eq. (7.25), it follows that:

$$s = \frac{g_1(T_{scr})}{g_0(T_{scr})} \Delta r_0 + \frac{g_2(T_{scr})}{g_0(T_{scr})} m^{-1} \Delta p_0 \equiv \beta \Delta r_0 + t_{eff} m^{-1} \Delta p_0. \tag{7.27}$$

The coefficient β in Eq. (7.27) is dimensionless, and the coefficient t_{eff} in Eq. (7.27) has the dimensions of time and characterizes the effective time during which the transformation of velocity modulation into density modulation occurs. Note that the use of Wronsky's theorem makes it possible to express the functions $g_1(t)$ and $g_2(t)$ and the parameters β and t_{eff} through the unperturbed solution $g_0(t)$ and the value of $D_{sp} = \dot{g}_0(0)$:

$$g_2(t) = g_0(t) \int_0^t [g_0(t')]^{-2}\, dt', \quad g_1(t) = g_0(t) - D_{sp} g_2(t),$$

$$t_{eff} = \int_0^{T_{scr}} [g_0(t')]^{-2}\, dt', \quad \beta = 1 - D_{sp} t_{eff}. \tag{7.28}$$

For the MEMSI with a two-electrode objective system, $T_{scr} = (L + 2l)/v_0$, where L is the distance from the objective to the screen. If $L \gg l$, we obtain:

$$M = \frac{L}{2l}, \quad \beta = \frac{1}{2}, \quad t_{eff} = \frac{3}{2} \tau_l = \frac{3l}{v_0}, \quad D_{sp} = \frac{v_0}{6l} = \frac{1}{2t_{eff}} \tag{7.29}$$

Equations (7.29) were first obtained for the MEM with a two-electrode objective system by means of pure geometrical (and not temporal) consideration of electron trajectories (Lukianov *et al.*, 1968a; Barnett and Nixon, 1967a,b).

The relationships,

$$\beta = \frac{1}{2}, \quad D_{sp} = \frac{1}{2t_{eff}} \tag{7.30}$$

are valid for the MEMSI of straight design with an arbitrary objective system and follow from the fact that the incident and reflected beams move in the same fields of lenses. Therefore the coefficient in the right-hand side of Eq. (7.12) is an even function of time t.

The above treatment was provided in the case of point source illumination. If the illumination source has a nonzero size r_2, then the coordinate of the point at the screen may be written as:

$$\boldsymbol{R} = M \cdot (\boldsymbol{r}_0 + \boldsymbol{s} + \boldsymbol{\rho}), \tag{7.31}$$

where $\boldsymbol{\rho}$ characterizes the displacement of the electron trajectory due to the velocity spread in the illumination beam:

$$\boldsymbol{\rho} = t_{eff} m^{-1} (\boldsymbol{p}_0 - m D_{sp} \boldsymbol{r}_0) \tag{7.32}$$

The spread over $\boldsymbol{\rho}$ may be interpreted as being due to the projection image of the specimen plane onto the screen from the source of finite sizes (Fig. 7.5). However, this interpretation, although it conveys the right picture of the influence of spread in $\boldsymbol{\rho}$, disregards the fact that the electron trajectories close to the surface of the specimen are parabolas and not straight lines, as is shown in Fig. 7.5. For the MEMSI with an arbitrary objective system the point of illumination source characterized by the coordinate $\boldsymbol{r}_2$ yields the related to the specimen displacement $\boldsymbol{\rho}$,

$$\boldsymbol{\rho} = -\frac{L}{L_2}\frac{\boldsymbol{r}_2}{M} \tag{7.33}$$

where L_2 is the distance from the source to the objective system.

According to Eq. (7.33), the related to the specimen size ρ of the projection source decreases when the magnification M increases. Therefore at a sufficiently high magnification M the finite size of the real illumination source no longer imposes limitations on the transverse resolving power of the MEMSI of straight design.

A detailed discussion of the contrast in the MEMSI will be presented later.

Focused Imaging Mode. In the MEM with focused images the trajectory of the reflected electron is also described by Eq. (7.18). This trajectory undergoes refractions in lenses and traverses the plane of the contrast aperture mounted at the cross-over; i.e., in the back focal plane of the objective lens. The moment t_c of the intersection of the plane of the cross-over by an electron is determined from the equation $g_1(t_c) = 0$ and therefore:

$$\boldsymbol{r}(t_c) \equiv \boldsymbol{r}_c = [\boldsymbol{p}_0 + \Delta\boldsymbol{p}_0(\boldsymbol{r}_0, \boldsymbol{p}_0, \epsilon)]\, m^{-1}\, \tau; \quad \tau = g_2(t_c) \tag{7.34}$$

The value $\tau = g_2(t_c)$ of the dimension of time characterizes the focusing properties of the objective lens. The value of τ may be estimated as $\tau \sim f/v_0 \sim t_c$ by the order of magnitude, where f is the focal length. In the focused mode, the optics sharply images the surface of the specimen onto the screen, i.e.,

$$g_2(T_{scr}) = 0, \quad \boldsymbol{R} = g_1(T_{scr}) \cdot (\boldsymbol{r}_0 + \Delta\boldsymbol{r}_0)$$

The image contrast formation in the focused mode is determined by the transparency function of the contrast aperture i.e., by the configuration of the aperture and its position with respect to the axis. We shall denote the transparency of the aperture as a function of momentum $\boldsymbol{p}$ with which the electron is reflected from the specimen as $T(\boldsymbol{p})$. Here, $\boldsymbol{p} = m\boldsymbol{r}_c/\tau$; besides, $T(\boldsymbol{p})$ may take on the values either 0 or 1.

The distribution of the electron flux in the beam over the values of $\boldsymbol{p}_0$ and ϵ will be denoted as $\mathcal{J}_0(\boldsymbol{p}_0, \epsilon)$ so that the flux per an area element $d^2\boldsymbol{r}_0$ of the surface of the specimen equals:

$$j_0\, d^2\boldsymbol{r}_0 = d^2\boldsymbol{r}_0 \cdot \int \mathcal{J}_0(\boldsymbol{p}_0, \epsilon)\, d^2\boldsymbol{p}_0\, d\epsilon \tag{7.35}$$

Introducing the coordinates of the screen $\boldsymbol{r}' = \boldsymbol{R}/M$ related to the specimen and the magnification $M = g_1(T_{scr})$, we obtain the expression for the current density on the viewing screen:

$$j(\boldsymbol{r}') = \int \mathcal{J}_0(\boldsymbol{p}_0, \epsilon)\, T(\boldsymbol{p}_0 + \Delta \boldsymbol{p}_0)\, \delta^{[2]}(\boldsymbol{r}' - \boldsymbol{r}_0 - \Delta \boldsymbol{r}_0)\, d^2 \boldsymbol{r}_0\, d^2 \boldsymbol{p}_0\, d\epsilon. \qquad (7.36)$$

Since the contribution of the terms $\backsim \Delta \boldsymbol{r}_0$ is usually small (see p. 275), it may be taken into account only in the first order. Thus we have:

$$j(\boldsymbol{r}_0) = \int \mathcal{J}_0(\boldsymbol{p}_0, \epsilon)\, T(\boldsymbol{p}_0 + \Delta \boldsymbol{p}_0) \left(1 - \frac{\partial \Delta x_0}{\partial x_0} - \frac{\partial \Delta y_0}{\partial y_0}\right) d^2 \boldsymbol{p}_0\, d\epsilon \qquad (7.37)$$

According to Eqs. (7.34)-(7.37) all information about the optics of the instrument, which is necessary for the solution of the problem of image contrast in the MEMFI, consists of (1) the distribution function $\mathcal{J}_0(\boldsymbol{p}_0, \epsilon)$ for the illumination beam, (2) the transparency function $T_1(\boldsymbol{r}_c) = T(\boldsymbol{p} = m\boldsymbol{r}_c/\tau)$, characterizing the configuration of the aperture and its position with respect to this axis, and (3) the value of τ from Eq. (7.34). If transparency $T(\boldsymbol{p})$ is known in the $\boldsymbol{p}$-coordinates, then even the value of τ is irrelevant. In the MEM with focused images described by Bok *et al.* (1971) the distributions $\mathcal{J}_0(\boldsymbol{p})$ and $T(\boldsymbol{p})$ were identical as a consequence of some design features of the instrument. The contrast in the MEMFI will be briefly discussed later.

Primary Modulation of Electron Beam by Microfields

In this section we shall calculate analytically the changes of the momentum $\Delta \boldsymbol{p}_0(\boldsymbol{r}_0, \boldsymbol{p}_0, \epsilon)$ and the coordinate $\Delta \boldsymbol{r}_0(\boldsymbol{r}_0, \boldsymbol{p}_0, \epsilon)$ obtained by the electron because of the action of electrical and magnetic microfields in the space over the specimen. The calculations are made in the first order in the amplitude of microfields; estimations from the Eqs. (7.5)-(7.8) show that it is a reasonable approximation even for the description of a strong contrast in the MEM. In this section the Hamiltonian form of the equations of motion is used. This allows us to write down the result in the most compact form even for the oblique incidence, i.e., for $\boldsymbol{p}_0 \neq 0$. Below, for electrical and for magnetic microfields, $\Delta \boldsymbol{p}_0$ will also be calculated through the equations of motion in a usual form.

In zeroth approximation the electron trajectory is given by Eqs. (7.13), and therefore in the first order of perturbation theory the equation for the change of the transverse momentum $\dot{\boldsymbol{p}}$ is:

$$\dot{\boldsymbol{p}} = -\frac{\partial V}{\partial \boldsymbol{r}} [\boldsymbol{r} = \boldsymbol{r}_0 + m^{-1} \boldsymbol{p}_0 t, \boldsymbol{p} = \boldsymbol{p}_0, z = z_0 + eE_0 t^2/2m] \qquad (7.38)$$

where $V(\boldsymbol{r}, z, \boldsymbol{p}, m\dot{z})$ is the Hamiltonian interaction of the electron with the microfields:

$$V = -e\varphi + \frac{e}{m}(p_x A_x + p_y A_y + m\dot{z} A_z) \tag{7.39}$$

(In Eq. (7.39) for V the terms $\backsim A^2$ were omitted.) Here $\varphi(x, y, z)$ is the scalar potential of an electrical microfield, and $A(x, y, z)$ is the vector-potential of a magnetic microfield, so that the three-dimensional vectors of microfields $\boldsymbol{E}$ and $\boldsymbol{B}$ are respectively equal to $\boldsymbol{E} = -\boldsymbol{\nabla}\varphi$, $\boldsymbol{B} = \text{rot}\,\boldsymbol{A}$.

Integration of Eq. (7.38) over time yields:

$$\Delta \boldsymbol{p}_0(\boldsymbol{r}_0, \boldsymbol{p}_0, \epsilon) = \left(\frac{\partial \Delta \Phi}{\partial \boldsymbol{r}_0}\right)_{\boldsymbol{p}_0, \epsilon} \tag{7.40}$$

$$\Delta \Phi(\boldsymbol{r}_0, \boldsymbol{p}_0, \epsilon) = -\int_{-\infty}^{+\infty} V\left(\boldsymbol{r}_0 + \frac{\boldsymbol{p}_0 t}{m}, z_0 + \frac{eE_0 t^2}{2m}, \boldsymbol{p}_0, m\dot{z} = eE_0 t\right) dt \tag{7.41}$$

The magnitude of $\Delta\Phi$ from Eq. (7.41) is the change of the classical action (in the sense of Hamilton's principle) due to the microfields; $\Delta\Phi$ in Eq. (7.41) is calculated in the first order of perturbation theory.* It could be shown that the change of the coordinate $\Delta \boldsymbol{r}_0$ may also be expressed as a partial derivative of the change of the action:

$$\Delta \boldsymbol{r}_0(\boldsymbol{r}_0, \boldsymbol{p}_0, \epsilon) = -\left(\frac{\partial \Delta \Phi}{\partial \boldsymbol{p}_0}\right)_{\boldsymbol{r}_0, \epsilon} \tag{7.42}$$

In the process of differentiation the dependence of $z_0(\boldsymbol{p}_0, \epsilon)$ according to Eq. (7.14) should be taken into account.

It can be shown that the contribution of terms $\backsim \Delta \boldsymbol{r}_0$ to the contrast of the electrical microfields is important exclusively for the MEMFI operating near its limit of resolution (see p. 275). Terms $\backsim \Delta \boldsymbol{r}_0$ for the magnetic microfields in the MEMSI are briefly discussed on p. 273.

Thus Eqs. (7.40)–(7.42) yield a unified description of the primary modulation of the electron beam (magnitudes $\Delta \boldsymbol{p}_0$ and $\Delta \boldsymbol{r}_0$) for the arbitrary values of $\boldsymbol{p}_0$ and ϵ when the beam is subject to the influence of both the electrical and magnetic microfields.

However, up to this point no use had been made of the fact that the microfields $\varphi(\boldsymbol{r}, z)$, $\boldsymbol{E}(\boldsymbol{r}, z)$, $\boldsymbol{A}(\boldsymbol{r}, z)$, $\boldsymbol{B}(\boldsymbol{r}, z)$ satisfy Laplace's equation in the space over the specimen, the boundary conditions on the surface of the specimen, and the condition of a decrease at $z \to +\infty$. It follows that any component of those fields in the entire space is completely determined by its values on surface $z = 0$ of the

*In the presence of magnetic fields the Hamilton's momentum $\boldsymbol{p}$ is connected with velocity $\boldsymbol{v}$ by $\boldsymbol{p} = m\boldsymbol{v} - e\boldsymbol{A}$; therefore, the right-hand side of Eq. (7.38) does not coincide with the usual expression for force $\boldsymbol{f} = -e\{\boldsymbol{E} + [\boldsymbol{vB}]\}$. However, since we assume $\boldsymbol{A} \to 0$ far from the specimen, the result [Eqs. (7.40)–(7.42)] turns out to be precisely equivalent to the integration of the usual force $\boldsymbol{f}(t)$.

specimen. There are many explicit expressions for this property; for example, the potential of a double layer. In this section an expansion into a two-dimensional Fourier integral will be used. The solution of Laplace's equation for $\varphi(\boldsymbol{r}, z)$:

$$\left(\frac{\partial^2}{\partial x^2} + \frac{\partial^2}{\partial y^2} + \frac{\partial^2}{\partial z^2}\right) \varphi(x, y, z) = 0$$

in this representation looks as follows:

$$\varphi(\boldsymbol{r}, z) = \int d^2 \boldsymbol{k}\, e^{i\boldsymbol{k}\cdot\boldsymbol{r} - |\boldsymbol{k}|z}\, \tilde{\varphi}(\boldsymbol{k}), \quad \tilde{\varphi}(\boldsymbol{k}) = (2\pi)^{-2} \int d^2 \boldsymbol{r}\, e^{-i\boldsymbol{k}\cdot\boldsymbol{r}} \varphi(\boldsymbol{r}, z = 0) \tag{7.43}$$

Analogous formulae are valid for any component of $\boldsymbol{E}$, $\boldsymbol{A}$, $\boldsymbol{B}$. Here $\boldsymbol{k} = (k_x, k_y)$ is the "wave vector" of a harmonical component of the field, $\lambda = 2\pi/|\boldsymbol{k}|$ is the "wavelength" of that component, and $\nu = \lambda^{-1} = |\boldsymbol{k}|/2\pi$ is the spatial "frequency."

In addition, the statical fields $\boldsymbol{E}$ and $\boldsymbol{B}$ are potential in vacuum; therefore, the total vector of the field $\boldsymbol{E}$ (or $\boldsymbol{B}$) may be established by defining only one function on the surface of the specimen. Potential $\varphi(\boldsymbol{r}, z = 0)$ will be used for the electrical microfield and the values of $B_z(\boldsymbol{r}, z = 0)$ will be used for the magnetic microfield, although some other components could also be used.

The values of the changes of the action $\Delta\Phi$, of the momentum $\Delta\boldsymbol{p}_0$ and of the coordinate $\Delta\boldsymbol{r}_0$ will also be expanded into the Fourier integral over the variable $\boldsymbol{r}_0$; for example,

$$\Delta\Phi(\boldsymbol{r}_0, \boldsymbol{p}_0, \epsilon) = \int d^2 \boldsymbol{k}\, e^{i\boldsymbol{k}\cdot\boldsymbol{r}_0}\, \Delta\tilde{\Phi}(\boldsymbol{k}, \boldsymbol{p}_0, \epsilon)$$

In this representation Eqs. (7.40) and (7.42) read as follows:

$$\Delta\tilde{\boldsymbol{p}}_0(\boldsymbol{k}, \boldsymbol{p}_0, \epsilon) = i\boldsymbol{k}\, \Delta\tilde{\Phi}; \quad \Delta\tilde{\boldsymbol{r}}_0(\boldsymbol{k}, \boldsymbol{p}_0, \epsilon) = -\left(\frac{\partial \Delta\tilde{\Phi}}{\partial \boldsymbol{p}_0}\right)_{\boldsymbol{k}, \epsilon} \tag{7.44a}$$

The solution of Laplace's equation in the form of a Fourier integral [Eq. (7.43)] makes it possible to calculate the change of the action $\Delta\Phi$ from Eq. (7.41):

$$\Delta\tilde{\Phi}(\boldsymbol{k}, \boldsymbol{p}_0, \epsilon) = \left(\frac{2\pi m e}{E_0 |\boldsymbol{k}|}\right)^{1/2} \cdot \left[\tilde{\varphi}(\boldsymbol{k}) + \frac{i}{m|\boldsymbol{k}|^2}(k_x p_{0y} - k_y p_{0x})\, \tilde{B}_z(\boldsymbol{k})\right]$$
$$\times \exp\left\{-|\boldsymbol{k}|\left(\frac{eU_B - \epsilon}{eE_0} + \frac{p_0^2}{2meE_0}\right) - \frac{(\boldsymbol{k}\cdot\boldsymbol{p}_0)^2}{2meE_0|\boldsymbol{k}|}\right\} \tag{7.44b}$$

The results obtained will be discussed here. Equations (7.40)-(7.42) and (7.44a,b) which solve the problem of determining $\Delta\boldsymbol{p}_0$ and $\Delta\boldsymbol{r}_0$ by the given microfields $\boldsymbol{E}$ and $\boldsymbol{B}$, contain a description of a number of effects, characterizing the interaction of the electron with the microfields at the transit time over the specimen. Some of those effects are given below.

The Fourier components of the microfields with a definite frequency $|k|$ decrease with the increase of height according to the law exp $(-|k|z)$; therefore all the magnitudes $\Delta\tilde{\Phi}$, $\Delta\tilde{p_0}$, $\Delta\tilde{r_0}$ contain a corresponding factor exp $(-|k|z_0)$. According to Eq. (7.14) the value of $z_0 = (eU_B - \epsilon + p_0^2/2m)/eE_0$ depends on momentum p_0, energy ϵ, and bias voltage U_B.

Equations (7.44) point to the exponential decrease of $\Delta\tilde{p_0}$ and $\Delta\tilde{r_0}$ with the increase of bias voltage. Therefore, increasing the bias voltage U_B and keeping fixed the optical parameters of the instrument, the operator can suppress the influence of the high frequency components and obtain a sort of blurred image formed only by the low frequency components of the microfield. Contrary to this, by reducing U_B, i.e., by placing the reversal point z_0 closer to the specimen, it is possible to achieve a greater influence of the high frequency components on the trajectory and on the contrast; the image seems to become sharper. In this case in order to suppress the low frequency components in the image the optical parameters of the apparatus should be changed, i.e., $t_{eff} \backsim \sqrt{l}$ should be reduced in the MEMSI. It should be remembered that the high frequency components correspond to the small-scale inhomogeneities, and low frequency components correspond to large-scale structures; the estimation $\delta x \sim |k|^{-1}$ can be made.

The spread over the energy ϵ leads to the spread $\delta z = \delta\epsilon/eE_0$ of values of the height of reflection. Therefore, the main part of the beam even for the best choice of U_B is reflected at the height $z_0 \gtrsim \delta\epsilon/eE_0$ from the specimen. That accounts for strong deterioration of the contrast produced by the microfields of the size $\delta x \lesssim \delta\epsilon/eE_0$, which in turn accounts for the practical limit of resolution of the MEM.

Two effects must be mentioned, which appear if $p_0 \neq 0$, i.e., when the incident trajectory is not vertical. First effect is the increase of the height of reflection of electrons z_0 with the increase of $|p_0|$. That leads to the decrease of modulation effect by a factor exp $\{-(|k|p_0^2/2meE_0)\}$ because of the exponential decrease of microfield with height. The second effect deals with the fact that at $p_0 \neq 0$ the trajectory of electron is not purely vertical, and electron "slides" past the transverse distribution of inhomogeneity. Therefore for a periodic inhomogeneity $\backsim$ exp $\{ikr\}$ the force acting on electrons oscillates in time

$$\backsim \exp\{i(kp_0)\,t/m\}.$$

That leads to the decrease of the modulation effect by a factor exp $\{-[(kp_0)^2/2meE_0|k|]\}$. Both effects are important for the microfields of sufficiently small size:

$$\delta x \sim |k|^{-1} \lesssim p_0^2/meE_0$$

The difference in the manifestations of these two effects consists in the following. The increase of the height of the reflection reduces the modulation independently of the mutual orientation of p_0 and k. The very point of reversal z_0

may in principle be brought closer to the specimen by a reduction of U_B, and in that way the effect of $\delta z_0 = \boldsymbol{p}_0^2/2meE_0$ may be compensated for. However, such compensation is possible only for the electrons with a given value of $\boldsymbol{p}_0^2/2m - \epsilon$.* The effect of sliding past the microfield is determined by the scalar product $(\boldsymbol{p}_0 \cdot \boldsymbol{k})$. Therefore, when one is obtaining images of microfields close to the one-dimensional ones, that effect may be eliminated if the specimen is illuminated by electrons with $\boldsymbol{p}_0 \perp \boldsymbol{k}$, i.e., by electrons moving along the structure instead of intersecting it.

In the MEMSI the value of $\boldsymbol{p}_0$ in the illumination beam is proportional to the distance $\boldsymbol{r}_0$ from the electrical center of the image: $\boldsymbol{p}_0 = mD_{\text{sp}}\,\boldsymbol{r}_0$. In this case the $z_0(\boldsymbol{r}_0)$-dependence [the "parabola of heights of reflection," Eq. (7.20)] and the sliding effect lead to a suppression of the contrast of small-scale structures far from the electrical center of the image (Fig. 7.6).

For the magnetic microfields the change of the action $\Delta\Phi$ may be expressed through the flux of magnetic induction "hitched up" by the parabolic trajectory during the motion over the specimen; for the transmission electron microscope this was pointed out by Wohlleben (1971).

Indeed, from Eqs. (7.39) and (7.41) the following equation may be derived:

$$\Delta\Phi = -\frac{e}{m}\int_{-\infty}^{+\infty}(A \cdot p)\,dt = -e\int(A \cdot dl) = -e\iint(\text{rot}\,A \cdot ds) = -e\iint(B \cdot ds) \tag{7.45}$$

[in Eq. (7.45) the three-dimensional vectors are understood]. It follows from Eq. (7.45) [or from Eq (7.44)] that for a vertically incident electron, i.e., at $\boldsymbol{p}_0 = 0$, the change $\Delta\boldsymbol{p}_0$ due to the effect of the magnetic microfields vanishes in the first order. The parabolic trajectory in such a case degenerates into two coinciding straight lines, and the magnetic flux "hitched up" by the trajectory equals zero. See p. 273 for a more detailed discussion of the modulation of the beam by magnetic microfields.

Image Contrast Formation of Electrical Microfields and Topographical Inhomogeneities

The potential distribution on the surface of the specimen was first observed by Orthuber (1948). It was found that the negatively charged areas appear dark on the image because those areas have a diverging effect on electron trajectories. In contrast, the positively charged areas act as converging lenses on electrons and they appear bright on the image. The topographical inhomogeneities on the

*In this connection it would seem to be attractive to produce such illumination in the MEM for which an approximate equality $\epsilon = \text{const} + \boldsymbol{p}_0^2/2m$ would hold. In this case all the electrons would have the same height of reflection z_0; however, such devices have not yet been realized experimentally.

specimen also distort the flat equipotential surfaces. Therefore, an analogous contrast (a dark spot with bright edges) is observed on the image of a hillock. The images of hollows are represented as bright spots. Experimental investigations by Bartz *et al.* (1956) on the sensitivity of the MEM operating in the shadow projection mode point to a possibility of revealing steps of 2.5 nm in height.

Theoretically the problem of image contrast formation of electrical microfields was first considered by Wiskott (1956a). Wiskott's approximation in the theory of image contrast formation in the MEMSI is explained below.

For the electrons vertically incident on the specimen, the change of the tangential velocity (momentum) subject to the action of microfields close to the surface of the specimen is calculated in the first order of perturbation theory. The further motion of the electron after reflection may be considered as a motion in the field of the immersion objective from the specimen to the screen in the absence of microfields, but with modified initial velocity.

In the shadow projection mode of the MEM the transverse coordinates of electrons hitting the screen obtain the displacement due to the action of microfields. It is useful to divide the screen coordinates $\boldsymbol{R}_{scr}$ by the magnification M; that yields the reduced screen coordinates $\boldsymbol{r}' = \boldsymbol{R}_{scr}/M$ which may be called "the coordinates of the screen related to the specimen". Denoting the reduced (i.e., "the related to the specimen") displacement by $\boldsymbol{s}$, one may write

$$x' = x + s_x(x, y), \quad y' = y + s_y(x, y) \tag{7.46}$$

The current density distribution $j(x', y')$ at the final screen is calculated on the basis of the conservation of current flow $j(x', y')\, dx'\, dy' = j_0\, dx\, dy$. Consequently we have:

$$j(x', y') = j_0 \cdot \left[\frac{\partial(x', y')}{\partial(x, y)}\right]^{-1} \tag{7.47}$$

Here j_0 is the current density distribution in the absence of perturbations. In a one-dimensional case when, for example, $s_y = 0$, the expressions (7.46) and (7.47) may be reduced to:

$$j(x') = j_0 \left[1 + \frac{ds_x(x)}{dx}\right]^{-1}, \quad x' = x + s_x(x) \tag{7.48}$$

The image contrast produced on the MEMSI screen is a result of redistribution of the density of electron trajectories. In the points of the screen, where the Jacobian of transformation $\partial(x', y')/\partial(x, y)$ is nullified, the current density tends to infinity provided that the electron beam is monoenergetic and originates from a point source. Intersection of electron trajectories reflecting from the adjacent parts of the specimen corresponds to the formation of caustic surfaces. When the caustic surfaces intersect the screen, the bright narrow lines (caustics) appear on the image.

Expression (7.47) for the current density of the MEMSI should be used in association with the law of coordinate transformation, Eq. (7.46). The latter shows that the imaging process in the shadow projection mode is accompanied by inhomogeneous scale distortions. In other words, the coordinate on the viewing screen of the MEMSI is connected not only with the position of the point on the specimen and magnification of the electron optical system but also with the perturbations, appearing because of the irregularities and the microfields on the surface of the specimen.

In a case of weak contrast an approximate expression for the current density on the screen may be written in first order in $\mathbf{s}$:

$$j(x, y) \approx j_0 \left[1 - \frac{\partial s_x}{\partial x} - \frac{\partial s_y}{\partial y}\right] \tag{7.49}$$

Thus, in order to calculate the contrast it is necessary to find the value of the change of transverse momentum (or the displacement of the electron trajectory) caused by the action of microfields on the surface of the specimen.

The equations of motion of the electron in the field of immersion objective E_0 in the presence of microfield $\varphi(x, y, z)$ are:

$$\frac{d^2 \mathbf{r}}{dt^2} = -\frac{e}{m} E(\mathbf{r}) = \frac{e}{m} \operatorname{grad} \left[(z - z_0) E_0 + \varphi(x, y, z)\right] \tag{7.50}$$

Here $E_0 = U_0/l$ is the field strength of the asymptotically uniform field of the objective, z_0 is the coordinate of the surface of the zero potential for an unperturbed field, i.e., the height of reflection of the monoenergetic electron beam, and $\varphi(x, y, z)$ is the potential appearing because of perturbations and microfields on the surface.

It will be assumed here that the value of perturbing microfields is smaller than that of the uniform fields of objective $(|\nabla\varphi| \ll E_0)$. In conformity with Wiskott the problem of the modification of the electron trajectory will be solved in the first order of perturbation theory with the value of $\nabla\varphi$ being a small parameter. In zero approximation the trajectories of the vertically incident electrons are:

$$x^{(0)}(t) = x = \text{const}, \quad y^{(0)}(t) = y = \text{const}, \quad z^{(0)}(t) = z_0 + eE_0 t^2/2m \tag{7.51}$$

From Eq. (7.50) the following expression is obtained for the change of the transverse momentum in the first order in the microfield:

$$\Delta p_x = e \int_{-\infty}^{+\infty} dt \, \frac{\partial \varphi(x, y, z(t))}{\partial x} = \sqrt{\frac{2em}{E_0}} \int_{z_0}^{\infty} \frac{\partial \varphi(x, y, z)}{\partial x} \cdot \frac{dz}{\sqrt{z - z_0}} \tag{7.52}$$

An analogous expression is valid for Δp_y.

Further concretization of the solution is connected with the fact that the potential $\varphi(x, y, z)$ in Eq. (7.52) satisfies Laplace's equation in the space over the specimen and boundary conditions on the surface.

The following method for the solution of the problem of image contrast has proven to be very fruitful (Sedov *et al.*, 1968a). The perturbing potential $\varphi(x, y, z)$ in that case was represented as a potential of an electrical double layer:

$$\varphi(x, y, z) = \frac{z}{2\pi} \iint\limits_{-\infty}^{+\infty} \frac{\varphi(x - \xi, y - \eta, z = 0)}{(\xi^2 + \eta^2 + z^2)^{3/2}} \, d\xi \, d\eta \tag{7.53}$$

where ξ and η are the integration variables. Equations (7.52) and (7.53) make it possible to connect the change of momentum $\Delta \boldsymbol{p}$ with the values of the perturbing potential $\varphi(x, y, z = 0)$ on the surface of the specimen. For $z_0 = 0$ we obtain:

$$\Delta p_x(x, y) = \left(\frac{2\pi em}{E_0}\right)^{1/2} \left[\Gamma\left(\frac{1}{4}\right)\right]^{-2} \iint d\xi \, d\eta \, (\xi^2 + \eta^2)^{-3/4} \, \frac{\partial \varphi(x - \xi, y - \eta, z = 0)}{\partial x}, \tag{7.54}$$

where $\Gamma(\frac{1}{4}) = 3.626$ is the value of the Eueler gamma function.

The related to the specimen displacement of the electron trajectory caused by the microfield is expressed by:

$$\boldsymbol{s}(x, y) = t_{\text{eff}} \, m^{-1} \, \Delta \boldsymbol{p}(x, y) = 3l \cdot (2meU_0)^{-1/2} \, \Delta \boldsymbol{p}(x, y) \tag{7.55}$$

where $U_0 = E_0 l$ is the accelerating voltage. In a one-dimensional case when $\varphi(x, y, z = 0)$ depends only on x, the related to the specimen displacement may be calculated from Eqs. (7.52) and (7.53) even for $z_0 \neq 0$:

$$s_x(x) = \frac{3l^{3/2}}{U_0\sqrt{2}} \int_{-\infty}^{+\infty} d\xi \, \frac{\sqrt{z_0 + \sqrt{\xi^2 + z_0^2}}}{\sqrt{\xi^2 + z_0^2}} \cdot \frac{\partial \varphi(x - \xi, z = 0)}{\partial x} \tag{7.56}$$

Equations (7.55) and (7.56) may be generalized for the MEMSI with an arbitrary objective system by substitution $t_{\text{eff}} \rightarrow (9ml/2eE_0)^{1/2}$.

Using Eqs. (7.46), (7.47), (7.54), and (7.55), it is possible to calculate the image contrast created by microfields of an arbitrary shape. In some particular cases analytical expressions may be obtained. One may consider, for example, the perturbing potential $\varphi(x, z = 0) = \pi^{-1} \varphi_0 \arctan(x/w) + \varphi_0/2$ which may serve as an approximation of the potential distribution on the surface of p-n junction. Here φ is the amplitude of the potential and w is the width of p-n junction. The solution of Laplace's equation for the potential of that kind is:

$$\varphi(x, z) = \frac{\varphi_0}{2} + \frac{\varphi_0}{\pi} \arctan \frac{x}{z + w} \tag{7.57}$$

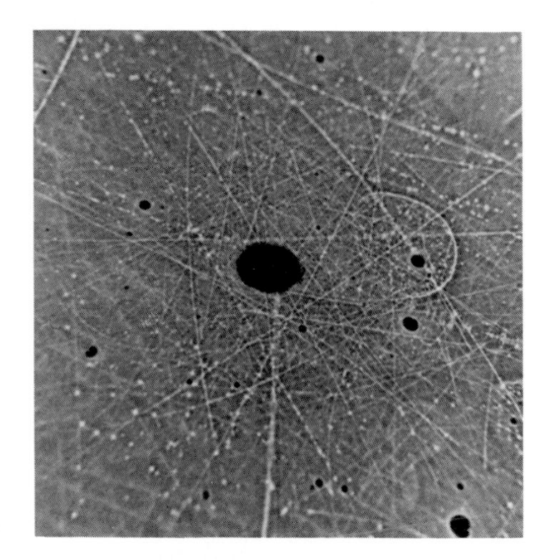
(a)

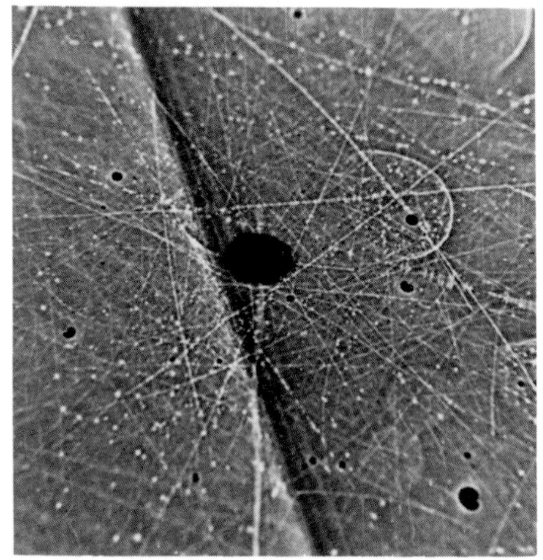
(b)

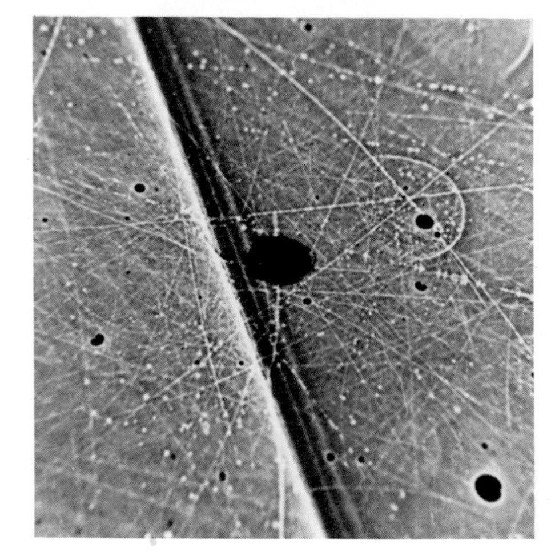
(c)

For the displacement of the electron trajectory related to the specimen and for the current density $j(x')$, the following explicit expressions were obtained (Sedov *et al.*, 1968a):

$$x' = x + s(x) = x + (z_0 + w)\,A\sqrt{\frac{1 + \sqrt{1 + \alpha^2}}{1 + \alpha^2}}, \quad \alpha = \frac{x}{z_0 + w},$$

$$\frac{j(x')}{j_0} = \left\{1 - A\,\frac{\alpha(2 + \sqrt{1 + \alpha^2})}{2(1 + \alpha^2)^{3/2}\sqrt{1 + \sqrt{1 + \alpha^2}}}\right\}^{-1},$$

$$A = \varphi_0\, t_{\text{eff}}\sqrt{e/mE_0(z_0 + w)^3} = 3\varphi_0(U_0\sqrt{2})^{-1}\,[l/(z_0 + w)]^{3/2}. \qquad (7.58)$$

The second expression for the constant A in Eq. (7.58) is given for the MEMSI with a two-electrode objective system. As an illustration, Fig. 7.6 shows the image of the semiconductor p-n junction with and without voltage applied.

Figure 7.7 represents the calculated curves of current density distribution due to the action of the field of p-n junction for the different values of the parameter A, i.e., for the different values of φ_0 or $(z_0 + w)$. It is easy to see from the curves that with an increase of φ_0 the brightness in the maximum also increases and that the maximum is shifted in the direction of the positively charged area of the specimen. At the value of $A = 2.404$, the intensity of the bright line increases infinitely and the caustic first appears at distance $x' \approx 3.63\ (z_0 + w)$ from the center of the p-n junction; it corresponds to the value of $\alpha = \tan(\pi/5) \approx 0.727$. The further increase of φ_0 leads to a split of the bright line, resulting in the appearance of two branches of caustics, which are shifted also in the direction of the positively charged areas.

The experimental oscillograms of brightness obtained in the MEM (Sedov *et al.*, 1968a), when the diffused silicon p-n junction was observed, demonstrated qualitative correspondence with the theoretical curves in Fig. 7.7.

As was mentioned above, the concave and convex topography inhomogeneities on the surface of the specimen produce contrast analogous to that of either negatively or positively charged areas, respectively. The local variation of the height Δh on the surface is approximately equivalent to the change of $\varphi = -E_0\,\Delta h$ of the local potential of the specimen. Thus, Eqs. (7.54) and (7.56) are applicable to the description of the image contrast created by the topography inhomogeneities when we replace $\varphi(x, y, z = 0) \Rightarrow -E_0\,\Delta h(x, y)$ (Fig. 7.8). As a result, an

---

Fig. 7.6 The MEMSI micrographs of semiconductor p-n junction. Accelerating voltage 10 kV; bias voltage 5 V, magnification 100X. The dark spot in the center corresponds to the hole (for primary and reflected beams) in the screen; other dark spots with bright rims are the images of local carbon contaminations. "Blurring" of the image of scratches at the periphery of the picture is due to the effects of sliding of electrons past the microfield and increased height of reflection. (a) Topography of the surface without applying voltage; (b) image of p-n junction with the voltage applied (φ_0 = 1 V); (c) φ_0 = 1.5 V.

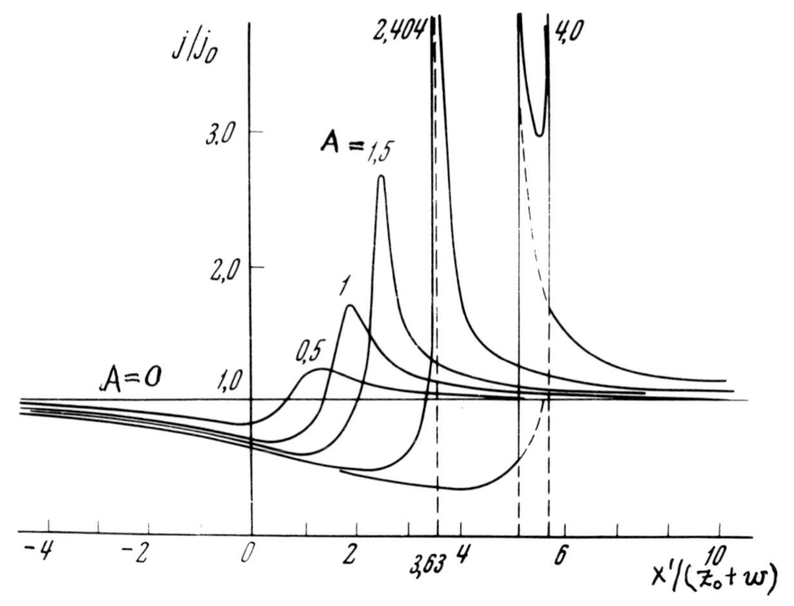

Fig. 7.7 Current density distribution on the screen of MEMSI for the potential of the type (7.57) (Sedov *et al.*, 1968a). (*Courtesy of N. N. Sedov.*)

empirical rule may be formulated by means of which the topographical contrast may be distinguished from that of potential profiles.

Indeed, when the accelerating voltage U_0 increases, the topographical contrast remains constant (the resolving power is assumed to be independent on the accelerating voltage). Contrary to that, the potential contrast decreases when the accelerating voltage increases. The effect of the decrease of the image contrast of electrical microfields when the accelerating voltage increases was examined in detail by Spivak and Lyubchenko (1959), and is known as the effect

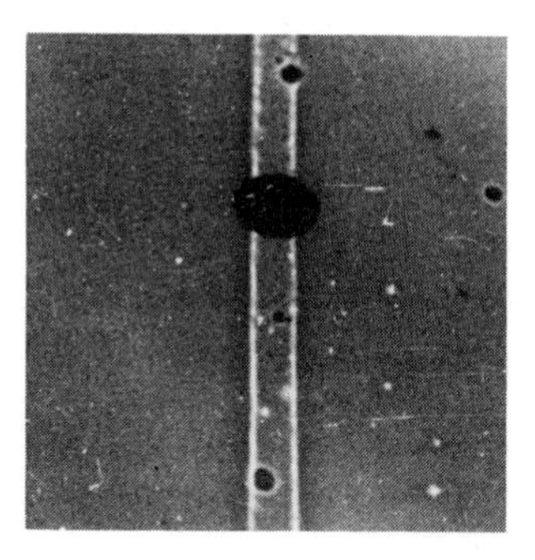

Fig. 7.8 The MEMSI micrograph of the rectangle concave inhomogeneity of a depth of 50 nm. Magnification 60X. Field strength of the uniform retarding field $E_0 = 2.10^6$ V/m. The image of each boundary of this inhomogeneity is similar to the image of potential "step"; compare with Fig. 7.6b.

of suppression of an electrical "microlens" by the "macrolens" of the accelerating field of the immersion objective.

For an estimation of MEMSI sensitivity to the potential step (or to the topographical step), the amount of contrast $|j - j_0| = 0.1\, j_0$ as a minimum detectable value will be assumed. From Eq. (7.58) it follows that such a value of the contrast is achieved at the $A = 0.24$. Thus, the minimum detectable height of a potential step $\varphi_{0\min}$ with the parameter of smoothness w at the height of reflection z_0 equals:

$$\varphi_{0\min} = 0.24\, t_{\text{eff}}^{-1}\, (mE_0/e)^{1/2}\, (z_0 + w)^{3/2} = 0.11\, U_0\, [(z_0 + w)/l]^{3/2} \qquad (7.59)$$

The image contrast produced by inhomogeneities of different shape was investigated by Brand and Schwartze (1963) on special equipment for simulation of electron trajectories. Comparing the experimental curves of current density distribution, produced by inhomogeneities of triangular, rectangular, and hemispherical configurations, the authors established that when the reflection occurs at some distance from the surface of the specimen there is a good agreement of experimental data with the theoretical calculations based on Wiskott's theory. When the reflection occurs at small distances, the agreement between the theory and the simulation results is worse. Data obtained by Brand and Schwartze (1963) are in full agreement with the conclusions by Wiskott (1956a).

The small-scale inhomogeneities are revealed at a close distance from the surface of the specimen, and a rough relief overshadows the fine one. The latter was illustrated by Heydenreich (1962). In that paper it was shown that simultaneous imaging of fine and rough structures on the cleavage surface of the crystal is impossible at the same value of bias voltage. Thus, if the image of fine details on the surface is desirable, care should be taken that the rough relief should not overshadow the fine details. That may be achieved by polishing the surface to perfection.

The similarity of mechanism of image contrast formation of potential and topographical steps was demonstrated in another work by Heydenreich (1966). It was shown that the images of cleavage steps consist of a double bright-and-dark line, the bright edge of which corresponds either to a deeper hollow or to a more positively charged area. When the optical images of the cleavage surface of NaCl, covered by a thin Ag film, were compared with the electron mirror images, it was established that the lines in the MEMSI images are shifted in the direction of hollows or of areas with greater potential. As a result, the images of pits are stretched and the images of hollows are shrunk. The scale distortion was clearly demonstrated by Barnett and England (1968), who had calculated the image contrast caused by sinusoidal potential distribution $\varphi(x,\ z = 0) = \varphi_1 \cos(kx)$. In that case the current density distribution on the final screen was found to be:

$$j(x')/j_0 = [1 - C_{\text{cl}} \cos(kx)]^{-1},\ x' = x - k^{-1}\, C_{\text{cl}} \sin kx,$$
$$C_{\text{cl}} = \varphi_1\, t_{\text{eff}}\, (2\pi e/mE_0)^{1/2}\, |k|^{3/2} \exp\{-|k|z_0\} \qquad (7.60)$$

It is convenient to represent Eq. (7.60) as an expansion into Fourier series in cos (nkx') (Barnett and England, 1968):

$$j(x')/j_0 = 1 + 2 \sum_{n=1}^{\infty} J_n(nC_{cl}) \cos(nkx'), \tag{7.61}$$

where $J_n(z)$ is a Bessel function of nth order. When $C_{cl} \geqslant 1$ the series is divergent, infinite peaks of brightness; i.e., caustic lines, appear on the image. Nevertheless, Eq. (7.61) gives the correct values of the different harmonic components of the current density distribution $j(x')$. The curves of the contrast described by Eqs. (7.60) and (7.61) of the classical theory are presented in Fig. 7.29 in the section on "Quantum Effects in MEM," where the results of the classical and wave-mechanical calculations will be compared.

It should be emphasized that such peculiar image "distortion" is directly associated with the mechanism of image contrast formation in the MEMSI, and is inherent in any instrument operating in a shadow projection imaging mode. In this connection one should take into account that in the MEMSI the sizes of the details on the images do not correspond to those on the specimen. The scale distortion in the imaging of topographical inhomogeneities was also discussed by Sedov (1970a).

The peculiarities of the image contrast formation in the MEMSI may be used for obtaining stereoscopic images. A specimen with inhomogeneous surface conductivity will now be considered. When the current flows through such a specimen, the more conductive areas are revealed as dark spots with bright edges against the background of surrounding less conductive areas. The observation of two images obtained from the same area of the surface at different currents gives an illusion of a three-dimensional effect (Mayer, 1957a), similar to that in the observation of a usual stereo-pair.

Spectral Transfer Function of MEM in Shadow Imaging Mode

A number of authors (Schwartze, 1964; Artamonov and Komolov, 1966; Barnett and Nixon, 1966; Bok *et al.*, 1968b) had applied the harmonical analysis to the problem of image contrast formation in the MEM. This approach made it possible to introduce the concept of the spectral transfer function of the MEM. In this approach the mirror electron optical system is considered as a filter similar to a usual radiotechnical filter with a definite frequency response.

In this approach the contrast $K(r)$ on the screen, $K(r) = [j(r) - j_0]/j_0$, and the potential $\varphi(r, z = 0)$ on the surface of the specimen are represented by a Fourier expansion:

$$\varphi(\boldsymbol{r}, z = 0) = \int \tilde{\varphi}(\boldsymbol{k}) e^{i\boldsymbol{k}\cdot\boldsymbol{r}} d^2 k, \quad K(\boldsymbol{r}) = \int \tilde{K}(\boldsymbol{k}) e^{i\boldsymbol{k}\cdot\boldsymbol{r}} d^2 k \tag{7.62}$$

Generally the contrast $K(r)$ is nonlinearly connected with the amplitude of perturbation $\varphi(r, z = 0)$ as has been shown above. However, if only linear terms are taken into account, from Eq. (7.49) it follows that:

$$K(r) \approx -\frac{\partial s_x}{\partial x} - \frac{\partial s_y}{\partial y} = -\frac{t_{\text{eff}}}{m}\left(\frac{\partial \Delta p_x}{\partial x} + \frac{\partial \Delta p_y}{\partial y}\right) \tag{7.63}$$

In this case the linear relation between the amplitudes of Fourier components of the potential $\tilde{\varphi}(k)$ and of the contrast $\tilde{K}(r)$ may be written:

$$\tilde{K}(k) = S_{\text{cl}}(k)\,\tilde{\varphi}(k) \tag{7.64}$$

Substitution of Eq. (7.44) into Eqs. (7.63) and (7.64) yields:

$$S_{\text{cl}}(k) = t_{\text{eff}}\left(\frac{2\pi e}{mE_0}\right)^{1/2} |k|^{3/2}\, e^{-|k|z_0} = \frac{3(\pi l)^{1/2}}{E_0} |k|^{3/2}\, e^{-|k|z_0} \tag{7.65}$$

where $S_{\text{cl}}(k)$ is the spectral transfer function of the MEMSI in the linearized classical problem, and the dimension of $S_{\text{cl}}(k)$ is $(\text{volt})^{-1}$. The second equality in Eq. (7.65) is written down for the MEMSI with a two-electrode objective system; all subsequent formulae are valid for an arbitrary objective system if substitutions $l^{1/2} \to t_{\text{eff}}\,(2eE_0/9m)^{1/2}$, $U_0 \to E_0 l$ are made. The dependence of $S_{\text{cl}}(k)$ on spatial frequency $\nu = |k|/2\pi$ is shown in Fig. 7.9. The maximum of function $S_{\text{cl}}(k)$ corresponds to the optimum spatial frequency $\nu_{\text{opt}} = |k_{\text{opt}}|/2\pi = 3/(4\pi z_0)$ (Barnett and Nixon, 1966). It means that the highest contrast for a given amplitude of micropotential φ_1 is achieved for the optimum period of sinusoidal perturbation $\lambda_{\text{opt}} = \nu_{\text{opt}}^{-1} = 4\pi z_0/3$.

Equation (7.65) is written down for the case of monochromatic illumination. The spectral transfer function for the case of nonmonochromatic beam may be obtained by averaging Eq. (7.65) over energy distribution of electrons with the

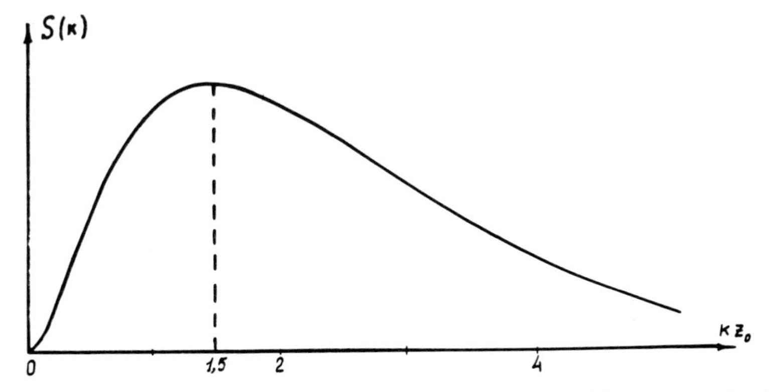

Fig. 7.9 Spectral transfer function of MEMSI. $|k|/2\pi$ is the spatial frequency and z_0 is the height of the reversal point.

relation $z_0 = (eU_B - \epsilon)/eE_0$ taken into consideration. Since the explicit form of the energy distribution is unknown, the following can only be pointed out. It may be assumed approximately that most of the electrons in the beam are reflected at the height $z_0 \sim \delta\epsilon/eE_0$ (where $\delta\epsilon$ is the energy spread), because if $eU_B \lesssim \delta\epsilon$, then most of the electrons hit the specimen and do not reach the screen. The optimum spatial period for the smallest reasonable value of U_B is determined by the energy spread and is equal in the order of magnitude to $\lambda_{opt} = 4\pi z_0/3 \sim 4\delta\epsilon/eE_0$. Thus, the spectral transfer function of the MEMSI at high frequencies may be limited because of the energy spread in the illuminating source (Artamonov and Komolov, 1966).

Artamonov *et al.* (1966) checked their theoretical predictions for the spectral transfer function by the experiments with a special model specimen. Those experiments showed satisfactory agreement between the theoretical and experimental data.

The use of the spectral transfer function makes it possible to obtain a linearized expression for the image contrast of an arbitrary microfield $\varphi(x, y, z = 0)$ on the surface of the specimen. It is convenient to write down separately these expressions for one- and two-dimensional cases:

$$K(x') = \int \varphi(x'', z = 0)\, G^{\mathrm{I}}_{\mathrm{cl}}(x' - x'')\, dx'',$$

$$K(x', y') = \iint \varphi(x'', y'', z = 0)\, G^{\mathrm{II}}_{\mathrm{cl}}\left(\sqrt{(x' - x'')^2 + (y' - y'')^2}\right) dx''\, dy'' \quad (7.66a)$$

where functions $G^{\mathrm{I}}_{\mathrm{cl}}(x)$ and $G^{\mathrm{II}}_{\mathrm{cl}}(r)$ describe linearized response of the contrast to a point source and are determined by the Fourier transform of the spectral transfer function:

$$G^{\mathrm{I}}_{\mathrm{cl}}(x) = (2\pi)^{-1} \int dk\, S_{\mathrm{cl}}(k) \exp\{ikx\},$$

$$G^{\mathrm{II}}_{\mathrm{cl}}(|\boldsymbol{r}|) = (2\pi)^{-2} \iint dk_x\, dk_y\, S_{\mathrm{cl}}(|\boldsymbol{k}|) \exp\{i\boldsymbol{k} \cdot \boldsymbol{r}\} \quad (7.66b)$$

The response functions to a point source have direct physical meaning, since such a point source may serve as a model of a potential produced by a local inhomogeneity on the surface of a conducting specimen. For example, the boundary conditions

$$\varphi_{\mathrm{I}}(x, z = 0) = -E_0\, Q_{\mathrm{eff}}\, \delta(x); \quad \varphi_{\mathrm{II}}(x, y, z = 0) = -E_0\, V_{\mathrm{eff}}\, \delta(x)\, \delta(y) \quad (7.67a)$$

on the surface of the specimen cause exactly the same perturbations as a hemicylindrical or hemispherical hillock on the conducting surface, respectively:

$$\varphi_{\mathrm{I}}(x, z) = E_0 z[1 - \pi^{-1} Q_{\mathrm{eff}} \cdot (x^2 + z^2)^{-1}];$$

$$\varphi_{\mathrm{II}}(x, y, z) = E_0 z[1 - (2\pi)^{-1} V_{\mathrm{eff}} \cdot (x^2 + y^2 + z^2)^{-3/2}]. \quad (7.67b)$$

The corresponding radii of the cylinder and of the sphere equal $R_{\mathrm{I}} = (Q_{\mathrm{eff}}/\pi)^{1/2}$, $R_{\mathrm{II}} = (V_{\mathrm{eff}}/2\pi)^{1/3}$. In the cases of an arbitrary local perturbation the expressions (7.67b) represent the most slowly decreasing part of the potential; i.e., they are the main terms of the asymptotic expansions of the exact solution of Dirichlet's problem. In other words, the coefficients Q_{eff} or V_{eff} in Eq. (7.67) may be considered to be characterizing the effective (in relation to the perturbation of equipotentials) cross section or the effective volume of a corresponding topographic perturbation on the surface.

Calculation of the integrals [Eq. (7.66b)] for a monoenergetic illumination yields:

$$G^{\mathrm{I}}_{\mathrm{cl}}(x) = \frac{9}{4}\,(l)^{1/2}\,E_0^{-1}\,z_0^{-5/2}\left(1+\frac{x^2}{z_0^2}\right)^{-5/4}\cos\left(\frac{5}{2}\arctan\frac{x}{z_0}\right),$$

$$G^{\mathrm{II}}_{\mathrm{cl}}(r) = \frac{45}{16}(l)^{1/2}\,E_0^{-1}\,z_0^{-7/2}\left(1+\frac{r^2}{z_0^2}\right)^{-7/4} P_{5/2}\left(\left[1+\frac{r^2}{z_0^2}\right]^{-1/2}\right), \qquad (7.68)$$

where $P_{5/2}(z)$ is the Legendre function of index 5/2. Figures 7.10 and 7.11 show curves $G^{\mathrm{I}}_{\mathrm{cl}}(x)$ and $G^{\mathrm{II}}_{\mathrm{cl}}(r)$. The half-width of these functions at half the maximum value equal $\Delta x^{\mathrm{I}} = 0.39z_0$ and $\Delta r^{\mathrm{II}} = 0.43z_0$, respectively. The half-width may be accepted as an estimation of the resolving power over the transverse coordinate in the MEMSI:

$$\Delta x_{\mathrm{cl}} \approx 0.4 z_0 \qquad (7.69)$$

For field strength $E_0 \sim 2.10^6$ V/m and for energy spread $\delta\epsilon \sim 1$ eV the value of $\Delta x_{\mathrm{cl}} \approx 0.4\delta\epsilon/eE_0$ corresponds to 200 nm approximately. Experimental values of the resolving power of the MEMSI over the transverse coordinates vary from 100 to 500 nm for different instruments.

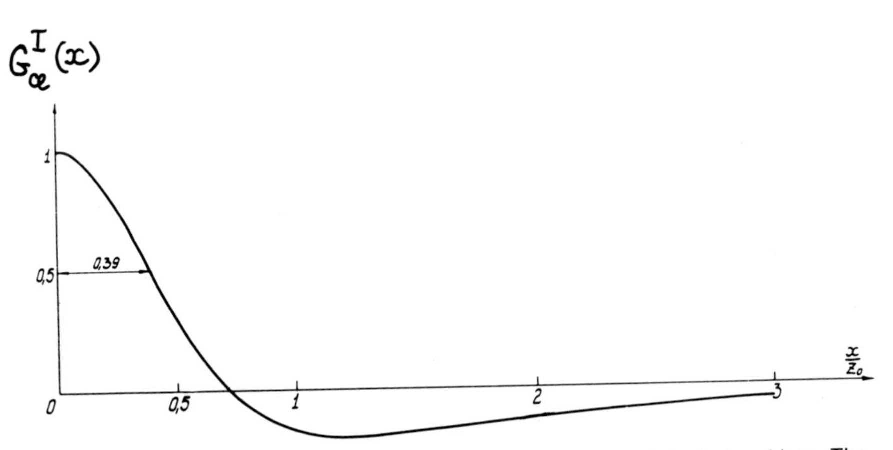

Fig. 7.10 Linear response to a point source for the one-dimensional classical problem. The curve is normalized to the value "1" at $x = 0$. z_0 is the height of the reversal point. *(Gvosdover and Zel'dovich, 1973.)*

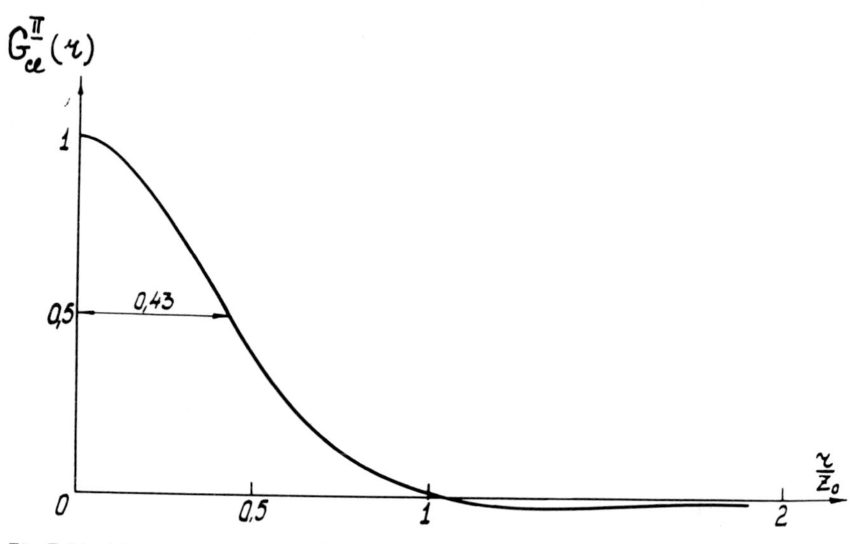

Fig. 7.11 Linear response to a point source for the two-dimensional classical problem. The curve is normalized to the value "1" at $r = 0$ where $r = \sqrt{x^2 + y^2}$. *(Gvosdover and Zel'dovich, 1973.)*

The minimum detectable values of Q_{eff} and V_{eff} corresponding to the value of contrast $K(x = 0) = 0.1$ or $K(r = 0) = 0.1$ are given by:

$$|Q_{\min}| = [10 E_0\, G_{\text{cl}}^{\text{I}}(0)]^{-1} = 0.044\, z_0^2 \cdot (z_0/l)^{1/2},$$

$$|V_{\min}| = [10 E_0\, G_{\text{cl}}^{\text{II}}(0)]^{-1} = 0.035\, z_0^3 \cdot (z_0/l)^{1/2} \tag{7.70}$$

The minimum detectable amplitude φ_1 of the sinusoidal potential (which corresponds to $|K| = 0.1$) for optimum spatial period equals:

$$|\varphi_{1\,\min}| = [10\, S_{\text{cl}}(k_{\text{opt}})]^{-1} \approx 0.046\, E_0\, z_0 (z_0/l)^{1/2} \tag{7.71}$$

The value $|\varphi_{1\,\min}|$ may be written in terms of height h_1 of the sinusoidal perturbation on the surface of the specimen: $h_1 = -\varphi_1/E_0$. For $l = 0.5$ cm, $E_0 = 2.10^6$ V/m, $z_0 = 1000$ nm ($U_B = 2$ volts) the minimum detectable amplitude φ_1 equals $|\varphi_{1\,\min}| = 1.3 \times 10^{-3}$ volt, and $h_{1\,\min} \approx 0.65$ nm and corresponds to a value $\lambda_{\text{opt}} = 4200$ nm being an optimum spatial period. Thus, the MEM can be said to have an "anisotropy" of resolving power: the minimum detectable height $h_1 \sim z_0 (z_0/l)^{1/2}$ is much smaller than the resolving power $\Delta x \sim z_0$ over the transverse coordinates.

It is of particular interest to compare the value $2\varphi_{1\,\min}$ of the double amplitude of a sinusoidal potential (corresponding to $|j - j_0| = 0.1\, j_0$) with value $\varphi_{0\,\min}$ from Eq. (7.59) which characterizes the height of the potential step producing the contrast $|j - j_0| = 0.1\, j_0$. Comparison of Eq. (7.59) for $w = 0$ and Eq. (7.71) yields $2\varphi_{1\,\min} = 0.8\, \varphi_{0\,\min}$. Hence, it can be said that spatial frequencies close to optimum make the main contribution to the image contrast formation of a rectangular potential or topographic step.

It follows from the formulae obtained that with a sufficiently monoenergetic illumination beam the sensitivity and resolving power of the MEMSI may be infinitely improved by the reduction of z_0, the latter being the height of reflection. However, this conclusion had been drawn on the basis of classical theory. Estimations from the Eqs. (7.9)–(7.11) and calculations from the section on "Quantum Effects in MEM" show that for a sufficiently small z_0, the sensitivity and resolving power of the MEMSI are limited by quantum-mechanical effects.

Magnetic Contrast

The possibility of visualizing magnetic microfields in the MEM was first demonstrated by Spivak *et al.* (1955), who had investigated a cobalt single crystal and an artificial specimen consisting of a number of magnetic and nonmagnetic plates. By a comparison of the mirror electron micrographs with the images obtained by the Bitter powder method, it was established that the bright areas on the micrographs correspond to those with maximum gradient of magnetic microfields.

The contrast of magnetic microfields of different origins (the domain structure of ferromagnetics, the fields recorded on magnetic tapes, the fields over the magnetic recording heads, and specimens with magnetization) was investigated by Mayer (1957b, 1958, 1959, 1961). Mayer gave a qualitative description of the magnetic contrast and formulated some criteria for distinguishing the image of magnetic structures from that of electric microfields and topographical inhomogeneities. These criteria are given below.

(1) Magnetic contrast is due to the interaction of the radial velocity component $\dot{r}$ with the component B_z of the magnetic microfield normal to the surface. As a result, the sensitivity of the MEMSI to magnetic fields is nullified in the optical center, where radial velocity equals zero and sensitivity increases from the center to the periphery.

(2) The areas of the specimen where $\Delta B_z/\Delta\varphi < 0$ (φ is the azimuthal angle) appear bright* on the image, and the areas appear dark if $\Delta B_z/\Delta\varphi > 0$. Inversion of magnetic contrast occurs when some selected area of the specimen is moved from one side relative to the electrical center of the image to another.

(3) The radial magnetic structures form better contrast in comparison with structures extended in azimuthal direction.

(4) The image of the secondary emission spot is distorted because of the influence of magnetic microfields. As an example, Figs. 7.12 and 7.13 show the electron mirror micrographs of the signals recorded on the magnetic tape and the image of microfield over the gap of a magnetic recording head.

*Contrary to Mayer, we use the right-hand coordinate system.

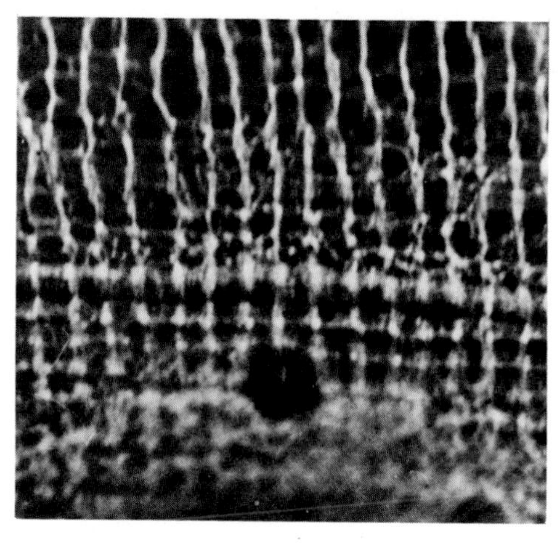

Fig. 7.12 The MEMSI micrograph of the sinusoidal magnetic field recorded on tape with a period of 42 μm. Magnification 85X. Electrical center lies below the bottom edge of the micrograph.

First quantitative estimation of magnetic contrast was made by Kranz and Bialas (1961). In conformity with Mayer these authors supposed that contrast is due only to the component $eB_z\dot{r}$ of the Lorentz force. On that assumption the linear relation between the current density distribution $j(r, \varphi)$ on the final screen and the first derivative of the normal component of magnetic microfield B_z was obtained:

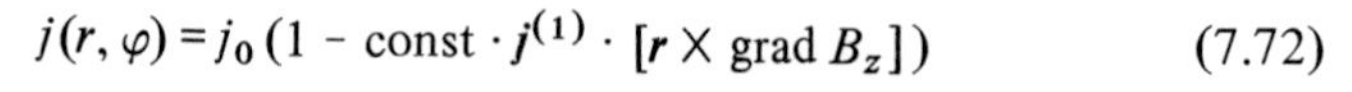

$$j(r, \varphi) = j_0 (1 - \text{const} \cdot j^{(1)} \cdot [\boldsymbol{r} \times \text{grad}\, B_z]) \qquad (7.72)$$

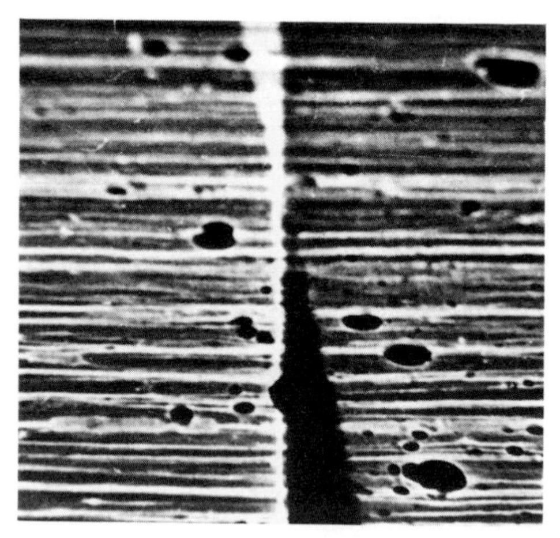

Fig. 7.13 The MEMSI micrograph of the gap of a magnetic recording head when the current flows through the excitation winding. Magnification 100X. Orientation of the gap on the specimen corresponds to vertical direction on this Figure. Horizontal lines correspond to mechanical scratches.

where $j^{(1)}$ is the unit vector of current density, and j_0 is the current density in the absence of perturbations.

It follows from Eq. (7.72) that the image contrast is proportional to $|\boldsymbol{r}|$ and is higher at the margins of the screen than in its center; the contrast reaches its maximum value when grad $B_z \perp \boldsymbol{r}$. It means that the radially extended structures are revealed better than the azimuthal ones. When a specimen with a magnetic pattern is moved through the electrical center, the image contrast of the magnetic microfield is reversed, i.e., bright areas become dark, because the term $j^{(1)} \cdot [r_x \text{ grad } B_z]$ changes the sign. Thus, Eq. (7.72) by Kranz and Bialas qualitatively describes the main features of the image contrast formation of magnetic microfields indicated first by Mayer. The problem of the image contrast formation of magnetic microfields was further discussed by Sedov *et al.* (1968b), Petrov *et al.* (1970a), and Gvosdover (1972a). A more detailed consideration of image contrast formation of magnetic microfields will be given here.

The equations of motion of the electron in the field of an immersion objective in the presence of perturbing magnetic microfield $\boldsymbol{B}(x, y, z)$ are:

$$\dot{p}_x = -e(v_y B_z - v_z B_y), \quad \dot{p}_y = -e(v_z B_x - v_x B_z), \quad \dot{p}_z = -e(v_x B_y - v_y B_x) + eE_0 \tag{7.73}$$

The image formation of magnetic microfield is connected with the change $\Delta \boldsymbol{p}$ of transverse momenta similar to the case of electrical microfields. In the Wiskott approximation, this change is determined in the first order of perturbation theory by the integral of Lorentz force:

$$\Delta p_x = -e \int_{-\infty}^{+\infty} (v_y B_z - v_z B_y)\, dt = \frac{e}{m} \int_{-\infty}^{+\infty} [eE_0\, t B_y(t) - p_{0y} B_z(t)]\, dt \tag{7.74}$$

[expression for Δp_y may be obtained from Eq. (7.74) by changing the general sign and permutation of indices $x \rightleftarrows y$]. The values of the field $\boldsymbol{B}(t)$ in Eq. (7.74) are taken at the points of unperturbed trajectory $(x(t), y(t), z(t))$.

In the case of a vertically incident electron beam (at $\boldsymbol{p}_0 = 0$), the second term in Eq. (7.74) vanishes. Function $t \cdot B_y(r, z_0 + eE_0 t^2/2m)$ is an odd function of time, and its integration gives zero. Thus the value of $\Delta \boldsymbol{p}$ is equal to zero when $\boldsymbol{p}_0 = 0$, since the contributions obtained before and after passing the reversal point have the same absolute values and opposite signs.

The change of momentum $\Delta \boldsymbol{p}$ appears only when $\boldsymbol{p}_0 \neq 0$ is taken into account. Here we shall consider only the terms of the first order in $\boldsymbol{p}_0$. Thus, it is sufficient to take the values of $B_z(t)$ on the line $(r, z(t))$, and for $B_x(t)$, $B_y(t)$ the terms of the first order in $\boldsymbol{p}_0$ must be taken into account; for example:

$$B_x(t) \approx B_x(x, y, z(t)) + \frac{p_{0x} t}{m} \frac{\partial B_x}{\partial x} + \frac{p_{0y} t}{m} \frac{\partial B_x}{\partial y} \tag{7.75}$$

Then the integration of both Eq. (7.74) and the analogous relation for Δp_y gives:

$$\Delta p_x = \gamma_{xx} p_{0x} + \gamma_{xy} p_{0y}, \quad \Delta p_y = \gamma_{yx} p_{0x} + \gamma_{yy} p_{0y} \tag{7.76}$$

where the components of the matrix $\gamma_{ik}(\boldsymbol{r})$ are expressed by:

$$\gamma_{xx} = 2\sqrt{\frac{2me}{E_0}} \int_{z_0}^{\infty} \frac{\partial B_y}{\partial x} \sqrt{z - z_0}\, dz,$$

$$\gamma_{xy} = 2\sqrt{\frac{2me}{E_0}} \int_{z_0}^{\infty} \frac{\partial B_y}{\partial y} \sqrt{z - z_0}\, dz - \sqrt{\frac{2me}{E_0}} \int_{z_0}^{\infty} \frac{B_z\, dz}{\sqrt{z - z_0}}$$

$$\gamma_{yx} = -2\sqrt{\frac{2me}{E_0}} \int_{z_0}^{\infty} \frac{\partial B_x}{\partial x} \sqrt{z - z_0}\, dz + \sqrt{\frac{2me}{E_0}} \int_{z_0}^{\infty} \frac{B_z\, dz}{\sqrt{z - z_0}}$$

$$\gamma_{yy} = -2\sqrt{\frac{2me}{E_0}} \int_{z_0}^{\infty} \frac{\partial B_x}{\partial y} \sqrt{z - z_0}\, dz \tag{7.77}$$

The matrix γ_{ik} may be separated in two parts:* $\gamma_{ik} = \gamma_{ik}^n + \gamma_{ik}^t$, associated with the contributions of normal (B_z) and tangential (B_x, B_y) components of the microfield, respectively. The contribution of the normal component is:

$$\gamma_{ik}^n = \begin{pmatrix} 0 & -\gamma_0 \\ \gamma_0 & 0 \end{pmatrix}, \quad \gamma_0(\boldsymbol{r}) = \sqrt{\frac{2me}{E_0}} \int_{z_0}^{\infty} \frac{B_z(\boldsymbol{r}, z)\, dz}{\sqrt{z - z_0}} \tag{7.78}$$

The transformation of vector $\boldsymbol{p}' = \boldsymbol{p}_0 + \hat{\gamma}^n \boldsymbol{p}_0$ corresponds to (at $|\gamma_0| \ll 1$) the rotation by the angle γ_0 as consequence of the Larmor precession with angular frequency $\omega \sim eB_z m^{-1}$ during the transit time in the microfield.

Such a model of the interaction of the electron beam with magnetic microfields was formulated by Sedov *et al.* (1968b), where the formula of the type of Eq. (7.78) was obtained. The treatment of image contrast formation of magnetic microfields with a consideration of both the normal and the transverse components of the field made by Gvosdover (1972a) showed that the contribution of the gradient of transverse components into $\Delta \boldsymbol{p}$ is of the same order as the contribution of the normal component of magnetic microfield. In that case $|\gamma_{ik}^t| \sim |\gamma_0|$, and the $\hat{\gamma}^t$-matrix corresponds to a more complicated transformation.

According to Eqs. (7.26b) and (7.27), the imaging onto a viewing screen when

*The conventional character of such a separation is connected with the fact that in the vacuum, where conditions rot $\boldsymbol{B} = 0$, div $\boldsymbol{B} = 0$ are satisfied, it is impossible to create microfields with only normal or only tangential components. Moreover, the values of the total vector $\boldsymbol{B}(\boldsymbol{r}, z)$ may be determined by $B_z(\boldsymbol{r}, z = 0)$ on the surface.

the expressions $t_{\text{eff}} = (2D_{\text{sp}})^{-1}$ and $\boldsymbol{p}_0 = mD_{\text{sp}}\boldsymbol{r}_0$ are taken into account, is described by:

$$(\boldsymbol{r}')_i = (\boldsymbol{r} + \boldsymbol{s})_i = (\boldsymbol{r})_i + \frac{1}{2} \sum_{k=1}^{2} (\boldsymbol{r})_k \cdot \gamma_{ik} \tag{7.79}$$

This relation shows that because of the electron optical properties of the objective system of the MEMSI of straight design the components of the γ_{ik} matrix, when projecting onto the screen, equal half of its initial value; the same holds for the deflection angle γ_0.

The vector of displacement $\boldsymbol{s}$ related to the specimen may also be represented as $\boldsymbol{s} = \boldsymbol{s}^n + \boldsymbol{s}^t$, where indexes n and t correspond to the contributions of the normal and the tangential components, respectively. Vector $\boldsymbol{s}^t$ satisfies the relation $\partial s_x^t/\partial x + \partial s_y^t/\partial y = 0$. For that reason it is possible to say that only the normal component of the field contributes to the contrast in the first order in the microfield:

$$j(r) \approx j_0 \left(1 - \frac{\partial s_x^n}{\partial x} - \frac{\partial s_y^n}{\partial y}\right) \tag{7.80}$$

Introducing the angle φ in a polar coordinate system with the center on the axis of the MEM [$\varphi = \arctan(y/x)$], from Eqs. (7.78)-(7.80) in the case of weak contrast we obtain:

$$j(r, \varphi)/j_0 \approx 1 - \frac{1}{2} \frac{\partial \gamma_0 (r \cos\varphi, r \sin\varphi)}{\partial \varphi} \tag{7.81}$$

where $\gamma_0(x, y)$ is defined by Eq. (7.78). Equation (7.81) obtained by Gvosdover (1972a) gives the quantitative generalization of the equation (7.72) by Kranz and Bialas (1961). Note that in the case of strong contrast (at $|r| \cdot |d\gamma_0/dr| \gtrsim 1$) the tangential components are as important as the normal component of the microfield.

One-dimensional microfields are of particular interest. Consider $B_x = B_x(x, z)$, $B_y = 0$, $B_z = B_z(x, z)$. Taking equality div $\boldsymbol{B} = 0$ into account, we obtain in that case:

$$x' = x - \gamma(x) \cdot y, \quad y' = y; \tag{7.82a}$$

$$j(x', y') = j_0 \left[1 - y \frac{d\gamma}{dx}\right]^{-1} \tag{7.82b}$$

where $\gamma(x) = 0.5\gamma_0(x)$ is the deflection angle obtained when the electron moves from the objective system to the screen. Thus, the contribution of both the tangential and the normal components into displacement s_x are compensated, and transformation [Eq. (7.82a)] cannot be represented as a pure rotation of vector $\boldsymbol{r}$.

As was mentioned above, the caustic formation in the case of strong contrast occurs when the Jacobian of the transformation $r \rightarrow r'$ is equal to zero. According to Eq. (7.82b) in the case of one-dimensional magnetic microfields, that occurs when $y\, d\gamma/dx = 1$. Thus, the equation of caustic in the screen coordinate system (x', y') may be represented as:

$$y'(x) = (d\gamma/dx)^{-1}, \quad x'(x) = x - (d\gamma/dx)^{-1}\, \gamma(x) \tag{7.83}$$

with x being a parameter.

Here is a more detailed consideration of an example of one-dimensional sinusoidal microfield with the lateral period $\lambda = 2\pi/k$ and with the amplitude B_0 on the surface. It will be shown that by means of that very simple model it is possible to explain the wedge-shaped structure observed by Spivak *et al.* (1963a) on the image of periodic magnetic microfields. In this case the solution of Laplace's equation for $\boldsymbol{B}$ gives $B_z(x, z) = B_0\, e^{-kz} \sin kx$, $B_x(x, z) = -B_0\, e^{-kz} \cos kx$. The deflection angle $\gamma(x)$ equals $\gamma(x) = 0.5\, \gamma_0(x) = A \sin kx$, where $A = 0.5\, B_0 (2\pi e/ mE_0\, k)^{1/2}\, e^{-kz_0}$. For the magnetic microfield of that type the explicit equation for the caustic lines in the screen coordinate system may be obtained (Petrov *et al.*, 1970b; Gvosdover and Petrov, 1972):

$$x' = \lambda \left[n \pm \frac{1}{2\pi} \arccos \frac{\lambda}{2\pi A y'} \right] \pm A y' \sqrt{1 - \left(\frac{\lambda}{2\pi A y'} \right)^2} \tag{7.84}$$

where n is an arbitrary whole number. The value of the wedge angle formed by the caustic lines is approximately equal to $2A$; i.e., it is twice the value of the amplitude of the deflection angle.

It is worth noting that similar contrast will appear if the one-dimensional periodic magnetic microfields of another configuration (e.g., triangle, rectangle) are observed. This phenomenon may be explained by the fact that because of strong damping of the microfield with the height over the specimen, higher harmonic components give a small contribution and the contrast is determined by the main sinusoidal harmonic of the microfield. Figures 7.14 and 7.15 show the micrograph of the domain structure on the prismatic plane of a cobalt single crystal (Spivak *et al.*, 1963a) and theoretically calculated caustic lines, which appear to be due to the sinusoidal magnetic microfield.

The calculation of image contrast caused by the magnetic microfield of the ferromagnetic stripe was made by Petrov *et al.* (1967). These authors obtained a satisfactory agreement between the experimentally measured local deflection angle and the calculated one when the magnetic field was approximated by a "step." When a ferromagnetic stripe is observed, the effects of the second order appear. The latter are revealed by narrowing the image of the ferromagnetic stripe on the screen. Effects of the second order in the amplitude of the magnetic microfield were discussed by Sedov *et al.* (1968b).

In order to verify the theoretical calculations of the image contrast based on the interaction of the radial velocity component of the electron with the normal component of magnetic microfield, experiments were carried out on the speci-

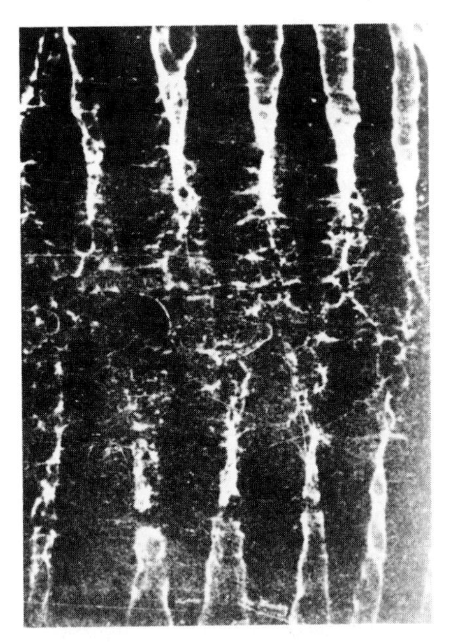

Fig. 7.14 The MEMSI micrograph of the domain structure on the prismatic plane of the cobalt single crystal. Bias voltage $U_B = 0$. Magnification 200X. (*Spivak* et al., *1963a; courtesy of G. V. Spivak.*)

men (wire carrying a current) with a known distribution of the magnetic microfield (Spivak *et al.*, 1968c). To guarantee better accuracy of the measured data, the stroboscopic mode of operation of the MEM was used. The reason for its use is that in the pulse mode, in comparison with the conventional mode, it is possible to have higher current through the specimen without causing its destruction. Thus, the value of $\boldsymbol{B}$ can be increased, resulting in an intensification of the contrast.

Figure 7.16 illustrates the "image" of the magnetic microfield of a wire carrying a current; it slightly resembles the picture of the magnetic field of a magnetic head. Figure 7.17 shows the oscillograms of the contrast, demonstrating that with an increase of the current the contrast also increases. These experiments have shown that an accuracy of the order of 20% can be attained. Thus, the up-to-date models of the image contrast formation of magnetic microfields in MEM yield both a qualitative and a quantitative description of the processes of depicting magnetic microfields.

In conclusion, we shall briefly discuss the contribution of the terms $\backsim \Delta r_0$ to the images of magnetic microfields in the MEMSI. The change of action $\Delta\Phi$ may be easily calculated in the first order in $\boldsymbol{p}_0$ from Eq. (7.45) for the magnetic flux "hitched up" by a trajectory:

$$\Delta\Phi(r_0, \boldsymbol{p}_0) = -\boldsymbol{p}_0 \cdot \Delta r_0(r_0) \tag{7.85a}$$

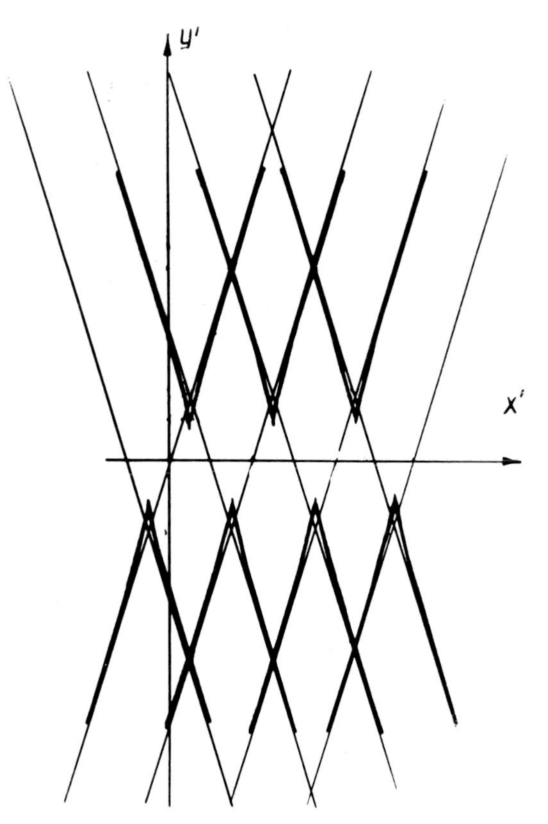

Fig. 7.15 A set of caustic lines corresponding to one-dimensional sinusoidal magnetic microfield; compare with Fig. 7.14. For clearness this Figure is drawn for a larger amplitude of the magnetic field.

where:

$$\Delta x_0(r_0) = -2\sqrt{\frac{2me}{E_0}}\int_{z_0}^{\infty} B_y(r_0, z)\sqrt{z - z_0}\, dz,$$

$$\Delta y_0(r_0) = +2\sqrt{\frac{2me}{E_0}}\int_{z_0}^{\infty} B_x(r_0, z)\sqrt{z - z_0}\, dz \qquad (7.85b)$$

Differentiation of Eq. (7.85a) over r_0 yields immediately the above expressions (7.77) with $\gamma_{ik} = -\partial(\Delta r_0)_i/\partial r_{0k}$. It should be noticed that the two-dimensional divergence of the vector Δr_0 equals zero: $\partial\Delta x_0/\partial x_0 + \partial\Delta y_0/\partial y_0 = 0$. Therefore, the terms $\backsim\Delta r_0$ do not contribute to the contrast in the first order.

The terms $\backsim\Delta r_0$ have the order of magnitude $|\Delta r_0| \sim \gamma\delta x$, where $\gamma \sim B(e\delta x/mE_0)^{1/2}$ is the rotation angle. "Wiskott" terms $\backsim\Delta \boldsymbol{p}_0$ result in the displacement $|s| \sim \gamma r_0$. Thus, at the areas far from the electrical center of the image (i.e., at

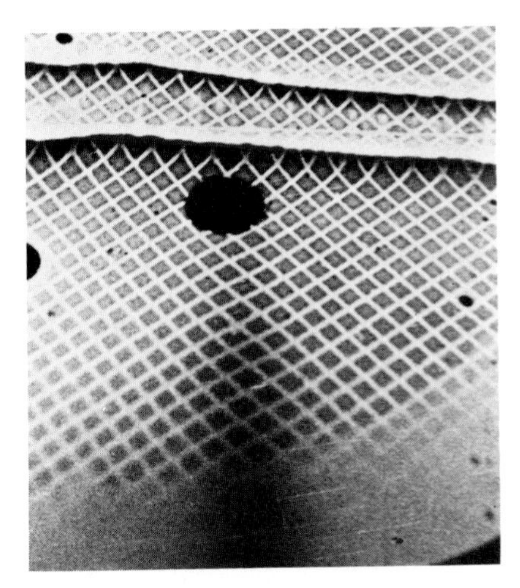

Fig. 7.16 Stroboscopic MEMSI micrograph of the magnetic microfield produced by a wire carrying a current. Dark spot in the center is the hole in the screen. Mesh period is 33 μm. Horizontal magnification 85X. Horizontal lines on the upper part are the cleavage steps on the mica covering a wire. Direction of the wire corresponds to the vertical direction on the image.

$|r_0| \gg \delta x$) the terms $\backsim \Delta r_0$ may be neglected in comparison with "Wiskott" terms $\backsim \Delta p_0$. The terms $\backsim \Delta r_0$ are more important close to the center, i.e., at $|r_0| \lesssim \delta x$. Besides, there may be a situation when the displacement is determined mostly by the terms $\backsim \Delta r_0$ because of large-scale components of microfields (i.e., with large δx), and the contrast is determined by the terms $\backsim \Delta p_0$ because of the small-scale microfields. Indeed, the displacement $s_{\Delta r} = 0.5\ \Delta r_0 \backsim B \cdot (\delta x)^{3/2}$ and the contrast $K_{\Delta p} \backsim B \cdot (\delta x)^{-1/2}$, and the above statement becomes plausible.

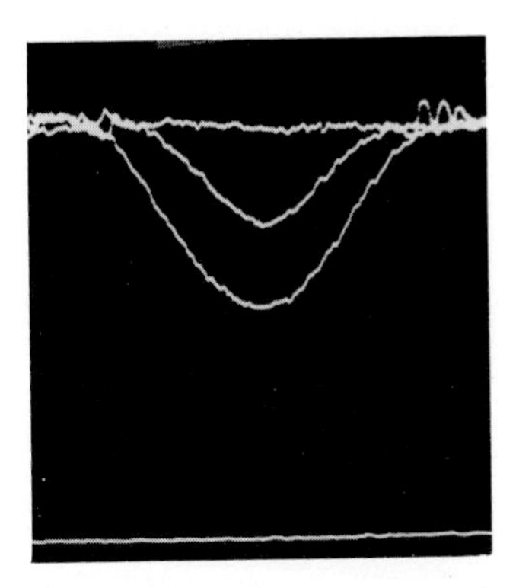

Fig. 7.17 Experimental oscillograms of the current density (corresponding to the image of Fig. 7.16) at different values of the amplitude of pulse current: 1–2 amp; 2–6 amp. The horizontal bright line corresponds to zero level of brightness (Spivak *et al.*, 1968b).

Sensitivity and Resolving Power of MEM in Focused Imaging Mode

Substitution of the expressions (7.40)-(7.42) for $\Delta \boldsymbol{p}_0$ and $\Delta \boldsymbol{r}_0$ into the equation (7.37) for the current density distribution on the final screen $j(\boldsymbol{r}')$ yields the total solution of the classical problem of the image contrast formation of electrical and magnetic microfields in the MEMFI. From the estimations made above (Eqs. (7.6), (7.8)), it follows that such calculations make it possible to describe even a strong contrast though the first order of perturbation theory was used to determine $\Delta \boldsymbol{p}_0$ and $\Delta \boldsymbol{r}_0$. In this section we shall not discuss the problem of strong contrast in the MEMFI; the reader can find it in the review by Bok *et al.* (1971). Here we shall only briefly formulate the problem of sensitivity and resolving power.

In linear approximation for the contrast, the transparency function of the aperture may be represented as an expansion with the accuracy of terms in the first order:

$$T(\boldsymbol{p}_0 + \Delta \boldsymbol{p}_0) \approx T(\boldsymbol{p}_0) + \frac{\partial T}{\partial \boldsymbol{p}_0} \cdot \Delta \boldsymbol{p}_0(\boldsymbol{r}_0, \boldsymbol{p}_0, \epsilon) \tag{7.86}$$

In the case of standard apertures with a sharp edge $[T(\boldsymbol{p}) = 1, \text{ or } T(\boldsymbol{p}) = 0]$, this expansion is not always valid since the derivative $\partial T/\partial \boldsymbol{p}_0$ contains a δ-function. The other possible procedure of the linearization consists in the transforming variables $\boldsymbol{p}_0 + \Delta \boldsymbol{p}_0 = \boldsymbol{p}'$ and in the further expanding function $\mathcal{J}_0(\boldsymbol{p}' - \Delta \boldsymbol{p}_0, \epsilon)$ and of the Jacobian of transformation into the series in $\Delta \boldsymbol{p}_0$. If function $\mathcal{J}_0(\boldsymbol{p}_0, \epsilon)$ is a smooth* one, then both methods of linearization mentioned above yield the same result. Therefore, we shall use a more simple equation (7.86). Thus we obtain in the first order that $j(\boldsymbol{r}_0) = j_0 + \delta j(\boldsymbol{r}_0)$, where:

$$j_0 = \int \mathcal{J}_0(\boldsymbol{p}_0, \epsilon) T(\boldsymbol{p}_0)\, d^2\boldsymbol{p}_0\, d\epsilon,$$

$$\delta j(\boldsymbol{r}_0) = \int \mathcal{J}_0(\boldsymbol{p}_0, \epsilon) \left\{ \frac{\partial T}{\partial \boldsymbol{p}_0} \cdot \Delta \boldsymbol{p}_0(\boldsymbol{r}_0, \boldsymbol{p}_0, \epsilon) - T(\boldsymbol{p}_0) \left(\frac{\partial \Delta x_0}{\partial x_0} + \frac{\partial \Delta y_0}{\partial y_0} \right) \right\} d^2\boldsymbol{p}_0 d\epsilon \tag{7.87}$$

The linearized spectral transfer function of the MEMFI for the electrical microfields equals:

*Contrary to that for the MEMFI from the work by Bok *et al.* (1971), both functions $T(\boldsymbol{p}_0)$ and $\mathcal{J}_0(\boldsymbol{p}_0, \epsilon)$ have the singularity at the same values of $\boldsymbol{p}_0$. Thus, the linearized description of the contrast given in this paragraph cannot be applied to such an instrument. In that case it is necessary to use general equations [(7.36), (7.37)]. As was mentioned in the review by Bok *et al.* (1971), the weak contrast in such an instrument $K \backsim |\Delta \boldsymbol{p}_0|$ and thus is not analytical in $\Delta \boldsymbol{p}_0$. That leads in particular to the fact that, for example, the image of a sinusoidal microfield will contain only the double and higher spatial frequencies.

$$S(k) = j_0^{-1} \int \mathcal{J}_0(p_0, \epsilon) \left(\frac{2\pi em}{E_0 |k|}\right)^{1/2} \exp\left\{-\frac{|k|}{eE_0}\left(eU_B - \epsilon + \frac{p_0^2}{2m} + \frac{(k \cdot p_0)^2}{2m|k|^2}\right)\right\}$$

$$\times \left[ik\frac{\partial T}{\partial p_0} - i\frac{2|k|(k \cdot p_0)}{meE_0} T(p_0)\right] d^2p_0 d\epsilon \quad (7.88)$$

In Eq. (7.88) terms $\backsim \partial T/\partial p_0$ are connected with effects $\backsim \Delta p_0$ and terms $\backsim T(p_0)$ are due to the effects $\backsim \Delta r_0$.

Consider the following example. Let the electron beam be monoenergetic and:

$$\mathcal{J}_0(p_0, \epsilon) = \text{const} \cdot \delta(\epsilon) \exp\left\{-\frac{p_0}{2q^2}\right\}$$

where q is the spread of transverse momenta in the illumination beam close to the specimen. For a circular aperture with the center on the axis of the system, the contrast of first order vanishes. For the knife edge aperture located precisely in the center of the cross-over, one has:

$$T(p_x, p_y) = \theta(p_x) = \begin{cases} 1, & p_x \geqslant 0 \\ 0, & p_x < 0 \end{cases}$$

In that case it follows from Eq. (7.88) that:

$$S(k) = \frac{2}{q}\left(\frac{me}{E_0}\right)^{1/2} \frac{ik_x}{|k|^{1/2}} \exp\left(-|k|\frac{U_B}{E_0}\right) \times \left[1 + \frac{q^2(2k_y^2 + k_x^2)}{meE_0|k|}\right]^{-1/2}$$

$$\cdot \left\{1 - \frac{2|k|q^2}{meE_0}\left[1 + \frac{q^2(2k_x^2 + k_y^2)}{meE_0|k|}\right]^{-1}\right\} \quad (7.89)$$

The first term in the braces in Eq. (7.89) corresponds to the terms $\backsim \Delta p_0$ and the second one to the terms $\backsim \Delta r_0$. From Eq. (7.89) it is clear that at $U_B = 0$ a considerable damping of the function $S(k)$ occurs at $|k| \gtrsim k_{\text{opt}} \sim (eE_0 m)/q^2$ when the effects of sliding and of increase of the height of reflection become essential. Besides, at $|k| \gtrsim k_{\text{opt}}$ the contributions of the terms $\backsim \Delta p_0$ and $\backsim \Delta r_0$ become of the same order of magnitude. As it was shown (see Eq. (7.3)), the value

$$\delta x_{\text{FI}} \approx \frac{q^2}{eE_0 m} \approx \frac{r_a^2 U_0}{f^2 E_0} \quad (7.90)$$

determines the transverse resolution of the MEMFI with a monoenergetic electron beam; here r_a is the size of the cross-over of the beam in the plane of contrast aperture and f is the focal length of the objective system. The substitution of values $r_a \sim 100\ \mu\text{m}$, $E_0 \sim 10^7 \text{V/m}$, $U_0 = 30$ kV, $f \sim 3$ cm yields $\delta x_{\text{FI}} \sim 30$ nm by the order of magnitude. The limit of resolving power connected with the energy spread of the electron beam is $\delta x \sim \delta\epsilon/eE_0$ and for $\delta\epsilon \sim 1$ eV the value of $\delta x \sim 100$ nm is obtained. In the review by Bok *et al.* (1971) the experimental value of the transverse resolving power $\delta x \sim 80$ to 100 nm had been reported for the MEMFI.

RECONSTRUCTION OF MICROFIELDS FROM IMAGE

The simplicity of the mechanism of primary modulation in the MEM (the change of transverse momentum under the influence of microfields) and the high reliability of the analytical calculations of that process allow us to formulate and solve theoretically and experimentally the problem of the reconstruction of microfields from the image contrast (Sedov, 1968). The essence of the problem may be illustrated by an example of a weak contrast of electrical microfields in the MEMSI.

The image contrast formation in the MEMSI may be considered in the linear approximation as a process of filtration of different Fourier components of micropotential $\tilde{\varphi}(k)$ or of microfield $\tilde{E}(k) = -ik\tilde{\varphi}(k)$, that is:

$$\tilde{K}(k) = C_1 e^{-|k|z_0} |k|^{3/2} \tilde{\varphi}(k) = -C_1 e^{-|k|z_0} \cdot \frac{i(k \cdot \tilde{E}(k))}{\sqrt{|k|}}, \qquad (7.91a)$$

$$\tilde{s}(k) = C_2 e^{-|k|z_0} \frac{ik}{\sqrt{|k|}} \tilde{\varphi}(k) = -C_2 e^{-|k|z_0} \frac{\tilde{E}(k)}{\sqrt{|k|}} \qquad (7.91b)$$

Here $\tilde{K}(k)$ and $\tilde{s}(k)$ are the Fourier transforms of the contrast $K(r)$ and displacement $s(r)$, respectively; C_1 and C_2 are some constants.

Relations (7.91a,b) show that the components of microfield with different values of spatial frequency $|k|/2\pi$ are transformed with a different coefficient into the displacement and contrast. The same equations (7.91a,b) or more accurate expressions from the section on "Image Contrast Formation" will be used below for determining microfields φ, E, B by the image.

The problem of reconstruction of microfields from the image will be stated more clearly if the structure of the relation between the contrast and the microfield is formulated once more. As it was described earlier, the transformation of microfields into contrast may be separated into two stages. In the first stage the transformation of microfields into the primary modulation of an electron beam occurs. For a wide range of the strength of microfield (with strength E satisfying $|E| \lesssim E_0$), the first transformation is linear up to a high accuracy, and according to Eq. (7.56) it is not local with respect to the transverse coordinate. In the second stage the transformation of velocity modulation into the modulation of the current density (i.e., the contrast) occurs on the screen. The second transformation is already nonlinear in the case of a strong contrast, yet it is of a local character for both the focused and the shadow imaging modes of the MEM.

The procedure for the reconstruction of microfields (i.e., the solution of the inverse problem of electron microscopy), which is discussed in this section, also falls into two stages. Firstly, from the contrast data we determine the displacement of electron trajectories or the change of the momentum versus the coordinate on the specimen. Secondly, from the obtained dependence $\Delta \boldsymbol{p}(\boldsymbol{r})$, we find the microfield at some height z_1 over the specimen by means of a linear transformation of the convolution type. In addition, a special class of wavelike microfields of great practical importance is considered. It turns out that for those microfields in a stroboscopic mode of operation the weak contrast is directly proportional to the amplitude of microfield, so that there is no need for any special procedures for solving the inverse problem.

Determination of Displacement in MEMSI

The simplest method for determining the related to the specimen displacement s of the electron trajectory consists in the measurement of the positions of the image of the point of a coordinate mesh previously deposited on the specimen.* The advantage of that method is that it makes possible the measurement of both components of the displacement vector $\boldsymbol{s} = (s_x(x, y), s_y(x, y)$. This method is also useful when the microfield under investigation may be switched on or off. The image of some small inhomogeneities of the specimen may also be used for that purpose.

Other methods for determining the displacement are based directly on the study of the current density distribution $j(x', y')$ on the final screen. Registration of only one function $j(x', y')$ makes it impossible in a general case to determine both components of the vector $\boldsymbol{s}$. Therefore, such methods are applicable only when the displacement vector $\boldsymbol{s}$ in some coordinate system has only one component.

For the Cartesian coordinates (x, y) that may be the case for the so-called one-dimensional microfields, when $\varphi(x, y) = \varphi(x)$ or $B_z(x, y) = B_z(x)$. As we have seen above in this case only x-component of the displacement exists: $s_x = s$; for the electrical microfield, s depends on x only, i.e., $s = s(x)$, and for the magnetic microfield, $s = y \cdot f(x)$. In such a "one-dimensional" problem the expression for the contrast is:

$$j(x', y') = j_0 \left[1 + \frac{\partial s(x, y)}{\partial x}\right]^{-1}; \quad y' = y, \quad x' = x + s(x, y) \tag{7.92}$$

Here and in the subsequent discussion we assume that there are no caustics.

If some point (x_1', y_1') is found on the screen for which $s(x_1', y_1') = 0$ can be assumed, and if Eq. (7.92) is integrated over dx' from x_1' up to some variable

*See also the section on "Electron Mirror Interferometer" where the interference fringes play the role of such a mesh.

value $x'(x)$, then we get:

$$s(x) = x'(x) - x_1' + j_0^{-1} \int_{x_1'}^{x'(x)} j(x'')\, dx'' \tag{7.93}$$

[the coordinate $y = y_1'$ in Eq. (7.93) is assumed to be fixed]. The graphical rule for the determination of the displacement (Sedov *et al.*, 1968a; Sedov, 1970b) from the integrated oscillogram of brightness follows from Eq. (7.93) (Fig. 7.18). In the above papers this rule was realized in the following way. The image was scanned with a constant velocity across the slit of a registration system. The electrical signal was integrated by analog integrator and recorded on the screen of the oscilloscope.

For convenience an oscillogram in the absence of the microfield under investigation was also recorded; however, that second oscillogram might be replaced by a straight line with the appropriate slope. The horizontal line in Fig. 7.18 characterizes the value of the displacement $s(x)$, and the x-coordinate of its intersection with the unperturbed straight line is equal to the value of the coordinate on the specimen x to which the obtained displacement must be attributed.

An analogous method for determining the displacement for a transmission electron microscope operating in a shadow projection mode was realized by Cohen and Harte (1969); this method might be described by the same equation (7.92). If it is known that the displacement $s(x_1) = s(x_2) = 0$ for some two points x_1 and x_2, then $s(x)$ can be reconstructed from the current density distribution on the screen $j(x')/j_0$ by a digital computation of the integral:

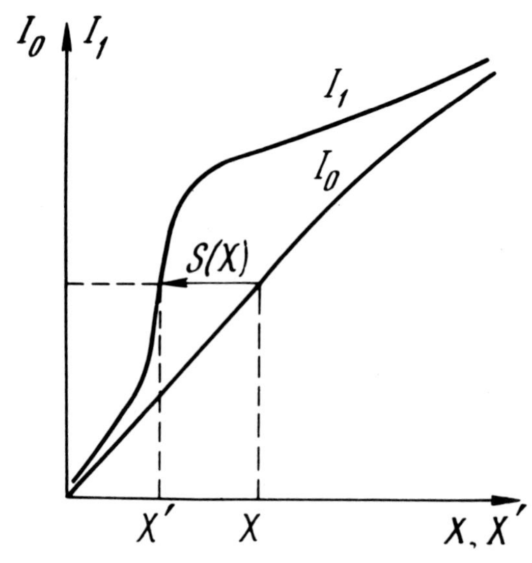

Fig. 7.18 On the graphical determination of the displacement of the electron by means of integrated curves of current density. (*Sedov* et al., *1968a; courtesy of N. N. Sedov.*)

$$f(\xi) = \alpha \int_{x_1}^{\xi} \left[1 - \frac{j(x')}{j_0}\right] dx' - (1 - \alpha) \int_{\xi}^{x_2} \left[1 - \frac{j(x')}{j_0}\right] dx' \quad (7.94a)$$

where α is an arbitrary real number. Function $s(x)$ is defined from Eq. (7.94a) parametrically:

$$s = f(\xi), \quad x = \xi - f(\xi) \quad (7.94b)$$

Equation (7.93) follows from Eq. (7.94) for $\alpha = 1$; in the paper by Cohen and Harte (1969) the values $\alpha = \frac{1}{2}$, $x_1 = -\infty$, $x_2 = +\infty$ were used. An analogous method was suggested for the case when $s(r)$ is directed along the radius-vector originating at some point of the (x, y)-plane (Sedov, 1971).

For the purpose of registering the displacement a method based on the introduction of a nonlinear feedback into the system making the image on the screen to scan over the slit was suggested by Gvosdover (1970). In this method the instantaneous shift of the image is determined by the integration of the signal $[j(t)]^{-1}$. The scanning over the specimen's coordinate turns out to be linear in this method: $x = ut$. The difference between that linear function and the instantaneous scanning voltage (expressed in units of the shift) equals the value of the displacement:

$$x = ut, \quad s(x) = x' - ut = -ut + u \int^{t} \frac{j_0}{j(t')} dt' \quad (7.95)$$

The described method is applicable to conventional MEMSI, although it has not been realized experimentally. The feedback method of automatic registration of the displacement may probably be easily realized in a scanning MEM with beam separation.

Reconstruction of Microfields from Displacement

Methods for displacement determination described above are well elaborated only for a one-dimensional case. Therefore reconstruction of microfields from the displacement will be considered in detail only for the one-dimensional problem; a general case of two-dimensional microfields may be treated in a similar way.

Equations (7.56), (7.78), and (7.82) yield the following expressions for the displacement related to the specimen in the case of reflection of an electron beam at a height z_0 over the specimen for the given electrical $(E_x(x, z = 0))$ or magnetic $(B_z(x, z = 0))$ one-dimensional microfield on the surface of the specimen:

$$s(x) = -t_{\text{eff}} \left(\frac{e}{mE_0}\right)^{1/2} \int_{-\infty}^{+\infty} d\xi E_x(x - \xi, z = 0) \frac{\sqrt{z_0 + \sqrt{z_0^2 + \xi^2}}}{\sqrt{z_0^2 + \xi^2}}, \quad (7.96a)$$

$$s(x,y) = -\frac{y}{2}\left(\frac{e}{mE_0}\right)^{1/2}\int_{-\infty}^{+\infty} d\xi\, B_z(x-\xi, z=0)\frac{\sqrt{z_0+\sqrt{z_0^2+\xi^2}}}{\sqrt{z_0^2+\xi^2}}, \tag{7.96b}$$

where for the magnetic microfield y is the distance of the straight line $y = \text{const}$ from the electrical center of the image.

The Fourier transformation of the relations (7.96) yields the spectral transfer function discussed above; see Eq. (7.91). The inversion of the relations (7.91) yields exponentially large values ($\sqrt{|k|}\exp\{|k|z_0\}$) of the coefficient of inverse transformation for $|k| \to \infty$. It means that the problem of reconstruction of the microfield on the surface $z = 0$ from the displacement measured at $z_0 > 0$ is an incorrect problem of mathematical physics. The solution of incorrect problems must be based on a special approach. Since the corresponding integrals are formally divergent, they should be regularized; and, what is most important, the values of these integrals are unstable in relation to the small errors encountered on the experimental and computational stages of work. The results themselves, which may be obtained by the regularization methods, depend on the errors of measurement and on the degree to which the microfields under investigation are *a priori* statistically undetermined. For further details, the reader is referred to the review by Turchin *et al.* (1970).

The difficulties of the formal divergence of the integrals which provide the solution of the inverse problem will be avoided in conformity with Sedov *et al.* (1968a,b) by means of a special "trick" of regularization. Suppose we are interested in the field $E_x(x, z_1)$, $B_z(x, z_1)$ at some height $z = z_1$ over the specimen. Then, the spectral coefficient of the inverse transformation is proportional to $|k|^{1/2}\exp\{-|k|(z_1 - z_0)\}$, and for $z_1 > z_0$ the formal convergence of the integrals which give the solution of the inverse problem is guaranteed. In the coordinate representation, the corresponding inverse transformation is:

$$E_x(x, z_1) = \frac{1}{4\pi}\left(\frac{E_0 m}{e}\right)^{1/2} t_{\text{eff}}^{-1}\, f(x, z_1, z_0);$$

$$B_z(x, z_1) = \frac{1}{2\pi y}\left(\frac{E_0 m}{e}\right)^{1/2} f(x, z_1, z_0) \tag{7.97}$$

where for $z_1 > z_0$:

$$f(x, z_1, z_0) = -\sqrt{2}\int_{-\infty}^{+\infty} d\xi\, s(x+\xi)\,\text{Re}\,[(z_1 - z_0 - i\xi)^{-3/2}] \tag{7.98}$$

Equations (7.97) and (7.98) are in themselves solutions of the inverse problem for $z_1 > z_0$. However, Eq. (7.98) does not allow us to obtain directly the result at $z_1 \to z_0$, since for $z_1 = z_0$ the integral [Eq. (7.98)] becomes formally divergent. One may overcome that difficulty taking into account that

$$\int_{-\infty}^{+\infty} \mathrm{Re}\,[(z_1 - z_0 - i\xi)^{-3/2}]\, d\xi = 0 \tag{7.99}$$

Adding the integral [Eq. (7.99)] multiplied by $s(x)\sqrt{2}$ to Eq. (7.98), we obtain the expression which is identically equal to Eq. (7.98). Function $s(x)$ must be differentiable, and because of that fact it is possible to obtain the limit of the corresponding expression at $z_1 \to z_0$. Thus,

$$f(x, z_0, z_0) = \int_{-\infty}^{+\infty} d\xi\, \frac{s(x + \xi) - s(x)}{|\xi|^{3/2}} \tag{7.100}$$

Equations (7.97) and (7.100) were employed in a number of papers for the determination of one-dimensional electrical and magnetic microfield [see references in the review by Lukianov *et al.* (1973)].

In spite of the convergence of the integral [Eq. (7.100)], calculation of it based on experimental data raises a number of difficulties. In particular the formulae obtained give a correct description of only those Fourier components the period $2\pi\,|\boldsymbol{k}|^{-1}$ of which is larger than the lateral resolving power.

In the experiments to be discussed, function $s(x)$ had practically been smoothed to improve the convergence of the integral [Eq. (7.100)] at $|\xi| \to 0$; that poor convergence is connected with the incorrectness of the problem. This fact led to a considerable loss of information about the high frequency components of the microfield.

The integral (7.100) should be calculated over the infinite interval, and the displacement had been measured experimentally within a finite interval only. The integral (7.100) converges relatively slow at $|\xi| \to \infty$. Therefore, it was necessary to approximate the values of $s(x)$ at $x \to \pm\infty$, basing the approximation on the data for the finite values of x and on the *a priori* information on the microfield under investigation.

As an illustration, Fig. 7.19 shows the image of electrical inhomogeneity accompanying the propagation of recombination waves in germanium, and Fig. 7.20 shows the distributions of the potential and field strength in the "active" region of the specimen (Gvosdover *et al.*, 1971). Figure 7.21 shows the distribution of the normal and tangential components of the magnetic microfield for the ring magnetic head. Considering the above-mentioned facts, the accuracy of the measurement of the main (i.e., with moderate values of spatial frequency) components of the field profile in the experiments mentioned above for values $E_x \sim 10^5$ V/m, $\delta x \sim 50\ \mu$m; $B_z \sim 100$ gauss, $\delta x \sim 20\ \mu$m was ~20%.

In conclusion it should be emphasized that the prospects of measuring microfields with the help of the MEM are very promising. They are founded on the simplicity and high reliability of the formulae which provide the solution of the

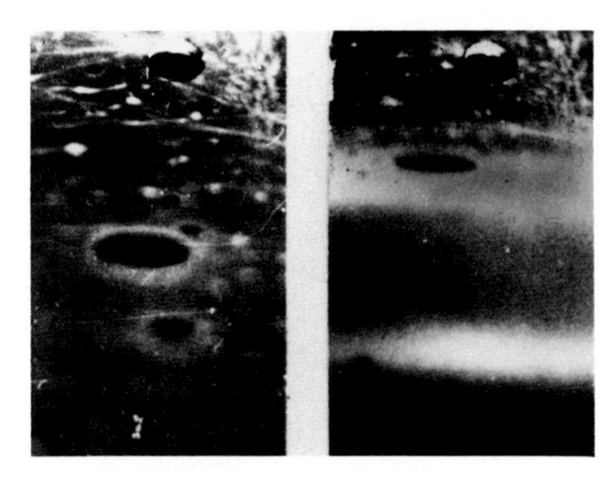

Fig. 7.19 Stroboscopic MEMSI micrographs of the surface of germanium specimen. (a) Topography of the surface; (b) image of electrical instability accompanying propagation of recombination waves. Magnification 50X. (*Gvosdover* et al., *1971*.)

direct problem. As to the concrete methods of the solution of the inverse problem, the "choice" of a microfield best fitting the experimental distribution of the current density on the screen seems to be most desirable. With such a method the microfield may be approximated by an expansion in terms of some system of functions or other; the optimum choice of this system depends on the noise of measurement and on the *a priori* information about the microfield (Turchin *et al.*, 1970). One of the advantages of such an approach is that it is

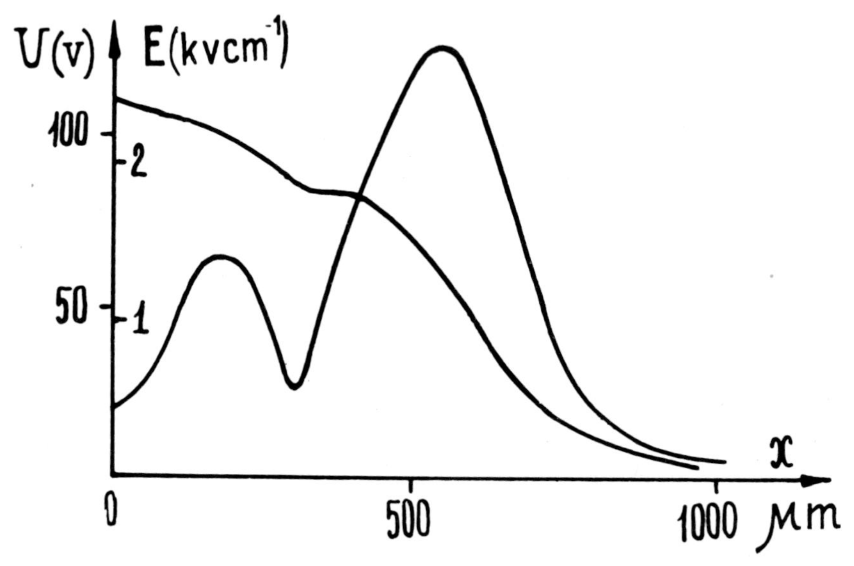

Fig. 7.20 Field strength (1) and potential distribution (2) in "active region" of the specimen represented in Fig. 7.19b (*Gvosdover* et al., *1971*.)

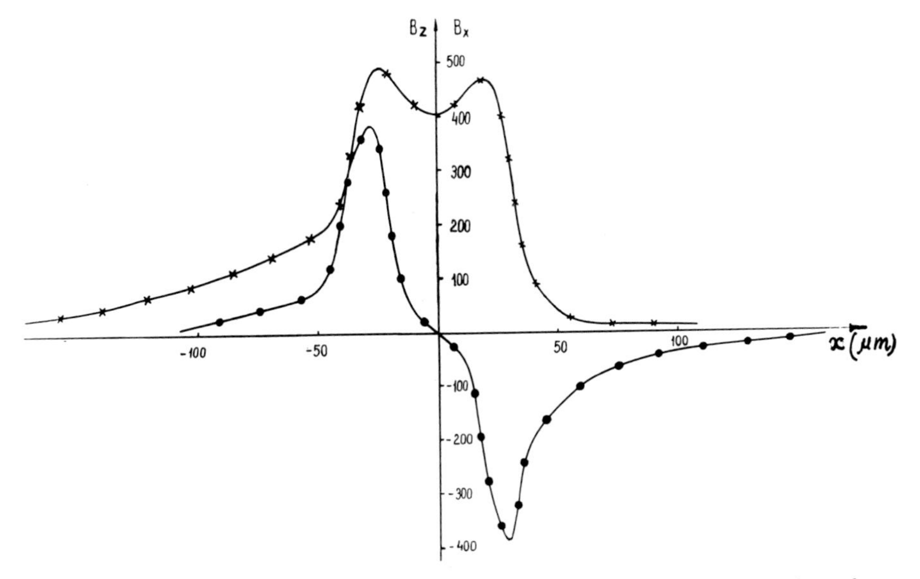

Fig. 7.21 Distribution of normal (*dots*) and tangential (*crosses*) components on the surface of ring magnetic head. The scale along the *x*-axis is microns, and along the *y*-axis is gauss.

not necessary to reconstruct the displacement from the contrast; the determination (in principle) of the function of two coordinates ($\varphi(x, y, z = 0)$ or $B_z(x, y, z = 0)$) straight from the current density distribution $j(x', y')$ is possible.

Depiction of Undulating Microfields

The MEMSI has been successfully used for the observation of piezoelectrical fields appearing on the surface of crystals when the acoustic (either bulk or surface) waves propagate through the crystal (Gvosdover *et al.*, 1968) (Fig. 7.22).

If the microfield varies periodically with time, then the displacement of electron trajectories and the image contrast are also time-dependent. The observation of the image may be provided either in a stroboscopic mode in which the image contrast corresponding to a definite phase of a periodical process is recorded, or in a conventional (nonstroboscopic) mode which reveals a time-averaged image contrast.

In the conventional mode of operation all the terms in the contrast linear in the amplitude of a harmonical microfield vanish as a result of time averaging. Therefore the expression for the contrast written with accuracy up to the terms of $\backsim s^2$ is of interest. For the derivation of such an expression one must take into account that the point of the screen $\boldsymbol{r}'$ in the MEMSI corresponds to the displaced point $\boldsymbol{r}$ on the specimen: $\boldsymbol{r} = \boldsymbol{r}' - \boldsymbol{s}(\boldsymbol{r})$. Here is the final expression for $j(x', y')$ with accuracy up to terms of $\backsim s^2$ (the details of the derivation may be

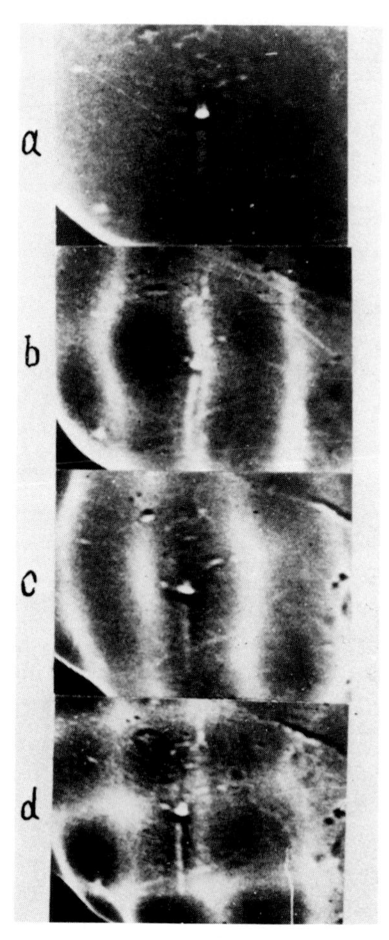

Fig. 7.22 Stroboscopic MEMSI micrographs of the surface of the CdS crystal. (a) Microtopography of the surface, with bright small spot in the center corresponding to the image of a pit; (b,c) images of piezoelectric fields on the surface corresponding to the bulk waves at the resonant frequency of 5 MHz; (d) image of the piezofield at nonresonant frequency 5.2 MHz. Magnification 15X. (*Gvosdover* et al., *1970*.)

found in the work by Gvosdover, 1972b):

$$j(x', y')/j_0 = 1 - \left[\frac{\partial s_x}{\partial x} + \frac{\partial s_y}{\partial y}\right]_{x', y'} + \left[s_x \frac{\partial^2 s_x}{\partial x^2} + s_y \frac{\partial^2 s_x}{\partial x \partial y} + s_x \frac{\partial^2 s_y}{\partial x \partial y} + s_y \frac{\partial^2 s_y}{\partial y^2}\right.$$

$$\left. + \left(\frac{\partial s_x}{\partial x}\right)^2 + \left(\frac{\partial s_y}{\partial y}\right)^2 + \frac{\partial s_x}{\partial x} \cdot \frac{\partial s_y}{\partial y} + \frac{\partial s_x}{\partial y} \cdot \frac{\partial s_y}{\partial x}\right]_{x', y'} \quad (7.101)$$

The values of all the functions $s(r, t)$ and their derivatives in the right-hand side of equality (7.101) should be taken at the point $r = r'$, i.e., at the "immovable" point r' of the screen which is of interest. In the one-dimensional case $s_y = 0$, $s_x = s(x)$ and the expression (7.101) may be reduced to:

$$j(x')/j_0 = 1 - \left[\frac{ds}{dx} - \frac{d}{dx}\left(s\frac{ds}{dx}\right)\right]_{x'} + 0(s^3) \tag{7.101a}$$

In the stroboscopic mode, the nonzero contrast may appear already because of the terms of the first order in microfields; in that case it is possible to observe both the traveling and standing waves of the piezopotential. The expression for the contrast caused by the piezofields of the surface Rayleigh-Lamb waves becomes especially simple in the stroboscopic mode. The potential of those waves on the surface:

$$\varphi(r, t) = \varphi_c(r) \cos(\omega t) + \varphi_s(r) \sin(\omega t) \tag{7.102}$$

satisfies the Helmholtz equation:

$$\left(\frac{\partial^2}{\partial x^2} + \frac{\partial^2}{\partial y^2}\right) \varphi_{c,s}(x, y) + k^2 \varphi_{c,s}(x, y) = 0 \tag{7.103}$$

where $k^2 = \omega^2/v_s^2$ and v_s is the velocity of the surface waves; its dependence on direction is disregarded. It follows from Eq. (7.103) that the solution of Laplace's equation corresponds to the profile $\exp\{-kz\}$ versus z-coordinate. Therefore, the expression for the contrast in the first order in φ is:

$$\frac{j(x, y, t)}{j_0} \approx 1 + t_{\text{eff}} \left(\frac{2\pi e}{mE_0}\right)^{1/2} k^{3/2} e^{-kz_0} \varphi(x, y, z = 0, t) \tag{7.104}$$

[Compare with Eqs. (7.64) and (7.65)].

Thus, the general two-dimensional undulating fields satisfying the Helmholtz equation (7.103) produce on the screen of a stroboscopic MEMSI a contrast proportional in the first order to the value $\varphi(r, z = 0, t)$ on the surface of the specimen. Because of its simplicity, Eq. (7.104) provides the solution of both the direct and inverse problems. The equation makes it possible to measure the amplitude of the piezopotential of the wave on the surface of a crystal (Fig. 7.23). The stroboscopic MEMSI had also been used for the investigation of the surface and bulk waves in the specimens irradiated by protons (Spivak *et al.*, 1972).

For the conventional (nonstroboscopic) mode of operation the linear terms in the contrast disappear. Calculation of the contrast with Eqs. (7.101) and (7.101a) and averaging over time lead to the following results. The traveling waves do not produce any contrast in the conventional mode (Fig. 7.24), and the standing waves produce a contrast with a spatial period λ' which is half the period λ of the microfield: $\lambda' = \lambda/2$. For example, for the standing wave

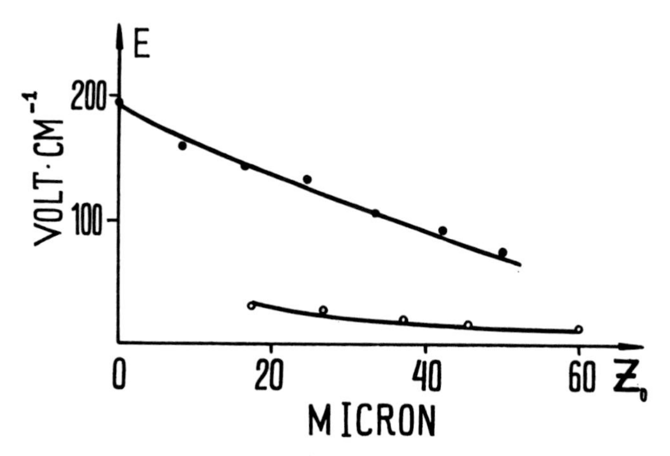

Fig. 7.23 Dependence of the field strength of the surface piezoelectric waves on the height over the surface of the specimen. Dots correspond to field strength distribution in a standing wave ($LiNbO_3$ delay line), and open circles correspond to traveling wave (quartz delay line). (*Gvosdover* et al., *1970.*)

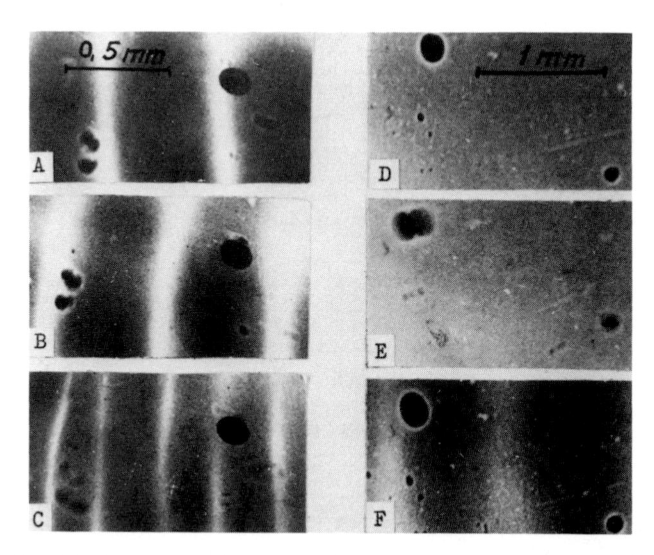

Fig. 7.24 The MEMSI micrographs of the surface waves on delay lines. (A,B,C) $LiNbO_3$ delay line; (D,E,F) quartz delay line; (A,B) stroboscopic images of the standing wave, where phase shift between images A and B is π; (C) the conventional mode image of the standing wave, with spatial period in the conventional mode image equal to half the period of the field; (D) microtopography of the surface; (E) image in conventional mode of operation with propagating traveling wave; (F) stroboscopic image of traveling wave. Magnification 30X (A,B,C); 18X (D,E,F). (*Gvosdover* et al., *1970.*)

$\varphi(x, t) = \varphi_1 \cos \omega t \cdot \cos kx$ the contrast in the stroboscopic and conventional modes equals, respectively (Gvosdover *et al.*, 1970):

$$j^{\mathrm{str}}(x', t) \approx j_0 [1 + C \cos(\omega t) \cos(kx')],$$
$$j^{\mathrm{conv}}(x') \approx j_0 [1 + 0.5\, C^2 \cos(2kx')] \qquad (7.105)$$

where $C = 3 E_0^{-1} (\pi l)^{1/2} k^{3/2} e^{-kz_0} \varphi_1$. The expressions (7.105) show that the use of conventional mode for $C \ll 1$ makes the contrast worse even for the standing waves (Fig. 7.24).

Equation (7.101) allows us to calculate the contrast of undulating microfield for the internal synchronous detection mode. This mode of operation is a modification of the stroboscopic one, but the illumination and observation in this mode is continuous. The image on the screen in this mode vibrates as a whole under the influence of harmonical signals of the frequency and the phase which are equal to those of the process under observation. Those signals must be applied to the deflection x and y plates (or coils), and their action may be described as additional displacement $\Delta \boldsymbol{s}$ which is independent of x and y:

$$\Delta \boldsymbol{s} = \{s_{xc}, s_{yc}\} \cdot \cos \omega t + \{s_{xs}, s_{ys}\} \sin \omega t \qquad (7.106)$$

In the absence of microfields, such displacements do not produce any contrast. In the presence of microfields of the type (7.102) and (7.103) and of the additional displacement (7.106), it is possible to observe the contrast of the first order in the microfield without strobing. For the calculation of such a contrast it is sufficient to use Eq. (7.101) and to take only the terms of the first order in the microfield after time averaging:

$$j(x', y')/j_0 = 1 + \frac{1}{2} t_{\mathrm{eff}} \left(\frac{2\pi e}{m E_0} \right)^{1/2} k^{3/2} e^{-kz_0} \cdot [(\boldsymbol{s}_c \cdot \boldsymbol{E}_c) + (\boldsymbol{s}_s \cdot \boldsymbol{E}_s)] \qquad (7.107)$$

where $\boldsymbol{E}_{c,s}$ is the undulating microfield on the surface of the specimen:

$$\boldsymbol{E}(x, y, z = 0, t) = \boldsymbol{E}_c(x, y) \cos \omega t + \boldsymbol{E}_s(x, y) \sin \omega t$$

Equation (7.107) shows that the contrast of the microfields [Eq. (7.103)] for the internal synchronous detection mode is proportional to the values of the microfield on the surface. Thus, for the image of such microfields in that mode of operation there is no need for complicated procedures for the solution of an inverse problem, even for a general two-dimensional case. As far as we know, such a mode of operation has not yet been realized experimentally.

QUANTUM EFFECTS IN MEM

The estimations from Eqs. (7.9)-(7.11) show that the lateral resolving power is limited by the wave mechanical effects in a different way for shadow projection and for focused imaging modes of the MEM. In the case of the shadow projection mode (MEMSI), the primary modulation of the monoenergetic electron beam introduces practically no limitations, and the quantum limit of resolving power is determined by the Fresnel diffraction of the beam on its way from the specimen to the "defocusing plane," i.e., to the screen. In the MEMFI with contrast aperture the quantum limit of resolving power is determined either by the Fraunhoffer diffraction on the aperture or by the effects of primary modulation of the beam. Only the second limit $\delta x \sim a = (\hbar^2/2meE_0)^{1/3}$ is inherent in the MEM, and for a given strength E_0 of the retarding field it cannot be reduced by any means.

The wave-mechanical problem of image contrast formation in the MEM was first brought up by Wiskott (1956b), where the above-mentioned extreme estimation $\delta x \sim a$ was obtained. Further study of quantum effects in the MEM was done by Bok (1968), Hermans and Petterson (1970), Petrov *et al.* (1970a,b), Bok *et al.* (1971), Lichte *et al.* (1972), Lenz (1972), and Gvosdover and Zel'dovich (1973).

In this section the quantum-mechanical formulae will be presented which describe the MEM in both modes of operation and which take all the above-mentioned effects into account; however, the aberrations will be disregarded. In the next paragraph will be presented a brief description of the general form of the wave function of electrons close to and far from the specimen. Far from the specimen this wave function satisfies the paraxial wave equation which is solved by means of the integral of Huygens-Fresnel type. Close to the specimen the wave function may be represented as a sum of incident and reflected waves except a very small area $|z - z_0| \lesssim (\hbar^2/2meE_0)^{1/3}$ in the vicinity of the reversal point.

The problem of the primary modulation is reduced to the calculation of the relation between the incident and the reflected waves; that problem will be solved later. The beam demodulation process is described by applying the general formula for the integral of Huygens-Fresnel type. The consideration of the quantum limit of resolving power of the MEMSI has been given earlier; p. 275 contains the same for the MEMFI. Finally, the description on p. 307 is devoted to the mirror electron interferometer (Lichte *et al.*, 1972), as well as the consideration of the resolving power of the MEM over the vertical z-coordinate.

Representation of Electron Wave Function

The monochromatic component (with the energy ϵ) of the electron beam may be characterized by the wave function $\Psi(r, z)$ which satisfies the stationary Schrödinger equation:

$$\left(\frac{\partial^2}{\partial x^2} + \frac{\partial^2}{\partial y^2} + \frac{\partial^2}{\partial z^2}\right)\Psi + \frac{2m(eU + \epsilon)}{\hbar^2}\Psi = 0 \tag{7.108}$$

The properties of the solution of that equation are quite different in different spatial areas of the MEM and are treated separately here. The potential $U(\mathbf{r}, z)$ close to the specimen may be represented as a sum of the potential of the uniform retarding field $U^{(0)}(z) = E_0 z - U_B$ and of the potential of perturbation $\varphi(\mathbf{r}, z)$.

Neglecting perturbation φ the equation (7.108) close to the specimen may be written as follows:

$$\left(\frac{\partial^2}{\partial x^2} + \frac{\partial^2}{\partial y^2} + \frac{\partial^2}{\partial z^2}\right) \Psi + \frac{(z - z_1)}{a^3} \Psi = 0 \tag{7.109a}$$

where z_1 is the height of the classical reversal point for a vertically incident electron with energy ϵ, i.e., $z_1 = (eU_B - \epsilon)/eE_0$. Parameter a,

$$a = (\hbar^2 / 2meE_0)^{1/3} \tag{7.109b}$$

has the dimension of length and characterizes the depth of penetration of an electron into the classically forbidden region in a uniform field with the strength E_0.

The solution of the equation (7.109a) describing the electron incident on the specimen with the transverse momentum $\mathbf{p}_0 = \hbar \mathbf{k}_0$ is (Landau and Lifschitz, 1963):

$$\Psi(\mathbf{r}, z) = \text{const} \cdot \Phi_A \left(\frac{z_0 - z}{a}\right) e^{i\mathbf{k}_0 \mathbf{r}} \tag{7.110}$$

Here $\Phi_A(x)$ is the Airy function:

$$\Phi_A(x) = (\pi)^{-1/2} \int_0^\infty \cos\left(ux + \frac{u^3}{3}\right) du \tag{7.111a}$$

and $z_0 = z_1 + p_0^2/2meE_0$ is the classical height of reflection of an electron with energy ϵ and momentum $\mathbf{p}_0$; see Eq. (7.14).

The asymptotic behavior of the solution [Eq. (7.110)] at $z - z_0 \gg a$ is of special interest; it corresponds to the sum of two terms:

$$\Psi(\mathbf{r}, z) = \text{const} \cdot \frac{1}{2} \left(\frac{z - z_0}{a}\right)^{-1/4} e^{i\frac{\pi}{4} + i\mathbf{k}_0 \mathbf{r}} \left[e^{-i\frac{2}{3}\left(\frac{z-z_0}{a}\right)^{3/2}} + e^{-i\frac{\pi}{2} + i\frac{2}{3}\left(\frac{z-z_0}{a}\right)^{3/2}} \right] \tag{7.111b}$$

where the first term stands for the incident wave and the second term for the reflected wave. Representation (7.111b) is a quasiclassical (WKB) asymptotic form of the solution of a Schrödinger equation (7.108). A WKB-approximation will be used for the description of the motion along the z-coordinate everywhere except the small vicinity $|z - z_0| \sim a$ of the point z_0.

It is clear that the incident and the reflected waves will still exist in the solution of Eq. (7.108) if the microfields are taken into account. However, in this case for the incident wave with a definite value of the "input" momentum $\boldsymbol{p}_0 = \hbar \boldsymbol{k}_0$ the reflected wave contains a number of momenta $\boldsymbol{p}' = \boldsymbol{p}_0 + \hbar \boldsymbol{k}$, where $\hbar \boldsymbol{k}$ is the change of momentum caused by the action of the microfield.

The description of the motion of electrons far from the specimen, i.e., in the optical system, calls for a consideration of Eq. (7.108) with the axially-symmetrical potential

$$U^{(0)}(z, |r|) \approx U^{(0)}(z, r = 0) - \frac{1}{4} r^2 \frac{\partial^2 U^{(0)}(z, 0)}{\partial z^2}$$

The quasiclassical WKB-approximation may be used to describe the motion along the axis (zeroth paraxial approximation). For this purpose it is useful to introduce "time" t calculated along the classical axial trajectory by the relation $p_z^{(0)}(z)\, dt = m dz$ similar to that in paraxial geometrical optics, where $p_z^{(0)}(z) = \pm\sqrt{2meU^{(0)}(z, 0)}$. In that approximation the following equation can be written:

$$\Psi(z, t) = (p_z^{(0)}(t))^{-1/2} \exp\left\{ i\hbar^{-1} \int^t p_z^{(0)}(t) \frac{dz}{dt}\, dt \right\} \Psi_1(t, \boldsymbol{r}) \quad (7.112)$$

The representation (7.112) of the wave function allows us to unite in a natural way the description of the incident and the reflected waves.*

Function $\Psi_1(t, \boldsymbol{r})$ describes the evolution of the transverse structure of the beam in the process of motion. Substituting Eq. (7.112) in Eq. (7.108) and neglecting second derivatives of $\Psi_1(t, \boldsymbol{r})$ over the "time" t, we obtain the equation for $\Psi_1(t, \boldsymbol{r})$:

$$i\hbar \frac{\partial \Psi_1}{\partial t} = -\frac{\hbar^2}{2m} \Delta_\perp \Psi_1 - e\left[\varphi(\boldsymbol{r}, z(t)) - \frac{1}{4} r^2 \frac{\partial^2 U^{(0)}}{\partial z^2} \right] \Psi_1 \quad (7.113)$$

where $\Delta_\perp = \partial^2/\partial x^2 + \partial^2/\partial y^2$. Equation (7.113) is equivalent to the nonstationary Schrödinger equation for the electron motion in two-dimensional space (x, y) in the presence of nonstationary potential

$$\left[\varphi(\boldsymbol{r}, z(t)) - \frac{1}{4} r^2 \frac{\partial^2 U^{(0)}}{\partial z^2} \right]$$

In the absence of perturbations $\varphi(\boldsymbol{r}, z)$, the solution of the paraxial equation in the plane $z = z_2 = z(t_2)$ (i.e., at the moment $t = t_2$) may be expressed by the values of the wave function at $t = t_1$ by means of the integral of the Huygens-Fresnel type (see also § 160 in the book by Glaser, 1952):

*The representation (7.112) fails to be correct in the vicinity of the reversal point $t = 0$, where $p_z(t) = eE_0 t$ falls to zero. However, at $z - z_0 \gg a$ the quasiclassical factor in Eq. (7.112) gives a correct description of the wave function including the phase factor: $(|t|)^{1/2} = \exp\{-i(\pi/2)\}(-|t|)^{1/2}$.

$$\Psi_1(r_2, t_2) = \frac{mB}{2\pi i\hbar} \int d^2 r_1 \, \Psi_1(r_1, t_1) \exp\left\{ i \frac{m}{2\hbar} (Ar_2^2 - 2Br_1 r_2 + Cr_1^2) \right\} \tag{7.114a}$$

In this integral, the constants A, B, and C of the dimension of inverse time may be expressed through the fundamental solutions $g_1(t_1, t_2)$ and $g_2(t_1, t_2)$ which describe the motion of an electron from the plane $z = z_1 = z(t_1)$ to the plane $z = z_2 = z(t_2)$ according to the laws of paraxial geometrical optics; compare with Eqs. (7.15) and (7.16). The expressions for A, B, and C are as follows:

$$A = g_2^{-1} \dot{g}_2, \quad B = g_2^{-1}, \quad C = g_2^{-1} g_1 \tag{7.114b}$$

where the point means the derivative over t_2.

Analysis made specially for the MEM by Gvosdover and Zel'dovich (1973) reveals that Eqs. (7.114a, b) are also applicable for describing the imaging of the wave function reflected from the specimen, i.e., for $z_1 = z_0$, $t_1 = 0$. In this case the value of the asymptotic coefficient in the "reflected part" of the wave function over the specimen should be taken as $\Psi_1(z_1, r_1)$, and the fundamental solutions $g_1(t)$ and $g_2(t)$ from Eqs. (7.15) and (7.16) should stand for $g_1(t_2, 0)$ and $g_2(t_2, 0)$. Modulation of the "reflected part" of the wave function by the microfields will be treated in the next section.

Primary Modulation of Wave Function by Microfields

The calculation of primary modulation of the electron beam will be provided in several stages each yielding the formulae with a different degree of accuracy (and respectively with a different degree of complexity).

Primary modulation may be calculated most easily for the vertically incident electron beam (zero paraxial approximation in the problem of primary modulation) subject to the action of electrical microfields. If the transverse size of a microfield δx (which coincides with the longitudinal size) is considerably greater than the "quantum" parameter $a = (\hbar^2/2meE_0)^{1/3}$, the primary modulation itself may be calculated with the help of the quasiclassical WKB-approximation. This is possible since the representation (7.112) for the wave function and the very quasiclassical equation (7.113) are not valid only within a small vicinity of the reversal point at $|z - z_0| \lesssim a$.

For the large scale microfields with $\delta x \sim \delta z \gg a$ one may neglect the contribution of the vicinity $|z - z_0| \sim a$ in comparison with the contribution of all the rest of the space occupied by the microfield. In this case in Eq. (7.113) $\partial^2 U^{(0)}/\partial z^2 = 0$ close to the specimen. Besides the term $\Delta_\perp \Psi_1$ in Eq. (7.113) may be neglected, and then it follows from Eq. (7.113) that:

$$\Psi_1(t_2, r) = \Psi_1(t_1, r) \exp\left\{ i\hbar^{-1} e \int_{t_1}^{t_2} \varphi(r, z(t)) \, dt \right\} \tag{7.115}$$

Here $t_1 < 0$ is the moment of time before the electron enters the microfield, $t_2 > 0$ is the moment after the electron leaves the microfield, and the dependence $z(t)$ is given by a classical equation (7.13).

Since the microfield decreases rapidly depending on the height (at a distance of $\delta z \sim \delta x$), the integral $\int \varphi \, dt$ may be taken at the interval $-\infty < t < +\infty$. In this case the phase from Eq. (7.115) equals the change of classical action $\Delta\Phi(\boldsymbol{r})$ due to the microfield [compare with Eq. (7.41)] divided by $\hbar$:

$$\Delta(\arg \Psi_1) = \hbar^{-1} \Delta\Phi(\boldsymbol{r}) = -\hbar^{-1} \int_{-\infty}^{+\infty} V(\boldsymbol{r}, z(t)) \, dt$$

$$= \hbar^{-1} \left(\frac{2me}{E_0} \right)^{1/2} \int_{z_0}^{\infty} \frac{\varphi(\boldsymbol{r}, z) \, dz}{\sqrt{z - z_0}} \tag{7.116}$$

The equation of the type of Eq. (7.116) and the treatment of primary modulation of the electron beam in the MEM as of phase modulation was first given by Hermans and Petterson (1970). The expressions (7.115) and (7.116) are already sufficient for the solution of a large number of problems in quantum theory of the MEM.

A more accurate (but still quasiclassical) consideration of the primary modulation must take into account the nonzero value of the transverse momentum $\boldsymbol{p}_0 = \hbar \boldsymbol{k}_0$ for the electrons incident upon the specimen. Detailed calculations yield the following expression for the "reflected part" of the wave function:

$$\Psi_{\text{ref}}(\boldsymbol{r}) = \text{const} \cdot \exp\left\{ i \frac{(\boldsymbol{p}_0 \cdot \boldsymbol{r})}{\hbar} + \frac{i}{\hbar} \Delta\Phi(\boldsymbol{r}, \boldsymbol{p}_0, \epsilon) \right\} \tag{7.117}$$

This formula is valid in the case of incident plane wave with momentum $\boldsymbol{p}_0$ for a general case of the action of electrical and magnetic fields. Here $\Delta\Phi(\boldsymbol{r}, \boldsymbol{p}_0, \epsilon)$ is the change of the classical action from Eq. (7.41) calculated for a definite value of $\boldsymbol{p}_0$.

Expressions (7.115)-(7.117) for the primary modulation reveal a close correspondence between quasiclassical and classical calculations. The quantum modulation of the phase of the beam with unchanged amplitude of the wave function corresponds to the classical modulation of the velocity of the beam with unchanged density. In that sense the assertion that at the stage of primary modulation there occur no essential differences between the classical and quantum approaches for microfields with size $\delta x \gg a$ is true. It follows from the earlier estimation that the sizes of microfields which are effectively imaged in the MEMSI are definitely much greater than the quantum parameter a. Therefore, the quantum (i.e., the diffraction) effects in the MEMSI may appear only at the stage of demodulation while the beam moves from the specimen to the screen.

Apart from being simple and clear, the quasiclassical description of primary modulation has one more important advantage, namely, the possibility of regarding the expression for the modulating phase factor (7.115) and (7.117) in case $\hbar^{-1}\Delta\Phi \gg 1$ as an infinite sum of the series of quantum-mechanical theory of perturbation:

$$\exp\{i\Delta\Phi/\hbar\} = 1 + (i\Delta\Phi/\hbar) + \frac{1}{2!}(i\Delta\Phi/\hbar)^2 + \cdots \tag{7.118}$$

where $\Delta\Phi$ is of the first order in amplitude of microfields. Therefore, when $\hbar^{-1}\Delta\Phi \gg 1$ the expressions of types (7.115) and (7.117) allow us to describe the contrast of the first order in microfield as well as a strong contrast.

Devising a quantum theory of the MEMFI requires an ability to calculate primary modulation of the wave function by a microfield of very small transverse size $\delta x \sim a$. In this case the quasiclassical WKB-approximation is definitely inapplicable, and therefore the Eq. (7.108) and wave functions of the type (7.110) should be resorted to. For the microfield with the size $\delta x \sim a$, an analytical solution of the problem of primary modulation had been obtained but only up to the first order of the quantum-mechanical perturbation theory. We shall formulate the final result, omitting the details of the calculations. For a given perturbation $\tilde{\varphi}(\boldsymbol{k})$ or $\tilde{\boldsymbol{B}}(\boldsymbol{k})$ with the wave vector $\boldsymbol{k}$ the strict quantum-mechanical solution of the problem leads to the multiplication of the quasiclassical expression of the amplitude of modulation of the first order by a factor:

$$\exp\left\{-2a^3|\boldsymbol{k}|(\boldsymbol{k}\cdot\boldsymbol{k}_0) - \frac{2}{3}|\boldsymbol{k}|^3a^3\right\} \tag{7.119}$$

A summary of all the results [Eqs. (7.115)-(7.119)] reveals that the reflected part of the wave function close to the specimen is modified by microfields in the following way:

$$\Psi_{\text{ref}}(\boldsymbol{r}, z=0) = \text{const}\cdot\exp\left\{i\frac{\boldsymbol{p}_0\cdot\boldsymbol{r}}{\hbar} + \frac{i}{\hbar}\Delta\Phi_q(\boldsymbol{r},\boldsymbol{p}_0,\epsilon)\right\} \tag{7.120a}$$

Here $\Delta\Phi_q$ is the change of the action corrected with the consideration of quantum effects:

$$\hbar^{-1}\Delta\Phi_q(\boldsymbol{r},\boldsymbol{p}_0,\epsilon) = \int d^2\boldsymbol{k}e^{i\boldsymbol{k}\cdot\boldsymbol{r}}\left[\tilde{\varphi}(\boldsymbol{k}) + i\frac{(k_xp_{0y} - k_yp_{0x})}{m|\boldsymbol{k}|^2}\tilde{B}_z(\boldsymbol{k})\right] \cdot R_q(\boldsymbol{k},\hbar^{-1}\boldsymbol{p}_0,\epsilon) \tag{7.120b}$$

In Eq. (7.120b) $R_q(\boldsymbol{k}, \boldsymbol{k}_0, \epsilon)$ is the spectral (i.e., dependent on the wave vector $\boldsymbol{k}$ of the microfield) coefficient of transformation of the potential of microfield $\varphi(\boldsymbol{r}, z=0)$ into the primary modulation of the phase $\hbar^{-1}\Delta\Phi_q$. This coefficient has the dimension volt^{-1} and equals

$$R_q(\boldsymbol{k}, \boldsymbol{k}_0, \epsilon) = \hbar^{-1} \left(\frac{2\pi me}{E_0}\right)^{1/2} |\boldsymbol{k}|^{-1/2} \exp\left\{-|\boldsymbol{k}|z_0 - \frac{a^3(\boldsymbol{k}\cdot\boldsymbol{k}_0)^2}{|\boldsymbol{k}|} - 2a^3|\boldsymbol{k}|(\boldsymbol{k}\cdot\boldsymbol{k}_0) - \frac{2}{3}a^3|\boldsymbol{k}|^3\right\} \quad (7.120c)$$

where:

$$z_0 = \frac{eU_B - \epsilon}{eE_0} + \frac{(\hbar \boldsymbol{k}_0)^2}{2meE_0} \quad (7.120d)$$

The results [Eqs. (7.119)-(7.120)] are valid also for the magnetic microfields (the deduction of that is omitted) and that point is reflected in Eq. (7.120b). However, it should be remembered that for the components of microfields $\tilde{\varphi}(\boldsymbol{k})$ and $\tilde{\boldsymbol{B}}(\boldsymbol{k})$ with $|\boldsymbol{k}| \sim a^{-1}$ the expression (7.120b) may be used only in the form of linear expansion:

$$\Psi_{\text{ref}} = e^{i\boldsymbol{k}_0\cdot\boldsymbol{r}}(1 + i\hbar^{-1}\Delta\Phi_q(\boldsymbol{r}, \hbar\boldsymbol{k}_0, \epsilon)) \quad (7.120e)$$

and all the subsequent calculations are valid only up to the first order in amplitude of perturbing microfield. It is worth noting that the use of the expression for the strength of the microfield $E_1 \sim |\boldsymbol{k}|\tilde{\varphi}(\boldsymbol{k})$ allows us to obtain the following estimation:

$$\hbar^{-1}\Delta\Phi_q \approx \sqrt{\pi}\,\frac{E_1}{E_0}(ka)^{-3/2} \quad (7.121)$$

where all exponential factors have been omitted. Estimation (7.121) shows that even a weak microfield ($E_1 \ll E_0$) may produce strong primary modulation in cases of perturbations of the size $\delta x \sim k^{-1} \gg a$.

The results obtained in this section may be illustrated by a schematic curve of $|\boldsymbol{k}|$-dependence of the spectral coefficient $R_q(\boldsymbol{k}, \hbar^{-1}\boldsymbol{p}_0, \epsilon)$ which connects the primary phase modulation with the potential of the microfield according to Eqs. (7.120) (Fig. 7.25). Figure 7.25 is out of scale and proportion.

The basic part of the curve is the solid line 1 corresponding to the law $\Delta\Phi \backsim \tilde{\varphi}\cdot|\boldsymbol{k}|^{-1/2}$, which is connected with the simple dependence of the transit time t_{tr} on the size of the microfield: $t_{\text{tr}} \backsim (\delta x)^{1/2} \sim |\boldsymbol{k}|^{-1/2}$. The first turn of the curve is the dashed line 2 representing the decrease of microfield depending on the height z_0 according to the law $\exp\{-|\boldsymbol{k}|z_0\}$. The height of reflection z_0 for the vertically incident electrons is determined either by the specially applied bias voltage ($z_0 = U_B/E_0$) or by the energy spread of electrons ($z_0 \sim \delta\epsilon/eE_0$). A decrease of U_B and $\delta\epsilon$ will result in a shift of that turn of the curve to the right. This turn of the curve is responsible for a strong chromatic dependence of the primary modulation by the microfield of the size $\delta x \lesssim \delta\epsilon/eE_0$. This point is a specific feature of the MEM; in the transmission electron

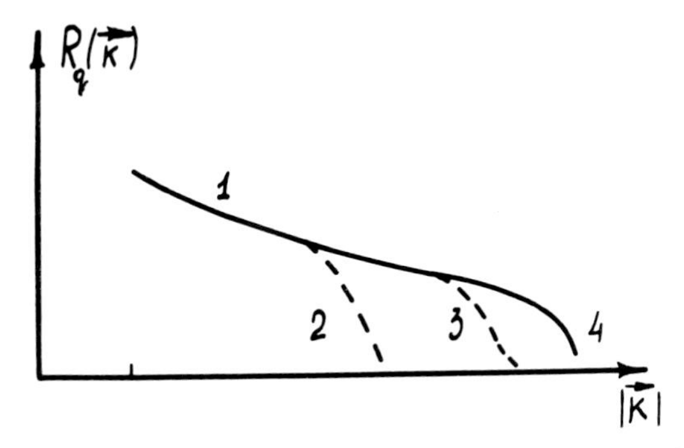

Fig. 7.25 Schematic graph of spectral coefficient of primary modulation in the MEM. The graph is out of scale and proportion. See explanations of parts 1, 2, 3, 4 in the text.

microscope the variations of energy due to the energy spread cause insignificant changes of the primary modulation.

Another turn of the curve is the dashed curve 3 which is connected with the lack of verticality for the unperturbed trajectory (classical effects of sliding past the microfield and the raise of height of reflection both appearing at $\boldsymbol{p}_0 \neq 0$). For the MEMSI there are no such effects close to the electrical center of the image, and for the suppression of these effects in the MEMFI the angular width of the illumination beam should be diminished.

The last high frequency turn of the curve is the solid line 4 which is situated at $|\boldsymbol{k}| \sim a^{-1} = (\hbar^2/2meE_0)^{-1/3}$. It is connected with the quantum nature of the electron and could be shifted (though not very much) to the larger values of $|\boldsymbol{k}|$ only by the increase of the strength of the field E_0; however, in this case the curve 1 as a whole would be diminished $\backsim |E_0|^{-1/2}$. The low frequency boundary of the curve ($|\boldsymbol{k}|_{\min} \sim l^{-1}$) is simply due to the fact that the size of the imaged field on the specimen in the MEM is definitely smaller than the distance l between the specimen and the nearest electrode (aperture).* The schematic curve in Fig. 7.25 together with the Eqs. (7.12) allow one to make a judgment about the potential of the MEM for the diagnostics of microfields.

Contrast and Quantum Limit of Resolving Power With Shadow Imaging

The process of demodulation in the MEMSI (i.e., the process of transformation of the primary phase modulation into an amplitude one) is quite similar to that of the phase image in a microscope of any other type operating in shadow projection imaging mode. In this discussion of the specific diffraction effects the spread of electrons over the momenta $\boldsymbol{p}_0$ and over the energy $\delta\epsilon$ will be disregarded. In this case the wave function of an electron may be assumed to be

*Probably a much stronger upper limit for the size of the imaged field must be imposed because of the aberrations.

coherent over the whole specimen.* A substitution of $\epsilon = \text{const} = \epsilon_0$, $\boldsymbol{p}_0 = mD_{sp}\boldsymbol{r}_0$ into the expression for the change of classical action $\Delta\Phi(\boldsymbol{r}_0, \boldsymbol{p}_0, \epsilon)$ should be made. In this case the modulating phase factor depends on $\boldsymbol{r}$ only:

$$\Psi_{\text{ref}}(\boldsymbol{r}) = \Psi_{\text{inc}}(\boldsymbol{r}) \exp\{i\hbar^{-1}\Delta\Phi(\boldsymbol{r})\} \tag{7.122}$$

where $\Delta\Phi(\boldsymbol{r}) \equiv \Delta\Phi(\boldsymbol{r}, mD_{sp}\boldsymbol{r}, \epsilon)$ and the function $\Psi_{\text{inc}}(\boldsymbol{r})$ equals

$$\Psi_{\text{inc}}(\boldsymbol{r}) = \text{const} \cdot \exp\left\{i \frac{m}{2\hbar} D_{sp}\boldsymbol{r}^2\right\} \tag{7.123}$$

The wave function of the type (7.123) corresponds to the classical relation $\boldsymbol{p} = mD_{sp}\boldsymbol{r}$ which is characteristic of the illumination in a shadow imaging mode.

If the size of the microfield is much smaller than the distance from the specimen to the nearest electrode, it may be assumed that the primary modulation occurs in the vicinity of the surface of the specimen. In this case the imaging to the screen may be calculated with the help of the Huygens-Fresnel integral [Eq. (7.114a)], where it is assumed that $t_1 = 0$, $t_2 = T_{scr}$. It is useful to introduce coordinates $\boldsymbol{r}' = \boldsymbol{R}_{scr}/M$, where $M = g_0(T_{scr})$ is the magnification of the MEMSI. Substitution of Eqs. (7.122) and (7.123) into Eq. (7.114) yields the value of the wave function at the point $\boldsymbol{R}_{scr} = M\boldsymbol{r}'$ of the screen:

$$\Psi(\boldsymbol{r}') = \frac{1}{4\pi i b^2} \int d^2\boldsymbol{r} \exp\left\{i\frac{(\boldsymbol{r} - \boldsymbol{r}')^2}{4b^2} + i\frac{\Delta\Phi}{\hbar}\right\}. \tag{7.124}$$

In Eq. (7.124) we have omitted the phase factor of the type $\exp\{iA\boldsymbol{r}'^2\}$ which does not change the contrast. The normalization coefficient in Eq. (7.124) is chosen in such a way that the current density on the screen is given by $j(\boldsymbol{r}') = j_0|\Psi(\boldsymbol{r}')|^2$. In Eq. (7.124) we have introduced the parameter b with dimensions of length:

$$b = \left(\frac{\hbar t_{\text{eff}}}{2m}\right)^{1/2} = \left(\frac{9\hbar^2 l}{8meE_0}\right)^{1/4} \tag{7.125}$$

where t_{eff} characterizes the effective time during which the velocity modulation is transformed in the MEMSI into the density modulation (see Eqs. (7.26), (7.27)). The second equality in Eq. (7.125) is written particularly for a MEMSI with a two-electrode objective system. Calculation of integral (7.124) in the stationary phase approximation yields the classical expression [Eq. (7.47)] for the current density on the screen.

The integral (7.124) with the first of the equalities (7.125) describes the deformation of the wave packet (of the phase-modulated wave function $\exp\{i\Delta\Phi(\boldsymbol{r})/\hbar\}$) in the free two-dimensional space during the time t_{eff}. For the

*Complete transverse coherence of the beam is assumed here to simplify the subsequent calculations. In fact for the validity of the results of this paragraph much weaker restrictions on the radius of coherence of the beam are necessary: $r_{coh} \gg b = (\hbar t_{\text{eff}}/2m)^{1/2}$.

MEMSI with a two-electrode objective system the integral (7.124) may also have the following interpretation. Let us introduce the wavelength $\lambda_0 = 2\pi\hbar/\sqrt{2meU_0}$, where U_0 is the accelerating voltage. Then the integral (7.124) describes the diffraction of the plane wave with wavelength λ_0 on the phase specimen $\hbar^{-1}\Delta\Phi(r)$ with subsequent propagation at a distance $z_2 = 3l$ from the specimen. It is clear from the interpretation that the resolving power in such a projection mode is about $\delta x \sim (\lambda_0 z_2)^{1/2} \sim (\lambda_0 l)^{1/2}$ (the parameter b from Eq. (7.125) equals $0.49 \cdot (\lambda_0 l)^{1/2}$).

An analysis of weak contrast will reveal quantitative determination of the resolving power of the MEMSI. In the linear in $\Delta\Phi$ approximation it follows from Eq. (7.124):

$$K(r') = \frac{j(r') - j_0}{j_0} = -\frac{1}{2\pi b^2} \int \frac{\Delta\Phi(r)}{\hbar} \cos\left(\frac{(r - r')^2}{4b^2}\right) d^2 r \qquad (7.126)$$

This linearized expression is applicable for the calculation of the image contrast formation of both the electrical and magnetic microfields.

Let us consider the contrast due to electrical microfields close to the optical center of the image in more detail. Here the sliding effect and that of the increase of height of reflection may be disregarded and the classical action may be calculated according to Eq. (7.116). At this point the spectral transfer function $S(k) = \tilde{K}(k)/\tilde{\varphi}(k)$ may be introduced just as was done in the classical theory. Then, from Eq. (7.120) it follows that:

$$S_q(k) = R_q(k, 0, \epsilon_0) \cdot S_D(k) \equiv \hbar^{-1} \left(\frac{2\pi me}{E_0}\right)^{1/2} |k|^{-1/2} \exp\{-|k| z_0\} \cdot 2 \sin(k^2 b^2) \qquad (7.127)$$

The first factor $R_q(k, 0, \epsilon_0)$ (volt^{-1}) is the spectral coefficient of transformation of the amplitude of micropotential (volt) into the dimensionless magnitude $\hbar^{-1}\Delta\Phi$, being the phase due to the primary modulation of a vertically incident monochromatic beam (see Eq. (7.120)). This factor incorporates all the specific features of the MEM as opposed to the instruments of other types. The second factor $S_D(k) = 2\sin(k^2 b^2)$ is dimensionless and characterizes the spectral coefficient of transformation of the phase modulation into modulation of intensity of the beam resulting from the Fresnel diffraction. This factor depends on only one parameter of the dimension of length $b = (\hbar t_{\text{eff}}/2m)^{1/2}$ and is typical for any instrument with defocused (shadow projection) imaging of a phase specimen.

The curves of the spectral transfer function for different values of the ratio z_0/b are given in Fig. 7.26. For $|k| \lesssim b^{-1}$ the quantum expression (7.127) coincides with the classical formula Eq. (7.65). If, in addition, the high frequencies are suppressed by a factor $\exp\{-|k| z_0\}$ the results of both quantum and classical consideration coincide throughout the range of values of $|k|$, where

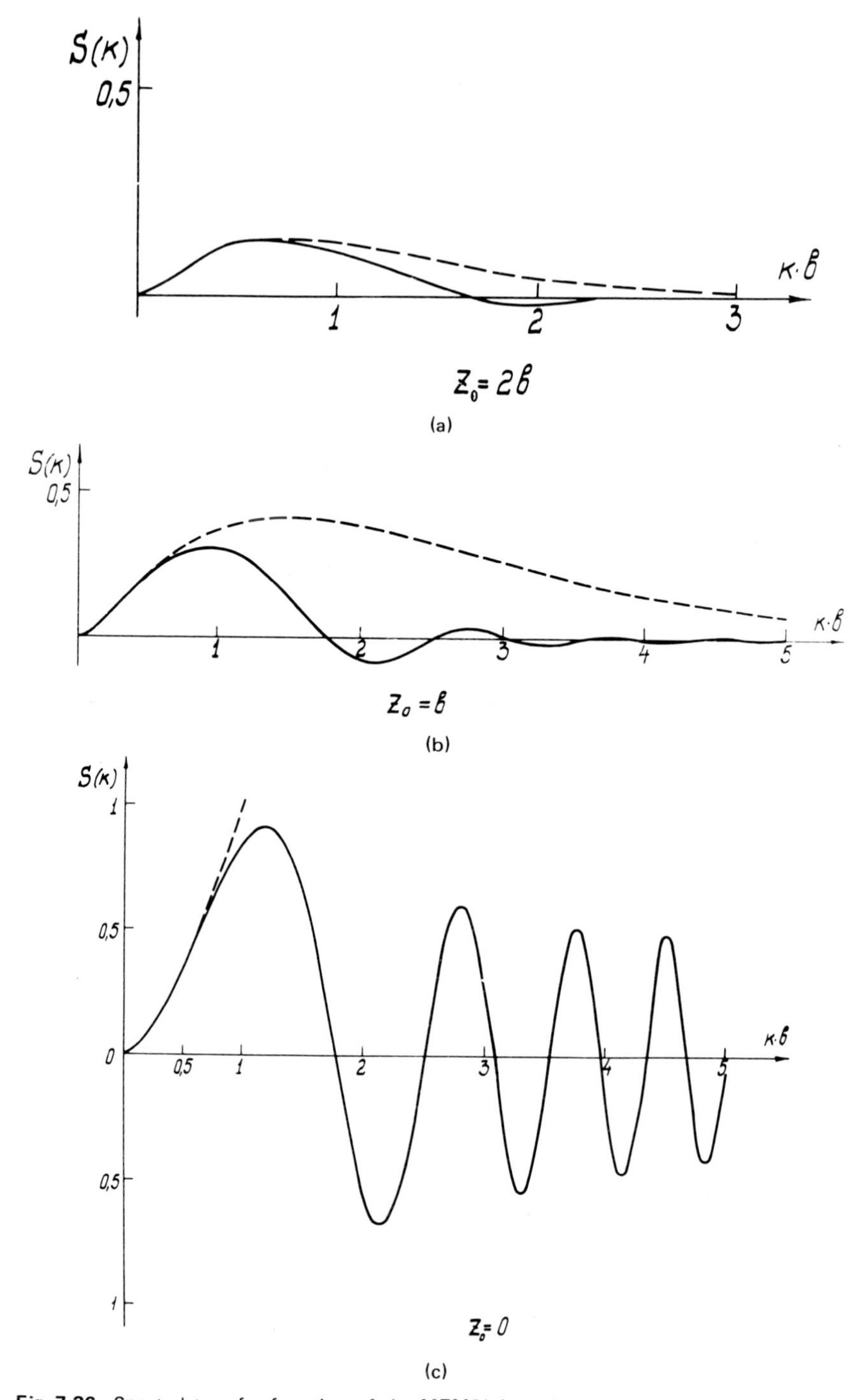

Fig. 7.26 Spectral transfer function of the MEMSI for different values of the height of reversal point z_0. Dashed curves correspond to the classical theory, and solid lines correspond to quantum theory. (a) $z_0 = 2b$; (b) $z_0 = b$; (c) $z_0 = 0$. (*Gvosdover and Zel'dovich, 1973.*)

$S(k)$ differs considerably from zero. This is the case for $z_0 \gtrsim 3b$ (Fig. 7.26a). When $z_0 \ll b$ and $|k| \gtrsim b$, the primary modulation causes a slow decrease $\backsim |k|^{-1/2}$, and the Fresnel diffractional demodulation causes rapid oscillations of the spectral transfer function (Fig. 7.26b,c). Thus, the effects of diffraction in weak contrast appear at $\delta x \sim k^{-1} \lesssim b$. In the case of strong contrast the effects of diffraction and interference may appear also at $\delta x > b$ (see below). The maximum of $S_q(k)$ (at $z_0 = 0$) corresponds to $k_{\text{opt}}|_{z_0=0} = 1.18b^{-1}$. In the general case $z_0 \neq 0$ the value of

$$k_{\text{opt}} \sim \min\,(b^{-1}, z_0^{-1})$$

defines approximately the resolving power in order of magnitude.

The function describing the response to a point source in the linear quantum problem is also of interest. The calculation for the one-dimensional case at $z_0 = 0$ yields*

$$G_q^{\text{I}}(x) = \left(\frac{me}{\hbar^2 E_0 b}\right)^{1/2} \pi^{1/2} \left(\frac{x}{b}\right)^{1/2} J_{-1/4}\left(\frac{x^2}{8b^2}\right) \sin\left(\frac{\pi}{8} - \frac{x^2}{8b^2}\right) \qquad (7.128)$$

where $J_{-1/4}$ is the Bessel function of index $-\frac{1}{4}$. Figure 7.27 represents the graph of this function. This function describes the result of the interference of the unperturbed wave and of the inhomogeneous cylindrical wave generated by a point source located at $x = 0, z = 0$. Figure 7.28 illustrates the analogy to the Fresnel diffraction on a phase specimen. However, the analogy is not complete, since according to Eq. (7.41) or Eq. (7.142) a point source of the type $\varphi(x, z = 0) = E_0 Q \delta(x)$ on the surface of the specimen leads to phase perturbations at $x \neq 0$ too.

The halfwidth of Eq. (7.128) at half the value at the point $x = 0$ equals:

$$\delta x_{z_0=0}^{\text{I}} = 1.25b = 1.25 \left(\frac{\hbar t_{\text{eff}}}{2m}\right)^{1/2} = 0.61\,(\lambda_0 l)^{1/2} \qquad (7.129)$$

[the last equality in Eq. (7.129) is given for a MEMSI with a two-electrode objective system]. In conformity with Gvosdover and Zel'dovich (1973) this value will be taken as a characteristic of the limit of resolving power of the MEMSI dependent on the diffraction effects; up to a coefficient ~ 1 the expression (7.129) coincides with the estimation in Eq. (7.9).

Strong contrast with diffraction taken into account is also of interest. For the sinusoidal microfield $\varphi = \varphi_1 \cos(kx)$ it is possible to obtain the expression for the contrast in the form of Fourier series:

$$j(x')/j_0 = 1 + 2 \sum_{n=1}^{\infty} J_n\left(\frac{C_{\text{cl}}}{k^2 b^2} \sin(nk^2 b^2)\right) \cos(nkx') \qquad (7.130)$$

*The expression for $G_q^{\text{I}}(x)$ at $z_0 \neq 0$ is rather cumbersome; see for reference Gvosdover and Zel'dovich (1973).

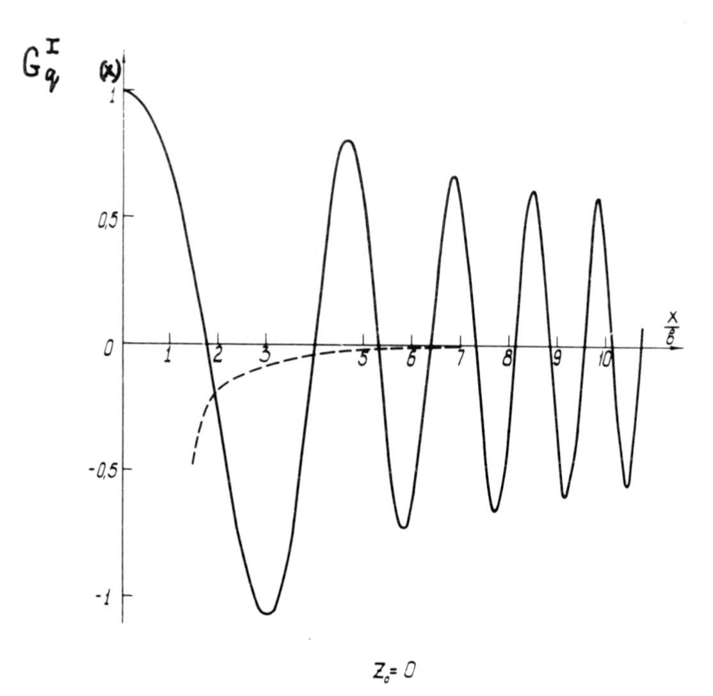

Fig. 7.27 Linear response to a point source for the one-dimensional quantum problem at $z_0 = 0$. Curve is normalized to the value "1" at $x = 0$ and dashed line is nonoscillating part of the response function corresponding to the classical case (Fig. 7.9) at $x \gg z_0$ (*Gvosdover and Zel'dovich, 1973.*)

The term C_{cl} in Eq. (7.130) is the classical parameter characterizing the strength of the contrast: $C_{cl} = \varphi_1 (2\pi e/mE_0)^{1/2} t_{eff} k^{3/2} \exp\{-kz_0\}$, for $C_{cl} \geqslant 1$ caustics appear on the image. The expression (7.130) has this advantage: its limit at $\hbar \to 0$ (or $b \to 0$) is just the classical expression, Eq. (7.61), for strong contrast. Figures 7.29a,b show the curves of contrast calculated from Eq. (7.130) for (a) $C_{cl} = 2$, $kb = 0.25$ and (b) $C_{cl} = 2$, $kb = 0.5$. In classical theory the infinite peaks of brightness should appear at this amplitude of microfield. The curves of current density (Fig. 7.29) clearly demonstrate how the peaks of brightness are smoothed because of diffraction effects.

Special attention should be paid to the additional maxima caused by the interference of the waves which come to the screen from different points of the specimen. Those interference maxima may produce a pattern with a very small-scale spatial structure. The period of this structure may be smaller than the size of perturbations on the specimen and even smaller than the limit of resolving power b obtained above. Therefore experimental investigation of resolving power of the MEMSI requires great care. Equation (7.130) for strong contrast in the MEMSI was obtained by Gvosdover and Zel'dovich (1973);

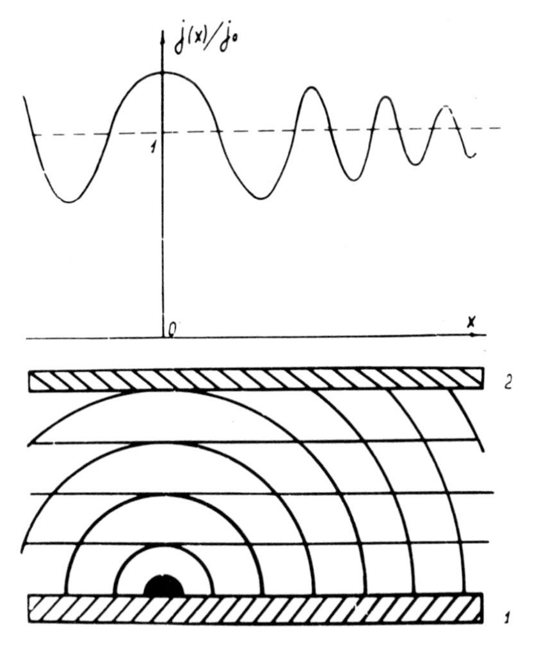

Fig. 7.28 Illustration to the interpretation of the response function to a point source as a result of interference of a plane unperturbed wave and cylindrical wave originating from the perturbation. 1, specimen; 2, equivalent location of the screen.

expressions of the type for another problem were first obtained by Warshalovich and Dyakonov (1971).

It follows from Eq. (7.129) that the diffraction limit of lateral resolving power of the MEMSI may be improved by diminishing t_{eff}, i.e., by changing optical parameters of the objective system. However, in that case the sensitivity of the MEM to microfields decreases. Loss of sensitivity to large-scale microfields is proportional to $t_{\text{eff}} \sim (\delta x_{\text{res}})^2$, which in itself is even an advantage since strong contrast caused by large-scale inhomogeneities hinders the observation of small-scale structures.

The minimum value of the amplitude of cosine microfield $\varphi_1 \cos kx$ producing contrast $|K| = 0.1$ is of interest. This minimum value may be obtained at $k = k_{\text{opt}} = 1.18\, b^{-1}$ and equals:

$$|\varphi_{1\,\text{min}}| = [10\, S\,(k_{\text{opt}})]^{-1} \approx 0.022 \left(\frac{\hbar^2 E_0}{meb}\right)^{1/2} \tag{7.131}$$

The value of φ_1 can be expressed in terms of perturbation of the height of the conducting specimen:

$$|h_{1\,\text{min}}| = |\varphi_{1\,\text{min}}|/E_0 \approx 5.10^{-3} \lambda_0^{3/4}\, l^{1/4} \tag{7.132}$$

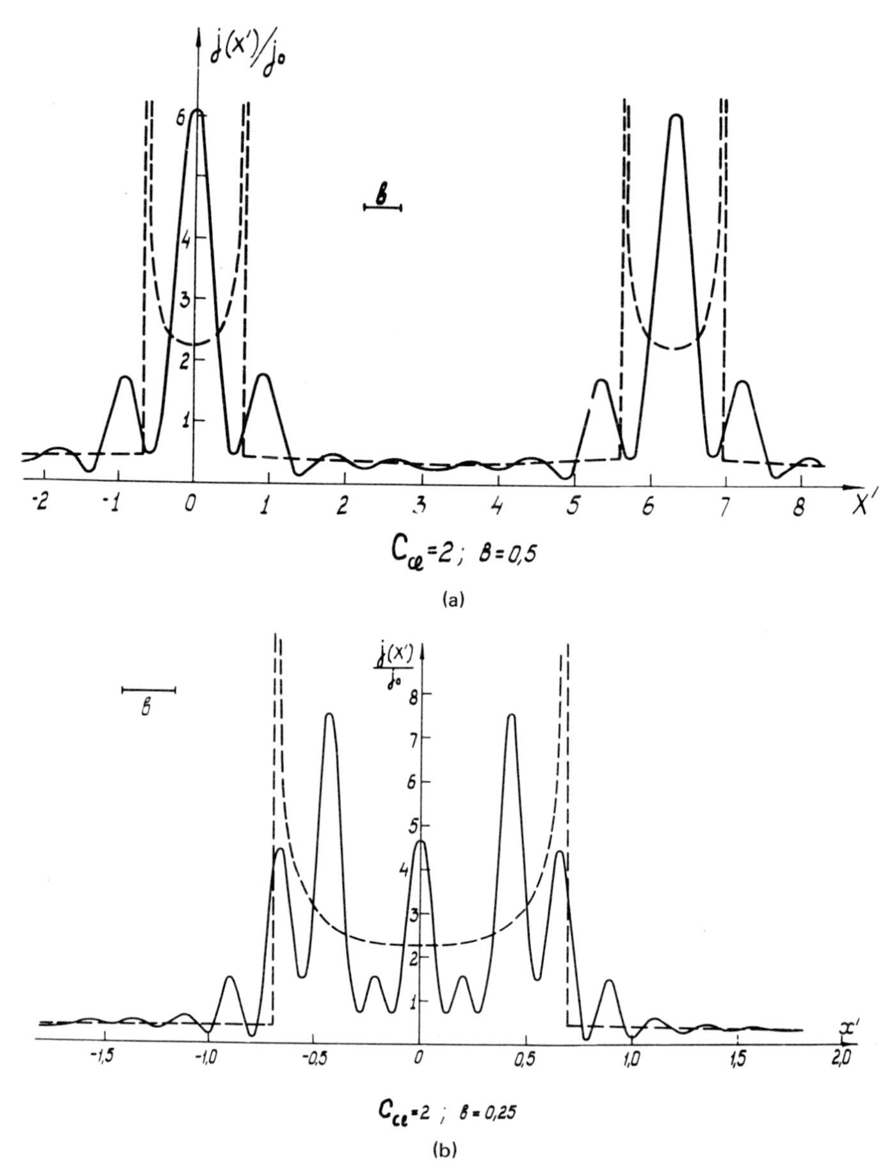

Fig. 7.29 Current density distribution due to the action of a sinusoidal microfield in the MEMSI. (a) $C = 2; b = 0.5; k = 1$. (b) $C = 2; b = 0.25; k = 1$. Dashed lines represent the current density distribution according to the classical theory (*Gvosdover and Zel'dovich, 1973.*)

[To make it more definite we have written down Eq. (7.132) for the case of an MEMSI with a two-electrode objective system].

A comparison of Eqs. (7.129) and (7.132) reveals an "anisotropy" of resolving power in the MEMSI: sensitivity to the vertical perturbations of the specimen structure is much better than the resolving power over the lateral coordinates. This statement is also valid in the case when resolving power and sensitivity are limited by pure classical effect, i.e., the energy spread of electrons. It also follows from Eq. (7.131) that improvement of the resolving power by diminishing t_{eff} leads to a decrease of sensitivity even to microfields with optimum spatial period ($S_{opt} \backsim (\delta x_{res})^{1/2}$ in accordance with curve 1 of Fig. 7.25).

Values V_{min} and Q_{min}, i.e., the volume and the cross section of the topography inhomogeneities of the surface of the specimen which cause contrast $|K| = 0.1$ (compare with classical calculation, Eqs. (7.70)) are equal to:

$$|Q_{min}| = [10 E_0 G_q^{I}(0)]^{-1} \approx 0.09 \left(\frac{b\hbar^2}{meE_0}\right)^{1/2},$$

$$|V_{min}| = [10 E_0 G_q^{II}(0)]^{-1} \approx 0.22 \left(\frac{b^3\hbar^2}{meE_0}\right)^{1/2} \qquad (7.133)$$

Let us make numerical estimations of the diffraction limit of the resolving power of the MEMSI. Substitution of the values $l \sim 5.10^{-3}$ m, $E_0 \sim 10^7$ V/m yields the following for a MEMSI with a two-electrode objective system:

$$\delta x \approx 1.25\, b \approx 100 \text{ nm}$$

A MEMSI having a resolving power of $\delta x \sim 80$ nm has been constructed by Ivanov and Abalmazova (1966). They used a five-electrode objective system with a value of parameter t_{eff} unknown to us; therefore, the coincidence of the numbers mentioned above may be accidental.

Contrast and Quantum Limit of Resolving Power With Focused Imaging

The wave-mechanical consideration of demodulation for the focused imaging mode of the MEM is based on the twofold use of Eq. (7.121); the plane of contrast aperture should be chosen as the plane for intermediate integration. The demodulation process itself in the MEMFI is quite similar to the demodulation of a phase image in a transmission electron microscope, and only the final formulae will be given here.

Up to unessential phase and constant factors the wave function in the coordinates $\boldsymbol{r}' = M^{-1}\boldsymbol{R}_{scr}$ related to the specimen may be expressed through the reflected part of wave function close to the specimen $\Psi_{ref}(\boldsymbol{r}, z = 0)$ by the relation:

$$\Psi(z_{scr}\, \boldsymbol{r}') = \int d^2 r\, \Psi_{ref}(\boldsymbol{r}, z = 0)\, \Pi\, (\boldsymbol{r}' - \boldsymbol{r}),$$

$$\Pi\, (\boldsymbol{r}' - \boldsymbol{r}) = (2\pi\hbar)^{-2} \int d^2 p\, \Gamma\, (\boldsymbol{p}) \exp\left\{i \frac{\boldsymbol{p}}{\hbar}(\boldsymbol{r}' - \boldsymbol{r})\right\} \qquad (7.134)$$

Here $\Gamma(\boldsymbol{p})$ is the amplitude transparency (in relation to the transparency of the wave function) of the contrast aperture at the point $\boldsymbol{r} = \boldsymbol{p}\tau/m$; see Eq. (7.34) about parameter τ. The current density of electrons at point $\boldsymbol{r}'$ of the screen is proportional to $\langle|\Psi(z_{\text{ser}}, \boldsymbol{r}')|^2\rangle$, where symbol $\langle\,\rangle$ stands for the averaging over the ensemble of the electrons incident on the specimen. Up to unessential factor we have:

$$j(\boldsymbol{r}') = \langle|\Psi(z_{\text{ser}}, \boldsymbol{r}')|^2\rangle = \iint d^2\boldsymbol{r}_1 d^2\boldsymbol{r}_2\ \Pi^*(\boldsymbol{r}' - \boldsymbol{r}_1)\,\Pi\,(\boldsymbol{r}' - \boldsymbol{r}_2)$$

$$\cdot\langle\Psi^*_{\text{ref}}(\boldsymbol{r}_1, z = 0)\Psi_{\text{ref}}(\boldsymbol{r}_2, z = 0)\rangle \quad (7.135)$$

The averaged bilinear combination of wave functions in the integral (7.135) is the density matrix. In our model of the illumination beam (7.35) it may be represented as an incoherent sum (as an integral over $d\boldsymbol{p}_0 d\epsilon$) of contributions due to the waves with different $\boldsymbol{p}_0$ and ϵ with the weight $\mathcal{I}_0(\boldsymbol{p}_0, \epsilon)$. Primary modulation of each wave is determined by Eqs. (7.120). Therefore:

$$\langle\Psi^*_{\text{ref}}(\boldsymbol{r}_1, z = 0)\ \Psi_{\text{ref}}(\boldsymbol{r}_2, z = 0)\rangle = \int \mathcal{I}_0(\boldsymbol{p}_0, \epsilon) d^2\boldsymbol{p}_0 d\epsilon\ \exp\left\{\frac{i}{\hbar}\,[\boldsymbol{p}_0(\boldsymbol{r}_2 - \boldsymbol{r}_1)\right.$$

$$\left. + \Delta\Phi_q(\boldsymbol{r}_2, \boldsymbol{p}_0, \epsilon) - \Delta\Phi^*_q(\boldsymbol{r}_1, \boldsymbol{p}_0, \epsilon)]\right\} \quad (7.136)$$

Equations (7.134)-(7.136) together with (7.120) give a complete solution of the image contrast problem (including the problem of strong contrast) for electrical and magnetic microfields in the MEMFI.

If we assume $\Gamma(\boldsymbol{p})$ to be real and $\Gamma(\boldsymbol{p}) = [T(\boldsymbol{p})]^{1/2}$, where $T(\boldsymbol{p})$ is the classical transparency of the aperture, then Eqs. (7.134)-(7.136) in the limit $\hbar \to 0$ turn† into classical expression (7.87).

The estimations Eqs. (7.2), (7.3) reveal the improvement of the resolving power and sensitivity of the MEMFI for narrower distribution of $\mathcal{I}_0(\boldsymbol{p}_0, \epsilon)$. Therefore, in the search of a quantum limit of resolving power of the MEMFI it will be assumed that $\mathcal{I}_0(\boldsymbol{p}_0, \epsilon) = \delta^{(2)}(\boldsymbol{p}_0)\,\delta\,(\epsilon)$, i.e., an ideal coherent illumination will be considered. In this case for the electrical microfield it follows from Eqs. (7.134)-(7.136) and (7.120) that in linear approximation the spectral transfer function equals:

$$S(\boldsymbol{k}) = \tilde{K}(\boldsymbol{k})/\tilde{\varphi}(\boldsymbol{k}) = R_q(\boldsymbol{k}, 0, \epsilon) \cdot S_D(\boldsymbol{k}) \equiv \hbar^{-1}\left(\frac{2\pi me}{E_0}\right)^{1/2} |\boldsymbol{k}|^{-1/2}$$

$$\cdot \exp\left\{-|\boldsymbol{k}|z_0 - \frac{2}{3}|\boldsymbol{k}|^3 a^3\right\} \cdot \frac{i[\Gamma^*(0)\Gamma(\hbar\boldsymbol{k}) - \Gamma(0)\Gamma^*(-\hbar\boldsymbol{k})]}{|\Gamma(0)|^2} \quad (7.137)$$

This expression [as well as Eq. (7.127) for the MEMSI] may be regarded as the product of two factors. The first is the spectral coefficient of primary

†It is interesting to note that the terms $\backsim \Delta\boldsymbol{r}_0$ in classical formulae correspond to the action $\Delta\Phi_q(\boldsymbol{r}_0, \boldsymbol{p}_0, \epsilon)$ in its imaginary part which is due to the quantum effects.

modulation R_q (volt^{-1}), and in Eq. (7.137) a more accurate expression for R_q than in Eq. (7.127) was used.

The second factor $S_D(k)$ is dimensionless; it characterizes the spectral coefficient of transformation of the phase modulation into the density modulation resulting from the diffraction on a contrast aperture. One can speak about Fraunhoffer diffraction since the aperture is placed at the focal plane of the objective system. Factor $S_D(k)$ essentially depends on the shape of the contrast aperture $\Gamma(\hbar k)$ and is typical for any instrument with focused imaging of a phase specimen.

Some general properties of the function $S_D(k)$ will be discussed here. First, since the amplitude transparency $\Gamma(\hbar k)$ satisfies definitely inequality $|\Gamma(\hbar k)| \leqslant 1$ than for $|S_D(k)|$ we have:

$$|S_D(k)| \leqslant 2\,|\Gamma(0)|^{-1} \tag{7.138}$$

and if the central beam is transmitted without a change of amplitude, i.e., if $|\Gamma(0)| = 1$, then* $|S_D(k)| \leqslant 2$. If the aperture is symmetrical with respect to the rotation at an angle 180° in the (x, y)-plane, i.e., if $\Gamma(p) = \Gamma(-p)$, then:

$$S_D(k) = 2\,\mathrm{Im}\,[\Gamma^*(\hbar k)/\Gamma^*(0)] \tag{7.139}$$

For example, the Zernike method of phase contrast microscopy (Born and Wolf, 1964) corresponds to $\Gamma(p) = 1$ at $|p| \neq 0$, $\Gamma(0) = i\sqrt{T_0}$, where T_0 is the transparency for the central beam. In this case $S_D(k) = 2\,(T_0)^{-1/2}$ for all $|k| \neq 0$.

For a usual aperture $\Gamma(p) = \Gamma^*(p)$, and $S_D(k)$ is determined by the degree of asymmetry of aperture with respect to the axis. For a symmetrical (with respect to $r \to -r$) aperture with $\Gamma = \Gamma^*$, contrast appears only in higher orders in microfields, and in the first order $S_D(k) = 0$. For an arbitrary asymmetrical aperture with size r_a (Fig. 7.30a) and with transparency $T = 1$ the demodulation coefficient $S_D(k)$ equals $\pm 2i$ at $|k| \lesssim m r_a/\hbar\tau$ and $S_D(k) = 0$ at $|k| \gtrsim m r_a/\hbar\tau$ (Fig. 7.30b). This imposes a limitation on the resolving power of the MEMFI because of Fraunhoffer diffraction on the contrast aperture:

$$\delta x_{\mathrm{res}} \gtrsim \frac{\hbar\tau}{m r_a} \sim \lambda_0 \frac{f}{r_a} \tag{7.140}$$

Here the estimation of $\tau \sim f v_0^{-1} = f \cdot (m/2eU_0)^{1/2}$ was used for parameter τ, and f is the focal length of the objective system. This is exactly the estimation (7.140) which had been obtained for the quantum limit of resolving power of the MEMFI in the review by Bok *et al.* (1971).

For an aperture of a sufficiently large size, however, the diffraction on it does not limit the resolving power any longer. In this case the resolving power is

*Note that the demodulation coefficient $S_D(k) = 2 \sin(k^2 b^2)$, for shadow imaging mode also satisfies $|S_D(k)| \leqslant 2$.

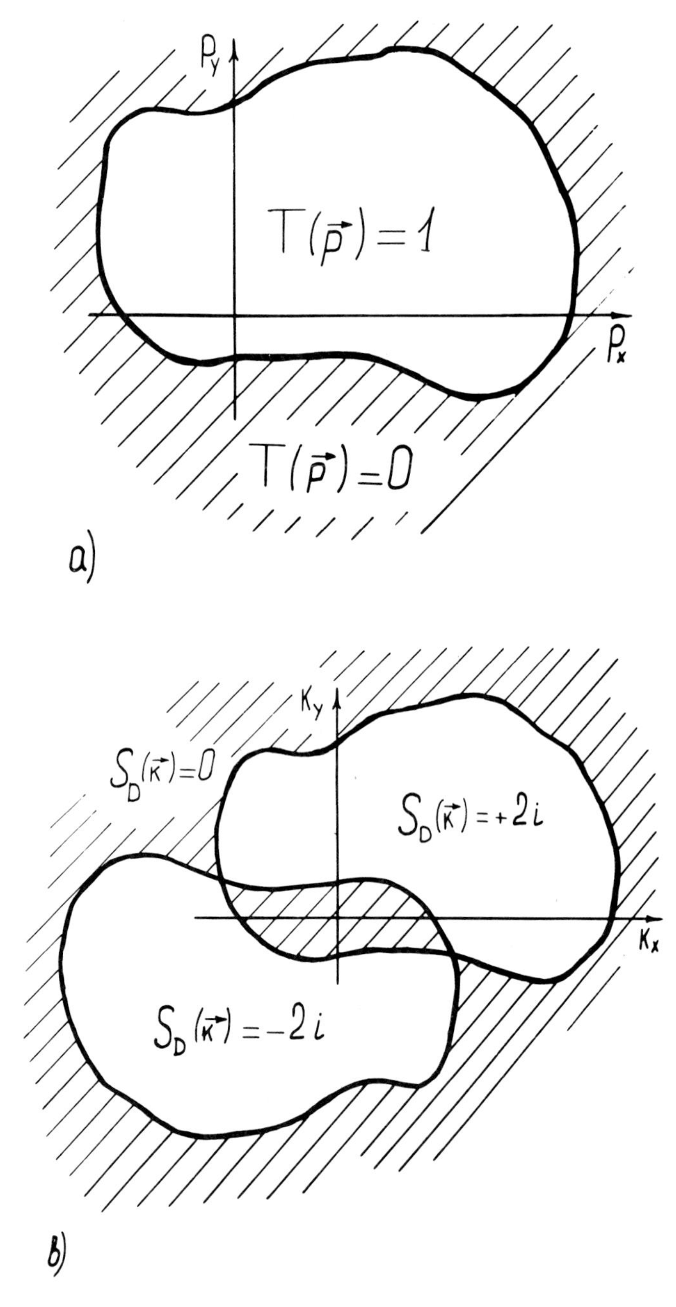

Fig. 7.30 (a) Transparency of the aperture $T(\boldsymbol{p})$ as a function of momentum $\boldsymbol{p}$; (b) spectral coefficient of demodulation $S_D(\boldsymbol{k})$ in MEMSI corresponding to Fig. 7.30a; $\boldsymbol{k} = \boldsymbol{p}/\hbar$.

determined either by spherical aberration* or (for an aberrationless objective system) by quantum-mechanical decrease of the primary modulation spectral function (curve 4 in Fig. 7.25). The latter limit of resolving power approximately equals:

$$\delta x_{\mathrm{res}} \gtrsim a = \left(\frac{\hbar^2}{2meE_0}\right)^{1/3} \tag{7.141}$$

A numerical evaluation of the parameter a gives the value $a \sim 1.5$ nm for $E_0 = 10^7$ volt/m being the field strength. It is interesting to note that the estimation obtained by Bok *et al.* (1971) also equals ~ 1.5 nm based on expression (7.140). That points to the fact that the size of an aperture chosen for a numerical estimation in the review by Bok *et al.* was optimum for the achievement of the resolving power for coherent illumination. As for the experimentally achieved resolving power in the MEMFI, it equals (Bok *et al.*, 1971) ~ 80 nm and is probably limited by the energy spread of electrons in the illumination beam.

Electron Mirror Interferometer

From the previous treatment it follows that microfields and topography inhomogeneities constitute a phase specimen. Various modes of operation account for the differences in the demodulation of the primary phase pattern. For the MEMSI it is a case of a Fresnel diffraction during the motion from the specimen to the screen. For the MEMFI it is a case of cutting-off a part of diffracted electrons by the contrast aperture; in the dark-field imaging mode those diffracted electrons are, on the contrary, transmitted. In principle, it is possible to use a mode similar to the Zernike method of phase-contrast microscopy (see the previous section).

An alternative to the Zernike method is the interference microscope. In this case an electrostatic (Möllenstedt and Düker, 1956) or some other biprism is mounted behind the objective system. The biprism produces an overlapping effect of two parts of the image on the screen. An interference fringe pattern appears in the part of the screen where both images overlap. The period of these fringes λ' is determined by the mutual angle of intersection α of two beams after the biprism: $\lambda' = \lambda_0/\alpha$, where λ_0 is the wavelength of the electrons accelerated to the voltage U_0. The contrast of the fringes depends on the degree of coherence of the illumination, i.e., on the size of the illumination source and also on the energy spread of electrons.

The variations of the phases of interfering rays depending on the coordinate

*Spectral coefficient of demodulation for a phase specimen with account of spherical aberration using a transmission electron microscope was treated by Hanszen (1970) (see also Hawkes, 1972, Chap. 3).

along the interference fringes are displayed by the transverse shift of the fringes. This allows us in principle to measure the variations of phase differences as functions of the coordinate along the fringes.

Mirror electron interferometer was first realized by Lichte *et al.* (1972) (Fig. 7.31). The micrographs of the interference pattern are shown in Fig. 7.32; the shift of the fringes is seen in the area where the fringes are intersected by the image of a topography inhomogeneity.

It should be noted that the shift of fringes is accompanied by the shift of the area where the beams overlap, i.e., by the shift of the field of interference. That is quite natural for an instrument operating in a shadow projection mode, since in that mode the gradient of the phase (i.e., of the classical action) leads to the shift of the coordinate of intersection of a classical trajectory with the screen plane.

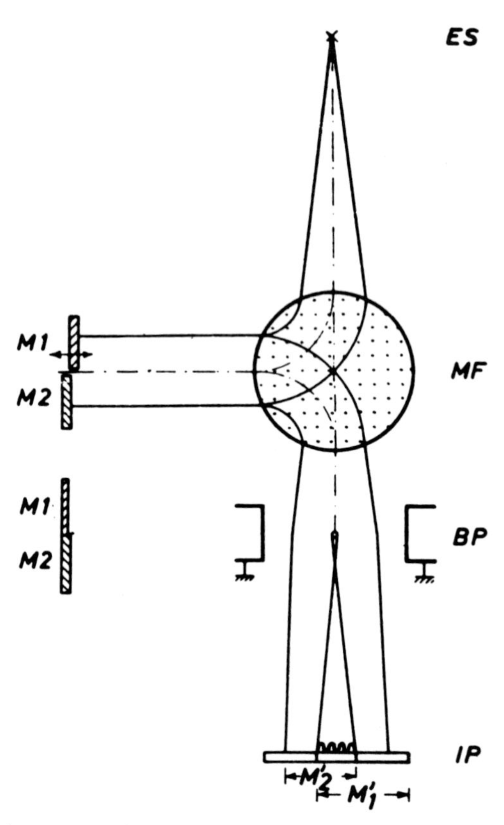

Fig. 7.31 Mirror electron interferometer of the Michelson type. Es, linear electron source, 100 nm in width, 25 keV; MF, magnetic deflection prism according to R. Castaing; M_1 and M_2, electron mirrors in different distances, realized by height steps on a mirror; BP, electron biprism; M_1' and M_2', coherent superpositions of the partial waves in the plane of observation IP. (*Lichte* et al., *1972; courtesy of G. Möllenstedt.*)

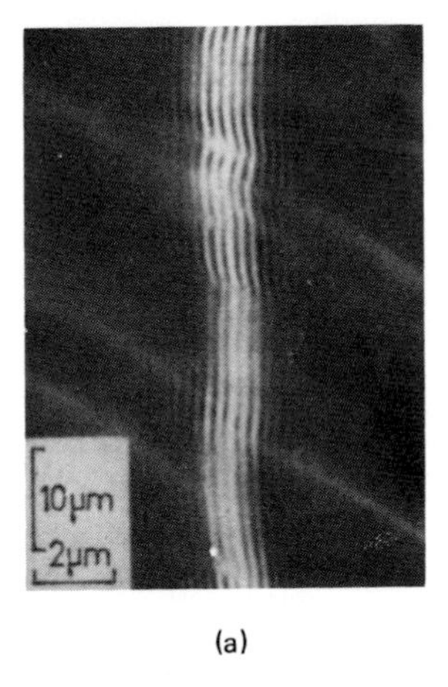

(a)

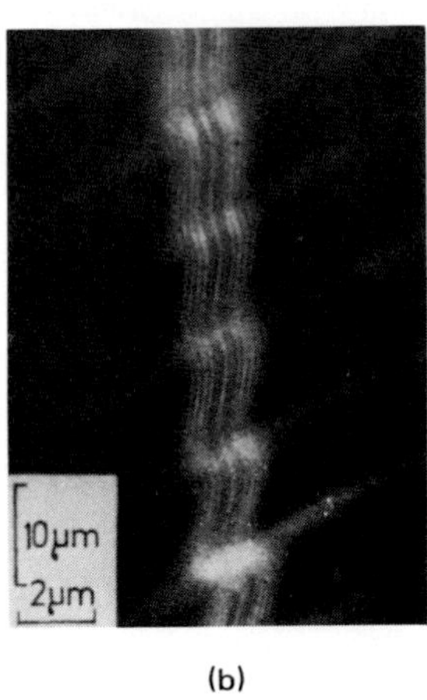

(b)

Fig. 7.32 (a) Object with several small disturbances ($U_B = 4$ *v*). (b) Object with grooves of varying depth and width (ca. 1 μm) drawn in Al-layer with a fine diamond point. The interference field is swept as a whole with decreasing contrast. (*Lichte* et al., *1972*; *courtesy of G. Möllenstedt.*)

The biprism may be said to have enabled those scientists to "deposit" a coordinate grating on the specimen; this grating is a set of fringes along one direction. The period of the fringes of the grating related to the specimen equals $\lambda' = 0.25$ μm (Fig. 7.31). The "classical" shift of electron trajectory may be measured by the shift of fringes. The minimum period λ' which can be used is determined by the size of the illumination source related to the specimen. It must be understood that the above "classical" interpretation is of limited applicability.

Let us make an estimation of the sensitivity of an electron mirror interferometer to the variations of potential φ_1 on the surface of the specimen or to the equivalent variations $h_1 = -\varphi_1/E_0$ of the height of the specimen. The essential difference between an MEM-interferometer and a light-optical interferometer consists of the following. The phase shift of a reflected beam in the MEM is due to the microfields, and it is picked up by an electron all along the distance it covers from the objective aperture to the specimen and back to the aperture. Microfield $\varphi_1(\boldsymbol{r}, z)$ satisfies Laplace's equation. Therefore, the perturbation of potential $\varphi_1(\boldsymbol{r}_0, z)$ along the ray $\boldsymbol{r}_0$ = const depends on the values of $\varphi_1(\boldsymbol{r}, z = 0)$ or $h_1(\boldsymbol{r})$ in each point $\boldsymbol{r}$ of the surface of the specimen. The expression for the phase $\hbar^{-1}\Delta\Phi(\boldsymbol{r})$ according to Eq. (7.116) is

$$\hbar^{-1}\Delta\Phi(x) = \hbar^{-1}(2me/E_0)^{1/2}\int_{-\infty}^{+\infty} dx'\varphi(x', z = 0)\,\mathrm{Re}\{[z_0 - i(x - x')]^{-1/2}\} \tag{7.142}$$

The schematic curve in Fig. 7.25 and the expressions (7.120) show that the maximum phase sensitivity of the electron mirror may be achieved for the perturbations with maximum lateral size δx about the value of distance l from the specimen to the aperture. For such large-scale perturbations ($\delta x \sim l$) from Eqs. (7.120) or (7.142) the following estimation may be obtained:

$$\Delta(\arg\Psi) = \hbar^{-1}\Delta\Phi \sim 2\pi\frac{\varphi_1}{\lambda_0 E_0} = 2\pi\frac{h_1}{\lambda_0} \tag{7.143}$$

On the basis of a similar relation first derived by Lenz (1972) a legitimate conclusion was drawn stating that an electron mirror interferometer makes it possible (in principle) to measure a very small height variation: $h_1 \sim \lambda_0 = 8.10^{-12}$ m; the latter corresponds to $U_0 = 25$ kV. However, that requires the interference of the electron rays which are about $\delta x \sim l \sim 1$ cm apart on the surface of the specimen. The given value of δx turns out to be the size of the area on the specimen to which such a small variation of height $h_1 \sim 8.10^{-12}$ m might be attributed. Thus, the value of vertical sensitivity $h_1 \sim 8.10^{-12}$ m corresponds to an extremely poor localization on the specimen. Apart from poor localization, the value 8.10^{-12} m is unrealistic since its realization requires a coherent illumination of the area of the specimen with the size $\delta x \sim 1$ cm; at present, this seems to be unattainable.

It follows from Eq. (7.142) that for the "step" potential of a smoothed type with a height of h_1 (i.e., $\varphi(x, z = 0) = E_0 h_1 \pi^{-1}\arctan(x/w)$) the phase modulation over point x equals:

$$\hbar^{-1}\Delta\Phi(x) = \frac{h_1}{\hbar}(2meE_0)^{1/2}\cdot\frac{x\sqrt{2}}{\sqrt{z_0 + w + \sqrt{x^2 + (z_0 + w)^2}}} \tag{7.144}$$

It will be assumed that the distance $|x|$ from the middle of the step is much greater than the parameter $(z_0 + w)$. The maximum obtainable phase difference corresponds to the maximum value $\pm|x|$ for which the illumination beam is coherent: $|x_{max}| \lesssim r_{coh}$. A demand that the phase shift should be larger or about 2π (shift not less than one fringe) yields a corresponding detectable value h_1:

$$h_1 = \lambda_0 \left(\frac{l}{\delta x}\right)^{1/2} \gtrsim \lambda_0 \left(\frac{l}{r_{\mathrm{coh}}}\right)^{1/2} \tag{7.145}$$

The size of coherence area r_{coh} is determined by the spread of lateral momenta of incident electrons: $r_{\mathrm{coh}} \approx h/\delta p_0$. According to the Eqs. (7.31)–(7.33) the value of δp_0 is proportional to the size r_2 of the illumination source: $\delta p_0 = 2r_2 m v_0/L_2$, where $v_0 = (2eU_0/m)^{1/2}$, and L_2 is the distance from the source to the two-electrode objective system. In the experiments by Lichte *et al.*, (1972) the minimum size of the source was 0.1 μm; assuming $L_2 \approx 20$ cm, the value $r_{\mathrm{coh}} \approx 4\ \mu$m is obtained. Indeed, in those experiments the rays which interfered on the screen were $\delta x \sim 2\ \mu$m apart on the surface of the specimen. When $l \sim 1$ cm is assumed, Eq. (7.145) yields the limit of sensitivity of the MEM interferometer: $h_1 \gtrsim 70\ \lambda_0 \approx 0.5$ nm; this limit corresponds to a one-fringe shift. The interferograms in Fig. 7.31 show the shift over several fringes. The order of magnitude of the height variations in that experiment was $\sim\delta h \sim$ 1000 nm; however, both the positive and the negative δh were close to each other on the specimen, so that their contributions might have been considerably compensated.

The energy spread of electrons is an important obstacle for the measurements of microfields by the MEM interferometer. The maximum number of interference fringes $N \sim eU_0/\delta\epsilon \gtrsim 10^4$ is large enough and is not important here. It is the extremely strong dependence of the phase of the reflected beam on energy ϵ which is an important obstacle. That is the chromatism of primary modulation which is a specific feature of the MEM.

It is interesting to compare the obtained figures with the vertical sensitivity of the MEMSI determined by the energy spread $\delta\epsilon$ of the electrons which illuminate the specimen. We impose a condition that the height of the reflection z_0 in the field E_0 equals $z_0 \approx 3\ \delta\epsilon/eE_0$. Assuming $\delta\epsilon \sim 1$ eV and $E_0 \approx 2.5.10^6$ V/m, the value $z_0 \sim 1.2\ \mu$m is obtained. Then, from the condition $|K| = 1$ for the contrast of a topography step in the MEMSI and Eqs. (7.58), the value $h_1 \gtrsim z_0 \cdot (z_0/l)^{1/2} \approx 1.3$ nm follows. The estimation $\varphi_1 = E_0 h_1 \sim 3$ mV corresponds to the above potential "step."

Thus, the MEM in a conventional shadow imaging mode has high sensitivity to microfields and topography inhomogeneities; the sensitivity may be improved even more by the application of the interference methods.

APPLICATIONS

Metals, Semiconductors, and Insulators

The potential distribution in thin metal films and semiconductors was first observed by Orthuber (1948). Examining a nickel cathode covered by a barium layer, Orthuber visualized the contact difference of potential between the nickel substrate and the barium as well as barium oxide layers.

The contact potentials were examined by Spivak *et al.* (1960). The secondary electron emitters, "L-cathode" and oxide cathode were subjected to the investigation. The oxide cathode was studied in mirror, thermo- and photo-emission imaging modes, which provided highly informative results. The qualitative evaluation of work function on the surface of patched emitters was provided by means of emission-mirror microscope (Spivak *et al.*, 1964a).

A number of experiments concerned with the visualization of contact fields on the surface of semiconductors, impurity segregation, dislocations, diffusion of impurities, and other phenomena were made by Igras and Warminski (1965, 1966, 1967, 1968).

The MEM was used for the examination of p-n junctions in semiconductors by Bartz and Weissenberg (1957), Spivak and Ivanov (1963), and Igras and Warminski (1965). The instrument was also used to study cleavage steps and different model specimens with "step"-potential distribution (Heydenreich, 1966). Recently, a number of papers on the application of the MEM for the investigation of integrated circuits have been published (Ivanov and Abalmazova, 1967; Igras and Warminski, 1968). Ivanov and Abalmazova (1967), for example, examined thin film capacitors. The stroboscopic mirror electron microscope was applied by Lukianov and Spivak (1966) in their observation of switching processes in semiconductor diodes.

Some other methods of quantitative measurements of microfields which are worth mentioning were suggested along with the visualization of p-n junctions and potential steps. For example, Sedov *et al.*, (1962) described the following method of potential distribution measurement. It is known that the secondary emission spot on the mirror image appears only in places where the potential of the surface of the specimen is greater or equal to that of the filament of the electron gun (including the contact difference of potentials). By measuring the displacement of the boundary of the secondary emission spot as a function of bias voltage, while fixed voltage is applied to the diode, a potential distribution on the surface can be obtained.

Another method of potential measurements was described by Guittard *et al.*, (1967a, 1968). The measurements of the current on the axis of the instrument at different bias voltages allowed the authors to establish that there is a linear dependence of the peak amplitude of the current on the surface potential. The linearity holds in the range up to 100 V of the measured voltages. The sensitivity of this method reaches maximum value corresponding to 0.1 mV, when the electron beam is reflected from the surface. The topographical resolution did not exceed 10 μm in these experiments. In the latest high vacuum models of the instrument (vacuum $\sim 10^{-11}$ torr) high stability and accuracy of measurements were attained. Application of that technique enabled the measurement of the contact difference of potentials between gold and aluminium on a specially prepared specimen, the potential of a band in a silicon doped by phosphorus; micro-circuits were also studied.

Another idea of potential measurements on the surface of the specimens is based on the solution of the inverse problem of electron optics applied to the electrical microfields (Sedov, 1968). By means of this method the field distributions in p-n junctions (Lukianov *et al.*, 1968b) and in metal-oxide structure were obtained. Accuracy of measurements of the order of 0.1 V with topographical resolution of 1 μm was obtained. The potential and the field strength accompanying the propagation of recombination waves in germanium were calculated with the use of that method (Gvosdover *et al.*, 1971). These experiments were carried out in the stroboscopic mode of operation of the MEM. The measured field strength was 10^4-10^5 V/m and was concentrated in an area of 50 μm in size.

It should be noted that the results of the measurements are strongly dependent on the carbon contaminations, polymerized by the action of either electron or ion beams. According to Schwartze (1962), when the pressure of residual gases is of the order of 10^{-4} torr, the ion current density is of the order of 5% of the electron current density.

The presence of contaminations in the column of the instrument leads to the polymerization of thin carbon films on the surface of the specimen. Because of the action of the electrons from the tail of Maxwellian distribution a local negative charge appears on the surface. As a result, dark spots with bright rims are observed on the screen. Those films under some conditions are positively charged by ion current. In that case bright stars corresponding to caustics appear on the image (Spivak *et al.*, 1964b). Lenz and Krimmel (1963) by means of digital computation found the shape of caustic surfaces which appear in the field of a point positive or negative charge. The results of their computations qualitatively coincide with the experimentally observed images of carbon contamination films. Thus, by means of the MEM the process of polymerization of thin films under electron or ion bombardment can be investigated, and the properties of those films may be examined.

Contaminations on the surface of the specimen under investigation lead to some errors of measurement of electrical and magnetic microfields. In that connection the observation of surface phenomena would be better provided in an ultrahigh vacuum electron mirror microscope with oil-free pumping. Undesirable effects caused by formation of carbon contaminations on the surface could probably be minimized in such instruments.

Mirror electron microscopy has also been applied in the study of domain structure of ferroelectrics. Most of the ferroelectrics examined in the MEM were insulators with 10^{12}-10^{14} $\Omega \cdot$cm resistivity. The observation of those high-resistivity specimens requires special preparations for investigation. To make the specimen conductive it is necessary to cover it with a thin metal or semiconductor film (usually by vacuum evaporation) to avoid excess charging of the surface.

The following ferroelectrics were examined in the MEM: barium titanate, triglycine sulfate (TGS), lead zirconate-titanate polycrystalline materials, and so on (Spivak *et al.*, 1959a, 1963b; English, 1968a,b; Someya and Kobayashi, 1971; Kobayashi *et al.*, 1972). Observations of surface topography and ferroelectric transition from the ferroelectric to the paraelectric state during the heating of crystals above the Curie point and cooling were made. The repolarization and the nucleation of domains were also examined. The use of the MEM made it possible to visualize the 180°-domain structure which is beyond the possibility of a polarizing light optical microscope.

The MEM was also used as photoconverters for conversion of infrared images into visible ones and as electron acoustical converters. The first photoconverter was devised by Orthuber (1948). The transducer was a thin PbS film which at the same time served as a mirror electrode. Threshold wavelength of the order of 1.8 μm was attained. A similar instrument made with a transducer containing a thin layer of 95 atomic percent of Se and 5 atomic percent of Bi had the sensitivity of 0.13 μA per μW per cm^2 at a 0.96 μm wavelength (Bates and England, 1969).

In the electron acoustical converter by Koch (1960), the mirror electrode was a quartz receiver mounted at the end window of the vessel filled with water where the specimen under investigation and the acoustic radiator were placed. The signal registered by the quartz receiver modulated the electron beam, and the image of the object situated in the vessel appeared on the viewing screen. It should be noted that the images obtained by Koch were blurred, which was due to the fact that the author was observing the averaged-in-time image of the surface potential and not the instantaneous one (compare with the Eqs. (7.105)).

The mirror electron microscope was used to study the oxide coating (Heydenreich and Vester, 1974) and the structures of the metal-oxide, metal-oxide-metal types (Ivanov and Abalmazova, 1967; Szentesi, 1972).

Magnetic Structures

The magnetic specimens were investigated mainly by Spivak *et al.* (1955–1972) and Mayer (1955–1961). Advances in the theory of image contrast formation and the development of new experimental techniques made possible not only the visual observation of magnetic structures but also the measurement of d.c., a.c. magnetic microfields by using the MEM.

The following areas were investigated by using the MEM: the domain structure in ferromagnetics (Spivak *et al.*, 1955, 1959b, 1961b, 1963a, 1964c, 1966; Mayer, 1957b, 1959, 1961), the magnetic recording (Mayer, 1958; Kuehler, 1960; Guittard *et al.*, 1967b), the fields over the magnetic recording heads (Spivak *et al.*, 1963c; Mayer 1961), the Abricosov vortex structure of superconductors (Wang *et al.*, 1966), and the superconductors in the intermediate state (Bostanjoglo and Seigel, 1967).

Together with the observation of magnetic microfields, some measuring techniques were proposed. In particular, for the estimations of the magnetic microfields over the gap of magnetic recording heads of different types the method of inverse problem solution was used (Sedov, 1968). The measurements of magnetic microfields on tapes recorded at audio- and video-frequencies were done by Rau *et al.*, (1970). An interesting method for intensifying the contrast of magnetic microfields recorded on magnetic tapes was suggested by Guittard *et al.*, (1967b). The essence of this technique consists in decentering the electron beam in a direction perpendicular to the direction of magnetic flux density. In that case the tangential velocity of the electron and the components of the Lorentz force arise, and contrast is intensified. The authors affirm that because of the increase of sensitivity of the MEM to magnetic microfields, the observation of magnetic signals through the metallized mica layer of 5 μm in thickness was possible without any essential deterioration of contrast. Thus, it is possible to make a device for read-out of signals recorded on magnetic tapes, and the latter may be placed outside a vacuum.

The use of the stroboscopic mode of operation of the MEM allowed the investigation of static magnetic fields, the observation of time-dependent periodic magnetic fields over magnetic recording heads (Spivak and Lukianov (1966b), and the measurement of frequency and phase responses, i.e., the dependence of the amplitude and phase of the magnetic field over the gap on the frequency at a fixed amplitude of current in the excitation winding (Rau *et al.*, 1970; Spivak *et al.*, 1971). The measuring technique is based on the following experimental result. When a sinusoidal or d.c. current flows in the excitation winding, it is possible to choose d.c. equivalent current producing the same d.c. field over the gap which would be the same as the field of a.c. current of fixed amplitude and frequency. In that case the electron optical image of the magnetic microfield (a typical dark wedge) and the oscillograms of current density distribution coincide. Taking the ratio of the d.c. equivalent current J to the fixed value of the a.c. current J_0 and keeping constant the phase of strob pulse, one could determine the magnetic field strength over the gap of the head in a wide (0–10 MHz) frequency range. The applications of MEM for the examination of magnetic structures are given in detail in a review by Petrov *et al.*, (1970c).

Biological Specimens

The mirror electron microscope was used for the examination of biological specimens by McLeod and Oman (1968). The specimen was the giant unicellular alga *Acetabularia crenulata.* For investigations a thin copper film was evaporated on the surface of the specimen. The authors attributed the observed conductivity patterns on the image to the existence of charging centers in the chloroplast. The authors also studied the activity of those centers in dynamics and showed that it tends to decrease with time after initial preparation of the cells.

Thus, with the help of the MEM it was shown that the charging centers exist in plant material and they could hold the charge for several seconds. The probable relation between the centers observed and the reactive centers, which are responsible for the transformation of electromagnetic energy into chemical energy, was discussed.

We are deeply grateful to Professor G. V. Spivak, whose pioneering works on mirror electron microscopy and whose attention, advice, and support served as a permanent source of inspiration in our work. The collaboration and fruitful discussions with Drs. A. E. Lukianov, V. I. Petrov, E. I. Rau, and N. N. Sedov from the Electron Optics Laboratory of Moscow State University are sincerely acknowledged.

References

Artamonov, O. M. (1968). A simple electron mirror microscope. *Proc. 4th Eur. Reg. Conf. Electron Microsc.* (Rome), Vol. 1, p. 97.

Artamonov, O. M., Guerasimova, N. B., Komolov, S. A. (1966). The experimental investigation of the operation of mirror electrical optical system. *Optiko-mechanich. promishl.* **33,** (12), 17.

Artamonov, O. M., and Komolov, S. A. (1966). On the potential relief mapping in the electron mirror. *Radiotechnika i Elektronika* **11,** 2186.

Barnett, M. E., and England, L. (1968). Distortions in shadow projection electron mirror images of periodic structures. *Optik* **27,** 341.

Barnett, M. E., and Nixon, W. C. (1966). Image contrast in mirror electron microscopy. *Proc. 6th Int. Cong. Electron Microsc.* (Kyoto), Vol. 1, p. 231.

Barnett, M. E., and Nixon, W. C. (1967a). A mirror electron microscope using magnetic lenses. *J. Sci. Instr.* **44,** 893.

Barnett, M. E., and Nixon, W. C. (1967b). Electrical contrast in mirror electron microscopy. *Optik* **26,** 310.

Bartz, G., Weissenberg, D., Wiskott, D. (1956). Ein Auflichtelektronenmikroskop. *Proc. 3rd Int. Conf. Electron Microsc.* (London), p. 395.

Bartz, G., and Weissenberg, G. (1957). Abbildung von p-n-Ubergängen in Halbleitern mit dem Auflichtelektronenmikroskop. *Naturwissenschaften* **44,** 229.

Bates, C. W., and England, L. (1969). An electron-mirror infrared image converter using vitreous selenium-bismouth photoconducting layers. *Appl. Phys. Lett.* **14,** 390.

Bethge, H., Hellgardt, J., and Heydenreich, J. (1960). Zum Aufban einfacher Anordnungen für Untersuchungen mit dem Elektronenspiegel. *Exp. Tech. Phys.* **8,** 49.

Bok, A. B. (1968). *A Mirror Electron Microscope.* Hooglanden Waltman, Delft.

Bok, A. B., Le Poole, J. B., and Roos, J. (1968a). A mirror electron microscope with focused image. *Proc. 4th Eur. Reg. Conf. Electron Microsc.* (Rome), Vol. 1, p. 103.

Bok, A. B., Le Poole, J. B., and Roos, J. (1968b). Contrast formation in a mirror electron microscope with focused images. *Proc. 4th Eur. Reg. Conf. Electron Microsc.* (Rome), Vol. 1, p. 101.

Bok, A. B., Le Poole, J. B., Roos, J., and De Lang, H. (1971). Mirror electron microscopy. In: *Advances in Optical and Electron Microscopy* **4,** 161. Academic Press, London and New York.

Born, M., and Wolf, E. (1964). *Principles of Optics.* Pergamon Press, London and New York.

Bostanjoglo, O., and Seigel, G. (1967). Electron mirror observation of superconductors in the intermediate state. *Cryogenics* **7**, 157.

Brand, U., and Schwartze, W. (1963). Modelliersuche zur Bildenstehung im Elektronenspiegel-Oberflächen-Mikroskop. *Exp. Tech. Phys.* **11**, 18.

Cline, J. E., Morris, J. M., and Schwartz, S. (1969). Scanning electron mirror microscopy and scanning electron microscopy of integrated circuits. *IEEE Trans. Electron Dev.* **ED-16**, 371.

Cohen, M. S., and Harte, K. J. (1969). Domain wall profiles in magnetic films. *J. Appl. Phys.* **40**, 3597.

Delong, A., and Drahos, V. (1964). Ein kombiniertes Emissions-Elektronenmikroskop. *Proc. 3rd Eur. Reg. Conf. Electron Microsc.* (Prague), Vol. A, p. 25.

English, F. L. (1968a). Electron-mirror-microscope analysis of surface potentials on ferroelectrics. *J. Appl. Phys.* **39**, 128.

English, F. L. (1968b). Domain nucleation on the surface of $BaTiO_3$. *J. Appl. Phys.* **39**, 2302.

Forst, G., and Wende, B. (1964). Zur Bildentstehung im Elektronenspiegelmikroskop. *Z. Angew. Phys.* **17**, 479.

Garood, J. R., and Nixon, W. C. (1968). Scanning mirror electron microscopy. *Proc. 4th Eur. Reg. Conf. Electron Microsc.* (Rome), Vol. 1, p. 95.

Glaser, W. (1952). Grundlagen der Elektronenoptik. Springer-Verlag, Wien.

Guittard, C., Babout, M., and Pernoux, E. (1968). Microscopie éléctronique à miroir: mesure des variations locales de potentiel. *Proc. 4th Eur. Reg. Conf. Electron Microsc.* (Rome), Vol. 1, p. 107.

Guittard, C., Babout, M., Savary, G., and Pernoux, E. (1967a). Amélioration dans la mesure des variations locales de potentiel au microscope électronique à miroir. *C. R. Acad. Sci., Paris* **264**, 592.

Guittard, C., Vassoille, R., and Pernoux, E. (1967b). Possibilité de realisation d'un lecteur de micropiste magnetique selon le principe du microscope électronique à miroir. *C. R. Acad. Sci.* **B264**, 924.

Gvosdover, R. S. (1970). The automatic registration of the displacement of electron trajectories for metrics of microfields in electron microscopy. *Radiotechnika i Elektronika* **15**, 2653.

Gvosdover, R. S. (1972a). The image contrast of magnetic microfields in mirror electron microscope. *Radiotechnika i Elektronika* **17**, 1697.

Gvosdover, R. S. (1972b). Image contrast formation of two-dimensional piezoelectric fields in mirror electron microscopy. *Proc. 5th Eur. Cong. Electron Microsc.* (Manchester), p. 504.

Gvosdover, R. S., Lukianov, A. E., Spivak, G. V., Ljamov, V. E., Rau, E. I., and Butilkin, A. I. (1968). The stroboscopic mirror electron microscopy observation of piezoelectric fields. *Radiotechnika i Elektronika* **13**, 2276.

Gvosdover, R. S., Karpova, I. V., Kalashnikov, S. G., Lukianov, A. E., Rau, E. I., and Spivak, G. V. (1971). The stroboscopic mirror electron microscopy observation of the recombination waves in germanium. *Phys. Stat. Sol.* (a) **5**, 65.

Gvosdover, R. S., Lukianov, A. E., Spivak, G. V., and Rau, E. I. (1970). The mirror electron microscopy of piezoelectric fields. *Proc. 7th Int. Cong. Electron Microsc.* (Grenoble, France), **1**, 199.

Gvosdover, R. S., and Petrov, V. I. (1972). The image contrast of one-dimensional sinusoidal magnetic microfield in mirror electron microscope. *Radiotechnika i Elektronika* **17**, 433.

Gvosdover, R. S., and Zel'dovich, B. Ya. (1973). The quantum theory of image contrast formation of electrical and magnetic microfields in mirror electron microscopy. *J. Microscopie* **17**, 107.

Hanszen, K. J. (1970). The optical transfer theory of the electron microscope; fundamental principles and applications. In: *Advances in Optical and Electron Microscopy* **4**, 1. Academic Press, London and New York.

Hawkes, P. W. (1972). *Electron Optics and Electron Microscopy.* Taylor and Francis Ltd., England.

Henneberg, W., and Recknagel, A. (1935). Zusammenhang zwischen Elektronenlinsen, Elektronenspiegel und Steuerung. *Z. Tech. Physik.* **16**, 621.

Hermans, A. J., and Petterson, J. A. (1970). A quantum mechanical treatment of the mirror electron microscope. *J. Eng. Math.* **4**, 141.

Heydenreich, J. (1962). Experimentelles zur Entstehung des Bildkontrastes im Elektronenspiegel Mikroskop. *Exp. Tech. Phys.* **10**, 346.

Heydenreich, J. (1966). Some remarks on the image formation in the diverging electron mirror microscope. *Proc. 6th Int. Cong. Electron Microsc.* (Kyoto), **1**, 233.

Heydenreich, Z., and Vester, J. (1974). Einsatz des Spiegel-Elektronenmikroskops zur Untersuchung dielektrischer dünner Schichten. *Kristall Technik* **9**, 107.

Hottenroth, G. (1937). Untersuchungen über Elektronenspiegel. *Ann. Phys.* (5) **30**, 689.

Igras, E., and Warminski, T. (1965). Electron mirror observations of semiconductor surfaces at low temperatures. *Phys. Stat. Sol.* **9**, 79.

Igras, E., and Warminski, T. (1966). Electron mirror observation of impurity segregations on a silicon surface. *Phys. Stat. Sol.* **13**, 169.

Igras, E., and Warminski, T. (1967). Electron mirror microscopic investigation of surface diffusion of lithium in silicon. *Phys. Stat. Sol.* **20**, K5.

Igras, E., and Warminski, T. (1968). Application of electron mirror microscope to semiconductor surface physics. *Proc. 4th Eur. Reg. Conf. Electron Microsc.* (Rome), Vol. 1, p. 109.

Ivanov, R. D., and Abalmazova, M. G. (1966). The high-resolution universal mirror electron microscope. *Izv. Akad. Nauk SSSR, Ser. Fiz.* **30**, 784.

Ivanov, R. D., and Abalmazova, M. G. (1967). The mirror electron observation of metal-semiconductor, metal-oxide thin film contacts. *Zs. Technich. Fiz.* **37**, 1351.

Kasper, E., and Wilska, A. P. (1967). An electron optical filter projection system with image intensification through storage. *Optik* **26**, 247.

Kelman, V. M., and Yavor, S. Ya. (1968). *Electron Optics.* Nauka, Leningrad.

Kobayashi, J., Someya, T., and Furuhata, Y. (1972). Application of electron-mirror microscopy to direct observation of moving ferroelectric domains of $Gd_2(MoO_4)_3$. *Phys. Lett.* **A38**, 309.

Koch, G. (1960). Ultraschallbildwandlung mit dem Elektronenspiegel. *Acoustica* **10**, 167.

Kranz, J., and Bialas, H. (1961). Uber die Entstehung des Bildkontrastes bei der Abbildung ferromagnetisher Bereiche im Auflicht-elektronenmikroskop. *Optik* **18**, 178.

Kuehler, J. D. (1960). A new electron mirror design. *IBM J. Res. Develop.* **4**, 202.

Landau, L. D., and Lifschitz, E. M. (1963). *The Quantum Mechanics. Nonrelativistic Theory.* Fizmatgiz, Moscow.

Leisegang, S. (1956). *Elektronenmikroskope.* Springer-Verlag, Berlin.

Lenz, F. (1972). Path differences in electron interferometers using mirrors. *Z. Physik* **249**, 462.

Lenz, F., and Krimmel, E. (1963). Bilder lokaler Aufladungen im Elektronen-Spiegelmikroskop. *Z. Phys.* **175**, 235.

Lichte, H., Möllenstedt, G., and Wahl, H. (1972). A Michelson interferometer using electron waves. *Z. Physik* **249**, 456.

Lukianov, A. E., Sedov, N. N., Spivak, G. V., and Gvosdover, R. S. (1968b). The electron mirror measurements of microfields. *Proc. 4th Eur. Reg. Conf. Electron Microsc.* (Rome), Vol. 1, p. 97.

Lukianov, A. E., and Spivak, G. V. (1966). Electron mirror microscopy of transient phenomena in semiconductor diodes. *Proc. 6th Int. Cong. Electron Microsc.* (Kyoto), Vol. 1, p. 611.

Lukianov, A. E., Spivak, G. V., Sedov, N. N., and Petrov, V. I. (1968a). The geometrical optics of mirror electron microscope with two-electrode immersion objective. *Izv. Akad. Nauk SSSR. Ser. Fiz.* **32,** 987.

Lukianov, A. E., Spivak, G. V., and Gvosdover, R. S. (1973). The mirror electron microscopy. *Uspekhi Fiz. Nauk.* **110,** 623.

Mayer, L. (1955). On electron mirror microscopy. *J. Appl. Phys.* **26,** 1228.

Mayer, L. (1957a). Stereo-micrographs of conductivity. *J. Appl. Phys.* **28,** 259.

Mayer, L. (1957b). Electron mirror microscopy of magnetic domains. *J. Appl. Phys.* **28,** 975.

Mayer, L. (1958). Electron mirror microscopy of patterns recorded on magnetic tape. *J. Appl. Phys.* **29,** 658.

Mayer, L. (1959). Electron mirror microscopy of magnetic stray fields on grain boundaries. *J. Appl. Phys.* **30,** 1101.

Mayer, L. (1961). Electron mirror microscopy. In: *Encyclopedia of Microscopy* (Clark, G. L., ed.), p. 316. Reinhold Publishing Corporation, New York.

McLeod, G. C., and Oman, R. M. (1968). The application of electron mirror microscopy to the examination of biological material. *J. Appl. Phys.* **39,** 2756.

Möllenstedt, G., and Düker, H. (1956). Beobachtungen und Messungen an Biprisma-Interferenzen mit Electronenwellen. *Z. Phys.* **145,** 377.

Ogilvie, R. E., Schippert, M. A., Moll, S. H., and Koffman, D. M. (1969). Scanning electron mirror microscopy. *Proc. 2nd Ann. SEM Symp.*, p. 425. IITRI, Chicago, Illinois.

Oman, R. M. (1969). Electron mirror microscopy. *Adv. Electronics Electron Phys.* **26,** 217.

Orthuber, R. (1948). Über die Anwendung des Elektronenspiegels zum Abbilden der Potentialverteilung auf Metallischen und Halbleiter-Oberflächen. *Z. Angew. Phys.* **1,** 79.

Petrov, V. I., Lukianov, A. E., and Gvosdover, R. S. (1970a). On the theory of magnetic contrast of one-dimensional microfields in mirror electron microscope. *Vestnik Mosk. Univers. Ser. Fiz., Astr.* **4,** 441.

Petrov, V. I., Lukianov, A. E., Gvosdover, R. S., and Spivak, G. V. (1970b). Some aspects of magnetic contrast in electron mirror microscope. *Microscopie Electronique. Proc. 7th Int. Conf. Electron Microsc.* (Grenoble), Vol. 2, p. 25.

Petrov, V. I., Lukianov, A. E., and Spivak, G. V. (1967). The solution of one problem of magnetic contrast in the mirror electron microscope. *Vestnik Mosk. Univers. Ser. III (Fiz., Astr.)* **8** (6), 102.

Petrov, V. I., Spivak, G. V., and Pavlyuchenko, O. P. (1970c). The electron microscopy of magnetic structures of bulk specimens. *Uspekhi Fiz. Nauk.* **102,** 529.

Rau, E. I., Lukianov, A. E., Spivak, G. V., and Gvosdover, R. S. (1970). The measurement of magnetic fields by mirror electron microscopy technique. *Izv. Akad. Nauk SSSR, Ser. Fiz.* **34,** 1539.

Schwartze, W. (1962). Origin and effects of ions in the electron-mirror surface microscope. *Čech. Čas. Fys.* **12,** 488.

Schwartze, W. (1964). "Feldkontrast" im Emissionsmikroskop und Elektronenspiegel. *Proc. 3rd Eur. Reg. Conf. Electron Microsc.* (Prague), Vol. A, p. 15.

Schwartze, W. (1966). Ein Elektronenspiegel. Oberflächenmikroskop. *Exp. Tech. Phys.* **14,** 293.

Schwartze, W. (1967). Elektronenspiegel. Oberflächenmikroskopie mit fokussierter Abbildung. *Optik* **25,** 260.

Sedov, N. N. (1968). The solution of the inverse problem of image contrast of microfields in electron microscopy. *Izv. Akad. Nauk SSSR, Ser. Fiz.* **32,** 1175.

Sedov, N. N. (1970a). The specific features of image formation in emission and mirror electron optical systems. *Izv. Akad. Nauk SSSR, Ser. Fiz.* **34,** 1529.

Sedov, N. N. (1970b). Théorie quantitative des systèmes en microscopie électronique à balayage, à miroir et à émission. *J. Microscopie* **9,** 1.

Sedov, N. N. (1971). The depiction of two-dimensional microfields in non-transmission electron microscopes. *Vestnik Mosk. Univers. Ser. III (Fiz., Astr.)* **12** (1), 106.

Sedov, N. N., Spivak, G. V., and Ivanov, R. D. (1962). Electron optical investigation of p-n junction in germanium and silicon. *Izv. Akad. Nauk SSSR, Ser. Fiz.* **26,** 1332.

Sedov, N. N., Lukianov, A. E., and Spivak, G. V. (1968a). The calculation of electric microfields by their image in mirror electron microscope. *Izv. Akad. Nauk SSSR, Ser. Fiz.* **32,** 996.

Sedov, N. N., Spivak, G. V., Petrov, V. I., Lukianov, A. E., and Rau, E. I. (1968b). The image contrast formation of magnetic microfields of an arbitrary type in mirror electron microscope. *Izv. Akad. Nauk SSSR, Ser. Fiz.* **32,** 1005.

Someya, T., and Kobayashi, J. (1971). Electron mirror microscopic observation of ferroelectric domains of $Ca_2Sr(C_2H_5CO_2)_6$. *Phys. Stat. Sol.* **4,** K 161.

Someya, T., and Watanabe, M. (1968). Development of electron mirror microscope. *Proc. 4th Eur. Reg. Conf. Electron Microsc.* (Rome), Vol. 1, p. 97.

Spivak, G. V., Antoshin, M. C., Lukianov, A. E., Rau, E. I., Ushakov, O. A., Akishin, A. I., Tokarev, G. A., Lyamov, V. E., and Bartel, I. (1972). The observation of propagation of elastic waves in piezoelectrics by means of scanning and mirror electron microscopes. *Izv. Akad. Nauk SSSR, Ser. Fiz.* **36,** 1954.

Spivak, G. V., Dubinina, E. M., Dyukov, V. G., Lukianov, A. E., Sedov, N. N., Petrov, V. I., Pavlyuchenko, O. P., Saparin, G. V., and Nevzorov, A. N. (1968a). The stroboscopic electron microscopy. *Izv. Akad. Nauk SSSR, Ser. Fiz.* **32,** 1098.

Spivak, G. V., Gvosdover, R. S., Lukianov, A. E., Sedov, N. N., Petrov, V. I., and Butilkin, A. I. (1968c). To the verification of the theory of magnetic contrast formation in mirror electron microscope. *Izv. Akad. Nauk SSSR, Ser. Fiz.* **32,** 1211.

Spivak, G. V., Igras, E., Pryamkova, I. A., and Zheludev, I. S. (1959a). On the observation of domain structure of barium titanate by means of electron mirror. *Kristallografiya* **4,** 123.

Spivak, G. V., and Ivanov, R. D. (1963). The mirror electron microscope and its application to the quantitative examination of semiconductors. *Izv. Acad. Nauk SSSR, Ser. Fiz.* **27,** 1203.

Spivak, G. V., Ivanov, R. D., Pavlyuchenko, O. P., and Sedov, N. N. (1963a). On the origin of image contrast formation in mirror, emission and scanning electron optical systems. *Izv. Akad. Nauk SSSR, Ser. Fiz.* **27,** 1139.

Spivak, G. V., Ivanov, R. D., Pavlyuchenko, O. P., Sedov, N. N., and Shvetz, V. F. (1963c). The visualization of magnetic audio recording field by means of electron mirror. *Izv. Akad. Nauk SSSR, Ser. Fiz.* **27,** 1210.

Spivak, G. V., Kirenski, L. V., Ivanov, R. D., and Sedov, N. N. (1961b). The development of mirror electron microscopy of magnetic microfields. *Izv. Akad. Nauk. SSSR, Ser. Fiz.* **25,** 1465.

Spivak, G. V., and Lukianov, A. E. (1966a). The observation and measurement of a.c. magnetic field of magnetic recording head in mirror electron microscope. *Izv. Akad. Nauk. SSSR, Ser. Fiz.* **30,** 803.

Spivak, G. V., and Lukianov, A. E. (1966b). The electron mirror microscope observation of p-n junction in pulsed mode. *Izv. Akad. Nauk SSSR, Ser. Fiz.* **30,** 781.

Spivak, G. V., Lukianov, A. E., and Abalmazova, M. G. (1964b). The observation of local contamination films in mirror electron microscope. *Izv. Akad. Nauk SSSR, Ser. Fiz.* **28,** 1382.
Spivak, G. V., Lukianov, A. E., Rau, E. I., and Gvosdover, R. S. (1971). Electron mirror measurements of magnetic fields over audio and video heads and tapes. *Intermag-71. IEEE Trans. Magnet.* **MAG-7,** 684.
Spivak, G. V., Lukianov, A. E., Toshev, S. D., and Koptsik, V. A. (1963b). On the observation of domain structure of triglycine sulfate by means of electron mirror. *Izv. Akad. Nauk SSSR, Ser. Fiz.* **27,** 1199.
Spivak, G. V., and Lyubchenko, V. I. (1959). On the resolving power of immersion objective in the presence of electrical and magnetic microfields on the cathode surface. *Izv. Akad. Nauk SSSR, Ser. Fiz.* **23,** 697.
Spivak, G. V., Pavlyuchenko, O. P., Ivanov, R. D., and Netischenskaja, G. P. (1964c). Die struktur des magnetfelds innerhalb der Domänewand mit Hilfe des Spiegelelektronenmikroskopes sichtbar gemacht. *Proc. 3rd Eur. Reg. Conf. Electron Microsc.* (Prague), Vol. A, p. 293.
Spivak, G. V., Pavlyuchenko, O. P., and Lukianov, A. E. (1966). The depiction of magnetic microfields in mirror electron microscope. *Izv. Akad. Nauk. SSSR, Ser. Fiz.* **30,** 813.
Spivak, G. V., Prilezhaeva, I. N., and Azovtsev, V. K. (1955). Magnetic contrast in electron mirror and observation of domains in ferromagnetics. *Dokl. Akad. Nauk SSSR* **100,** 965.
Spivak, G. V., Pryamkova, I. A., Fetisov, D. V., Kabanov, A. N., Lazareva, L. V., and Shilina, L. I. (1961a). The mirror electron microscope for the investigation of surface structures. *Izv. Akad. Nauk SSSR, Ser. Fiz.* **25,** 683.
Spivak, G. V., Pryamkova, I. A., and Igras, E. (1959b). On the observation of ferromagnetic and ferroelectric domains by means of electron mirror. *Izv. Akad. Nauk SSSR, Ser. Fiz.* **23,** 729.
Spivak, G. V., Pryamkova, I. A., and Sedov, N. N. (1960). On the electron optical contrast formation during observation of "patch fields" on the surface of emitters. *Izv. Akad. Nauk SSSR, Ser. Fiz.* **24,** 640.
Spivak, G. V., Schischkin, B. B., Lukianov, A. E., Mitschurina, K. A. (1964a). Über das quantitative Studium der Emitter mittels eines Hochvakuumemissionsmikroskopes. *Proc. 3rd Eur. Reg. Conf. Electron Microsc.* (Prague), Vol. A, p. 109.
Spivak, G. V., Shakmanov, V. V., Petrov, V. I., Lukianov, A. E., and Yakunin, S. I. (1968b). On the use of deflection plates gates in stroboscopic electron microscopes. *Izv. Akad. Nauk SSSR, Ser. Fiz.* **32,** 1111.
Szentesi, O. I. (1972). Stroboscopic electron microscopy at frequencies up to 100 MHz. *J. Phys.* **E5,** 563.
Turchin, V. F., Kozlov, V. P., and Malkevich, M. S. (1970). The use of methods of mathematical statistics for the solution of incorrect problems. *Uspekhi Fiz. Nauk,* **102,** 345.
Vassoille, R., Guittard, C., Pernoux, E., and Bernard, R. (1970). Monochromatisation et mesure de l'etalement energetique d'un faisceau d'electrons. *Microscopie Electronique. Proc. 7th Int. Conf. Electron Microsc.*, Vol. 1, p. 179.
Vertzner, V. N., and Chentzov, Yu. V. (1963). The scanning-mirror electron microscope. *Pribory i Techn. Eksper.* **N 5,** 180.
Wang, S. T., Challis, L. J., and Little, W. A. (1966). An electron mirror microscope study of the Abricosov vortex structure of superconductors. *Proc. 10th Int. Conf. Low Temperature Physics* (Moscow, USSR), Vol. 2A, p. 364.
Warshalovitch, D. A., and Dyakonov, M. I. (1971). Quantum theory of optical frequency modulation of an electron beam. *Zh. Eksper. Teor. Fiz.* **60,** 90.

Wiskott, D. (1956a). Zur Theorie des Auflicht-Elektronenmikroskops. I. Geometrische Elektronenoptik in der Umgebung des Objekts. *Optik* **13**, 463.

Wiskott, D. (1956b). Zur Theorie des Auflicht-Elektronenmikroskops. II. Wellenmechanische Elektronenoptik in der Umgebung des Objects. *Optik* **13**, 481.

Wohlleben, D. (1971). Magnetic phase contrast. In: *Electron Microscopy in Material Science*, p. 713. Academic Press, New York.

Zworykin, V. K., Morton, G. A., Ramberg, E. G., Hillier, J., and Vance, A. W. (1945). *Electron Optics and the Electron Microscope*. John Wiley & Sons, New York.

8. ELECTRON MICROSCOPY OF BANDED MAMMALIAN CHROMOSOMES

Gary D. Burkholder

Department of Anatomy, College of Medicine, University of Saskatchewan, Saskatoon, Saskatchewan, Canada

---

INTRODUCTION

In recent years, several methods have been developed which produce differentially staining segments (bands) along metaphase chromosomes. These methods include Q-banding, C-banding, G-banding, and R-banding (for reviews see: Miller *et al.*, 1973; Hsu, 1974; Schnedl, 1974; Dutrillaux and Lejeune, 1975; Ruzicka, in this volume). The impact of these banding methods on the study of human and other mammalian chromosomes has been monumental, for each chromosome has its own particular pattern of banding, and consequently every chromosome in the complement can be unequivocally identified. Not only have these methods found widespread applicability in the identification of normal and rearranged chromosomes, but they have also stimulated additional probing of the structural organization of chromosomes.

Virtually all of the work on chromosome banding has been performed at the light microscope level of resolution, and the bands at this magnification appear to be rather gross features of chromosomes. Although electron microscopy can provide a wealth of information not attainable with the light microscope, few attempts have been made to study banded chromosomes at high resolutions (Comings *et al.*, 1973; Bahr *et al.*, 1973; Green and Bahr, 1975). Such high resolution studies are important for two reasons: (1) They may reveal previously unidentifiable bands, that is, bands not resolved by light microscopy. This may permit the construction of more accurate chromosome banding maps, particularly in humans, and may be useful in more precisely pinpointing the exact location of breaks and exchanges in aberrant chromosomes. (2) They may improve our knowledge of the structural basis of chromosome banding and provide some insight into the mechanism of banding.

There are two conventional methods of observing chromosomes by electron microscopy: thin sectioning and whole mount electron microscopy. Since chromosomes are relatively large structures, thin sectioning only permits the examination of a wafer of the chromosome in one section, and this method has provided little information on chromosome structure or organization. The most common method for the ultrastructural analysis of metaphase chromosomes is the technique of whole mount electron microscopy (Gall, 1963; DuPraw, 1965; Hayat and Zirkin, 1973), in which unfixed cells are spread on the surface of distilled water. This method has proved to be very useful and has provided most of our current-day knowledge of chromosome structure (DuPraw, 1965, 1970; Comings and Okada, 1972). Several investigators have used whole mount electron microscopy to study chromosome banding (Comings *et al.*, 1973; Bahr *et al.*, 1973; Green and Bahr, 1975); however, this method commonly produces various degrees of stretching and/or dispersion of the chromosomes and chromatin fibers, which can be a distinct disadvantage for investigations of chromosome banding since the banding methods are normally performed on condensed and fixed chromosomes.

An ideal method for ultrastructural studies of chromosome banding should have the following attributes: (1) The methods of chromosome preparation and banding, for electron microscopy, should be the same as those already established and routinely used for light microscopy. (2) The same chromosome should be observable by both light and electron microscopy so that comparative studies of the banding patterns can be performed at the two levels of resolution. (3) The method should, as far as possible, preserve the normal chromatin organization of the chromosomes as it is known from whole mount electron microscopy.

A method of observing chromosome bands at high resolution, fulfilling all of these criteria, has been developed (Burkholder, 1974, 1975). This chapter describes the details of this method. For a related method, the reader is referred to Ruzicka in this volume.

COLLECTION AND PREPARATION OF METAPHASE CELLS

Method.

1. Accumulate metaphase cells with colcemid (0.05 μg/ml), 10 min to 4 hr.
2. Collect metaphase cells by selective detachment, or alternatively, collect all of the cells from the culture by trypsinization or scraping.

3. Pellet the cells by centrifugation at 250 *g*, 3–5 min.
4. Add 0.075 M KCl to the cell pellet and incubate at 37°C for 5–15 min.
5. Fix the cells in three changes of methanol–glacial acetic acid (3 : 1), 20 min each.
6. Suspend the cells in a small volume of 6 : 1 methanol–glacial acetic acid.

Comments. Any type of actively dividing mammalian cell may be employed as a source of metaphase chromosomes for banding and electron microscopy. I have used Chinese hamster Don cells, human amnion, and HeLa cells. The cells are grown as monolayer cultures in glass Blake bottles and are used during the logarithmic phase of cell growth to ensure an ample supply of metaphase cells.

Colcemid (Ciba), which arrests cells at metaphase and causes chromosome contraction, is added to each log phase culture in a final concentration of 0.05 μg/ml for a period of 10 min to 4 hr, depending on the cell type. The chromosomes of amnion and HeLa cells are more readily contracted by colcemid than those of Chinese hamster cells, and consequently require only a short exposure time, or no colcemid treatment at all. Overcontracted chromosomes should be avoided since they band very poorly.

It is usually advantageous to collect metaphase cells by the method of selective detachment (Terasima and Tolmach, 1963; Stubblefield and Klevecz, 1965), rather than harvesting the whole culture. Although this procedure requires the use of larger numbers of cells (usually 6–8 Blake bottle cultures are used for one chromosome harvest), it considerably increases the percentage of metaphases in the final preparation, thereby facilitating scanning of the grids. Following the exposure to colcemid, each culture is repeatedly shaken against the palm of the hand to dislodge the loosely attached metaphase cells into the medium. The proportion of detached cells can be monitored by examining each culture with an inverted microscope.

The method of cell harvesting is the conventional technique used in many laboratories to prepare metaphase chromosomes for light microscopy (Moorhead *et al.*, 1960; Hsu, 1972), and involves treatment of the cells with a hypotonic solution (0.075 M KCl) followed by fixation in methanol–glacial acetic acid (3 : 1). After fixation, the quality of the harvest may be monitored by placing a drop of the cell suspension on a clean glass slide and allowing it to air-dry. By phase contrast microscopy, the chromosomes should appear well-spread, flat, and free of background cytoplasm. Finally, for electron microscopy, the cells are suspended in a small quantity of diluted fixative made from six parts methanol and one part glacial acetic acid. In some cases, the cells may be left in this dilute fixative overnight in a refrigerator and chromosome preparations can be made the following day. This prolonged fixation often results in better chromosome spreads and flatter chromosomes.

PREPARATION OF CHROMOSOMES FOR ELECTRON MICROSCOPY

Method.

1. Make Formvar films by dipping standard glass microscope slides into 0.5% Formvar made in 1,2-dichloroethane.
2. Coat the films with carbon, using a vacuum evaporator.
3. Score the edges of the Formvar-carbon film with a scalpel, breathe heavily on the film so it becomes moistened with a layer of condensation, and lower the slide into a tray of distilled water so that the film separates from the slide and floats on the surface of the water (Fig. 8.1A).
4. Place a stiff wire screen (approximately 3 X 4 cm) on the bottom of the tray of water, and arrange 75-mesh grids on top (7–9 grids per screen).
5. Put a drop of the cells, suspended in 6:1 methanol–glacial acetic acid, on top of the floating film and air-dry (Fig. 8.1B).
6. Using forceps, lift the wire screen, with grids, out of the water so that the Formvar-carbon film is spread over the grids (Fig. 8.1C).
7. Dry at room temperature overnight.

Comments. This method is basically the standard method used in the preparation of Formvar films and their transfer to grids (Hayat, 1970), except that the films are carbon-coated before they are floated on water, and the specimens are placed on the films before the latter are attached to grids. A black developing tray filled with distilled water is useful for this procedure since the dull black background facilitates visualization of the floating films and the air-drying of the cells onto the films.

When a drop of cells, suspended in methanol–glacial acetic acid, is placed on Formvar-carbon films attached to glass slides (glass substrate), the acetic acid in the fixative causes the film to crack and partially disintegrate. This problem is avoided when water is used as a substrate for the Formvar-carbon film. The fluid substrate protects the floating film from the destructive effect of the fixative in some unknown manner.

Carbon coating is essential for the protection of the Formvar films since Formvar films alone will be destroyed by the fixative regardless of whether they are on glass or water substrates. The degree of carbon coating is quite critical. If the Formvar is sufficiently coated with carbon, the drop of cell suspension does not flow evenly over the surface of the film, and drys too slowly to produce good chromosome spreading. On the other hand, if the Formvar is too heavily carbon-coated, the film is very difficult to separate from the glass slide. The latter problem is somewhat alleviated by precleaning the slides in a 1:10 dilution of 7X detergent (Linbro Chemical Co., New Haven, Conn.). A moderately heavy coating of carbon is the best. The degree of coating is judged by including a small piece of white filter paper, partially covered by the slides, in the evaporation chamber. The exposed part of the filter paper appears to be a

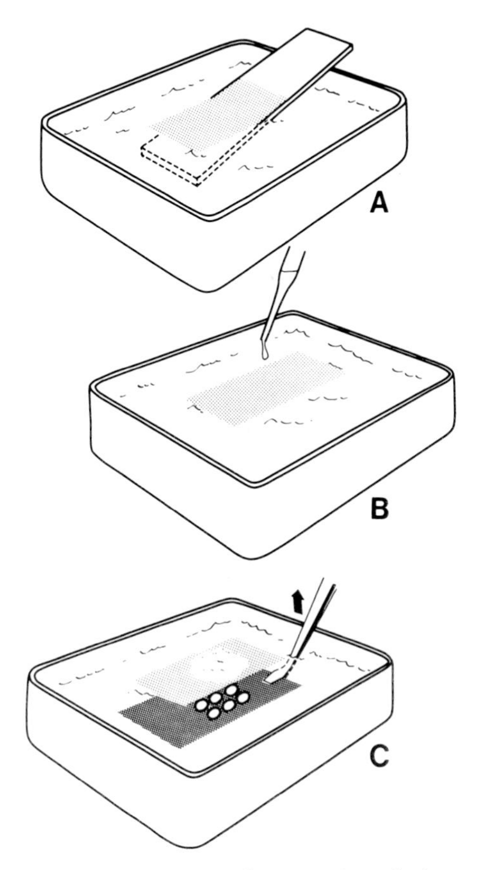

Fig. 8.1 Diagram illustrating the method of preparation of chromosomes for electron microscopy. (A) Formvar-carbon films, made on glass slides, are floated on the surface of a tray of water. (B) A drop of the cells, suspended in 6:1 methanol: glacial acetic acid, is placed on the film and allowed to air-dry. (C) Grids are arranged on a stiff wire screen placed on the bottom of the tray, and the screen is then lifted out of the water so that the film, bearing the chromosome preparations, is spread over the grids.

sandy-brown color when the correct carbon thickness has been attained. I use a Balzers high vacuum freeze etch unit (Model BA36OM) for carbon coating, but any carbon evaporator will do. The films should be transferred to water immediately after carbon-coating; otherwise they become difficult to separate from the glass.

The reduction in the acid content of the fixative (by using 6:1 instead of 3:1) prior to spreading on the Formvar-carbon films is a necessary part of the procedure. Although the carbon-coated Formvar is impervious to 3:1, this fixative does not spread thinly over the film but remains in a compact drop which dries very slowly.

When the droplet of cells in the 6:1 fixative is placed on the floating film, the solution should spread evenly over the film. Often the film will move wildly about on the surface of the water as part of the fixative flows off the edge. After the cells have air-dried, which usually takes ~1 min, the film will appear wrinkled in the area where the cells are located; however, this wrinkling disappears when the films are attached to the grids. It is also possible to flame-dry the cells onto the film by igniting the solution, but I have found it difficult to subsequently band flame-dried chromosomes.

The film, supporting the air-dried cells, may be picked up on a microscope slide and examined in a wet condition by phase contrast microscopy in order to determine the density of the metaphase cells, the degree of chromosome spreading, and the flatness of the chromosomes. Optimal metaphase spreads are essential in order to achieve good chromosome banding for electron microscopy; therefore the films should be monitored before they are attached to grids, thus avoiding the use of preparations which might later prove useless for banding. After examination, the film may be refloated on the water provided that it has not been allowed to dry onto the slide.

If the cells are too dense in the air-dried preparation, the suspension may be diluted by the addition of more 6:1; if they are too sparse, the suspension may be centrifuged and the pellet resuspended in a smaller volume of the fixative. Considerably cytoplasmic background around the chromosomes usually indicates that the hypotonic treatment was not of sufficient duration. If the chromosomes are widely scattered, and the chromosome complements are incomplete, the hypotonic treatment was probably too long. If the chromosomes appear shiny and not flat, the cells may be improperly fixed.

CHROMOSOME BANDING

G-Banding

Method.

1. Remove individual grids from the screen using jeweler's forceps. Check the quality of the chromosome preparation by placing the grid on a glass slide and examining it with phase contrast optics.
2. Float a grid with the chromosomes face-down on 0.1% trypsin (Difco) made in Ca^{2+}- and Mg^{2+}-free balanced salt solution. The length of time of treatment can only be determined by trial and error. Start with a 1-min treatment and proceed from there.
3. Rinse the grid in two changes of 70% ethanol, followed by two changes of 100% ethanol. Air-dry.
4. Check the degree of banding by phase contrast microscopy. Alternatively, the chromosomes may be stained with Giemsa and then examined. If the chromosomes are not sufficiently banded, the grids may be re-exposed to trypsin.

Comments. There are a large variety of methods which can be used to reveal G-bands. These include: (1) incubation at 60°C in a salt solution (ASG technique) (Sumner *et al.*, 1971); (2) denaturation in NaOH followed by incubation in salt solution or phosphate buffer at 60°-65°C (Drets and Shaw, 1971; Schnedl, 1971); (3) staining chromosomes in Giemsa at pH 9.0 (Giemsa 9 staining) (Patil *et al.*, 1971); (4) pretreatment with trypsin (Seabright, 1971; Wang and Fedoroff, 1972), α-chymotrypsin (Finaz and de Grouchy, 1971), pepsin (Lee *et al.*, 1972), pronase (Finaz and de Grouchy, 1972), urea (Kato and Yosida, 1972), guanidine-HCl (Kato and Moriwaki, 1972), or sodium dodecyl sulfate (Kato and Moriwaki, 1972). These methods generally involve some kind of pretreatment of the chromosomes prior to staining with Giemsa; however, it has also been reported that G-bands may sometimes be observed in fixed but otherwise untreated chromosomes (McKay, 1973; Yunis and Sanchez, 1973). The addition of Actinomycin-D or other agents into cultures for a few hours before harvest can also induce G-banding without any postfixation treatment (Hsu *et al.*, 1973).

The trypsin method is a fast and easy-to-use technique; however, a number of laboratories have had difficulties reproducing this method. It appears that several factors are critical for the successful preparation of G-banded chromosomes with trypsin, and each laboratory has to adjust the basic methodology to fit the local conditions in order to achieve good banding. One key to success is to have optimal chromosome preparations to begin with. The chromosomes should be well spread and free of cytoplasmic background, and should appear flat by phase contrast microscopy, rather than shiny. The age of the preparations has also been found to be a critical factor in determining the quality of the chromosome bands. For slide preparations, some investigators recommend that banding be done soon after preparation; others let the slides age for several days. I usually store the film preparations in a dry place for 2-3 days before banding, and have found that good banding can be obtained for several days thereafter. Preparations two to several weeks old do not band very well.

The trypsin solution is also critical. I use the crude powdered trypsin (1:250) made by Difco Laboratories, Detroit, Michigan, and have found that it works as well as highly purified enzyme. A stock solution of 5% trypsin, made in Puck's balanced salt solution (Ca^{2+}- and Mg^{2+}-free), is stored frozen. To make a working solution (0.1%), 0.75 ml of the stock solution is mixed with 40 ml of a 1:1 mixture of Puck's BSS and versene (0.02% EDTA in saline) and adjusted to pH 8.0 with sodium bicarbonate. This working solution is used for banding at room temperature and may be stored in the refrigerator for several days without noticeable deterioration. It may sometimes be convenient to stabilize the temperature of the trypsin at 0°C on ice, or at 37°C on a slide warming tray, particularly if the ambient air temperature is quite variable. It is convenient to use 10-ml plastic disposable beakers to hold the solutions for banding and staining the grid preparations.

The time of exposure to the trypsin solution, to obtain optimal banding, can only be determined by trial and error. Start with a one-minute treatment, and monitor the extent of banding by placing the dry grid loosely on a microscope slide and examining it by phase contrast microscopy. Clear bands seen with phase contrast optics usually represent overtreated chromosomes when stained with Giemsa. It is often useful to use two or three test grids initially, staining with Giemsa after each treatment until the correct exposure time is determined. If the chromosomes appear undertreated (uniform in density or staining), the grid may be re-exposed to trypsin for additional time; if the chromosomes are overtreated (appear ghostlike), the time of exposure should be reduced. It may be necessary to vary the trypsin concentration to obtain optimal banding. Lowering the trypsin concentration will slow down the reaction so that longer exposures will be necessary; however, it is far easier to control the banding reaction with longer exposure times than with very short ones. Once good banding has been obtained, additional grids may be treated using the same conditions.

C-Banding

Methods. There are two methods of producing C-bands for electron microscopy:

(1) NaOH Treatment. Invert the grids onto the surface of 0.07 N NaOH for 5-30 sec, rinse vigorously in two changes of 70% and two changes of 100% ethanol, and air-dry.

(2) Overtrypsinization. If one prolongs the trypsin treatment beyond the time required to produce optimal G-bands, C-bands will be produced. The length of time for this treatment can only be ascertained by trial and error. After trypsinization, the grids are rinsed in 70% and 100% ethanol and air-dried.

Comments. The classical method of C-banding, which demonstrates constitutive heterochromatin, was developed by Arrighi and Hsu (1971) and involves several steps: pretreatment of chromosome preparations with 0.2 N HCl for 30 min, rinsing in distilled water, treatment with 0.07 N NaOH for 2 min, rinsing in ethanol, overnight incubation in a salt solution at 65°C, and staining with Giemsa. Hsu (1974) has recently recommended that the NaOH solution be made as a mixture of 1 part 0.07 N NaOH and 6 parts of 2 x SSC (1 x SSC is 0.15 M NaCl, 0.015 M trisodium citrate), and it has been reported that barium hydroxide octahydrate (5% aqueous) may be substituted for NaOH (Sumner, 1972). For electron microscopy, NaOH treatment alone is sufficient to produce C-banding.

As with G-banding, the success of the C-banding procedure hinges on the use of good chromosome preparations. The time of exposure to NaOH is very critical, and an experimental series of grids should be treated for various lengths of time to determine the optimal exposure. The concentration of NaOH may be changed, if necessary. C-banding may readily be observed by phase contrast microscopy of unstained preparations, or by ordinary optics after staining with Giemsa. Overtreatment results in swelling and distortion of the chromosomes and a hollow, ghostlike appearance.

C-banding may also be produced by overtrypsinization. With prolonged trypsin treatment, the G-bands gradually disappear and are replaced by C-bands. At some intermediate stages, both G- and C-bands may be observed on the chromosomes. Often one can observe C-bands when trying to establish the conditions for G-banding. This is a sure indication that the chromosomes are overtreated if one is really trying to produce G-bands.

CHROMOSOME STAINING

Methods.

(1) Giemsa Staining. Float the grid, chromosome side down, on Giemsa solution (Fisher Scientific Giemsa, diluted 1 part to 20 parts with Sörensen's buffer, pH 6.8) for 5 min. Rinse the grid in distilled water and air-dry.

(2) Uranyl Acetate Staining. Float the grid on 2% aqueous uranyl acetate (filtered through three layers of Whatman #1 filter paper) for 10 min, then rinse in 50%, 70%, 95%, and 100% ethanol, approximately 1 min in each. Rinse in two changes of isoamyl acetate, 10 min each, and finally air-dry.

Comments. For light microscopy, the chromosomes may be examined in an unstained state using phase contrast optics or by ordinary optics after Giemsa staining. Giemsa staining is the preferential method, since the chromosomes and bands appear more sharply defined. For electron microsopy, the chromosomes may be left unstained, or stained with Giemsa or uranyl acetate. The optimal method for electron microscopy is uranyl acetate staining; however, some of the banded chromosome preparations appear almost equally clear when examined in an unstained state. Giemsa staining increases the electron density of the chromosomes, and the bands do not appear as sharply defined as with the other methods.

If comparisons are to be made between the light microscopic and electron microscopic appearance of the same chromosomes, it is probably advantageous to examine and photograph the chromosomes with the electron microscope first, and subsequently stain the grids with Giemsa for light microscopy. Staining the chromosomes with uranyl acetate does not interfere with subsequent Giemsa staining.

CONCLUDING REMARKS

Representative examples of G-banded chromosomes, visualized by electron microscopy, are shown in Figs. 8.2–8.5. Untreated control chromosomes appeared to be uniformly dense structures, and chromatin fibers were not visible,

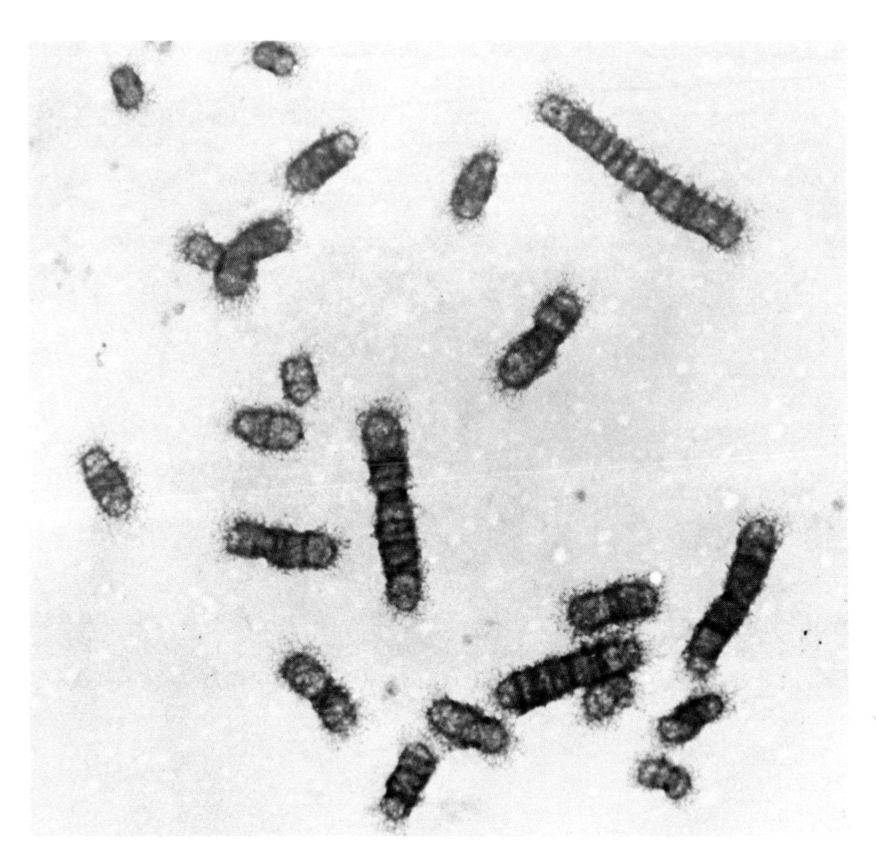

Fig. 8.2. A survey electron micrograph of a complete Chinese hamster chromosome complement, G-banded with trypsin. Unstained preparation. ×2,408. (*From Burkholder, G. D. (1974).* Nature **247**, *292.*)

presumably because they were very compactly organized within the fixed chromosomes. Trypsinized preparations readily revealed G-bands, and at low magnifications whole chromosome complements could be observed (Fig. 8.2), making studies of the total complement possible. The bands appeared to be electron-dense regions which were often more finely resolved than they were by light microscopy. The G-banded chromosomes showed various degrees of dispersion of the chromatin fibers, depending on the length of trypsinization (Figs. 8.3 and 8.4). Mildly treated chromosomes showed little or no dispersion of chromatin fibers, while after longer treatments the chromatin fibers appeared to spew out of the chromosome. This dispersion of the chromatin fibers affected the interbands first, with the bands remaining relatively intact (Fig. 8.3). With longer trypsinization, extensive dispersion was seen in both band and interband regions; however, the bands still remained relatively condensed (Fig. 8.4).

Very prolonged trypsinization completely removed all of the chromatin, and the chromosomes appeared as ghostlike structures.

It appears, from these results, that chromatin dispersion is not a prerequisite for G-banding, since bands can be clearly seen before any chromatin dispersion occurs. The bands, however, represent areas of packed chromatin fibers which are relatively more resistant to unfolding and dispersion than the interbands. These results are consistent with the hypothesis that the bands are inherent structural features of chromosomes and not artifactual by-products of the pretreatment or staining reaction. It is postulated that the particular types of

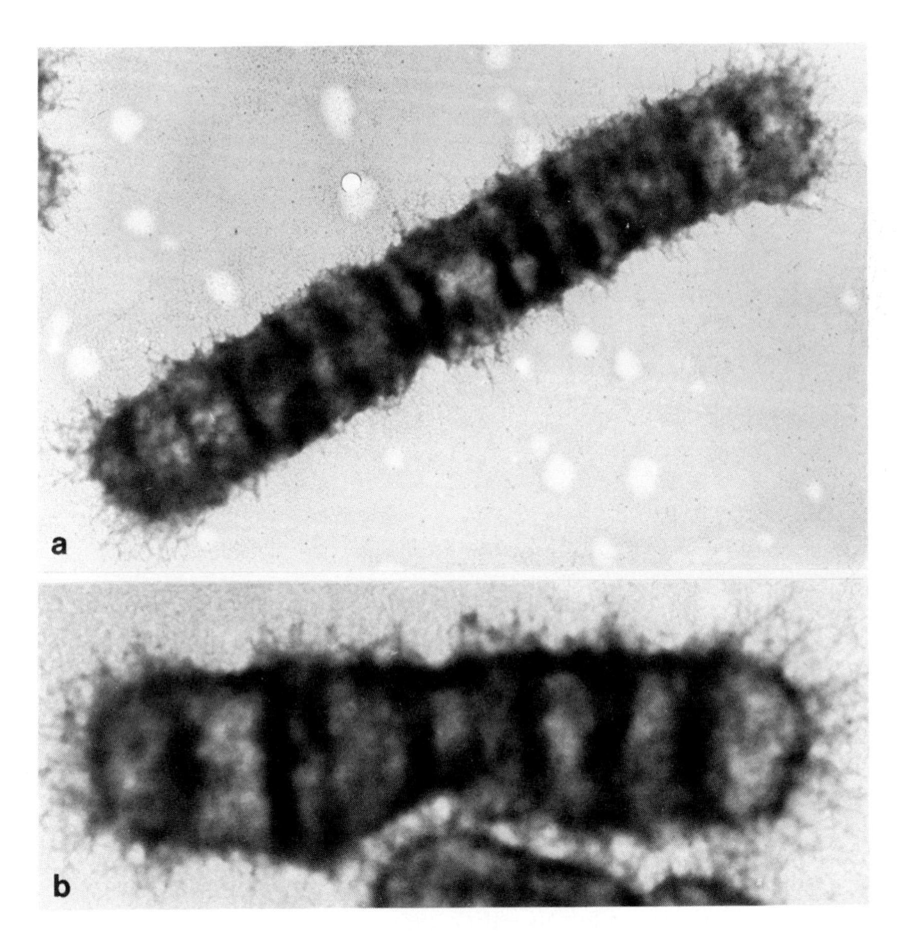

Fig. 8.3. G-banded Chinese hamster chromosomes. (a) Chromosome showing slight dispersion of the chromatin fibers, primarily from interband regions. Unstained preparation. ×9,180. *(From Burkholder, G. D. (1975).* Exp. Cell Res. **90**, *269*.) (b) Chromosome showing slight dispersion of the chromatin fibers from both band and interband regions. Unstained preparation. ×11,610.

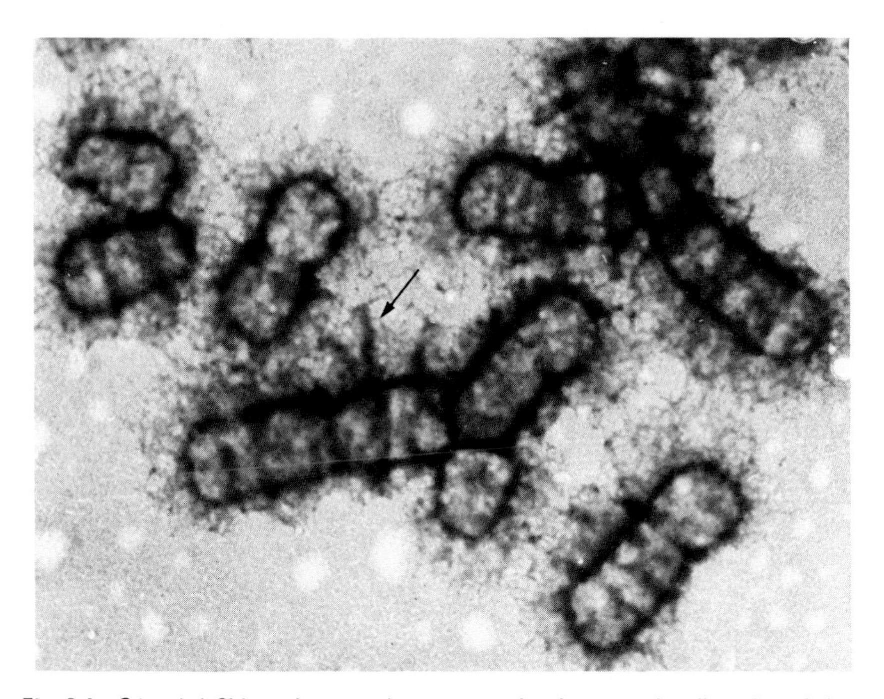

Fig. 8.4. G-banded Chinese hamster chromosomes showing extensive dispersion of chromatin fibers from both band and interband regions. Chromatin from the band regions, although displaced outside the chromosome, retains a condensed appearance (*arrow*). Unstained preparation. ×5,412.

proteins associated with the band regions may render these areas more resistant to dispersion than the interband regions.

In many of the trypsin-treated preparations, the chromatin fibers were clearly visible (Fig. 8.5), and many of these fibers were comparable in diameter (~250 Å) to those seen by whole mount electron microscopy (DuPraw, 1965, 1970; Comings and Okada, 1972). Fixation in acetic-methanol and subsequent trypsinization did not appear to have an adverse effect on the chromatin fiber morphology, although it did produce an apparent stickiness of the chromatin fibers since adjacent fibers were often fused together.

C-banding was clearly seen after either NaOH treatment (Fig. 8.6) or overtrypsinization (Fig. 8.7). Short NaOH or trypsin treatment produced G-bands, while C-bands resulted from a longer treatment which disrupted the relatively delicate G-bands. The formation of G- and subsequently C-bands is therefore a sequential process depending on the degree of treatment. Different exposure times produced different degrees of chromatin extraction, and it is clear from the electron micrographs (Figs. 8.6 and 8.7) that the mechanism of C-banding involves the selective loss of non-C-band chromatin, while the constitutive heterochromatin remains largely intact and electron-dense. It seems reasonable that the particular proteins associated with the DNA, or the degree of binding of proteins to DNA, in constitutive heterochromatin may make this chromatin resistant to dispersion and extraction by trypsin or NaOH.

These results clearly indicate that this method of chromosome preparation for electron microscopy is useful for probing the structural organization of banded chromosomes, particularly the organization of chromatin within the band and interband regions. Since the same methods of chromosome preparation (hypotonic treatment, fixation in methanol-glacial acetic acid) and chromosome banding are used for both light and electron microscopy, the two levels of resolution are strictly comparable: the electron microscope reveals the fine details of what is grossly observed at the lower level of resolution. The effect of any type of treatment or staining reaction on fixed preparations of metaphase chromosomes can potentially be analyzed at the ultrastructural level using this method, and there are several immediate applications of the method relating to chromosome structure. These applications include: (1) A comparison of the fine structure of chromosomes banded by different G-band techniques—do these different methods (controlled heating, denaturation-renaturation, high pH

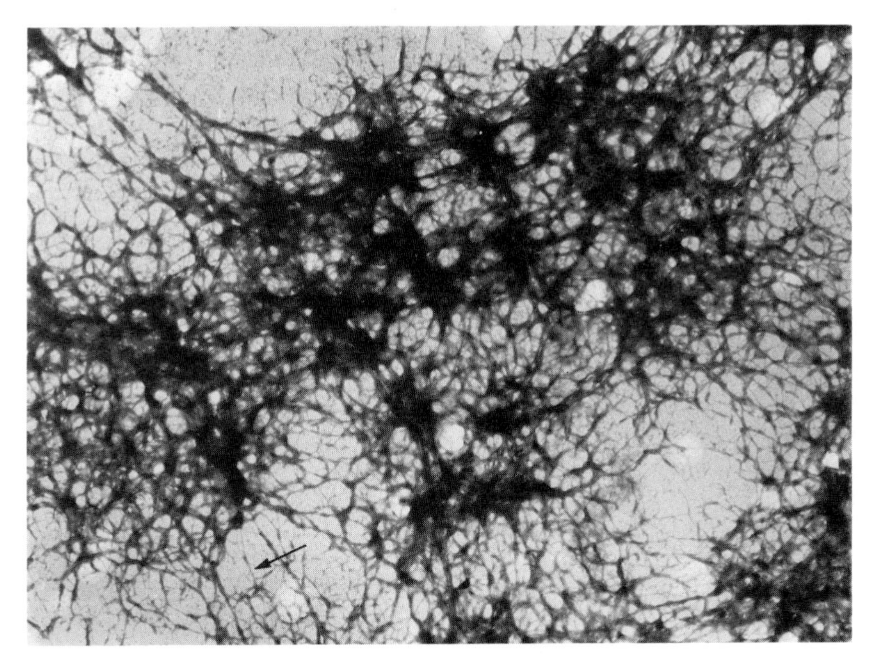

Fig. 8.5. Parts of several G-banded Chinese hamster chromosomes with well-defined chromatin fibers. The chromatin fibers are highly packed together in the band regions, less dense in the interbands. Single chromatin fibers are about 26.5 nm in width (*arrow*). Uranyl acetate staining. ×9,247.

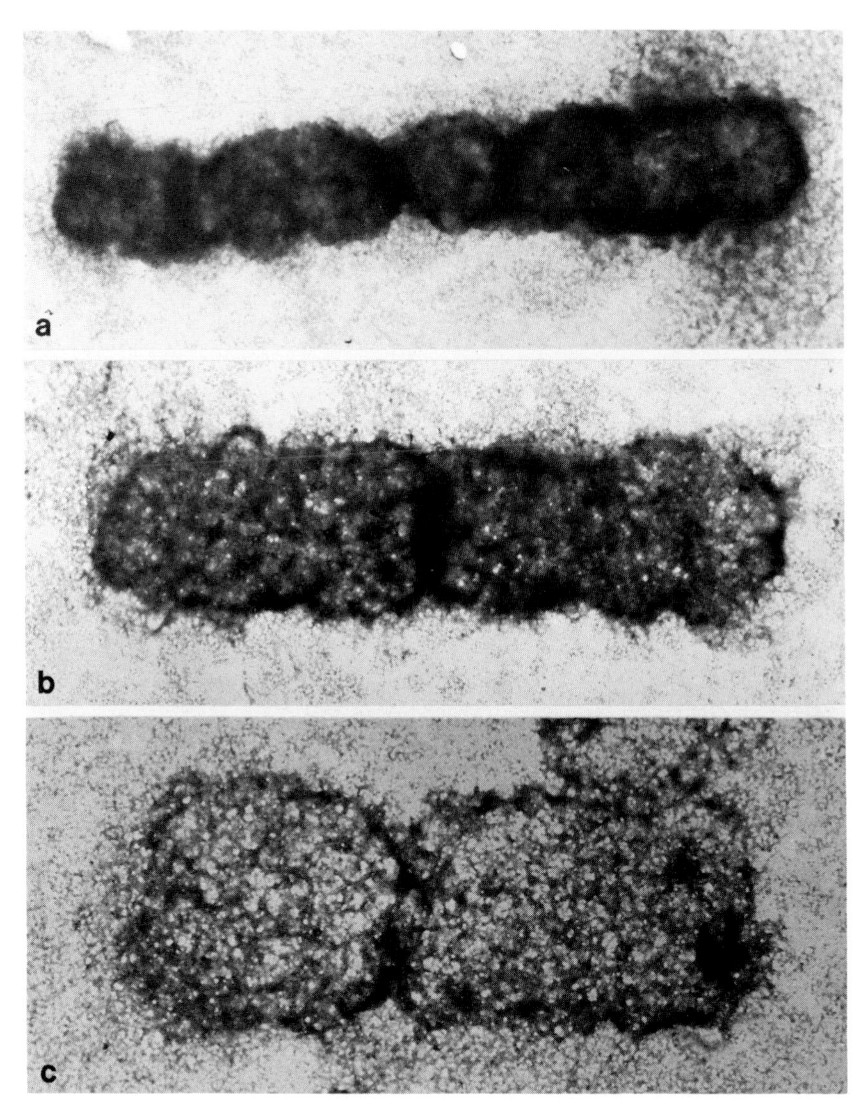

Fig. 8.6 (a)–(c). C-banded Chinese hamster chromosomes, produced by NaOH treatment. The constitutive heterochromatin in the centromere regions remains intact and resistant to extraction, while varying amounts of non-C-band chromatin are extracted. In (a), the bands seen in the chromosome arms probably correspond to G-bands. (a)–(c) Uranyl acetate staining. (a,b) ×10,560. (c) ×19,000.

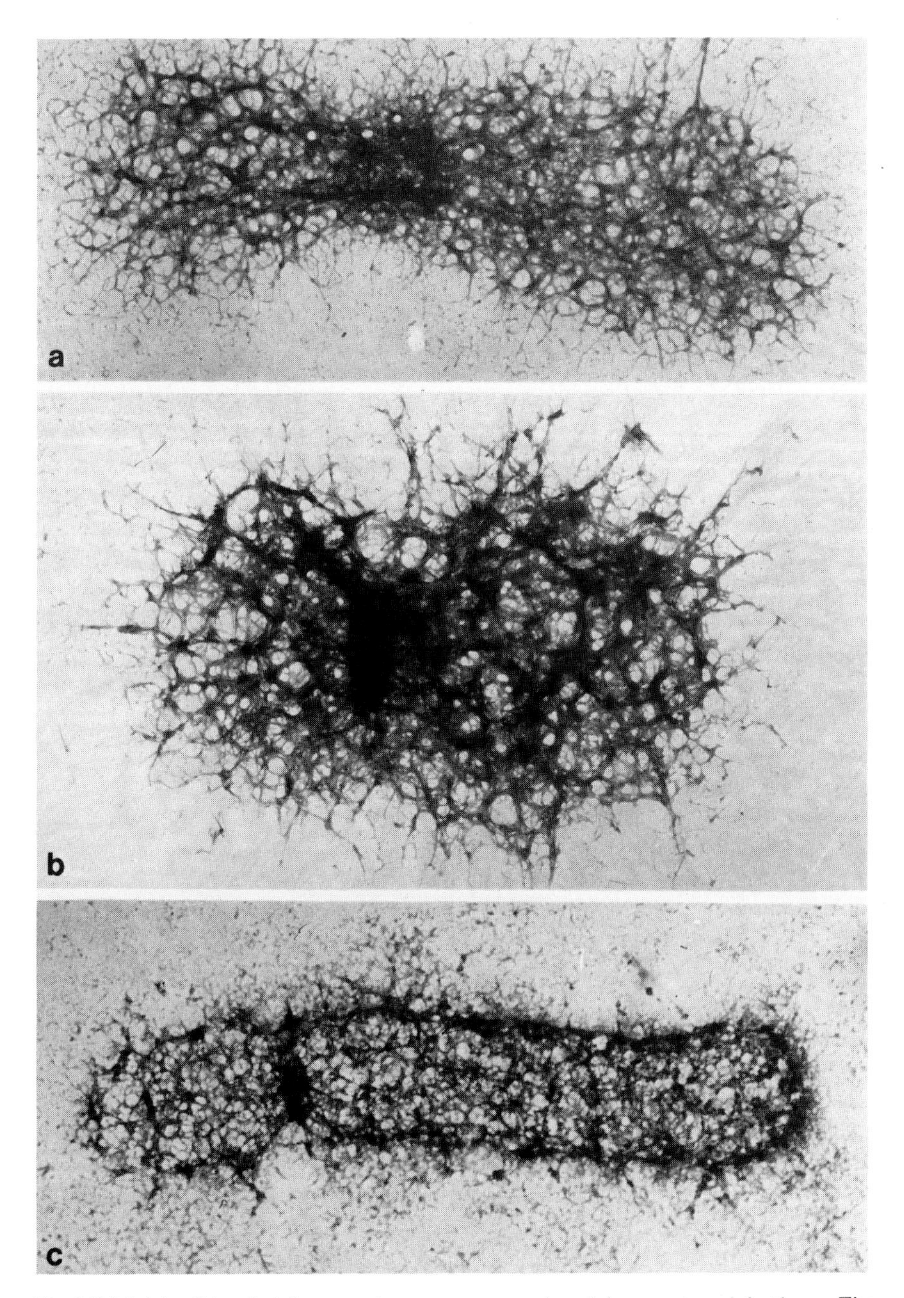

Fig. 8.7 (a)–(c). C-banded human chromosomes produced by overtrypsinization. The centromeric constitutive heterochromatin is highly condensed and resistant to extraction. Variable amounts of chromatin are extracted from the chromosome arms. (a)–(c) Uranyl acetate staining. (a) ×19,000. (b) ×17,280. (c) ×17,550.

staining, proteolytic enzymes or other chemicals) all produce a common ultrastructural appearance? (2) An investigation of the organization of chromatin in R-banded chromosomes—how does the structure of the reverse bands compare with that of the G-bands and interbands? (3) A study of the differentially stained sister chromatids produced by heat treatment of metaphase chromosomes after two rounds of DNA replication in the presence of 5-bromodeoxyuridine (Perry and Wolff, 1974; Korenberg and Freedlender, 1974)—what structural differences exist between the unifilarly-substituted (darkly stained) chromatid and the bifilarly-substituted (lightly stained) chromatid?

The normal human karyotype has been extensively studied by light microscopy of banded chromosomes, and chromosome maps based on the banding patterns have been formulated (Paris Conference, 1971). The origin of structurally abnormal chromosomes can also be determined with a high degree of accuracy since the banding pattern in each chromosome segment remains a constant feature of that segment. This has been of immense importance in clinical cytogenetics, since the accuracy of genetic counseling is often dependent upon the accuracy with which chromosomes can be identified. Considerable advantage might be gained by analyzing the human karyotype at a higher level of resolution. Can the bands visible by light microscopy be resolved as finer entities at the ultrastructural level? If finer landmarks can indeed be recognized in banded chromosomes by electron microscopy, then high resolution maps of human chromosomes would produce an even greater degree of specificity in characterizing chromosome anomalies.

In this regard, some evidence has been obtained in the Chinese hamster that additional chromosome bands may be revealed by electron microscopy (Burkholder, 1975). When the same G-banded metaphase chromosome was observed by both light and electron microscopy, some of the light-microscopic G-bands were represented by two or more ultrastructural bands. The number of bands seen in metaphase chromosomes by electron microscopy appeared to approach the increased number of bands generally seen in prometaphase chromosomes by light microscopy. It is well known that several of the prometaphase bands may amalgamate during chromosome contraction and appear as a single band at metaphase, and my results indicate that such fused bands at the light microscope level may be recognized as several discrete bands by electron microscopy. An even greater number of bands may be revealed by electron microscopy of the elongated prometaphase chromosomes.

The high degree of resolution obtained with banded Chinese hamster chromosomes suggests that this method holds particular promise for a similar high-resolution investigation of human chromosomes, particularly if prometaphase chromosomes are examined. It appears likely that the increased resolution provided by the electron microscope will reveal previously unidentifiable structural details in human chromosomes, so that more accurate chromosome maps, based on the ultrastructural G-bands, can be constructed. Such maps would provide a basis for the study of structurally abnormal human chromosomes, and would permit the mapping of break points and exchanges with an even greater degree of precision than is presently possible at the light microscope level.

This work was supported by grant number MA-5125 from the Medical Research Council of Canada. The author is a Medical Research Council Scholar.

References

Arrighi, F. E., and Hsu, T. C. (1971). Localization of heterochromatin in human chromosomes. *Cytogenetics* **10,** 81.

Bahr, G. F., Mikel, U., and Engler, W. F. (1973). Correlates of chromosomal banding at the level of ultrastructure. In: *Chromosome Identification–Techniques and Applications in Biology and Medicine* (Caspersson, T., and Zech, L, eds.). Academic Press, New York.

Burkholder, G. D. (1974). Electron microscopic visualization of chromosomes banded with trypsin. *Nature* **247,** 292.

Burkholder, G. D. (1975). The ultrastructure of G- and C-banded chromosomes. *Exp. Cell Res.* **90,** 269.

Comings, D. E., Avelino, E., Okada, T. A., and Wyandt, H. E. (1973). The mechanism of C- and G-banding of chromosomes. *Exp. Cell Res.* **77,** 469.

Comings, D. E., and Okada, T. A. (1972). Electron microscopy of chromosomes. In: *Perspectives in Cytogenetics* (Wright, S. W., Crandall, B. F., and Boyer, L., eds.), p. 223. Charles C. Thomas, Springfield, Illinois.

Drets, M. E., and Shaw, M. W. (1971). Specific banding patterns of human chromosomes. *Proc. Nat. Acad. Sci.* **68,** 2073.

DuPraw, E. J. (1965). Macromolecular organization of nuclei and chromosomes: a folded fibre model based on whole mount electron microscopy. *Nature* **206,** 338.

DuPraw, E. J. (1970). *DNA and Chromosomes.* Holt, Rinehart and Winston, Inc., New York.

Dutrillaux, B., and Lejeune, J. (1975). New techniques in the study of human chromosomes: methods and applications. *Adv. Human Genet.* **5,** 119.

Finaz, C., and de Grouchy, J. (1971). Le caryotype humain après traitement par l'α-chymotrypsine. *Ann. Génét.* **14,** 309.

Finaz, C., and de Grouchy, J. (1972). Identification of individual chromosomes in the human karyotype by their banding pattern after proteolytic digestion. *Humangenetik* **15,** 249.

Gall, J. G. (1963). Chromosome fibers from an interphase nucleus. *Science* **139,** 120.

Green, R. J., and Bahr, G. F. (1975). Comparison of G-, Q-, and EM-banding patterns exhibited by the chromosome complement of the Indian muntjac, *Muntiacus muntjak*, with reference to nuclear DNA content and chromatin ultrastructure. *Chromosoma* **50,** 53.

Hayat, M. A. (1970). *Principles and Techniques of Electron Microscopy: Biological Applications*, Vol. 1. Van Nostrand Reinhold Company, New York and London.

Hayat, M. A., and Zirkin, B. R. (1973). Critical point-drying method. In: *Principles and Techniques of Electron Microscopy: Biological Applications*, Vol. 3 (Hayat, M. A., ed.). Van Nostrand Reinhold Company, New York and London.

Hsu, T. C. (1972). Procedures for mammalian chromosome preparations. In: *Methods in Cell Physiology*, Vol. 3 (Prescott, D. M., ed.). Academic Press, New York and London.

Hsu, T. C. (1974). Longitudinal differentiation of chromosomes. *Ann. Rev. Genet.* **7**, 153.

Hsu, T. C., Pathak, S., and Shafer, D. A. (1973). Induction of chromosome crossbanding by treating cells with chemical agents before fixation. *Exp. Cell Res.* **79**, 484.

Kato, H., and Moriwaki, K. (1972). Factors involved in the production of banded structures in mammalian chromosomes. *Chromosoma* **38**, 105.

Kato, H., and Yosida, T. H. (1972). Banding patterns of Chinese hamster chromosomes revealed by new techniques. *Chromosoma* **36**, 272.

Korenberg, J. R., and Freedlender, E. F. (1974). Giemsa technique for the detection of sister chromatid exchanges. *Chromosoma* **48**, 355.

Lee, C. L. Y., Welch, J. P., and Winsor, E. J. T. (1972). Banding patterns in human chromosomes: production by proteolytic enzymes. *J. Hered.* **63**, 296.

McKay, R. D. G. (1973). The mechanism of G and C banding in mammalian metaphase chromosomes. *Chromosoma* **44**, 1.

Miller, O. J., Miller, D. A., and Warburton, D. (1973). Application of new staining techniques to the study of human chromosomes. *Prog. Med. Genet.* **9**, 1.

Moorhead, P. S., Nowell, P. C., Mellman, W. J., Battips, D. M., and Hungerford, D. A. (1960). Chromosome preparations of leukocytes cultured from human peripheral blood. *Exp. Cell Res.* **20**, 613.

Paris Conference. (1971). *Standardization in Human Cytogenetics.* Birth Defects: Original Article Series, **VIII**, 7 (1972). The National Foundation, New York.

Patil, S. R., Merrick, S., and Lubs, H. A. (1971). Identification of each human chromosome with a modified Giemsa stain. *Science* **173**, 821.

Perry, P., and Wolff, S. (1974). New Giemsa method for the differential staining of sister chromatids. *Nature* **251**, 156.

Schnedl, W. (1971). Banding pattern of human chromosomes. *Nature, New Biol.* **233**, 93.

Schnedl, W. (1974). Banding patterns in chromosomes. *Int. Rev. Cytol. Suppl.* **4**, 237.

Seabright, M. (1971). A rapid banding technique for human chromosomes. *Lancet* **II**, 971.

Stubblefield, E., and Klevecz, R. (1965). Synchronization of Chinese hamster cells by reversal of colcemid inhibition. *Exp. Cell Res.* **40**, 660.

Sumner, A. T. (1972). A simple technique for demonstrating centromeric heterochromatin. *Exp. Cell Res.* **75**, 304.

Sumner, A. T., Evans, H. J., and Buckland, R. A. (1971). New technique for distinguishing between human chromosomes. *Nature, New Biol.* **232**, 31.

Terasima, T., and Tolmach, L. J. (1963). Growth and nucleic acid synthesis in synchronously dividing populations of HeLa cells. *Exp. Cell Res.* **30**, 344.

Wang, H. C., and Fedoroff, S. (1972). Banding in human chromosomes treated with trypsin. *Nature, New Biol.* **235**, 52.

Yunis, J. J., and Sanchez, O. (1973). G-banding and chromosome structure. *Chromosoma* **44**, 15.

9. EQUIDENSITOMETRY: SOME NEUROBIOLOGICAL APPLICATIONS

L. T. Ellison and D. G. Jones

Department of Anatomy and Human Biology, University of Western Australia, Nedlands, Australia

INTRODUCTION

Equidensitometry, the study of lines or areas of equal tonal density within an image, is one of several techniques to have emerged in recent years as a result of developments in the field of image analysis. It aroused general scientific interest during the 1950s following Krug and Lau's initial publication (1952) in which they drew attention to the methods available for producing equidensities and their potential advantages in the analysis of images. Subsequently equidensitometry was adopted by a number of scientific and technological researchers, faced with the common problem of interpreting images with varying degrees of spread (Lau and Krug, 1968). In the process, methods were refined and diversified, three principal approaches being employed: (a) *photographic*, e.g., the two-layer method (Lau and Krug, 1968; see also Schmalbruch and Kamieniecka, 1974), the Sabattier effect (Kind, 1954; Lau and Krug, 1968), and the *Agfacontour* film (Ranz, 1970); (b) *physical*, e.g., the density-relief method (Lau, 1958; Lau *et al.*, 1958), large area photometry with dark ground illumination (Lau and Hess, 1960), and the scattered light method (Lau and Johannesson, 1934); (c) *electronic*, e.g., the *equidensitometre* (Jobin and Yvon, 1934), the equidensograph (Schusta, 1954), the Tech-Op/Joyce-Loebl isodensitometer (Miller *et al.*, 1964), and scanning microscope and television equidensitometers (Lau and Krug, 1968).

Two types of equidensitometry have developed from the original equidensity concept—line equidensities and area equidensities. Of these, line equidensities are used to depict the density distribution of the original photograph as a series of "contour lines," while area equidensities yield a complete photometric coverage of the density distribution of the original photograph. Each assesses density in a different way; the line equidensities providing general information on the shape, size, and density gradients of the object illustrated in the original photograph, and the area equidensities giving local density values to the various tones of the original photograph.

Area equidensities include color equidensities (Lau and Krug, 1968; Ranz and Schneider, 1971), as well as "large area photometry" (Lau and Hess, 1960), and screen equidensities. In each case, density is coded by converting its value at any point of the original photograph into a visually recognizable quantity, for instance, a color or a screen symbol. The value of the density is then determined by the shade of the color or by the conformation of the screen symbol.

Equidensitometry has been successfully used in the study of interferograms and spectrograms, and in the fields of astrophysics, illumination engineering, hydrology, photography, and radiation physics. Biological applications have included its use in radiation therapy (Keller, 1964; Rakow, 1965; Tsien and Robbins, 1966), and in the analysis of X-rays (Rakow, 1959; Rakow and Zapf, 1962; Heidenreich *et al.*, 1964). A recent attempt has been made in this laboratory to use equidensitometry in synaptic ultrastructural studies (Nolan and Jones, 1973, 1974). Some of these techniques, particularly the production of screen equidensities, have been modified by the present authors, in order to provide standard procedures for routine investigations. Furthermore, new methods have been introduced to permit the formation of line equidensities and families of equidensities from high power electron micrographs. This feat had not previously been accomplished owing to the granularity of the enlarged micrographs. Some of these results are reported in a preliminary way in this chapter.

EQUIDENSITOMETRY: PRODUCTION TECHNIQUES

As mentioned above, equidensities have been produced using photographic, physical, and electronic methods. Different applications favor different approaches, although the recent advances in electronic techniques make this particular approach the most flexible and efficient. Access to the appropriate electronic equipment is, however, beyond the resources of many small laboratories, and in these circumstances equidensity production by means of photographic methods is a viable alternative.

Nevertheless, photographic techniques do have some shortcomings. These include a lack of flexibility (electronic methods by contrast permit changes of equidensity position, width, and number by means of simple controls), and the time taken to acquire skill in using the film. These limitations have to be balanced against the potential value of equidensity measurements in any intended project. The scale of the project itself also has to be considered: large-scale projects render photographic methods unwieldy and time consuming, whereas small-scale projects do not warrant the expenditure required in obtaining the appropriate electronic equipment. Photographic methods are viable therefore, even today in small-scale projects. Given such a project and the adoption of a photographic equidensity technique, one of the most convenient approaches is that provided by the "Agfacontour professional" film, a film specifically designed for the production of equidensities.

The Agfacontour Film

Basic Principles. The *Agfacontour* film possesses two emulsion layers on a polyester base (Fig. 9.1). One of these is a conventional silver chloride emulsion (2); the other a thin silver bromide emulsion (1). The two emulsions operate at slightly different speeds, with the result that three exposure conditions can be defined: (a) high exposure, with reduction of both emulsions, (b) intermediate exposure, with reduction only of the faster bromide emulsion, and (c) low exposure, with no reduction taking place. It is evident from Fig. 9.1 that during development the fully exposed emulsions (beneath "a") undergo further reduc-

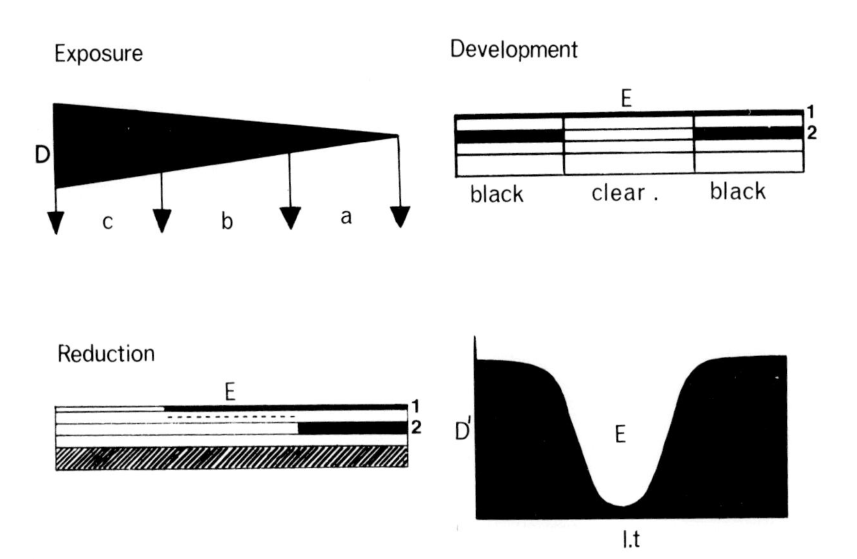

Fig. 9.1 Basic properties of the *Agfacontour* film. For details, see text. With development the portion of the film exposed to E, the critical exposure, remains clear while surrounding areas are blackened. The resulting density profile, shown in the graph of D^1 against I.t, constitutes a characteristic curve of the film. Note that D^1 represents the density of blackened portions of the *Agfacontour* film, whereas D refers to the density of a filter, wedge or photographic transparency.

tion with visible blackening of the film (blue-black, characteristic of *chemical* development). In addition, the underexposed emulsions (beneath "c") are developed *physically* (a reaction to sulfite present in the developer), and this process also results in blackening of the film (brown-black). Finally the emulsions with intermediate exposure (beneath "b"), react in an unexpected fashion; some chemical darkening of the reduced bromide emulsion occurs. It is, however, too thin for this effect to be very noticeable, and the bromide ions (dotted line in Fig. 9.1) released during reduction of the bromide emulsion prevent physical blackening of the adjacent underexposed chloride emulsion. The film therefore remains clear, and the intermediate exposure condition is subsequently referred to as the "critical exposure" (E). Provided that standard development conditions are maintained, E is an invariant property of the *Agfacontour* film.

Exposure (I.t). The critical exposure (E) may be obtained at any point on the surface of the *Agfacontour* film by manipulating two factors: (a) the exposure (I.t) or overall amount of light, and (b) the density (D) or resistance to the passage of that light (Fig. 9.1). Variations in the exposure may be introduced in a number of different ways. First the light intensity (I) may be modified by varying the filament current of the light source (V), the aperture setting of the lens system (F), or the amount of light filtration present (Y). Second, the time of exposure (t) may be altered, the effect of increased exposure time being the same as that of increased light intensity (I). Variations in the density are brought about by interrupting the light rays at either the back focal plane of the lens system, such, for instance, as occurs in the projection of a negative, or the front focal plane as, for example, in contact printing.

If the exposure I.t is set at the critical level E so as to produce a clear *Agfacontour* film copy, and then a grey filter D is placed between the light source and the film, the latter will now be underexposed ($\log I.t/D < E$) and consequently blue-black. In order to reproduce the conditions (E) necessary for the production of clear film the general exposure (I.t) will have to be increased until the ratio $\log L/D$ is again equal to E.

Transparencies. Replacing the grey filter by a continuous tone grey wedge (black-grey-clear) results in a portion of the wedge, approximately equal in density to the original filter, being copied as a clear area onto the *Agfacontour* film. The remainder is copied black as a result of either over- or underexposure. When positive prints of the exposed *Agfacontour* film are made, the clear area is printed black, and the adjacent blue- or brown-black regions are printed white.

The two factors—exposure (I.t) and density (D) are related to each other and to the critical exposure (E) by the formula: log I.t/D = E. Therefore the position (P) of the black band on the positive print which corresponds with the portion of the grey wedge (D) transmitting the critical amount of light, is given by the density value: D = log I.t/E. Increased exposure results in a black band corresponding in position to the darker portions of the wedge (the ratio log I.t/D is constant) and conversely decreased exposure results in reproduction of lighter portions of the wedge. The actual relationship between exposure and density reproduced is given by the formula: $t_2 = t_1 \times 10^{D_2 - D_1}$. This equation assumes that the light intensity (I) is constant, so that any given density (D) may be reproduced by varying the exposure time (t).

Replacement of the grey wedge by a black and white transparency results in a situation in which some densities pass the critical amount of light, while others transmit either too much or too little light. Different density regions are reproduced, as with the wedge, by varying the exposure time (t). If the exposure (I.t) remains constant, as occurs in the production of any one film copy, then it can be seen from the formula, D = log I.t/E, that clear regions of the film representing points of critical exposure (E) must also represent points of equal density (D) in the original transparency. These areas are therefore, by definition "equidensities."

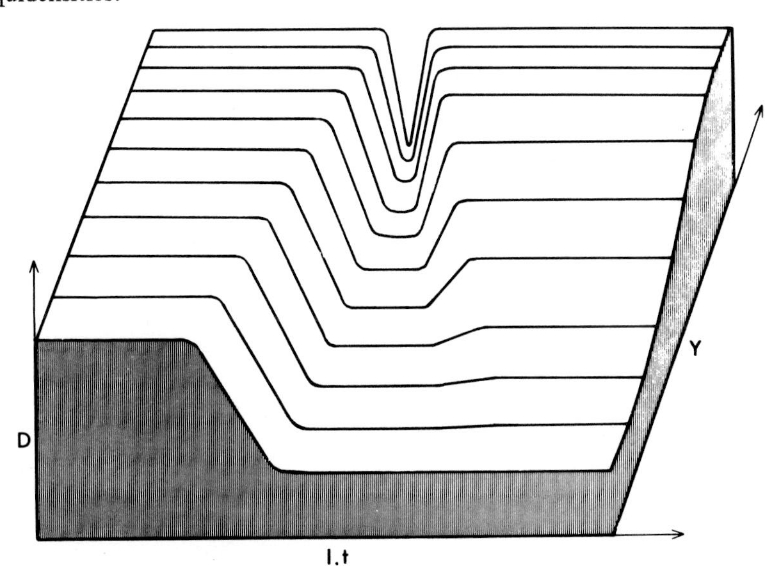

Fig. 9.2 The effect of Yellow Filtration (Y) on the characteristic curve of the *Agfacontour* film is demonstrated by plotting D^1 against I.t. The foremost curves are those for blue light, while those in the background are for green light. An increasingly narrow trough is evident towards the rear and this indicates a corresponding decrease in equidensity width. It is associated with a higher density to the right of the trough, which occurs as a result of reduction and *chemical* development of the chloride emulsion. The minimum exposures T.1 and T.2 (see text) are represented respectively by the left and right edges of the trough.

Yellow Filters. Placing a yellow filter between the light source and the *Agfacontour* film reduces the difference in the speed of the two emulsions. This lessens the size of the intermediate or "critical" exposure, and narrows the range of densities (D) reproduced as an equidensity (Fig. 9.2). It also affects the total amount of light reaching the film (I.t) and therefore changes the position (P) of the equidensity. These alterations are brought about in the following way. The minimum exposure (T.1) necessary to reduce the bromide emulsion is most sensitive to changes in the intensity of blue light, whereas the minimum for the chloride emulsion (T.2) is most reactive to green light. As the time T.1 is less than T.2, the use of a yellow color filter, which eliminates blue light while admitting green, lessens the difference (T.2 - T.1) by increasing T.1 at a faster rate than T.2. The fact that the light is filtered however, diminishes the overall light intensity and causes a shift in the position (P) of the equidensity to a less dense region within the transparency.

Production of Line Equidensities

Equidensity Width and Position. Equidensity *width* refers to the range of densities present in the original transparency and reproduced as an equidensity, while equidensity *position* refers to the average numerical value of the densities reproduced. The former is varied by suitable color filtration (see page 344), while the latter depends upon the total illumination (I.t). An equidensity may therefore be regarded as being composed of two components: (a) a plus and minus error term or variance which is related to the width (W) of the equidensity and the degree of yellow filtration (Y), and (b) a mean or positional value (P), which is determined by the conditions of exposure. If, for instance, the position (P) of an equidensity remains unchanged with respect to a neighboring equidensity, and the width (W) of both equidensities is reduced, gaps appear between the equidensities. The equidensities may also be separated by reducing the exposure of one equidensity, or increasing that of the other. In practice a combination of the two factors, (W) and (P), operates to control the degree of separation or overlap between equidensities.

Equidensities of Higher Order. Copy of the original transparency onto *Agfacontour* film constitutes a first order line equidensity (Figs. 9.1, 9.3b, and 9.7). Equidensities of the second order are produced by re-exposure of the first order equidensity onto *Agfacontour* film, a process which results in the first line equidensity being outlined by two clear zones. The outlining occurs in the following way: first, as a result of the high contrast of the silver chloride emulsion, only a thin grey region is present at the borders of the first order equidensity. Second, a portion of this grey region admits the critical amount of light (E) along the whole length of the border, and is therefore copied onto the *Agfacon-*

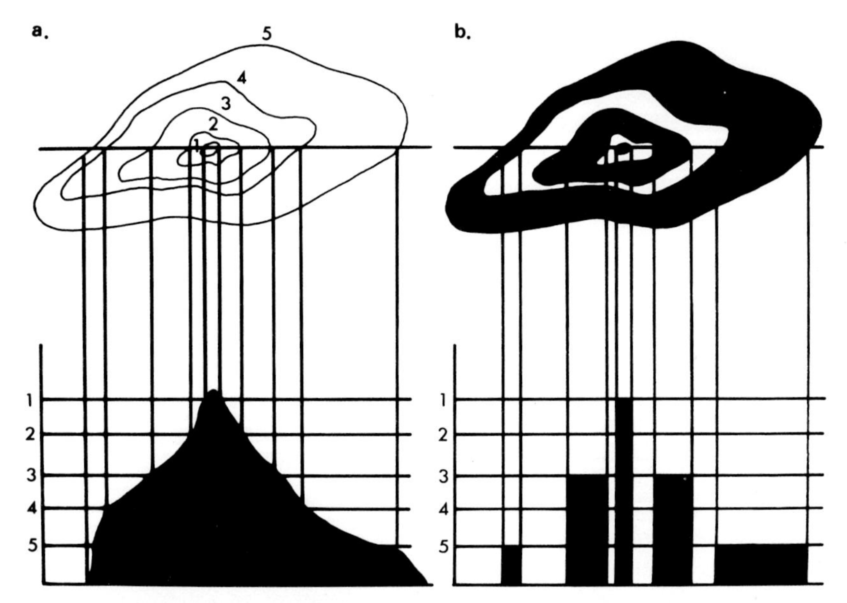

Fig. 9.3 The production of families of equidensities. The contour map at (a) is obtained by copying the three first order equidensities at (b) onto *Agfacontour* film to produce second order equidensities. The first order equidensities are initially produced separately and then placed in register for subsequent copying. Beneath each family of equidensities a density profile has been constructed. The first (a) is a "smooth" reconstruction, while the second (b) reconstructs only the densities represented by the corresponding equidensities.

tour film as a clear line. Subsequent printing results in two black lines outlining the first equidensity (Figs. 9.3a, 9.6, and 9.7). By repeating this process, equidensities of any desired order may be obtained. Usually however, the image complexity, the inability to resolve very fine paired lines, and the outlining of statistical grain variation (page 352) means that there is no practical value in proceeding beyond a limited number of stages.

Families of Equidensities. Families of equidensities may be produced by mounting a number of equidensities in register. First order equidensities may be used if the original density gradients are sufficiently high, but usually the second or third order equidensities derived from two or three well spaced first order equidensities, are more effective. In both instances the position (P) and width (W) of the first order equidensities are carefully calculated so that the resulting contour lines will be neatly spaced thereby giving a balanced overall impression of the spatial density distribution of the original tones. The second method, involving higher order equidensities, is particularly useful in evaluating small density gradients. With these, the first order equidensities are spread over a large area, but when they are recopied onto *Agfacontour* film, thin line equidensities appear at their borders, the resulting contours throwing the density gradients into relief (Fig. 9.3).

Successful production of "contour-line" equidensities from electron micrographs requires that the picture be as "smooth" as possible before any first order equidensities are produced, a situation seldom encountered in practice. A useful device for achieving this required "smoothness" is to defocus the original transparency when making the *Agfacontour* film copy. This results in visual loss of electron grain and object irregularities without at the same time losing any significant amount of optical information (see McKechnie, 1973, for a discussion of the effect of defocus on resolution). A more general view of the object is then obtained, a situation which is often advantageous in studying complex structures.

Production of Screen Equidensities

The Printing Screen. A typical printing screen consists of a square lattice of linear grey bands, expanded at their points of intersection to form dark grey globules. The bands merge gradually into the lighter regions at the center of the lattice squares. Figure 9.4 shows a series of prints of an enlarged screen (100 lines/in actual size) with varying exposure. This gives some idea of the distribution of densities within the screen. The first picture of the series was made with a short exposure and therefore illustrates the shape of the lighter densities of the screen. Subsequent pictures show the growth of the diamond shaped symbols with increasing exposure as regions of higher density are recruited. The final picture shows the nodal points of the screen (maximum density) as small white dots.

Screen Equidensities. Screen symbols are in fact line equidensities of the screen density patterns. They are therefore equivalent to black copies of the grey diamonds of Fig. 9.4 (1), or black copies of the grey fringes surrounding the patterns in each of the remaining pictures in Fig. 9.4. Some of the principal screen symbols are illustrated in Fig. 9.5. The extreme exposure values are represented by the diamond shaped symbol (electron opacity) and the dot shaped symbol (electron translucency), while intermediate exposures are represented by varying types of grid pattern.

The width of the equidensities is controlled as before (page 343) by using color filters. This affects the clarity but not the conformation of the screen symbols. In contrast to the line equidensities, only a single exposure is necessary to produce the complete range of tonal densities available from a negative film. As previously pointed out the exposure at any point is determined by the density of the original negative at that point, so that a range of exposures is presented to the screen for translation into screen symbols on the *Agfacontour* film. Because the screen symbol conformation depends upon the exposure, the resulting array of screen symbols reflects the values of the original tonal densities.

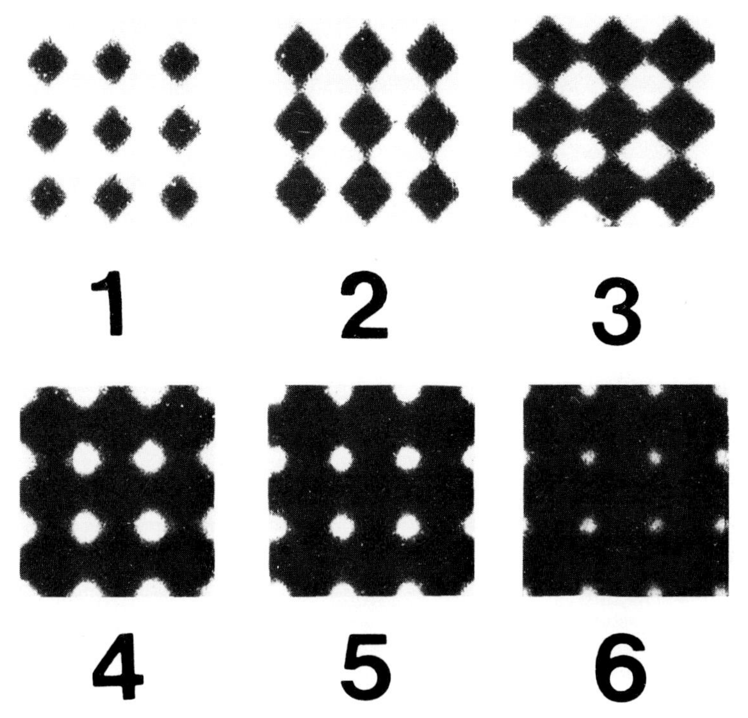

Fig. 9.4 This shows a series of photographs of a printing screen with increasing exposures from 1-6. The pictures were produced by enlarging a 100 lines/in screen through an optical enlarger onto photographic printing paper.

Tonal Range. Screens are available in a variety of tonal ranges, the appropriate choice depending upon the tonal range of the negative film to be copied. Screens with a high tonal range will only reproduce a few symbols from a low contrast negative, while screens with a low tonal range will only reproduce parts of a high contrast negative. The matching of screen and negative is therefore important. In our work to date we have been unable to obtain a high contrast screen to match our negatives (densities up to 2 or more density units), and so we have attempted to find methods of reducing the tonal contrast of the original negatives.

The method adopted in our current series of screen equidensities consisted of making double copies of the original 36-mm negative, and then subsequent enlargement to produce 5″ X 4″ negative transparencies with a maximum density of one density unit. Two procedures have been employed to obtain the enlarged, low contrast negatives: (1) enlargement to produce a full tone positive film copy, followed by contact printing with reduced exposure to lessen the contrast, and (2) contact printing with reduced exposure followed by enlargement. As anticipated, both procedures gave similar results although for simplicity and economy the latter is preferred.

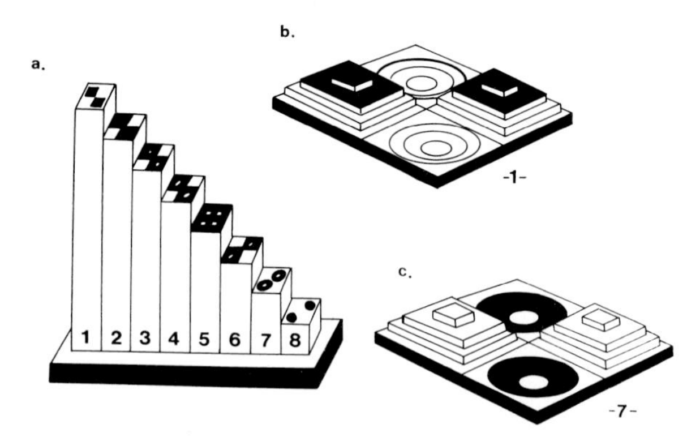

Fig. 9.5 Screen symbol configurations. (a) Columns 1–8 represent diminishing densities, and the screen symbol corresponding to each density level is shown at the top of the column. The symbols have been loosely labelled for reference: (1) small diamond, (2) diamond, (3) studded diamond, (4) pregrid, (5) grid, (6) postgrid, (7) circle, and (8) dot. Note that these discrete configurations are taken from a *continuum* of screen symbol shapes, and also that the direction of the symbols changes at density level 5. In (b) and (c) the symbols from columns 1 and 7, respectively, are shown enlarged, and in addition the actual densities of a small portion of the printing screen are depicted. In (b) the two elevated squares representing dark regions of the screen are seen to give rise to the screen symbols, whereas in (c) the two depressions representing light regions of the screen are responsible for the formation of the screen symbols. The reason for the previously noted change in the *direction* of the symbols is therefore, readily apparent.

In addition to these procedures, some alternative solutions to the problem of reducing negative tonal contrast should be mentioned. First the original negative may be lightened either by shortening the exposure to the electron beam or by shortening the subsequent development time, although both may result in loss of resolution. In addition, the electron image may be defocused and so blur both the higher and lower densities. This method is questionable however, as the effect of defocus on electron images may not be the same as its effect on light images, and the electron grain still present requires further optical defocus to obtain a "smooth" picture. Another method consists of optically defocussing an enlarged copy of the original negative to produce a visually "smooth" picture, with further defocus to reduce the contrast to the required level. The general effect of defocus is to lose visual information, although the original resolution is not lost. It may in fact be regained using line equidensities (Heidenreich *et al.*, 1964).

APPLICATIONS OF LINE EQUIDENSITIES

Line equidensities simplify the density distribution of a photograph by presenting it in the form of a contour map. In some cases this is advantageous because of the simplicity of presentation of the image data, while in others there is in addition the benefit of a clear border for measurement. In this section some of the applications outside the field of neurobiology are first examined to discover the types of situation best suited to line equidensity evaluation, and also to elucidate some of the limitations of line equidensities.

Interferograms

The line equidensity approach was extensively developed and tested in conjunction with the study of interferograms. In these, the width and spacing of the interference fringes relate to physical characteristics of the test object, e.g., surface curvature, surface roughness, film thickness, etc. Measurement of width and spacing of fringes is, therefore, of critical importance, although as a result of the sinusoidal nature of the image intensities in cross-section, width refers to the intensities in the immediate vicinity on either side of the intensity maxima, and spacing to the distance between intensity maxima or regions of a particular intensity on a given side of the maxima. The accuracy of these measurements is improved by the use of a densitometer or by using equidensities, the latter often proving to be more efficient in practice (Lau and Krug, 1968).

Interferograms are the end products in a number of procedures: (a) the testing of reflecting surfaces; (b) the measurement of crystal steps and the thickness of thin films (Tolansky, 1948); (c) the measurement of reduced silver present in the photographic emulsion (Lau, 1958; Lau *et al.*, 1958); (d) experiments using a transparent wedge with interference derived from reflected rays (Schult, 1927); and (e) experiments in which plexiglass or synthetic resin models are placed under mechanical stress while transmitting light (Schwieger and Haberland, 1955, 1956).

Spectra and Radiation Studies

The line equidensity approach has also been used in spectrophotometry, although in this application the wavelength of the light is also important. The additional information has been obtained in a number of ways: (a) conversion into a two-dimensional image before production of equidensities (Kramer, 1951; Lau, 1953; Lau and Krug, 1968); (b) conversion to interference patterns before production of equidensities (Schmidt, 1959, 1960); and (c) direct evaluation by means of Sabattier equidensities (Schröter, 1958). In the latter instance, the equidensities, which may be as high as the third order, can be used to deduce the granular velocities of the solar spectrum from say the wavelength variations, and the elements represented.

Two further applications of the line equidensity approach have also been investigated. These are firstly the production of isodose curves in clinical radiation therapy (Kölle *et al.*, 1956; Rakow, 1959, 1965), and secondly the evaluation of X-rays. The equidensity contour lines correspond closely to lines of equal intensity in the particular type of radiation employed, for instance alpha, beta, gamma, or X-rays (Lau and Krug, 1968).

Photomicrographs

Initial attention was focused on the information retrieval aspect of equidensitometry. Consequently some of its early applications included an increased resolution of over-enlarged or de-focused photographs (Lau and Krug, 1968), and a more objective assessment of the information content of resolution-test-photographs (Heidenreich *et al.*, 1964).

The evaluation of photomicrographs by means of line equidensities has been utilized by Banig and Mieler (1964) to obtain improved recognition of aberrant chromosomes. They used equidensities of the third order to produce families of equidensities, with additional information on gradients being obtained from the local thickness and spacing of equidensity contours (see also Lau and Krug, 1968). More recently, Nolan and Jones (1974) found that families of equidensities did not significantly improve the evaluation of electron-opaque structures in the junctional region of glutaraldehyde-PTA stained synapses, although line equidensities obtained by taking first order equidensities of a lithographic (high contrast) copy of the original micrograph, permitted readier differentiation of dense substructures from less dense background material.

Lau and Krug (1968) have defined the "limit of equidensitometry" as the order (i.e., first order, second order, etc.; see page 346) at which deviations in the contour line are no longer due to changes in the structure of the object, but instead are due to chance variations in the density of the original, such fluctuations being termed "statistical grain variation." The factors involved in the production of random density variations in electron micrographs are considered in detail by Röhler (1967).

Briefly, each electron strike results in a cluster of photographic grains. The clusters are in turn randomly positioned in such a way that the average number of clusters per unit area follows a Poisson distribution. Other factors which may contribute to the level of background noise in electron micrographs are (a) the presence of impurities, both in the tissue and in the embedding media, (b) the effect of any carbon supporting film which may be present, (c) the effect of uneven stain distribution, and (d) the presence of irregularities resulting from poor fixation. The high power electron micrograph with its prominent electron grain, possesses therefore, a higher level of background noise than for instance, an interferogram, and theoretically it will not be capable of giving rise to high order equidensities or even clear first order equidensities. The result of the direct application of contour-line equidensity methods is in fact a confusingly complex picture as illustrated in Fig. 9.6a and c. In addition, the objects of interest are often contained within a very narrow density range, while the accompanying low density gradients make the formation of "contour-line" equidensities difficult.

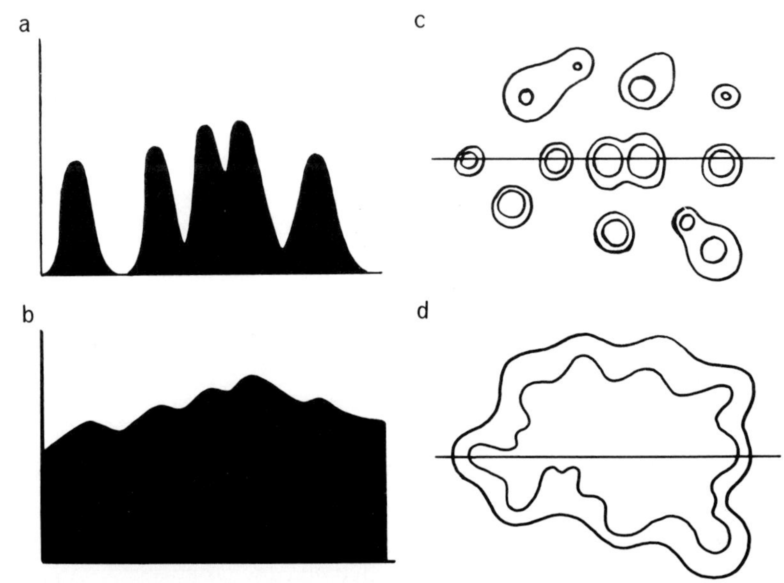

Fig. 9.6 An illustration of the effect of background noise on the production of line equidensities. In (a) the spikes represent a density profile of part of a noisy electron micrograph, while (c) demonstrates that the formation of two second order equidensities results in an outlining of individual grains. Although the spacing of these individual outlines gives some idea of the form of the original object, the image complexity is a drawback. This can be considerably reduced if the original picture is rendered "smooth" as in (b) by defocusing before equidensity production. The object outline (d) can then be more readily discerned, although the indentations still present in the border mean that background noise can never be entirely eliminated using this method.

Successful application of the line equidensity method requires the reduction of background noise to a low level before producing the contour-line equidensities (Fig. 9.6b and d).

Nolan and Jones (1974) in fact avoided "the critical exposure of the original onto *Agfacontour* film" by using an intermediate lithographic copy of the original. The effect of this procedure is illustrated in Fig. 9.7. Reference to Fig. 4 in Nolan and Jones' paper reveals however, that this method does not reduce the background grain of the photograph. There seems, if anything, to be some enhancement of the grain, but along with this effect there is an improvement in the recognition of the electron-opaque regions of the original photograph. In this respect, further exposure onto *Agfacontour* film does little more than outline the dense areas, already emphasized by exposure onto lithographic film. This outlining procedure introduces an element of confusion, because both the original dense regions and original light regions are shown as being light, with the result that reference to the original micrograph is required in order to distinguish

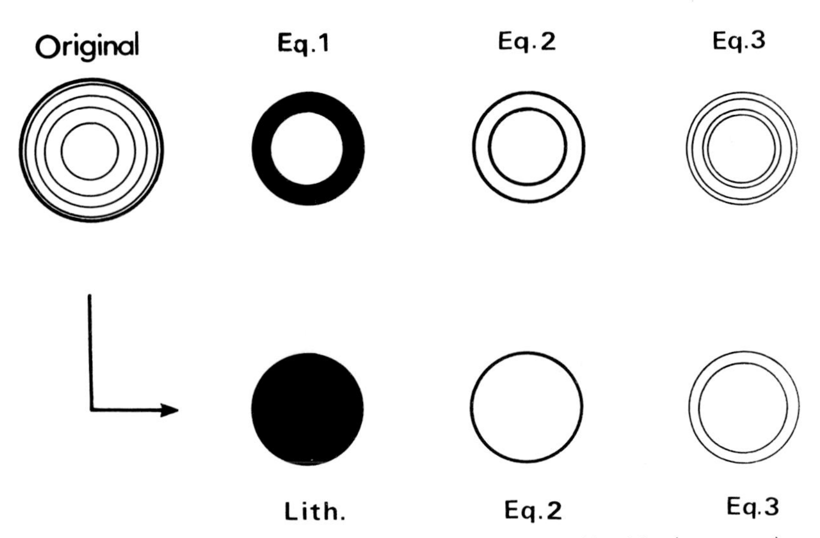

Fig. 9.7 This diagram depicts first, second, and third order equidensities (upper row) produced serially from an original profile. A direct lithographic copy (arrowed) is also used to produce first and second order equidensities (lower row). The latter procedure was used by Nolan and Jones (1974) to obtain line equidensities of the first order from an original electron micrograph of a synaptic junction.

electron-opaque from electron-translucent regions. From this we may conclude that, although this technique is useful in highlighting electron-opaque regions of the original micrograph, all the information necessary for such an appraisal is present in the lithographic copy. Furthermore, because of the excessive background noise there is only a negligible gain in accuracy of measurement with further copying onto *Agfacontour* film.

APPLICATIONS OF AREA EQUIDENSITIES

Area equidensities differ from line equidensities in the following respects: (a) complete photometric coverage of the photograph takes place, the image being represented graphically as columns of density (Fig. 9.5) rather than as a series of contour lines (Figs. 9.3, 9.7); (b) tonal density coding is needed to distinguish between adjacent regions of different density, the most commonly encountered types being either colors or screen symbols. Each of these achieves the same purpose, although the former is technically more efficient, whereas the latter is aesthetically more appealing.

Screen Equidensities

Screen equidensity patterns were originally produced by Lau and Hess (1960) using a printing screen and dark ground illumination. The process was called "large area photometry" and a number of applications were found, for example, in astronomy, optics of the atmosphere, illumination engineering, and microscopy (Lau and Krug, 1968). Interest in the screen technique was revived with the introduction of the *Agfacontour* film. Suitable exposure through a screen onto the film results in an array of symbols similar to those seen in large area photometry. Some of the subsequent applications are presented in the paper by Ranz and Schneider (1971), and include: a study of aerial photographs of agricultural fields; a study of aerially produced thermal maps of cold water springs (color equidensities were used in this case, although screen equidensities would have permitted similar measurements) and a study of satellite photographs of cloud formations. Biological studies based on the same principle include the identification of cells and organelles in thrombi, the identification of functionally distinct muscle bundles (Schmalbruch and Kamieniecka, 1974), and the identification of areas of mineral deposition in bone from their density distribution in X-rays (Konermann, 1971).

The essence of such studies is two-fold: first, the detection of different quantities (e.g., cereal crops at different stages of maturation, or water at different temperatures) identifiable by the shade of grey representing them in the photograph, and second, definition of the extent of these quantities by the corresponding extent of their equidensities. These principles suggest a direct application to electron microscopy, namely, the detection of specific substances in sectioned material using equidensitometric labeling techniques. Other more feasible applications of screen equidensitometry are given on page 363 and include stereological density measures and equidensitometric digitization of micrographs.

Screen Equidensities and Synapses

Before undertaking further discussion of the potential applications of equidensitometry to electron microscopy, it would be wise to examine a typical micrograph of the junctional region of a PTA-stained synapse in order to illustrate essential synaptic features and to provide a basis for evaluating the screen equidensities depicted in Figs. 9.9–9.13. Two synaptic junctions are shown in Fig. 9.8; a micrograph (a) and a sectional diagram (b). Together they illustrate the principal features of synaptic junctions when prepared under nonosmicated conditions (see Jones, 1975). Dense projections (dp), cleft densities (cd), and a postsynaptic thickening (pt) are present, while a diffuse presynaptic network pervades part of the presynaptic terminal above the cleft. A network-like profile is also visible in the postsynaptic cytoplasm. In some cases the presynaptic network is apparently continuous with the spikes (sp) of the dense projections, and this continuity may correspond to the presence of vesicles closely applied to the dense projections (Jones *et al.*, 1976; see also Akert *et al.*, 1969 and Pfenninger *et al.*, 1969 for a discussion of the "presynaptic vesicular grid" concept). Several possible vesicle sites (ves) have been outlined in Fig. 9.8 to illustrate this point. The postsynaptic network possesses globular shaped contours (glob), closely

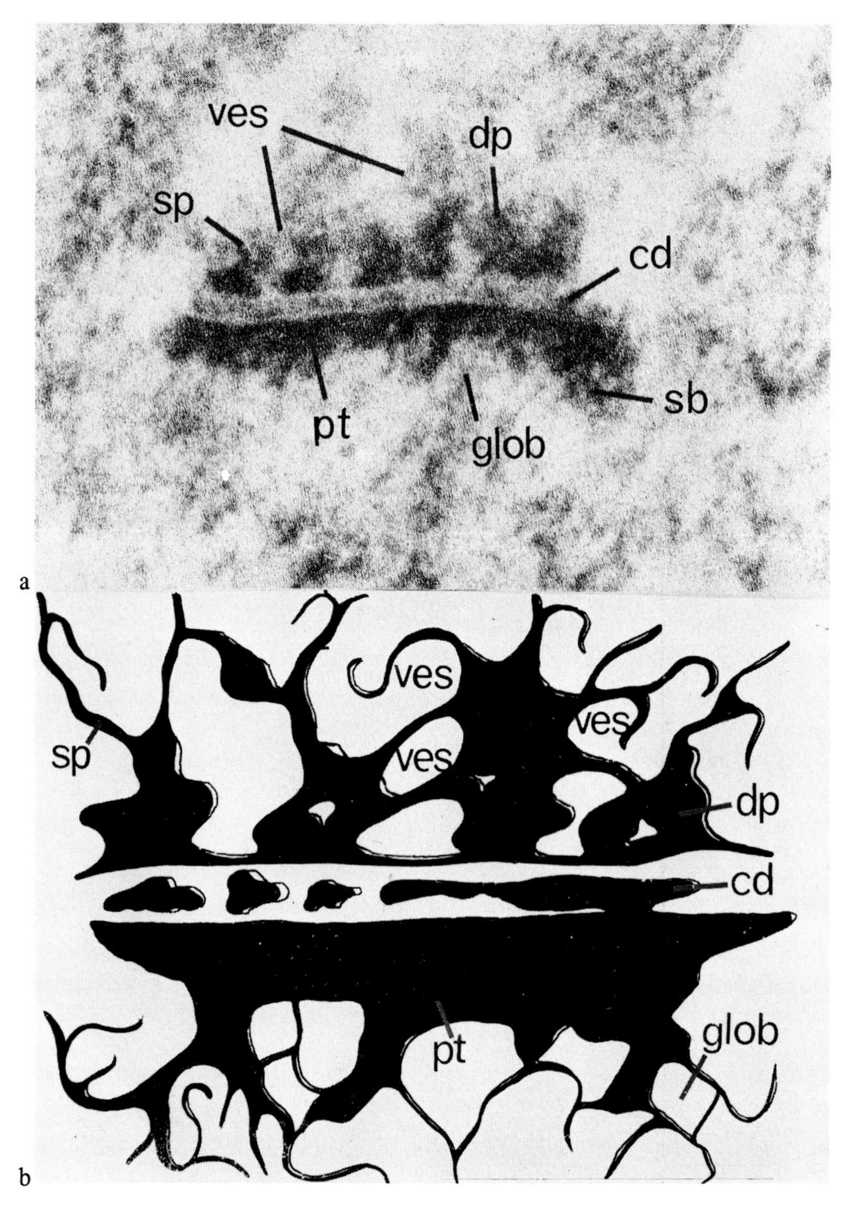

Fig. 9.8a A micrograph of a synaptic junction from rat cerebral cortex, prepared under non-osmicated conditions and stained with PTA. It depicts the essential features found in mature synaptic junctions: dense projections (dp), cleft densities (cd), a postsynaptic thickening (pt), spikes (sp), possible vesicle sites (ves) outlined by strands of the presynaptic network, a subjunctional body (sb), and indeterminate globular shapes (glob) in the postsynaptic cytoplasm. X 170,000.

Fig. 9.8b This diagram duplicates most of the features of the junction in Fig. 9.8a in order to demonstrate the basic two-dimensional nature of a literal interpretation of a micrograph. The effect of section thickness (single dark line) is shown in this instance as adding depth to an already existing object profile. Three possible vesicle sites (ves) are indicated to emphasize the close relationship between vesicles, dense projections (dp), spikes (sp), and network strands. Remaining labelling as in Fig. 9.8a.

associated with the postsynaptic thickening. A notable feature of these and other micrographs is the presence of electron grain, a factor which contributes to the elevated background noise of the electron micrograph.

Figure 9.9 inset is a micrograph of a highly magnified junction, possessing a high level of background noise. It has been selected in order to demonstrate the performance of the direct (i.e., without defocus) screen equidensity technique under adverse conditions. This is shown in Fig. 9.9, which highlights a single dense projection (dp). Cleft densities (cd) are situated immediately subjacent to the dense projection, and a portion of the postsynaptic thickening is present at the lower border.

Some elimination of background noise is achieved by using the screen technique provided that the area covered by each screen symbol is sufficient to average the positive and negative density fluctuations of the noise in the original photograph. This average, represented by the configuration of the screen symbol, estimates the signal (i.e., object) density. The result may be displayed as a three-dimensional histogram or density surface, in which screen symbol values, reflecting signal density, are presented in the form of columns of varying height (Fig. 9.5). Interpretation of the symbols is facilitated by consulting the scheme outlined in Fig. 9.5. They are approximately labeled in this Figure. The symbols of the actual screen equidensities in Fig. 9.9–9.13 are more easily seen if a ruler is placed along a diagonal line and moved slowly across the plates.

Figure 9.10 depicts a screen equidensity of a complete synaptic junction. Interest centers on the diamond-shaped symbols representing regions of high electron opacity (numbered 1-5), and also upon the electron translucent areas (6) representing the positions of the terminal membrane of the presynaptic region and of a neighboring neuron. A discontinuous bar (7) is situated between two of the electron translucent areas. The bar may be due to the presence of intercellular material. In connection with this it is notable that the entire cleft region possesses a higher density level than that of the adjacent border membranes, even though the two are in continuity. The screen symbols representing this higher density level are mainly diamond, studded, and grid symbols. This high density level is at least partly attributable to the presence of cleft densities (4), but may also be due to the effect of specimen tilt, a situation remedied by using a goniometer tilt stage.

The paramembranous densities exhibit little substructural detail, so that the general picture to emerge is one of isolated regions of high electron density (1-5) constituting a number of "centers" each of which is surrounded by a fringe (8) of material of varying but lesser density. An outer mantle (9) extends to the perimeter of the picture, and consists of small dots and circles. The border of the paramembranous densities is for the most part clearly shown (10), although strands of material apparently continuous with the densities are in places seen projecting into the adjacent cytoplasm (11). In addition, the region between adjacent densities is in one case clear (12) and in another occupied by material of intermediate density (13). The regions of highest density correspond in

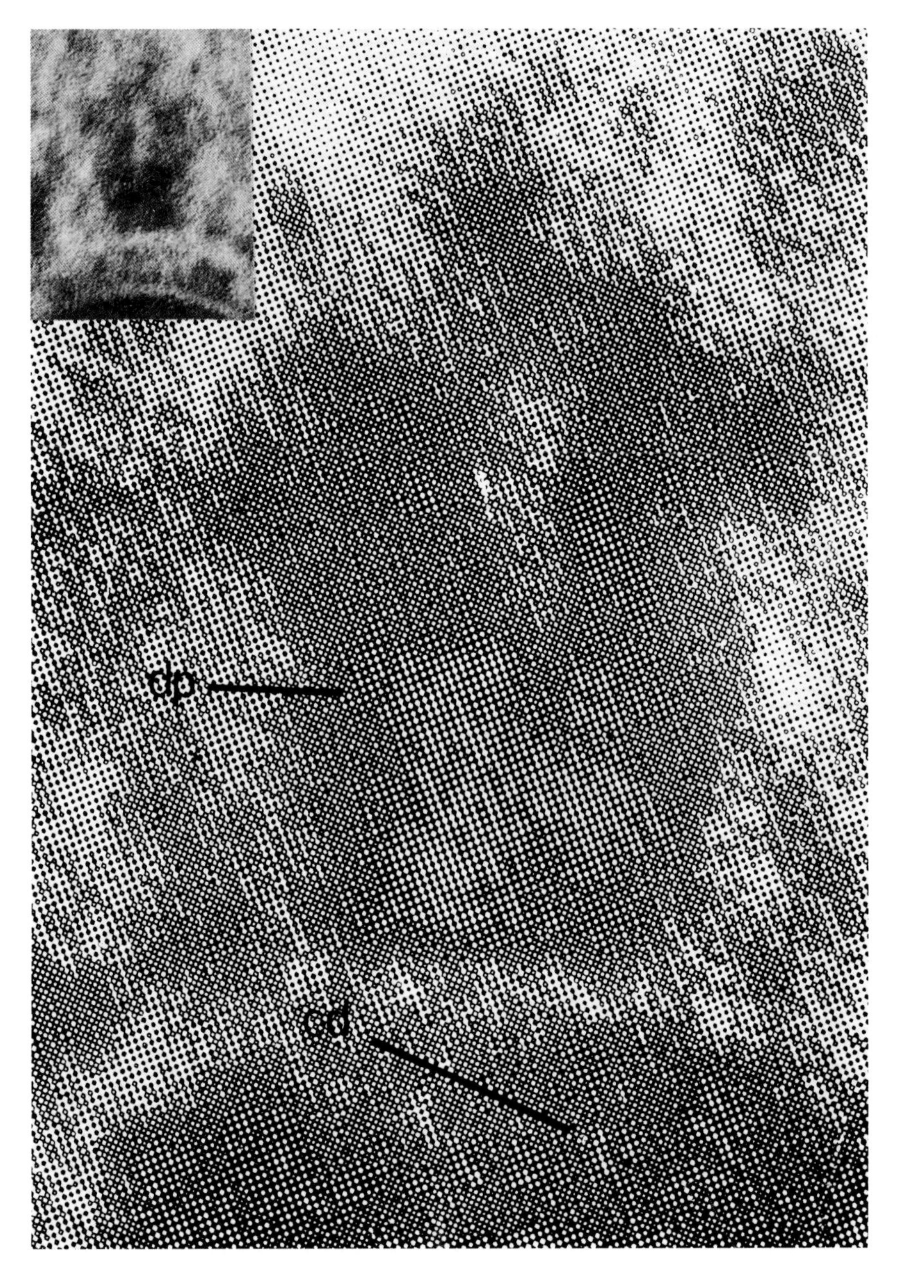

Fig. 9.9 Screen equidensity of an enlarged single dense projection (dp). Also shown are a number of the cleft densities (cd). The small crescent shaped region of diamond symbols to the right of the lower border is part of the postsynaptic thickening. X 1,800,000. *Inset*: Conventional micrograph of the same dense projection. X 180,000.

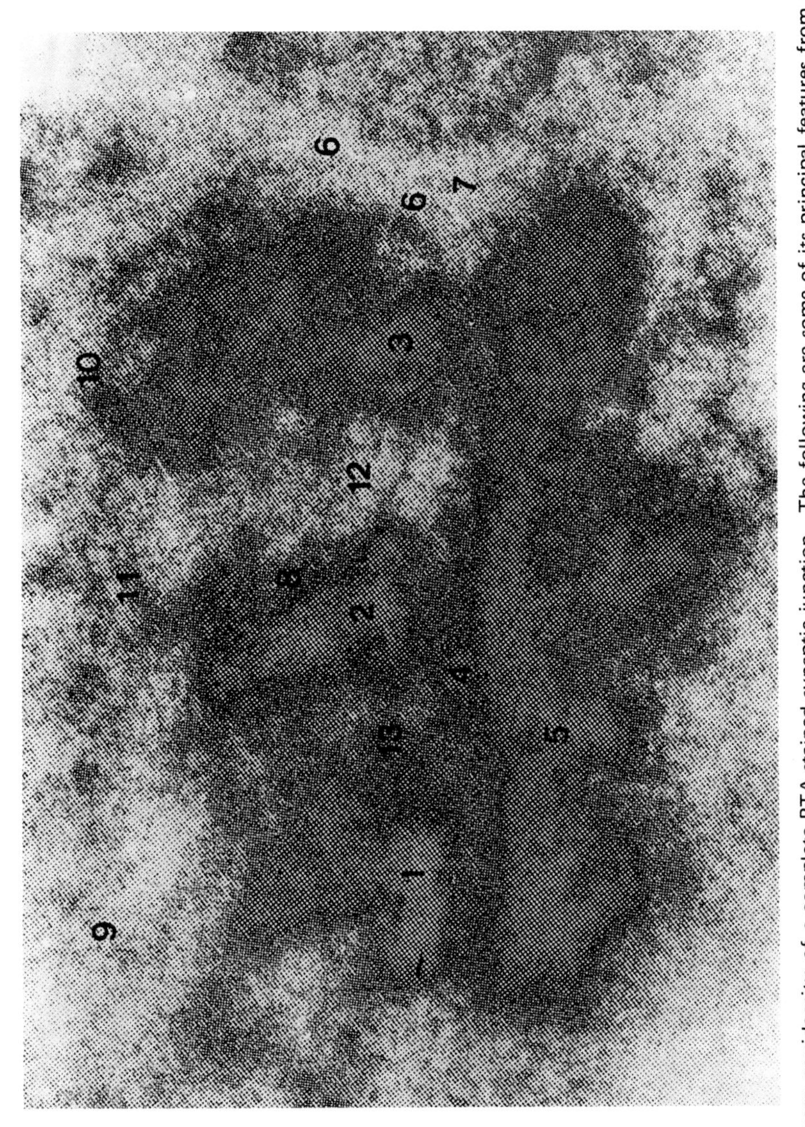

Fig. 9.10 Screen equidensity of a complete PTA-stained synaptic junction. The following are some of its principal features from a descriptive angle: dense projections (1–3), cleft density (4), postsynaptic thickening (5), membranous borders (6), electron dense bar (7), fringe of intermediate density (8), electron translucent background (9), well defined border of dense projection (10), a cytoplasmic strand continuous with a dense projection (11), electron translucent region between dense projections (12), material of intermediate density between dense projections (13). X400,000.

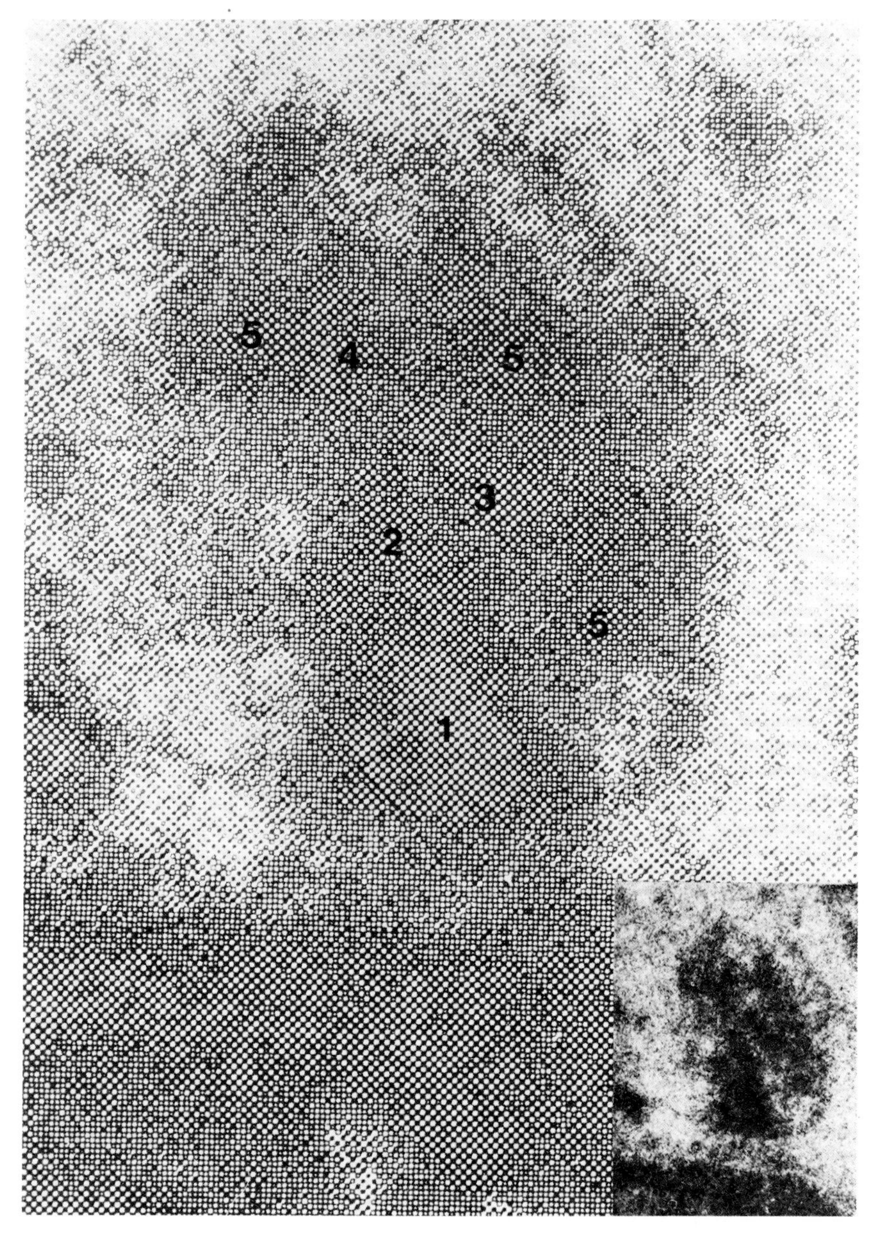

Fig. 9.11 Screen equidensity of the dense projection on the right of the junction in Fig. 9.10. Illustrated is a core of electron dense material (see also Fig. 9.14c): base (1), vertical column (2), branches (3 and 4), terminal globules (5). ×900,000. *Inset*: Conventional micrograph of the same dense projection. ×270,000.

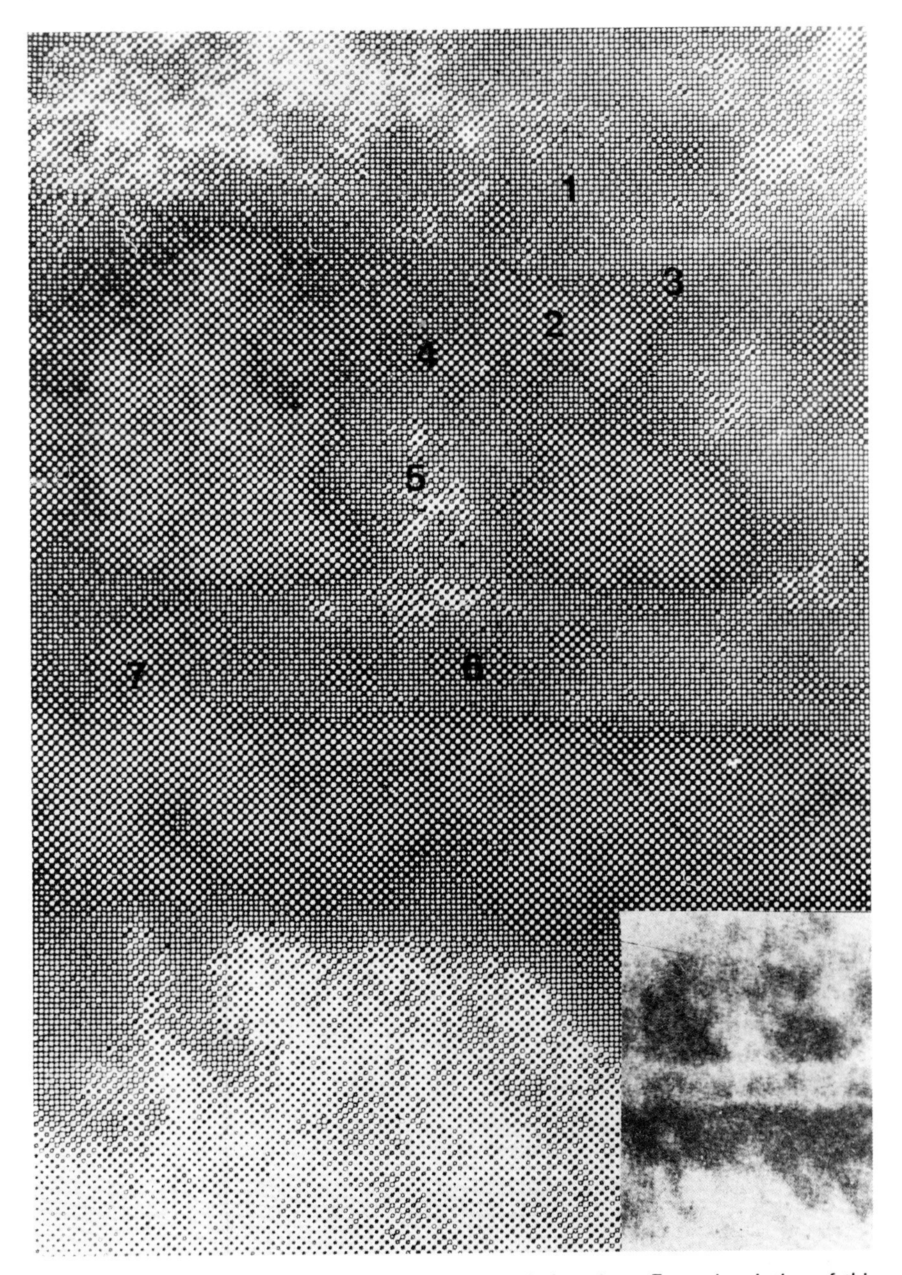

Fig. 9.12 Screen equidensity of portion of a synaptic junction. For a description of this equidensity, see text. ×900,000. *Inset*: Corresponding electron micrograph. ×270,000.

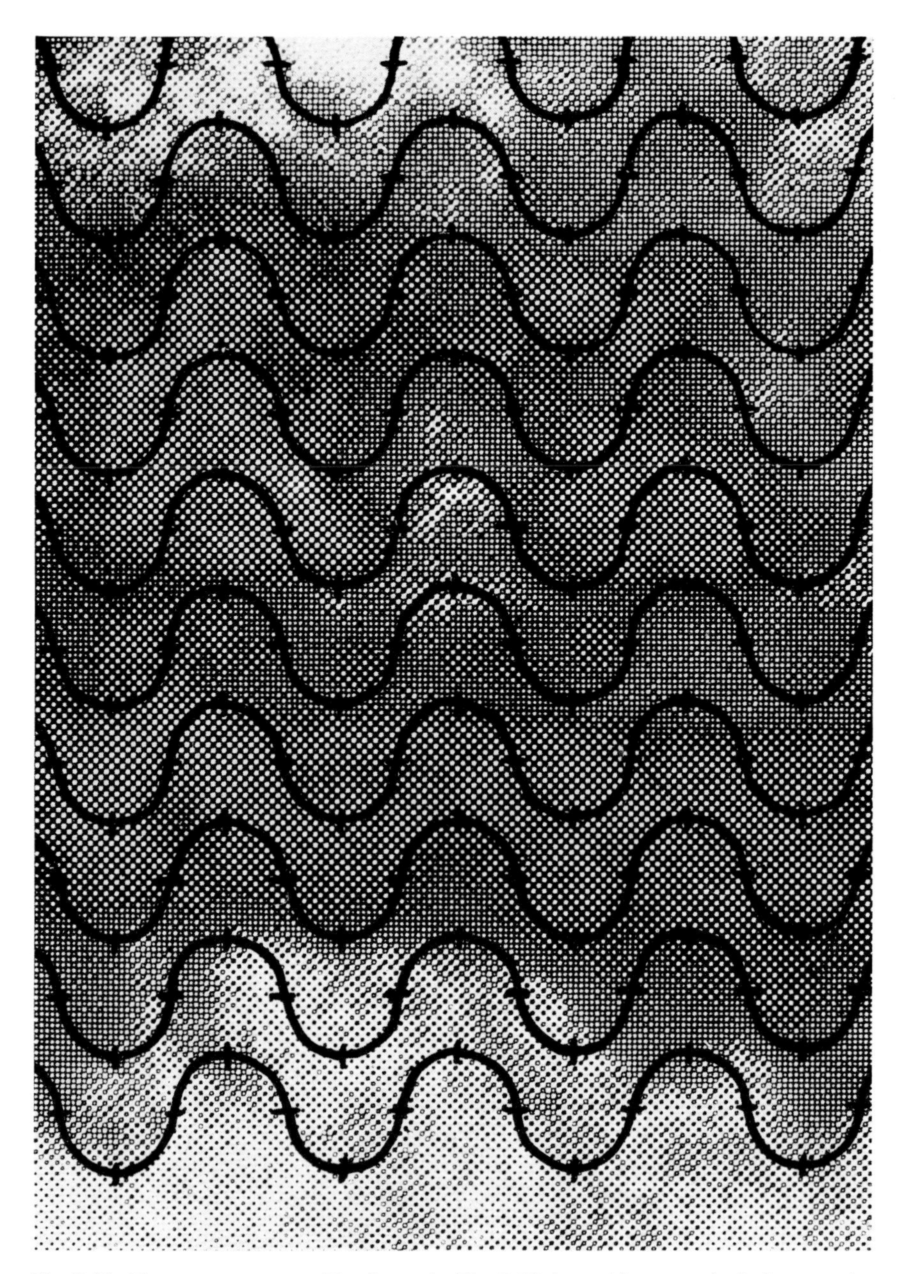

Fig. 9.13 The same screen equidensity as in Fig. 9.12, but with a stereological test probe superimposed. Several types of stereological data may be obtained: (1) point counts—the screen symbol immediately to the lower left of each test point (i.e., intersection of wavy line and small "cross-lines") being arbitrarily chosen for data recording. The actual density value of the screen symbol may be obtained from a calibration chart (page 367; also Fig.

position to density projections (1-3), cleft densities (4) and the postsynaptic thickening (5).

Figure 9.11 is an enlargement of the dense projection on the right of Fig. 9.10, and it shows some of the substructural detail latent within this profile. The shape of the regions containing the diamond-shaped symbols is illustrated in Fig. 9.14 (c). There is a core (1) of highest density close to the base of the dense projection. This is the area in which the diamond-shaped symbols are lightly shaded. (This apparent anomaly, lighter symbols representing higher densities, is explained by the fact that symbols at very high and very low densities are both seen as small dots.) Radiating from this central core is a narrow column (2), which then branches to the right (3) and left (4), and terminates in a series of globules (5).

Figure 9.12 illustrates the effect of uneven illumination on the differentiation of the screen symbols. The original micrograph is shown in the inset. The upper portion of the picture contains a crescent-shaped light region (1) which passes through the top of the right hand dense projection (2) and appears to cut it off at this point. In fact the dense projection continues into the crescent-shaped region because the large gradient (3) of the border can be seen in both regions. Some other notable features of this junction include the apparent continuity between dense projections (4), this continuity suggesting a possible vesicle site (5). The cleft is clearly displayed and contains small dense globules (6) and a density which is evidently continuous with the postsynaptic thickening (7). The predominantly higher density level of the cleft as compared with the background (lower portion of the picture) is also noticeable and indicates there may be two types of dense materials present in the cleft: (a) the cleft densities, and (b) a diffuse intercellular substance.

In Fig. 9.13, a stereological test probe has been superimposed on the screen equidensity shown in Fig. 9.12. This demonstrates the way in which stereological procedures may be used in conjunction with equidensitometry to obtain quantitative density data (see page 368).

EQUIDENSITOMETRY—POTENTIAL APPLICATIONS

In this section an examination is made of some of the potential uses of equidensity techniques in the following fields: (a) precision of measurement, (b) border determination, (c) digitalization of micrographs, (d) stereology of densities, (e) three-dimensional reconstruction.

9.14b). The results might be presented as a histogram depicting the frequencies of the density values; (2) intersection counts—the actual intersection taking place between a border (seen as a sudden transition between one symbol type and another) and one of the wavy lines of the test probe. Such counts would provide a measure of the complexity of dense bodies found in the micrograph. By using the appropriate stereological formulae, these point and intersection counts can be converted into three-dimensional quantities.

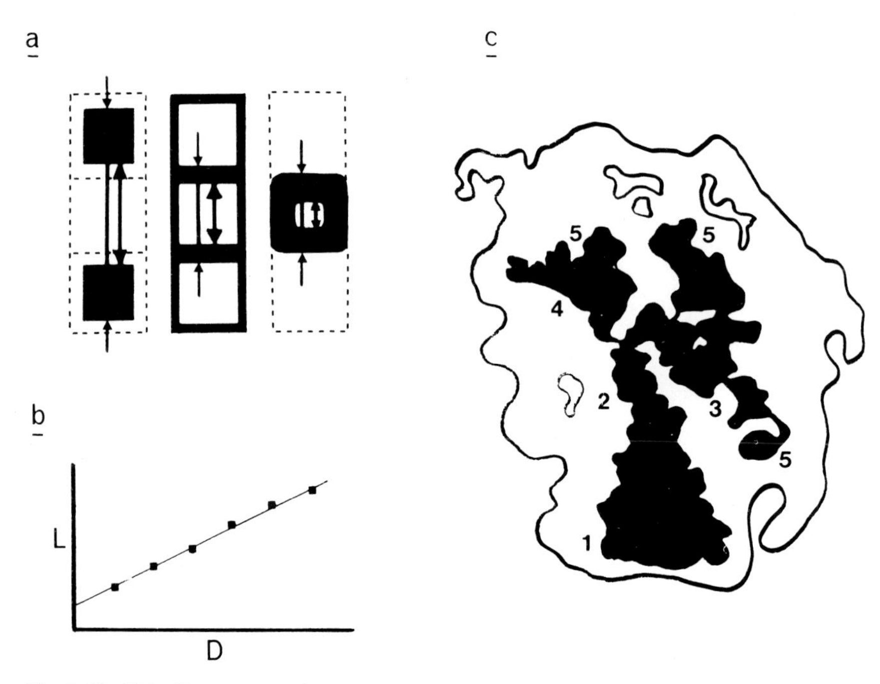

Fig. 9.14 This illustrates a theoretical basis for measurement of the numerical value of screen symbols. Three screen symbol configurations are depicted in (a): small diamond, grid and circle; and in each case the dimensions of the symbol are estimated by the sum (L) of the lengths of the two lines indicated by the arrows. The sum (L) is correlated with the density (D), which the screen symbol represents, and this relationship is illustrated graphically in (b). (C) is a diagram of the dense projection in Fig. 9.12, showing the central core of diamond shaped symbols. The key to the numbers 1–5 is given on page 363.

Measurement Precision

The success of line equidensities in increasing the precision of measurements of, for instance, interference fringes and spectrograms, indicates that with appropriate precautions, similar methods might be used to increase the accuracy of measurements of dense projections and other diffuse synaptic structures. Caution, however, needs to be exercised in view of some of the negative findings of a number of researchers when attempting to use line equidensities to improve the evaluation of complex asymmetrical density distributions (Lau and Krug, 1968).

Usually when taking measurements of a particular parameter in a group of objects, it is desirable to reduce variation between "like" objects in order to facilitate the detection of differences between groups belonging, for instance, to different age classes or to different treatment categories. Precision of measurement, brought about by the exclusion of extraneous sources of variation, is desirable at all stages–from the experimental animal to the final electron micrograph. Unwanted variation is avoided by controlling all factors other than those actually under investigation. This principle applies also to the measurement of micrograph component parameters, a process which often involves the introduction of random or even biased errors into the final data.

The contribution of several types of measurement error to the total variance of a parameter, for instance the "dense projection base width," may be analyzed as follows. Subjective measurement error depends upon (a) the clarity of the structure, and (b) the degree of personal bias present at the time of measurement. Personal bias may, in turn, be subdivided into (a) preconceptions involving the shape of the dense projections (for example, whether triangular, circular, or irregular), and (b) preconceptions concerning the degree of homogeneity expected in the parameter measurements. Experience has shown that subjects who were primed with, for instance, the information that dense projections are essentially triangular in shape, produce larger "base width" measurements than those who are told they are circular. Likewise, those informed that dense projections may be highly variable in size and shape produce measurements with higher variance than the measurements of those told that dense projections are equivalent in size and shape. Clearly a more objective method of assessing parameter measurements would eliminate some of these errors. Objective measurement is however, dependent upon the appropriate choice of a border criterion.

Border Criteria

The border has finally to be defined as a density function in terms of position, and in the case of indistinct object outlines, a number of borders are equally satisfactory depending upon the particular border criteria adopted. An immediate limitation imposed by the equidensity method is that the border must represent places of equal density, and therefore the problem reduces to one of deciding at which density level to place the border. Several alternative solutions are offered: (a) successively superimposing a series of photographically obtained contour lines to obtain a visual best-fit solution; (b) adjusting the density level of the border line by giving it a value which is a proportion of the maximum density of the object being outlined, and (c) standardizing the density level of the border line in each of a number of different micrographs, provided that the overall photographic density level remains the same.

Both (b) and (c) are arbitrary procedures and, while offering the possible advantage of complete objectivity, they also provide for unwanted variation in the form of illumination changes, uneven section thickness, or inconsistencies in the original photographic development. The first procedure (a), on the other hand, offers greater flexibility at the expense of some subjective errors. Minor adjustments to the position of the border can be made in individual cases, and this flexibility is advantageous in many situations. A number of synaptic ultrastructural components, for instance the dense projections, are still only visual "objects" in the sense that they possess no functional label, or definitive chemical structure. Hence their border is based on visual observation of density rather than on the distribution of any chemical substance.

Placement of a border is complicated by the fact that each micrograph may be interpreted in a number of different ways. Figure 9.15 illustrates two- and three-dimensional representations of a dense object. In (a) it has been shown as a two-dimensional slice in a plane parallel to the direction of the electron bream; the resulting micrograph (c) is shown graphically as a density profile. In Fig. 9.16 the same profile is taken and by a reverse process used to reconstruct a symmetrical estimate of the shape of the original object. Horizontal parallel lines represent the contour-line equidensity levels (a, c), while perpendicular lines drawn through the points of intersection of the horizontal lines and the density profile, pass through the horizontal axis and in doing so give rise to the actual contour-line equidensities themselves. At each stage a three-dimensional interpretation is shown beside the two-dimensional picture (b, d).

This example clearly illustrates the ambiguity of density information taken

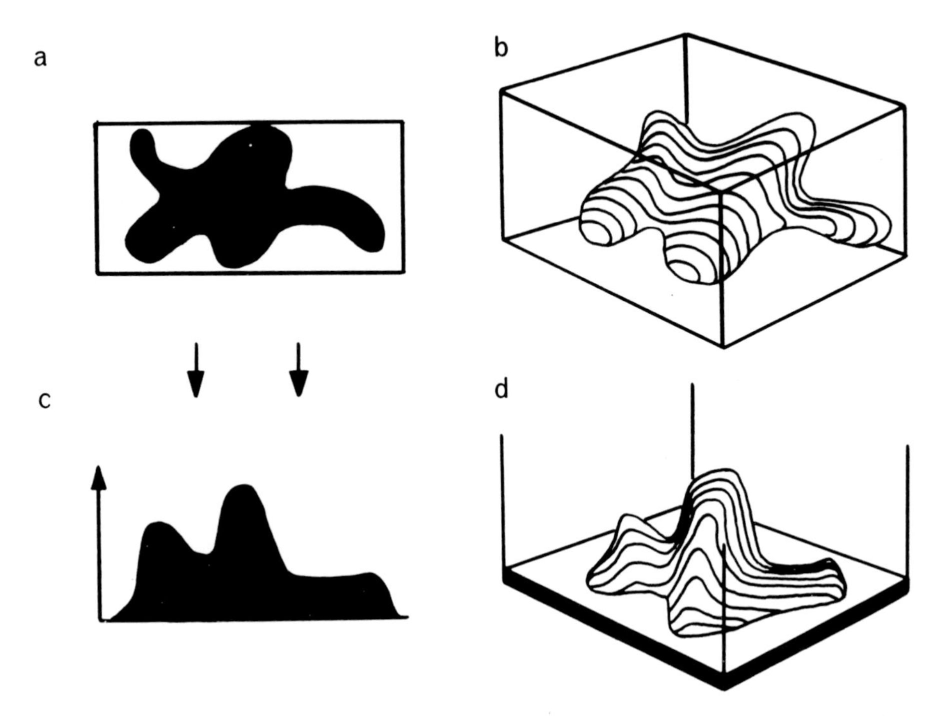

Fig. 9.15 A dense object embedded in a section is illustrated both in two (a) and three (b) dimensions. Its corresponding micrograph is shown as a photographic density profile, also in two (c) and three (d) dimensions.

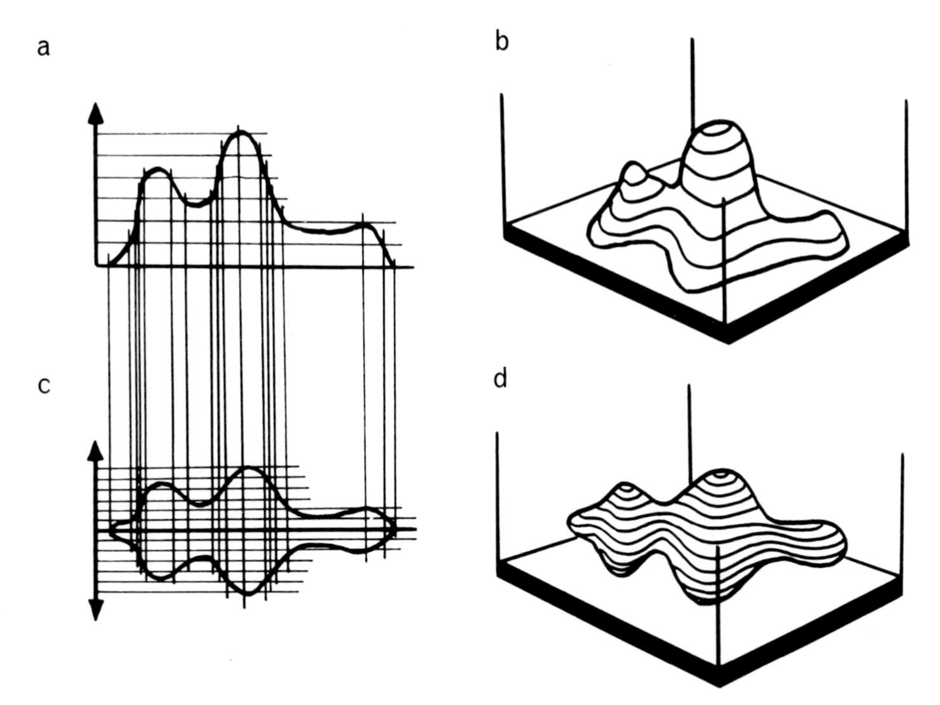

Fig. 9.16 An electron micrograph is depicted as a photographic density profile in two (a) and three (b) dimensions. It is the same profile shown in Figs. 9.15c and d, although horizontal lines representing a family of equidensities are also included. These are used to produce a symmetrical reconstruction of the object pictured in the micrograph in two (c) and three (d) dimensions. The reconstructions may be compared with the original object (Figs. 9.15a and b), to determine the degree of resemblance.

from a single micrograph. Attempts to reduce this uncertainty include statistical (page 368) and analytical (page 369) methods. Both these methods require a further technique for the assessment of densities in a micrograph. Such a technique is provided in the following section. While cumbersome in comparison with, for instance, a scanning microdensitometer, it may nevertheless give satisfactory results on a small scale.

Equidensitometric Digitization of Micrographs

On page 354 it was shown that the screen symbols reflect the corresponding density values of the original micrograph. Quantification of this relationship gives rise to the possibility of using screen symbols to measure original density values, and provided that a reasonably rapid method of reading the symbols is available, complete digitalization of the picture may be undertaken.

Figure 9.14 illustrates graphically the relationship between light intensity and the conformation of the screen symbol. It can be seen that the different symbol configurations are correlated with the sum of the lengths of the two arrows in each case. Direct measurement of the density value of the symbols is possible therefore, although such a procedure would involve enlargement of the symbols and possibly recopying onto *Agfacontour* film to further clarify their borders. In practice, screen symbols are most conveniently produced at the magnification shown in Figs. 9.9–9.13. The symbols are ~1 mm apart, and the width of their borders in comparison with their actual size is large, so that accurate measurement is difficult. It is nevertheless possible by comparing the symbols with other symbols of the same size produced using a continuous grey wedge, to obtain a close approximation of the density value represented by the symbol and also to evaluate each symbol rapidly in succession.

Potential applications of this approach are numerous, and many of these—which are also of interest to the biologist—are concerned with improving the evaluation of electron micrographs or reconstructing the actual three-dimensional electron opacities of the original object. For a summary of the procedures and applications involving digitalization of micrographs we refer the reader to one of the reviews of computerized electron microscopy, such as that of Frank (1973).

Stereology of Equidensities

Stereology is a science concerned with the statistical determination of three-dimensional quantities, such as volume or surface area (see review by Weibel and Bolender, 1973). Other aspects of structure, including shape and size, may also be found using these methods, although the details of such applications in the case of complex objects are not currently known. Some of the principal advantages of stereology when compared with conventional morphometry include its straight-forward "yes" or "no" type of answers to such questions as: "does a test probe point fall within a given region of the photograph?" or "does the test probe line intersect a given border of a structure within the photograph?" Additional advantages of stereology stem from its simple formulas for the conversion of quantities from two-dimensions to three, and its relative freedom from shape criteria (volume and surface area are shapeless quantities). Morphometry, by contrast, deals with linear measurements from point to point. Its parameters involve *a priori* assumptions as to the shape of the object, and complex formulas are demanded for the transition from two-dimensions to three.

The simultaneous development of both stereology and equidensitometry in recent years has, to the best of our knowledge, not yet resulted in any synthesis of ideas from the two fields. Point counting of densities appears nevertheless to offer a new approach to the quantification of micrographs. Density is a dimensionless quantity, and as such need not be characterized by definite shape or size. A border is implied however, and this permits point or intersection counts to be made. At the same time, the numerical values of the densities at these points, or on either side of the boundaries, are evaluated by means of screen equidensities or calibrated contour-line equidensities. The resulting data can be processed and presented in the form of a stereometric model (e.g., Weibel, 1972). Such a model could be used, for example, to detect gross chemical changes in a tissue brought about by different experimental conditions, or to compare the "densitometric composition" of one type of tissue with that of another. Parameters expected to be significant in making such comparisons are: (a) the mean density of the tissue, although rigorous standardization procedures would need to be adopted before conclusions were accepted as valid, (b) the volume proportion of each density level in relation to the total density, and (c) the surface area of the interfaces between discrete density levels, giving some idea of the complexity of the density patterns present in the tissue.

Three-Dimensional Reconstruction

As three-dimensional reconstruction is one of the ultimate aims of the morphologist, we will deal briefly with some of the main approaches to this problem and the potential contribution of equidensitometry to each. Two of the more conventional approaches are those of the morphometrist and of the stereologist. Morphometric models, for instance, the synaptic junction of Akert *et al.* (1969), have been constructed to account both for observed structural features and measured structural parameters. Stereometric models of a whole synapse have not to our knowledge been published, although some stereological parameters have been used, for instance in the synapse studies of Vrensen and De Groot (1973).

The recent approach to three-dimensional reconstruction employing a number of projections of an object tilted about an axis, appears to offer satisfactory complete reconstructions if a resolution limit is accepted. Two different methods of calculating the three-dimensional structure of the original object from its projections may then be used—the Fourier theory projection theorem as used by De Rosier and Klug (1968) and the "real space" algebraic equations as presented by Crowther *et al.* (1970) and implemented using an iterative method by Gordon *et al.* (1970). A recent account of the iterative and Fourier methods with extensive references to pertinent literature can be found in the papers by Zwick and Zeitler (1973) and Zeitler (1974). The relevance of equidensitometry to these reconstruction techniques lies in its ability to digitalize a micrograph, providing the density information necessary for the reconstruction of the original electron opacities. The actual reconstruction may then be carried out using computer facilities.

Much current research is being carried out with a view to finding reliable and efficient holographic reconstruction techniques. The similarity between holograms and interferograms suggests that an application of equidensitometry may also be found to this rapidly expanding field.

We wish to acknowledge grants from the Australian Research Grants Committee and the Nuffield Foundation toward the support of this and related work.

REFERENCES

Akert, K., Moor, H., Pfenninger, K., and Sandri, C. (1969). Contribution of new impregnation methods and freeze-etching to the problems of synaptic fine structure. *Prog. Br. Res.* **31,** 223.

Banig, T., and Mieler, W. (1964). Die Anwendung der Äquidensitometrie zur verbesserten Wiedergabe von Chromosomen. *Z. Kinderheilk* **90,** 54.

Crowther, R. A., De Rosier, D. J., and Klug, A. (1970). The reconstruction of a three-dimensional structure from projections and its application to electron microscopy. *Proc. Roy. Soc. A.* **317,** 319.

De Rosier, D. J., and Klug, A. (1968). Reconstruction of three dimensional structures from electron micrographs. *Nature* **217,** 130.

Frank, J. (1973). Computer processing of electron micrographs. In: *Advanced Techniques in Biological Electron Microscopy* (Koehler, J. K., ed.), p. 215. Springer-Verlag, Berlin.

Gordon, R., Bender, R., and Herman, G. T. (1970). Algebraic reconstruction techniques (ART) for three-dimensional electron microscopy and X-ray photography. *J. Theor. Biol.* **29,** 471.

Heidenreich, J., Bethge, H., and Ruess, U. (1964). Zur objektiven Auswertung von Auflösungstest-Aufnahmen. *Proc. 3rd. Europ. Reg. Conf. Electron Microscopy, Prague.* **1,** 125.

Jobin, M. M. and Yvon. (1934). Equidensitomètre. *Rev. Opt. théor instrum.* **13,** 179.

Jones, D. G. (1975). *Synapses and Synaptosomes: Morphological Aspects.* Chapman and Hall, London.

Jones, D. G., Reading, L. C., Dittmer, M. M., and Ellison, L. T. (1976). A critical evaluation of the relationship between the presynaptic network, synaptic vesicles, and dense projections in central synapses. *Cell Tiss. Res.* **169,** 49.

Keller, H. L. (1964). Filmdosimetrie in der Strahlentherapie. *Radiologe* **4,** 272.

Kind, E. G. (1954). Photographische Herstellung von Äquidensiten. *Arbeitstagung Optik, Jena* 117.

Kölle, W., Eichhorn, H. J., and Degenhardt, K. H. (1956). Die Herstellung von Isodosentafeln mittels einer photographischen Methode. *Probl Erg. Biophys Strachlenbiol, Leipzig,* 205.

Konermann, H. (1971). Quantitative Bestimmung der Matercalverteilung nach Röntgenbildern des Knochens mit einer nenen photographischen Methode. *Z. Anatomie.* **134,** 13.

Kramer, W. (1951). Ein optisches Registrierphotometer. *Z. Naturf.* **6a,** 658.

Krug, W., and Lau, E. (1952). Die Äquidensitometrie, ein neues Meßverfahren für Wissenschaft und Technik. *Feingerätetechn.* **1,** 391.

Lau, E. (1953). Die Äquidensiten in der Verwendung für spectroskopische Photometrie. *Exp. Techn. Phys.* **1,** 199.

Lau, E. (1958). Schwärzungsplastik und ihre Anwendungen, *Tagesbericht II. Internat. Koll Hochsch Elektrotech, Ilmenan 1957,* 46.

Lau, E., and Hess, G. (1960). Photographische groß flächenphotometrie. *Bild. Ton.* **13,** 71.

Lau, E., and Johannesson, J. (1934). Das Optimum der Detailwiedergabe der photographischen Schichten. *Z. Phys.,* **35,** 505.

Lau, E., Kind, E. G., and Roose, G. (1958). Photometry of photographic plates by interference microscopy. *Mon. Tech. Rev.* **2,** 118.

Lau. E., and Krug, W. (1968). *Equidensitometry.* Focal Press, London.

McKechnie, T. S. (1973). The effect of defocus on the resolution of two points. *Optica Acta* **20,** 253.

Miller, C. S., Parsons, F. G., and Kofsky, I. L. (1964). Simplified two-dimensional microdensitometry. *Nature* **202,** 1196.

Nolan, T. M., and Jones, D. G. (1973). Morphometry of synaptic ultrastructure using equidensitometry. *Amer. J. Anat.* **138,** 527.

Nolan, T. M., and Jones, D. G. (1974). Equidensitometric analytical techniques applied to the study of synaptic ultrastructure. *J. Neurocytol.* **3**, 327.

Pfenninger, K., Sandri, C., Akert, K., and Eugster, C. H. (1969). Contribution to the problem of structural organization of the presynaptic area. *Brain Res.* **12**, 10.

Rakow, A. (1959). Zur Anwendung von Äquidensitometrie in der Radiologie. *Dtsch. Gesdwes.* **14**, 382.

Rakow, A. (1965). Der Wert der photographischen Bestimmung der Dosisverteilung für die Bestrahlungsplanung. *Radiobiol. Radiother.* **6**, 69.

Rakow, A., and Zapf, K. (1962). Zur Äquidensitometrie und ihrer Bedeutung für die Bildanalyse elekronenmikroskopischer Aufnahmen biologischer Objekte. *Microskopie* **17**, 217.

Ranz, E. (1970). Agfacontour, a new film for simplified production of equidensities. *Visual* **8**, 49.

Ranz, E., and Schneider, S. (1971). Progress in the application of Agfacontour equidensity film for Geo-scientific photo interpretation. *Proc. 7th Intern. Symp. on Remote Sensing of Environment. 17-21 May*, p. 779. Institute of Science and Technology, University of Michigan.

Röhler, R. (1967). Informationstheorie in der Optik. In: *Optik und Feinmechanik in Einzeldarstellungen*, vol. **6** (Gunther, W., ed.), p. 171. Verlagsgesellschaft.

Schmalbruch, H., and Kamieniecka, Z. (1974). Fiber types in the human brachial biceps muscle. *Expt. Neurol.* **44**, 313.

Schmidt, H. (1959). Erfahrungen mit dem Schwärzungsplastikverhafren. *Optik* **16**, 538.

Schmidt, H. (1960). Photographische Photometrie von Sternspektren mit dem Interferenzmikroskop. *Veröff Sternw Babelsberg.* **13**, part 1.

Schröter, E. H. (1958). Chromosphäre Strukturen in den Balmerlinien. *Z. Astrophys.* **45**, 68.

Schult, E. (1927). Intensitätmessungen an Interferenzerscheinungen nebst Untersuchung stehender Lichtwellen. *Annln Phys.* (Leipzig) **82**, 1025.

Schusta, J. (1954). Elektronische Herstellung der Äquidensiten. *Arbeitstagung Optik, Jena*, 123.

Schwieger, H., and Haberland, G. (1955). Die Anwendung der Äquidensiten in der Spannungsoptik. *Bauplanung Bautechn* **9**, 71.

Schwieger, H., and Haberland, G. (1956). Die photographische Darstellung von Schubgleichen unterhalb der 1. Isochromatenordnung bei spannung-stoptischen Versuchen. *Z. angew Phys.* **8**, 350.

Tolansky, S. (1948). *Multiple-beam Interferometry of Surfaces and Films.* Oxford.

Tsien, K., and Robbins, R. (1966). The photographic representation of isodose patterns by the application of the Sabattier effect. *Brit. J. Radiol.* **39**, 1.

Vrensen, G., and De Groot, D. (1973). Quantitative stereology of synapses: a critical investigation. *Brain Res.* **58**, 25.

Weibel, E. R. (1972). A stereological method for estimating volume and surface of sarcoplasmic reticulum. *J. Microscopy* **95**, 229.

Weibel, E. R., and Bolender, R. P. (1973). Stereological Methods. In: *Principles and Techniques of Electron Microscopy: Biological Applications*, Vol. 3 (Hayat, M. A., ed.), p. 237. Van Nostrand-Reinhold Company, New York and London.

Zeitler, E. (1974). The reconstruction of objects from their projections. *Optik* **39**, 396.

Zwick, M., and Zeitler, E. (1973). Image reconstruction from projections. *Optik* **38**, 550.

AUTHOR INDEX

SUBJECT INDEX